AF807981

WITHDRAWN

Substituent Effects in Radical Chemistry

NATO ASI Series

Advanced Science Institutes Series

A series presenting the results of activities sponsored by the NATO Science Committee, which aims at the dissemination of advanced scientific and technological knowledge, with a view to strengthening links between scientific communities.

The series is published by an international board of publishers in conjunction with the NATO Scientific Affairs Division

A Life Sciences B Physics	Plenum Publishing Corporation London and New York
C Mathematical and Physical Sciences	D. Reidel Publishing Company Dordrecht, Boston, Lancaster and Tokyo
D Behavioural and Social Sciences E Engineering and Materials Sciences	Martinus Nijhoff Publishers The Hague, Boston and Lancaster
F Computer and Systems Sciences G Ecological Sciences	Springer-Verlag Berlin, Heidelberg, New York and Tokyo

Series C: Mathematical and Physical Sciences Vol. 189

Substituent Effects in Radical Chemistry

edited by

Heinz G. Viehe

Zdenek Janousek

and

Robert Merényi

University of Louvain,
Louvain-la-Neuve, Belgium

D. Reidel Publishing Company

Dordrecht / Boston / Lancaster / Tokyo

Published in cooperation with NATO Scientific Affairs Division

Proceedings of the NATO Advanced Research Workshop on
Substituent Effects in Radical Chemistry
Louvain-la-Neuve, Belgium
January 20-24, 1986

Library of Congress Cataloging in Publication Data

Substituent effects in radical chemistry.

(NATO ASI series. Series C, Mathematical and physical sciences; vol. 189)
Proceedings of the NATO Advanced Research Workshop on Substituent Effects in Radical Chemistry, Louvain-la-Neuve, Belgium, January 20–24, 1986.
"Published in cooperation with NATO Scientific Affairs Division."
Includes index.
1. Radicals (Chemistry)—Congresses. I. Viehe, Heinz Günter. II. Janousek, Zdenek. III. Merényi, Robert. IV. NATO Advanced Research Workshop on Substituent Effects in Radical Chemistry (1986 : Louvain-la-Neuve, Belgium) V. Series: NATO ASI series. Series C, Mathematical and physical sciences; vol. 189.
QD471.S86 1986 541.2′24 86–17895
ISBN 90–277–2340–0

QD
471
.S86
1986

Published by D. Reidel Publishing Company
P.O. Box 17, 3300 AA Dordrecht, Holland

Sold and distributed in the U.S.A. and Canada
by Kluwer Academic Publishers,
101 Philip Drive, Assinippi Park, Norwell, MA 02061, U.S.A.

In all other countries, sold and distributed
by Kluwer Academic Publishers Group,
P.O. Box 322, 3300 AH Dordrecht, Holland

D. Reidel Publishing Company is a member of the Kluwer Academic Publishers Group

All Rights Reserved
© 1986 by D. Reidel Publishing Company, Dordrecht, Holland.
No part of the material protected by this copyright notice may be reproduced or utilized in any form or by any means, electronic or mechanical, including photocopying, recording or by any information storage and retrieval system, without written permission from the copyright owner.

Printed in The Netherlands

CONTENTS

CONTRIBUTORS

Arnold Donald D., Department of Chemistry, Dalhousie University, Halifax, Nova Scotia, Canada B3H 4J3

Aurich Hans G., Fachbereich Chemie der Philipps-Universität, Hans-Meerwein Strasse, D-3550 Marburg, Germany

Bachi Mario D., The Weizmann Institute of Science, Department of Organic Chemistry, 76100 Rehovot, Israel

Balaban A.T., Str.Poenaru Bordea 6, Apt13, R-70502 Bucharest Roumania.

Barton Sir Derek, Institut de Chimie des Substances Naturelles, CNRS, F-91190 Gif-sur-Yvette, France

Beckhaus Hans-Dieter, Institut für Organische Chemie und Biochemie, Universität Freiburg, Albertstrasse 21, D-7800 Freiburg 1, Germany

Citterio Attilio, Politecnico di Milano, Istituto di Chimica Piazzo L. de Vinci 32 I-20133 Milano, Italy

Clark Timothy, Friedrich-Alexander Universität Erlangen-Nürnberg, Institut für Organ. Chemie, Henkestrasse 42, D-8520 Erlangen, Germany

Costantini M., Rhône-Poulenc Industries, Centre de Recherches de St. Fons, BP62, F-69190 Saint-Fons, France

Creary Xavier, Department of Chemistry, University of Notre Dame, Notre Dame IN 46556 U.S.A.

Crozet Michel P., Université de Droit, d'Economie et des Sciences Naturelles d'Aix-Marseille III. CNRS VA 109, Rue H. Poincaré, F-13397 Marseille Cedex 13, France

Döpp Dietrich, Universität Duisburg, Fachgebiet Organische Chemie Postfach 101629, D-4100 Duisburg 1, Germany

Fischer Hanns, Physikalisch-Chemisches Institut der Universität Zürich, Winterthurerstr.190, 8057 Zürich, Switzerland

Fossey Jacques, L.E.S.C.O.-C.N.R.S. 2 A 8, rue Henri Dunant, BP 28, F-94320 Thiais, France

Giese Bernd, Technische Hochschule Darmstadt, Institut für Organische Chemie, Petersenstrasse 22, D-6100 Darmstadt, Germany

Griller David, National Research Council of Canada, Division of Chemistry, Ottawa, Canada K1A OR9

Ingold Keith U., National Research Council of Canada, Division of Chemistry, Ottawa, Canada K1A OR9

Jackson Richard A., School of Chemistry and Molecular Sciences, University of Sussex, Brighton BN1 9QJ, Great Britain

Janousek Zdenek, Université de Louvain, Laboratoire de Chimie Organique, Place Louis Pasteur, 1, B-1348 Louvain-la-Neuve, Belgium

Katritzky Alan R., University of Florida, Department of Chemistry, Gainesville, FL Florida 32611, U.S.A.

Koch Tad H., University of Colorado, Department of Chemistry and Biochemistry, Boulder Co 80309-0215, U.S.A.

Korth H.G., Institut für Organische Chemie, Universität Essen, GHS, D-4300 Essen 1, Germany

Kost Daniel, Ben Gurion University of the Negev, Department of Chemistry, Beer-Sheva 84105, Israel

Leroy George, Université de Louvain, Laboratoire de Chimie Quantique, Place Louis Pasteur, 1, B-1348 Louvain-la-Neuve, Belgium

Louw Robert, Center for Chemistry and the Environment Gorleaus Labs, The State University of Leiden, P.O. Box 9502, NL-2300 RA Leiden, The Netherlands

Maillard Bernard, CNRS VA 35, Laboratoire de Chimie Organique et Organométallique, Université de Bordeaux 1, 351 Cours de la Libération, F-33405 Talence-Cedex, France

Merényi Robert, Université de Louvain, Laboratoire de Chimie Organique, Place Louis Pasteur, 1, B-1348 Louvain-la-Neuve, Belgium

Minisci Francesco, Politechnico di Milano, Dipartimento di Chimica, Piazza L. da Vinci, 32, I-20133 Milano, Italy

Neumann Wilhelm P., Universität Dortmund, Otto-Hahn-Strasse, Postfach 50 05 00, D-4600 Dortmund 50, Germany

Rüchardt Christoph, Institut für Organische Chemie und Biochemie, Albertstrasse, 21, D-7800 Freiburg 1 , Germany

Stella Lucien, Université de Droit, d'Economie et des Sciences Naturelles d'Aix-Marseille III, Laboratoire de Chimie Organique B, Rue Henri Poincaré, F-13397 Marseille Cedex 13, France

Sustmann Reiner, Institut für Organische Chemie, Universität-GHS Essen Universitätsstrasse 5, D-4300 Essen 1, Germany

Tedder Lord John M., Department of Chemistry, University of St. Andrews, St. Andrews Fife, KY16 9ST, United Kingdom

Tiecco Marcello, Istituto di Chimica Organica, Facolta di Farmacia, Universita di Perugia , I-06100 Perugia, Italy

Timberlake Jack W., University of New Orleans, Department of Chemistry, Lakefront, New Orleans, LA 70122, U.S.A.

Tordo Paul, Université de Provence - Structure et Réactivité des Espèces Paramagnétiques, C.N.R.S. UA 126, Rue H. Poincaré, F-13397 Marseille Cedex 4, France

Vertommen Luc, AKZO, Research Laboratories, Velperweg, 76, NL-6824 BM Arnhem, The Netherlands

Viehe Heinz Gunter, Laboratoire de Chimie Organique, Université de Louvain, Place Louis Pasteur, 1, B-1348 Louvain-la-Neuve, Belgium

von Ragué Schleyer Paul, Friedrich-Alexander-Universität Erlangen-Nurnberg, Institut für Organische Chemie, Henkestrasse 42,D-8520 Erlangen, Germany

Walton John C., Department of Chemistry, University of St. Andrews, St. Andrews Fife, KY16 9ST, United Kingdom

PREFACE

Respectably old radical chemistry which plays a major role in life-processes, both desired (breathing, ...) and non-desired (inflammatory diseases, ageing,...) has been gaining new youth in the past decade.

Modern spectroscopy and other physical methods, recent advances in computational methods as well as impressive and mechanistically well-understood syntheses have led to a number of spectacular developments in the field of radical chemistry. The impact of these achievements will reach far beyond the field of organic chemistry, for example into biology and medicine.

New facts and concepts in this rapidly expanding field deserved discussion among a number of leading experts present at the Workshop both for the information and constructive criticism. This happened during a particular brainstorming session and some of the impressions and reflections exchanged are recorded in the Epilogue.

Indeed the topic chosen seems to be quintessential for the whole radical chemistry : understanding and exploiting (single and multiple) substituent effects on radicals.
We had the honour to organise this Workshop because some years ago, in collaboration with L. Stella, we became involved in radical stabilising effects when both a donor and an acceptor (captor) groups are present on a C-radical. This leads to an enhanced stabilisation which we called the captodative effect and we have endeavoured the first extensive investigation of this effect on carbon centered radicals. Nevertheless such an effect was intimated as early as in 1952 by Dewar, but it was not until about 20 years later when Katritzky provided some first experimental results and called this effect "merostabilisation". Similar effect on nitrogen-centeres radicals has been discovered and well documented by Balaban under the name "push-pull" stabilisation. There has not yet been reached a complete agreement as to the extent of such an effect but by now more leading groups are active in this field. Owing to this concern it came to light that even the effect of a single substituent may still be subject to controversy. This explains why the organization of such a meeting was of paramount importance.

Thanks to a generous grant from NATO Scientific Affairs Division such a meeting among almost 50 specialists became possible at

Louvain-la-Neuve (Belgium). The lectures and posters were streamlined on the following topics :
- computations concerning both closed and open-shell species;
- kinetic and thermodynamic phenomena;
- ESR techniques;
- devising radical stabilising scales;
- new radical reactions in synthesis.

Many points have been clarified and those which still manage to escape our full understanding will stimulate the participants and, as we hope, the readers of this book to help to solve them.

H.G. Viehe,
Z. Janousek,
R. Merényi,

Louvain-la-Neuve, 1986.

A THEORETICAL APPROACH TO SUBSTITUENT EFFECTS IN RADICAL CHEMISTRY

G. Leroy, D. Peeters, M. Sana* and C. Wilante
Laboratoire de Chimie Quantique
Université Catholique de Louvain
Place L. Pasteur, 1,
B-1348 Louvain-la-Neuve, Belgium

ABSTRACT. In this paper we make a survey of various aspects of radical chemistry to which we have made some contribution. Our Localized Molecular Orbital (LMO) approach to the electronic structure and react vity of chemical species allowed us to generalize the Linnett theory and to describe the mechanism of various processes involving free radicals. We have also proposed a general definition of the concept of stabilization energy applying to any compound. Using this concept, we have been able to systematically estimate the thermodynamic stabilization of free radicals and to carry out a critical analysis of former definitions of the stabilization energy of these species. Furthermore, we have shown that the thermochemistry and, eventually, the activation energies of radical reactions can be rationalized in terms of stabilization energies. In this prospect, we have proposed an original interpretation of the concept of kinetic stabilization and we have made interesting predictions concerning the thermochemistry and the kinetic of radical recombination reactions. We have also devised a semi-empiri cal method for estimating the barrier heights of atom transfer reactio So, we have been able to systematically determine the kinetic paramete of these processes using the transition state theory with an Eckart tunnelling correction. Our barrier height modelling allowed us to rationalize the substituent effects on the activation barriers of hydrogen transfer reactions leading to carbon centered radicals. Besides, the kinetic parameters of various hydrogen atom additions to substituted ethylenes have been calculated at the ab initio level. On the whole, we have found that, in spite of their large stabilization energy, captodative radicals are generally transient because of the relatively small stabilization energy of their dimer. On the other han the recombination of dicaptor species would be made difficult by the large destabilization of their dimer. Finally, didonor radicals, would always be obtained at lower rates than the captodative and dicaptor species but nothing except steric effects would hinder their dimerization.

* Maître de Recherches du Fonds National Belge de la Recherche Scientifique.

H. G. Viehe et al. (eds.), Substituent Effects in Radical Chemistry, 1–48.
© 1986 by D. Reidel Publishing Co.

1. INTRODUCTION

In a previous series of papers, we have successively proposed :
i) An original procedure for determining the electronic structure of any chemical species (1). Its generalization to open-shell systems (2) allowed us to describe the distribution of electrons of each spin (α and β electrons) in free radicals and so to obtain chemical formulas equivalent to the corresponding Linnett structures (3,4).
ii) A general definition of the stabilization energy of chemical species based on the concept of standard bond energy term (5-7). In this way, we have been able to rationalize the thermochemistry of any physical or chemical process involving neutral or ionic species in terms of stabilization energies (8-11).
iii) A semi-empirical method for the estimation of activation barriers and kinetic parameters of atom transfer reactions (12-14). This procedure allowed us to improve our first ab initio results concerning these processes (15) and to rationalize the effects of substituents on the barrier heights of hydrogen abstraction reactions from methane derivatives.
Moreover, our various contributions to the theoretical approach to static and dynamic properties of free radicals have been described in some specialized (16-20) or review articles (21-23).
In this paper, we shall be more particularly interested in the effects of substituents on the structure, the stability and the reactivity of free radicals. Our theoretical approach to this problem shall essentially be based on the various methods and concepts cited above.

2. THE ELECTRONIC STRUCTURE OF FREE RADICALS

The electronic structure of free radicals can be analysed in terms of distributions of α and β electrons (Linnett structures) and spin properties. Such an analysis has been performed at various theoretical levels and the results obtained with different basis sets are quite similar. Thus, our preceding conclusions relating to the electronic structure of the carbon centered species and the prototypes of various series of radicals remain generally valid (9,21). As shown elsewhere (8,9), they lead to a detailed classification of free radicals. We summarize below the main conclusions deduced from former and more recent results.

2.1. The distribution of α and β electrons in free radicals

In carbon-centered free radicals, π-donor substituents remove the unpaired electron from the α carbon atom and the C_α-substituent bond has a three-electron bond character. On the other hand, π-acceptor substituents draw the unpaired electron towards the bond C_α-substituent which has also a three-electron bond character. Finally, a pair of substituents of opposite polarity delocalizes the unpaired electron and the bonds between C_α and the substituents have both a three-electron bond character. These conclusions are summarized in the formulas given in figure 1. They are in complete agreement with the results of geometry optimization and the corresponding spin density maps (21). A more precise representation of the α and β electron distribution in the aminocyanomethyl radical, obtained at the 6-31 G level, is shown in figure 2. It confirms the qualitative description of the captodative radical considered in figure 1.

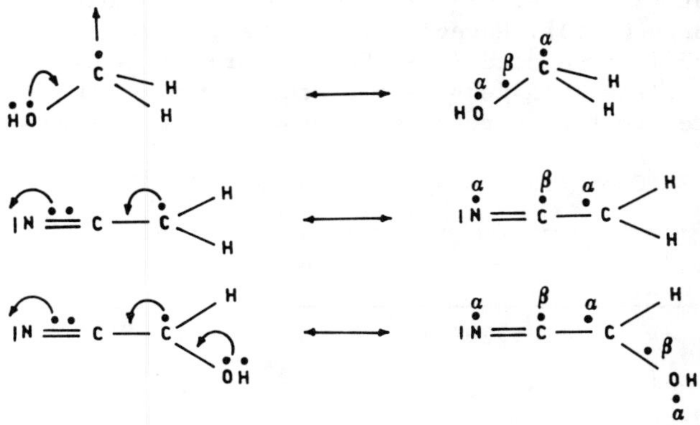

Figure 1. Electronic structure of some carbon-centered free radicals.

In summary, the unpaired electron of π carbon-radicals is generally delocalized, this effect being particularly important in captodative species, but, in σ carbon-radicals, it is more centered on the C_α atom. In nitrogen- and oxygen-centered free radicals, the effects of substituents on the unpaired electron are qualitatively the same as in carbon-centered species. The particularly large effects of NH_2 is attested by the typical three-electron bonds in hydrazyl and nitroxide radicals.

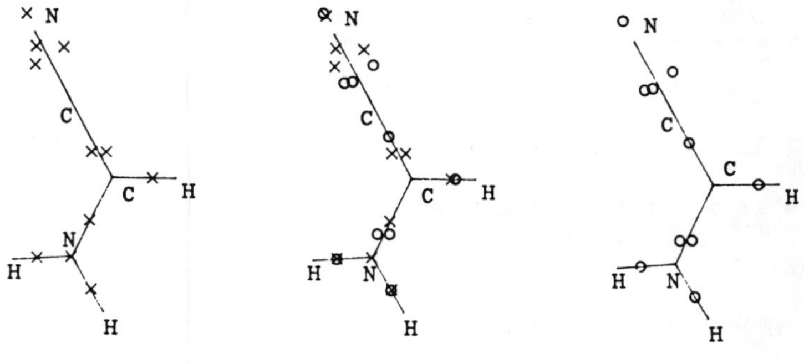

Figure 2. Distribution of α and β electrons in aminocyanomethyl radical (x = α ; o = β).

2.2. Spin properties

Spin properties calculated at the Unrestricted Hartree-Fock (UHF) level using Gaussian Type Orbitals (GTO's) are not very reliable. Indeed UHF wave functions do not represent pure doublet spin states. Moreover, gaussian basis sets are very inadequate near the nuclei because they do not satisfy the cusp condition. Consequently they fail to predict spin densities at the nuclei [$\rho_s(N)$] and the corresponding hyperfine splitting constants [a(N)]. However, in general, the sign of the spin density is correctly predicted (21). Furthermore, as shown by the data listed in table I, the spin densities at the C_α nucleus [$a(^{13}C_\alpha)$] are roughly correlated to the corresponding hyperfine splitting constants.

Table I. Comparison of $\rho_s(C_\alpha)$ and $a(^{13}C_\alpha)$.

R$^\cdot$	$\rho_s(C_\alpha)^a$	$a(^{13}C_\alpha)$, G
$\dot{C}H_3$	0.229	38.34 (24)
$\dot{C}H_2CH_3$	0.249	39.07 (24)
$\dot{C}H(CH_3)_2$	0.262	41.30 (24)
$\dot{C}H_2OCH_3$	0.280	47.20 (25)
$\dot{C}H_2OH$	0.283	47.40 (25)
$\dot{C}H(OH)_2$	0.366	101.10 (26)
$\dot{C}HF_2$	0.435	148.80 (27)
$\dot{C}F_3$	0.619	271.60 (27)

[a] 4-31 G results

The following relation is easily deduced :

$$a(^{13}C_\alpha) = -120.96 + 625.22\ \rho_s(C_\alpha) \qquad r=0.995 \qquad [1]$$

One observes also a good correlation between the theoretical π-spin densities of the α carbon atom and those estimated using the Mc Connell relation (28) :

$$a(H_\alpha) = 23.04\ \rho_s^\pi(C_\alpha) \qquad [2]$$

Indeed, the data of table II lead to the next equation :

$$\rho_s^\pi(C_\alpha)\text{theor.} = -0.189 + 1.174\ \rho_s^\pi(C_\alpha)\text{emp.} \qquad r=0.980 \qquad [3]$$

It is seen that the unpaired electron is more delocalized in the hydroxycyanomethyl radical. This captodative effect (29) is expected to be particularly important in the aminocyano- and aminoethynyl radicals for their hyperfine splitting constants $a(H_\alpha)$ are respectively equal to 14.9 and 14.7 G (32).

Table II. Theoretical and empirical values of $\rho_s^\pi(C_\alpha)$.

$R^\cdot$	$a(H_\alpha)$	$\rho_s^\pi(C_\alpha)$ emp.	$\rho_s^\pi(C_\alpha)$ theor.
$\dot{C}H_3$	23.04 (24)	1	1
$\dot{C}H_2CN$	20.88 (30)	0.906	0.855
$\dot{C}H(CN)_2$	19.18 (31)	0.832	0.769
$\dot{C}H(CN)(OH)$	18.10 (30)	0.786	0.757

We want to recall here that, even in the case of planar radicals (the only species for which they could be valid), the Mc Connell types of relations :

$$a(H_\alpha) = Q_\alpha \rho_s^\pi(C_\alpha) \qquad [4]$$

and

$$a(H_\beta) = Q_\beta \rho_s^\pi(C_\alpha) \qquad [5]$$

are very approximate. This is partly due to the lack of precision of experimental data and partly to the crudeness of the spin polarization and hyperconjugation models on which they are based. Indeed, if equations [4] and [5] were strictly verified, the ratio $a(H_\beta)/a(H_\alpha)$ should be constant in the series of planar α-substituted ethyl radicals. The data listed in table III show that this ratio is actually far from being constant. The correlation between $a(H_\beta)$ and $a(H_\alpha)$ is in fact relatively poor :

$$a(H_\beta) = -1.753 + 1.294\, a(H_\alpha) \qquad r=0.949 \qquad [6]$$

Table III. Comparison of $a(H_\alpha)$ and $a(H_\beta)$ in planar $CH_3-\dot{C}HX$ radicals.

$R^\bullet$	$a(H_\alpha)$, G	$a(H_\beta)$, G	$a(H_\beta)/a(H_\alpha)$
$CH_3-\dot{C}H_2$	22.38 (24)	26.87 (24)	1.200
$CH_3-\dot{C}H-COOCH_3$	20.30 (33)	24.90 (33)	1.227
$CH_3-\dot{C}H-COOH$	19.80 (33)	24.80 (33)	1.252
$CH_3-\dot{C}H-CN$	19.65 (30)	23.05 (30)	1.173
$CH_3-\dot{C}H-COCH_3$	18.82 (34)	22.29 (34)	1.184
$CH_3-\dot{C}H-CHO$	18.10 (25)	21.50 (25)	1.188

From the mean value of the ratio $a(H_\beta)/a(H_\alpha)$ and the generally assumed value of Q_α, we can easily estimate the proportionality constant Q_β of equation [5] :

$$\langle a(H_\beta)/a(H_\alpha)\rangle = \frac{Q_\beta}{Q_\alpha} = 1.204 \qquad [7]$$

$$Q = 23.04 \times 1.206 = 27.74 \text{ G} \qquad [8]$$

Thus, we can evaluate $\rho_s^\pi(C_\alpha)$ in any planar methyl or ethyl substituted radical using one of the following equations :

$$a(H_\alpha) = 23.04 \ \rho_s^\pi(C_\alpha) \qquad [9]$$

or

$$a(H_\beta) = 27.74 \ \rho_s^\pi(C_\alpha) \qquad [10]$$

Some typical results obtained in this way are collected in table IV.

Table IV. Empirical values of $\rho_s^\pi(C_\alpha)$ in planar $\dot{C}XYZ$ radicals

X	Y	Z	$a(H_\alpha)$	$a(H_\beta)$	$\rho_s^\pi(C_\alpha)$
H	H	H	23.04 (24)	-	1
CH_3	H	H	-	26.87 (24)	0.969
H	H	$COOCH_3$	22.60 (33)	-	0.981
CH_3	H	$COOCH_3$	-	24.90 (33)	0.898

Table IV. (continued)

X	Y	Z	$a(H_\alpha)$	$a(H_\beta)$	$\rho_s^\pi(C_\alpha)$
H	H	COOH	21.31 (33)	-	0.925
CH_3	H	COOH	-	24.80 (33)	0.894
H	COOH	COOH	21.20 (35)	-	0.920
CH_3	COOH	COOH	-	25.30 (35)	0.912
H	H	CN	20.88 (30)	-	0.906
CH_3	H	CN	-	23.05 (30)	0.831
H	CN	COOH	19.81 (30)	-	0.860
H	H	$COCH_3$	19.50 (33)	-	0.846
CH_3	H	$COCH_3$	-	22.29 (34)	0.804
H	CN	CN	19.18 (31)	-	0.832
H	H	CHO	19.00 (36)	-	0.825
CH_3	H	CHO	-	21.50 (25)	0.775
H	H	C≡CH	18.90 (37)	-	0.820
H	OH	CN	18.10 (30)	-	0.786
CH_3	OH	CN	-	17.88 (30)	0.645
H	OH	COOH	16.80 (35)	-	0.729
CH_3	OH	COOH	-	17.10 (38)	0.616
H	NH_2	CN	14.90 (32)	-	0.647
H	NH_2	C≡CH	14.70 (32)	-	0.638
CH_3	SEt	CN	-	15.00 (39)	0.541
H	NH_2	COOH	11.77 (40)	-	0.424
CH_3	OH	$COCH_3$	-	13.33 (34)	0.481

It is seen that the combined action of an acceptor and a donor substituent particularly decreases the π spin density on the radical center.

3. THE STABILITY OF FREE RADICALS

The concept of stability is widely used in chemistry but, until very recently, it did not have received any unambiguous definition. Let us first notice that this concept may be used to characterize either an isolated species without making reference to any actual chemical process, or a given thermodynamical system generally containing one or several constituents capable to undergo a chemical reaction. It is well known that such a system is said to be thermodynamically stable if the corresponding standard free enthalpy change is largely positive :

$$\Delta G_o^o = \Delta H_o^o - T\Delta S_o^o \gg 0 \qquad [11]$$

where ΔH_o^o and ΔS_o^o are respectively the enthalpy and entropy changes of the chemical process under consideration. When this condition is fulfilled, the system is chemically inert and the corresponding reaction, impossible.

From a kinetic point of view, the stability of a compound may be defined by its lifetime under given experimental conditions. The rate constant or, restrictively, the activation energy of a chosen reaction of the considered compound can also be used. This type of stability is related to the concept of kinetic stabilization which has been carefully analised by Griller and Ingold (41), in the case of free radicals. These authors have introduced various adjectives for characterizing the kinetic stabilization of these species. A given radical is said to be transient if its lifetime is comparable to that of methyl radical under the same experimental conditions. In solution, the rate constant for the process by which the transient radical decays approaches the diffusion-controlled limit. A radical is persistent if its lifetime is significantly greater than that of methyl, again under the same experimental conditions. Finally, a pure radical is kinetically stable if it can be handled and stored without special precautions.

Therefore, it is seen that both thermodynamic and kinetic stability (stabilization) characterize a thermodynamical system which, in the present case, is a pure radical capable to undergo a so-called decay reaction. Thus, the decay rate constant can be used to measure the kinetic stabilization of the considered radical. However, if the decay process exclusively consists in the dimerization of the radical, the kinetic stabilization of the latter can be measured by the dimerization rate constant or, restrictively, the corresponding activation energy if the entropy contribution is assumed to be constant. Let us now define an intrinsic stability which does not depend on any chemical transformation of the species under consideration.

3.1. The stabilization energy of chemical species.

For defining the stabilization energy of a chemical species independently of any transformation we shall use the complementary notions of standard bond energies and reference compounds. Many experimental facts suggest to assign to chemical bonds characteristic energies transferable from one compound to another. So, for example, the thermoneutrality of the following reactions :

$$CH_3\text{-}CH_2\text{-}CH_2\text{-}CH_3 \;+\; CH_3\text{-}CH_3 \;\rightarrow\; 2\; CH_3\text{-}CH_2\text{-}CH_3$$

and

$$C_6H_{12} + 6\; CH_3\text{-}CH_3 \;\rightarrow\; 6\; CH_3\text{-}CH_2\text{-}CH_3$$

leads to write the corresponding heats of reactions as a sum of bond contributions :

$$\Delta H^o_o = -\sum_i k_i N_i [\, N_{CC} E(C\text{-}C) + N_p E(C\text{-}H)_p + N_s E(C\text{-}H)_s]_i = 0 \qquad [12]$$

where k_i is equal to +1 or -1 depending on whether it refers to a product or a reactant and N_i represents the stoechiometric coefficient of the considered species. So we define typical bonds (C-C, primary C-H, secondary C-H) having each characteristic properties (particularly their energy) transferable from one alkane to another. For this reason, these bonds are considered as standard ones. Equation [12] amounts to assume that the heat of atomization of each compound under consideration may be simply written as a sum of standard bond energy terms :

$$\Delta H^o_a = \sum N_{AB} E^o_{AB} \qquad [13]$$

which leads to a usual expression of the heat of reaction :

$$\Delta H^o_o = \sum_i k_i N_i (\Delta H^o_a)_i \qquad [14]$$

Linear alkanes, cyclohexane and, in general, all the species obeying the above formalism will be considered as normal or reference compounds being, by definition, neither stabilized nor destabilized. The additive form of equation [13] allowed us to calculate the standard bond energies given in table V and the specific bond increments listed in table VI (7,9,23).

Any compound having a heat of atomization which does not obey equation [13] shall be considered as stabilized or destabilized. Thus, we may logically define the thermodynamical stabilization energy of a chemical species by the following relation :

$$SE^o = \Delta H^o_a - \sum N_{AB} E^o_{AB} \qquad [15]$$

The stabilization energy shall be positive or negative depending on whether the corresponding species is stabilized or destabilized.

The data of tables V and VI can be used to calculate either the heats of atomization of normal compounds or, otherwise, the stabilization energies of chemical species, as illustrated in table VII.

Table V. Standard bond energies (kcal mol^{-1}, 298.15 K)

E(C-C)	85.05	E(C-N)	76.61	E(C-F)	109.50
$E(C-H)_p$	98.25	$E(N-H)_2$	91.31	$E(C-H)_p^F$	97.79
$E(C-H)_s$	97.50	$E(N-H)_1$	90.29	$E(C-H)_s^F$	97.26
$E(C-H)_t$	96.78	$E(C-H)_p^N$	96.75	$E(C-H)_t^F$	96.56
E(C=C)	136.77	$E(C-H)_s^N$	96.27	E(N-F)	64.70
$E(C_d-C)$	88.92	$E(C-H)_t^N$	95.87		
$E(C_d-H)_2$	100.24	E(N-N)	47.88	E(C-Cl)	85.56
$E(C_d-H)_1$	99.72	E(S-S)	61.99	$E(C-H)_p^{Cl}$	96.72
E(C≡C)	184.50	$E(C-H)_p^{SS}$	97.71	$E(C-H)_s^{Cl}$	96.38
$E(C_t-C)$	93.93	$E(C-H)_s^{SS}$	97.27	$E(C-H)_t^{Cl}$	95.83
$E(C_t-H)$	103.51				
		$E(C-NO_2)$	287.42	E(C-Br)	71.50
E(C-O)	91.33	$E(NO_2)$	210.81	$E(C-H)_p^{Br}$	97.11
E(O-H)	108.12	$E(C-H)_p^{NO_2}$	96.59	$E(C-H)_s^{Br}$	96.71
$E(C-H)_p^O$	95.86	$E(C-H)_s^{NO_2}$	95.79	$E(C-H)_t^{Br}$	96.78
$E(C-H)_s^O$	95.49	$E(C-H)_t^{NO_2}$	95.71		
$E(C-H)_t^O$	95.35	E(N=O)	119.35	E(C-I)	55.87
E(O-O)	46.14	E(N-O)	47.12	$E(C-H)_p^I$	97.73
$E(C-H)_p^{OO}$	95.79			$E(C-H)_s^I$	97.51
$E(C-H)_s^{OO}$	95.35	E(C-CN)	297.62	$E(C-H)_t^I$	97.42
E(C=O)	160.17	E(C≡N)	203.69		
$E(C_{CO}-H)_2$	100.24	$E(C-H)_p^{CN}$	98.59	E(N=N)	107.68

Table V. (Continued)

$E(C_{CO}-H)_1$	101.33	$E(C-H)_s^{CN}$	98.14
$E(_{CO}-C)$	93.68	$E(C-H)_t^{CN}$	98.23
$E(C-H)_p^{CO}$	98.25		
$E(C-H)_s^{CO}$	97.41	$E(C-S)$	71.63
$E(C-H)_t^{CO}$	96.33	$E(S-H)$	85.30
$E(COOH)$	383.96	$E(C-H)_p^{S}$	97.67
		$E(C-H)_s^{S}$	97.35
		$E(C-H)_t^{S}$	96.75

Table VI. Standard bond increments (kcal mol^{-1}, 298.15 K)

$\Delta(C_{CO}-C)$	8.63	$\Delta(C-H)_p^{Cl}$	-1.53
$\Delta(C_d-C)$	3.87	$\Delta(C-H)_s^{Cl}$	-1.12
$\Delta(C_t-C)$	8.88	$\Delta(C-H)_t^{Cl}$	-0.95
$\Delta(C_{CN}-C)$	8.88		
		$\Delta(C-H)_p^{Br}$	-1.14
$\Delta(C-H)_p^{O}$	-2.39	$\Delta(C-H)_s^{Br}$	-0.79
$\Delta(C-H)_s^{O}$	-2.01	$\Delta(C-H)_t^{Br}$	0.00
$\Delta(C-H)_t^{O}$	-1.43		
		$\Delta(C-H)_p^{I}$	-0.52
$\Delta(C-H)_p^{N}$	-1.50	$\Delta(C-H)_s^{I}$	0.01
$\Delta(C-H)_s^{N}$	-1.23	$\Delta(C-H)_t^{I}$	0.64
$\Delta(C-H)_t^{N}$	-0.91		

Table VI. (Continued)

		$\Delta(C\text{-}H)_p^{CO}$	0.00
$\Delta(C\text{-}H)_p^{OO}$	-2.46	$\Delta(C\text{-}H)_s^{CO}$	-0.09
$\Delta(C\text{-}H)_s^{OO}$	-2.15	$\Delta(C\text{-}H)_t^{CO}$	-0.45
$\Delta(C\text{-}H)_p^{CN}$	0.34	$\Delta(C\text{-}H)_p^{S}$	-0.58
$\Delta(C\text{-}H)_s^{CN}$	0.64	$\Delta(C\text{-}H)_s^{S}$	-0.15
$\Delta(C\text{-}H)_t^{CN}$	1.45	$\Delta(C\text{-}H)_t^{S}$	-0.03
$\Delta(C\text{-}H)_p^{NO_2}$	-1.66	$\Delta(C\text{-}H)_p^{SS}$	-0.54
$\Delta(C\text{-}H)_s^{NO_2}$	-1.71	$\Delta(C\text{-}H)_s^{SS}$	-0.23
$\Delta(C\text{-}H)_t^{NO_2}$	-1.07		
$\Delta(C\text{-}H)_s^{F}$	-0.24		
$\Delta(C\text{-}H)_t^{F}$	-0.22		

Table VII. Typical heats of atomization (kcal mol^{-1},298.15 K)

Compound	$\Sigma N_{AB}E^o_{AB}$	ΔH^o_a	SE^o
$nC_{18}H_{38}$	5155.35	5155.00 (42)	∿0
$(tC_4H_9)_2CHCH(tC_4H_9)_2$	5176.41	5115.19 (43)	-61.22

It is seen that the heat of atomization of normal octadecane is practically equal to the sum of standard bond energies while tetratertiobutylethane appears to be a highly destabilized compound, obviously owing to very large steric repulsions between the tertiobutyl groups.

Our definition of thermodynamical stabilization energy is absolutely general for its applies to any molecule, free radical and positive or negative ion (7,9,11,23). Moreover, the concept of stabilization energy achieves a synthesis of former disparate notions such as hyperconjugation, resonance, ring strain and steric repulsion energies. Finally, as it does not explicitly refer to any chemical reaction, it may be taken as an intrinsic property of the considered species.

3.2. The stabilization energy of free radicals

For determining the stabilization energy of any species using equation [15] one needs to know its heat of atomization and therefore its heat of formation. As the heats of formation of many chemical species, specially reactive intermediates, are still experimentally not determined, we briefly recall here the semiempirical procedure for estimating these quantities.

3.2.1. The semiempirical procedure for determining ΔH_f^o. Unknown heats of formation can be estimated by using an interesting semiempirical procedure proposed by Pople and coworkers (97). This is briefly summarized below. Heats of reaction can be theoretically calculated at any temperature from the total energies of the partners and the frequencies of their normal modes of vibration. However, it is well known that the results obtained at the SCF level are not very accurate, essentially owing to correlation errors on total energies which only accidentally cancel. The Pople et al. procedure is based on the concept of isodesmic reaction. By definition, in this type of process, all the bonds are conserved (in number and nature) which leads to approximate cancellation not only of correlation errors but also ZPE and thermal corrections. Thus, the corresponding energy change at 0 K, theoretically obtained at the SCF level, is approximately equal to the experimental heat of reaction at 298.15 K

$$\Delta H_o^o(0) \simeq \Delta H_o^o(298.15) = \sum_i k_i N_i [\Delta H_f^o(298.15)]_i \qquad [16]$$

So, we can use a semiempirical procedure for estimating unknown heats of formation. It consists simply in introducing experimental heats of formation of reference compounds in the above equation as explicitly described elsewhere (7,9). The reliability of the results obtained by this procedure essentially depends on three factors :
i) The quality of the SCF energies. Except where otherwise specified we have used total energies calculated at the ab initio SCF 4-31 G level with full geometry optimization using the gradient techniques of the Gaussian series of programs (98). These energies are generally referred as 4-31 G/4-31 G.
ii) The choice of the isodesmic reaction. The best choice corresponds to a process where the properties of all the bonds, particularly their length, are conserved. It is the reason why we have now abandonned the reaction with ethane for determining the heat of formation of

carbon centered free radicals. We try to systematically use the reactions which respect the above condition. For example, the conservation of the bond lengths is reasonably respected in the following isodesmic process:

$$H_2N - \dot{C}H - CN + \dot{C}H_3 \rightarrow H_2N - \dot{C}H_2 + \dot{C}H_2 - CN$$

$$1.374 \quad 1.378\ (\text{Å}) \qquad\qquad 1.393 \qquad\qquad 1.377\ (\text{Å})$$

Moreover, the heat of reactions is very small (1.43 kcal mol^{-1}) which is another criterion of isodesmicity. On the other hand, in the reaction with ethane :

$$H_2N - \dot{C}H - CN + CH_3 - CH_3 \rightarrow \dot{C}H_2 - CH_3 + H_2N - CH_2 - CN$$

$$1.374 \quad 1.378 \qquad 1.526\ (\text{Å}) \quad 1.496 \qquad 1.472 \quad 1.445\ (\text{Å})$$

the bond lengths are really not conserved and the thermal effect is very large (19.28 kcal mol^{-1}).
iii) The heats of formation of the reference compounds. For the last few years, the recommended values of the experimental heats of formation of several free radicals have known very large modifications. The aminomethyl radical is a typical example which deserves to be mentioned. Until 1982, the value of ΔH_f^o (298) recommended by Mc Millen and Golden was 35.7 kcal mol^{-1} (44). It was based on the work of Griller and Lossing (45). On the other hand, in 1977, Colussi and Benson had found the value of 37 kcal mol-1 (46) which becomes 34.2 kcal mol^{-1} if a more recent value of the heat of formation of benzyl radical is used [47 kcal mol^{-1} (47)]. Finally using two different experimental methods Grela and Colussi respectively obtain 31.1 (48) and 30.3 kcal mol^{-1} (47). For lack of a better choice we shall adopt a mean value of 33 kcal mol^{-1} for this quantity.

The stabilization energies of a series of carbon centered free radicals are collected in table VIII. They have been derived either from the corresponding experimental heats of formation when these were available or from semiempirical heats of formation generally calculated at the 4 - 31 G/4 - 31 G level using the isodesmic reaction with $\dot{C}H_3$ and the heats of formation of reference species collected in table IX. The total energies used in this work are given in reference (18) or listed in table X.

Table VIII. Thermodynamical properties of carbon centered radicals.

$R^\cdot$	$\Delta H_f^o(298.15)^a$	$SE^o(298.15)^a$
$\dot{C}(CN)_3$	134.90^b	-5.16
$\dot{C}H(CH_3)_2$	19.2 (50)	1.82

Table VIII. (Continued).

$R^{\cdot}$	$\Delta H_f^o(298.15)$[a]	$SE^o(298.15)$[a]
$CH(CN)_2$	92.54[b]	3.34
$\dot{C}H_2COOH$	-58.64[c]	3.43
$\dot{C}(CH_3)_3$	9.4 (50)	3.70
$\dot{C}H(CN)(OH)$	18.86[b]	5.83
$CH_3\dot{C}HNH_2$	24.61[c]	7.89
$CH_3\dot{C}HCN$	50.1 (44)	8.35
$\dot{C}H(OH)_2$	55.23[b]	8.73
$\dot{C}H(NH_2)_2$	33.79[b]	10.21
$\dot{C}H_2C{\equiv}CH$	81.50 (54)	10.56
$\dot{C}H(OH)(COOH)$	106.03[c]	10.82
$NH_2\dot{C}(CN)_2$	89.35[b]	12.08
$CH_2{=}CH\dot{C}(CN)_2$	102.86[b]	12.84
$CH_3\dot{C}HC{\equiv}CH$	72.4 (54)	13.18
$\dot{C}H_2CH{=}CH_2$	39.1 (44)	13.21
$\dot{C}H(NH_2)(CN)$	54.97	14.96
$CH_3\dot{C}HCH{=}CH_2$	30.4 (44)	15.43
$CH_2{=}CH\dot{C}(OH)_t(CN)_c$	25.67[b]	15.93
$CH_3\dot{C}(NH_2)(CN)$	46.47[c]	16.08
$CH_2{=}CH\dot{C}HOH$	-4.03(tr.);-4.09(cis.)[b]	16.10(tr.);16.16(cis.)
$CH_2{=}CH\dot{C}(OH)_c(CN)_t$	24.79[b]	16.81
$CH_2{=}CH\dot{C}HCN$	66.07(tr.);66.57(cis.)[b]	17.19(tr.);16.69(cis.)
$\dot{C}H(NH_2)(C{\equiv}CH)$	79.25[c]	17.81
$\dot{C}H(NH_2)(COOH)$	-69.92[c]	19.95

Table VIII. (Continued).

$R^{\cdot}$	ΔH_f^o (298.15)[a]	SE^o (298.15)[a]
$CH_2CH\dot{C}(OH)_2$	-52.54[b]	20.07
$\dot{C}H(CH=CH_2)_2$	49.0 (44)	20.31

[a] In kcal mol^{-1}
[b] 4-31 G/4-31 G results.
[c] 6-31 G/6-31 results.

Table IX. Reference thermodynamical data (kcal mol^{-1},298.15 K)

Radical	ΔH_f^o	SE^o	BDE (C-H)
$\dot{C}H_3$	35.1 (44)	-2.65	105.09 (49)
$\dot{C}H_2CH_3$	28.0 (50)	-0.50	100.48 (49)
$\dot{C}H_2F$	-7.8 (44)	-2.06	101.10 (51)
$\dot{C}H_2OH$	-6.2 (44)	2.53	93.97 (49)
$\dot{C}H_2OCH_3$	-2.8 (44)	3.44	93.28 (49)
$\dot{C}H_2NH_2$	33.0[a]	7.53	90.60 (49)
$\dot{C}H_2Cl$	28.3 (44)	-2.60	100.00 (49)
$\dot{C}H_2CN$	58.5 (44)	6.60	92.90 (52)
$\dot{C}H_2CHO$	3.0 (53)	4.66	94.83 (49)
$\dot{C}H_2CHCH_2$	39.1 (44)	13.21	86.62 (49)
$\dot{C}H_2C_6H_5$	47.0 (47)	11.61[c]	87.10 (49)
$\dot{C}H_2CCH$	81.5 (54-55)	10.56	89.21 (49)

[a] See text.
[b] References correspond to ΔH_f^o(R-H)
[c] This SE^o does not contain the resonance energy of the phenyl ring.

Table X. Total energies (a.u.)

Species	Energy
$\dot{C}H_3$	- 39.54666[a]
$\dot{C}H_2NH_2$	- 94.55252[a]
$\dot{C}H_2CN$	-131.25116[a]
$\dot{C}H_2C\equiv CH$	-115.21505[a]
$\dot{C}H_2COOH$	-227.08096[a]
$\dot{C}H(OH)(COOH)$	-301.90693[a]
$\dot{C}H(NH_2)(COOH)$	-282.10145[a]
$\dot{C}H(NH_2)(C\equiv CH)$	-170.22114[a]
$\dot{C}H_2CH_3$	- 78.56862[a]
$CH_3\dot{C}HNH_2$	-133.57717[a]
$CH_3\dot{C}HCN$	-170.27466[a]
$CH_3\dot{C}(NH_2)(CN)$	-225.28359[a]
$\dot{C}H_2-CH=CH_2$	-116.30561[b]
$CH_2=CH-\dot{C}(CN)(H)$ (t)	-207.90840[b]
$CH_2=CH-\dot{C}(H)(CN)$ (c)	-207.90760[b]
$CH_2=CH-\dot{C}(OH)(H)$ (t)	-191.05007[b]
$CH_2=CH-\dot{C}(H)(OH)$ (c)	-191.05016[b]
$CH_2=CH-\dot{C}(CN)(CN)$	-299.49554[b]

Table X. (Continued)

Species	Energy
$CH_2=CH-\dot{C}$ ↗ OH ↘ OH	-265.80311[b]
$CH_2=CH-\dot{C}$ ↗ CN (t) ↘ OH (c)	-282.64990[b]
$CH_2=CH-\dot{C}$ ↗ OH (t) ↘ CN (c)	-282.64850[b]
CH_3COOH	-227.70112[a]
$CH_3C\equiv CH$	-115.82353[a]
CH_3CN	-131.86523[a]
H	-0.49823[a]
C_2H_4	-78.00446[a]
$CH_2=CHNH_2$	-133.01574[a]
$CH_2=CHCN$	-169.69392[a]
$CH_2=C(NH_2)(CN)$	-224.69983[a]

[a] 6 - 31 G/6 - 31 G results

[b] 4 - 31 G/4 - 31 G results

The difference between the results of table VIII and those previously described essentially come from the chosen isodesmic process and reference data. In table XI, we summarize the influence of substituents on the stabilization energy of carbon centered free radicals. The results corresponding to a strict additivity of the effects of substituents are also given in the same table (in brackets).

It is seen that captodative radicals generally have a larger stabilization energy than the sum of the stabilization energies of the corresponding monosubstituted species. Hydroxycyanomethyl radical is a typical exception to this rule unless its experimental heat of atomization is inaccurate. Moreover, CH_3 behaves as a donor group and

Table XI. Influence of substituents on the stabilization energies of $\dot{C}HXY$ radicals (kcal mol^{-1}, 298.15 K)

	H	CH_3	CN	C≡CH	COOH	OH	NH_2	$CH=CH_2$
H	-2.65	-0.50	6.60	10.56	3.43	2.53	7.53	13.21
CH_3	-0.50	1.82 (-1.00)	8.35 (6.10)	13.18 (10.06)	- (2.93)	- (2.03)	7.89 (7.03)	15.43 (12.71)
CN	6.60	8.35 (6.10)	3.34 (13.20)	- (17.16)	- (10.03)	5.83 (9.13)	14.96 (14.13)	16.94[a] (19.81)
C≡CH	10.56	13.18 (10.06)	- (17.16)	- (21.12)	- (13.99)	- (13.09)	17.81 (18.09)	- (23.77)
COOH	3.43	- (2.93)	- (10.03)	- (13.99)	- (6.86)	10.82 (5.96)	19.95 (10.96)	- (16.64)
OH	2.53	- (2.03)	5.83 (9.13)	- (13.09)	10.82 (5.96)	8.73 (5.06)	- (10.06)	16.13[a] (15.74)
NH_2	7.53	7.89 (7.03)	14.96 (14.13)	17.81 (18.09)	19.95 (10.96)	- (10.06)	10.21 (15.06)	- (20.74)
$CH=CH_2$	13.21	15.43 (12.71)	16.94[a] (19.81)	- (23.77)	- (16.64)	16.13[a] (15.74)	- (20.74)	20.31 (26.42)

[a] Mean value calculated from the stabilization energies of the transoid and cisoid configurations.

CH_2=CH either as a donor or a captor depending on whether it is opposed to a captor or a donor group respectively. The best captodative couples are ($CH=CH_2$; OH), (CN, NH_2), (C≡CH, NH_2) and (COOH, NH_2). For the latter, the so-called extrastabilization energy (56) is of about 9 kcal mol^{-1}. Let us also point out the large stabilization energy of the species $\dot{C}H(CH=CH_2)_2$ which however is smaller than twice the allyl stabilization energy.

3.2.2. A critical review of various methods for estimating stabilization energies of free radicals.

1° The methods based on thermochemical and kinetical data.

We shall first compare our definition of radical stabilization to former ones successively proposed by Szwarc (57), Benson (58) and Rodgers et al (59). According to Szwarc, the resonance energy of an unsaturated radical ($R_\pi\dot{C}H_2$) is the heat of reaction of the following isodesmic process :

$$R_\pi\dot{C}H_2 + CH_4 \rightarrow R_\pi CH_3 + \dot{C}H_3$$

We may write :

$$\Delta H^o_o = BDE(C\text{-}H)_{CH_4} - BDE(C\text{-}H)_{R_\pi CH_2} \qquad [17]$$

This approach has been generalized to any type of radicals and it is the reason why the R-H bond dissociation energies have been widely used to estimate the unpaired electron delocalization energies of these species.
The definitions of Benson and Rodgers et al. only differ from the previous one by the choice of the isodesmic reaction, CH_4 being replaced in this process by $R_\pi CH_3$ or C_2H_6 respectively. It is easy to demonstrate that BDE(R-H) cannot generally be used for estimating the stabilization energy of a free radical.Indeed, BDE(R-H) can be explicitly written as :

$$BDE(R\text{-}H) = \Delta H^o_a(RH) - \Delta H^o_a(R^\cdot) \qquad [18]$$

or, according to equation [15]

$$BDE(R\text{-}H) = \sum_{RH} N_{AB}E^o_{AB} - \sum_{R^\cdot} N_{AB}E^o_{AB} + SE^o(RH) - SE^o(R^\cdot) \qquad [19]$$

and, finally

$$BDE(R\text{-}H) = \Delta\Sigma N_{AB}E^o_{AB} + \Delta SE^o \qquad [20]$$

In the particular case of carbon centered species, the term $\Delta\Sigma N_{AB}E^o_{AB}$ is equal to the standard energy of a given C-H bond and it generally varies from one molecule to another. Then, strictly speaking, BDE(C-H) does not linearly depends on ΔSE^o and, a fortiori, on $SE^o(R^\cdot)$ only.

Table XII. Bond dissociation energies, BDE(C-H), of substituted methanes (kcal mol^{-1};298.15 K)

RH	ΔH_f^o(RH)	SE^o(RH)	SE^o(R$^\cdot$)	BDE(R-H)
$H-C(CN)_3$	114.78 (7)	-35.03	-5.16	72.22
$H-CH_2(CH_3)_2$	-24.83 (49)	-0.27	1.82	96.13
$H-CH_2(CN)_2$	63.5 (49)	-13.40	3.34	81.14
$H-CH_2COOH$	-103.27 (49)	0.00	3.43	96.73
$H-C(CH_3)_3$	-32.41 (49)	0.83	3.70	93.92
H-CH(OH)(CN)	-11.48 (18)	-7.19	5.83	82.44
$H-CH(CH_3)(NH_2)$	-11.35 (49)	-0.72	7.89	88.06
$H-CH(CH_3)(CN)$	12.3 (60)	0.00	8.35	89.90
$H-CH(OH)_2$	-93.5 (61)	6.06	8.73	90.37
$H-CH(NH_2)_2$	-1.19 (11)	2.15	10.21	87.08
$H-CH_2C{\equiv}CH$	44.39 (49)	0.00	10.56	89.21
$H-C(NH_2)(CN)_2$	75.26 (7)	-20.24	12.08	66.19
$H-CH_2CH{=}CH_2$	4.88 (49)	-0.22	13.21	86.62
$H-CH(NH_2)(CN)$	27.94 (18)	-2.41	14.96	79.13
$H-CH(CH_3)(CH{=}CH_2)$	-0.2 (49)	0.00	15.43	82.70
$H-CH(OH)(CH{=}CH_2)$	-29.55 (49)	-1.91	16.13	77.59
$H-CH(NH_2)(COOH)$	-93.72 (49)	-1.27	19.95	75.10
$H-CH(CH{=}CH_2)_2$	25.25 (49)	-0.73	20.31	75.75

The data of table XII actually lead to the following correlation :

$$BDE(C-H) = 96.527 + 0.929[SE^o(RH) - SE^o(R^\cdot)] \quad r=0.969 \qquad [20]$$

So, a small BDE(C-H) may correspond either to a very stabilized radical such as $\dot{C}H(NH_2)(COOH)$ or to a very destabilized molecule as $CH(CN)_3$. All the intermediate cases can obviously be encountered

The thermochemistry and the kinetics of the thermolysis of azo compounds has also been used to estimate the relative stabilities of carbon centered free radicals (62-63).
Assuming the following stoechiometric equation for this process

$$R\text{-}N{=}N\text{-}R \rightarrow 2R^{\cdot} + N_2$$

it is easy to derive the explicit expression of the corresponding heat of reaction :

$$\Delta H_o^o = \sum_{AZO} N_{AB}E_{AB}^o - 2\sum_{R^{\cdot}} N_{AB}E_{AB}^o - E^o(N{\equiv}N) + \Delta SE^o \qquad [21]$$

where SE^o represents the difference between the stabilization energies of the radicals and the azo compound :

$$\Delta SE^o = SE^o(azo) - 2SE^o(R^{\cdot}) \qquad [22]$$

One may easily verify that the contribution of the three first terms of equation [21] is not constant in a series of azo compounds. Thus we cannot expect to find a strict linear relation neither between ΔH_o^o and ΔSE^o nor, a fortiori, between ΔH_o^o and $SE^o(R^{\cdot})$. Nevertheless, using the data collected in table XIII, one observes a good correlation between ΔH_o^o and $SE^o(R^{\cdot})$:

$$\Delta H_o^o = 34.748 - 2.198\ SE^o(R^{\cdot}) \qquad r=0.995 \qquad [23]$$

Moreover, as an Evans-Polanyi relation does actually exist for these reactions :

$$E_a = 26.868 + 0.620\ \Delta H_o^o \qquad r=0.997 \qquad [24]$$

a good correlation is also observed between E_a and $SE^o(R^{\cdot})$:

$$E_a = 48.395 - 1.361\ SE^o(R^{\cdot}) \qquad r=0.992 \qquad [25]$$

Table XIII. Heats of reaction and activation energies of the thermolysis of azo compounds (kcal mol^{-1}, 298.15K)

Compound	ΔH^o_o	E_a	$SE^o(R^\cdot)$
$C_6H_5N{=}NC_6H_5$	62.7 (68)	66.44 (64)	-11.17 (44)
$CF_3N{=}NCF_3$	-	57.3 (65)	10.98 (44)
$CH_3N{=}NCH_3$	38.0 (69)	51.2 (66)	-2.65
$C_2H_5N{=}NC_2H_5$	34.03 (70)	48.5 (66)	-0.50
$nC_3H_7N{=}NC_3H_7$	33.33 (70)	45.7 (67)	0
$(CH_3)_2CHN{=}NCH(CH_3)_2$	29.89 (70)	45.31 (66)	1.82
tBuN=NtBu	27.50 (70)	43.0 (67)	3.70
$NCC(CH_3)_2N{=}NC(CH_3)_2CN$	10.53 (71)	33.4 (67)	12.12 (44)
$(CH_2CH)C(CH_3)_2N{=}NC(CH_3)_2(CHCH_2)$	-8.40 (67)	22.4 (62-63)	19.61 (44)

However, it is worth noticing that relations [23] and [25] do not hold when the stabilization energy of the azo compound is large. The species $CF_3N{=}NCF_3$ is a characteristic example of this type of species :

$$SE^o(CF_3N{=}NCF_3) \simeq 2SE^o(CF_3H) \simeq 2 \times 21.26 \text{ (49)} \qquad [26]$$

The corresponding activation energy (57.3) obviously does not obey equation [25]. We may also anticipate that neither ΔH^o_o nor E_a for the thermolysis of the azo compound $(NH_2)(CN)_2CN{=}NC(CN)_2(NH_2)$ would be directly related to the stabilization energy of the corresponding radical.

On the other hand, it is easy to demonstrate that a rotational barrier is exactly equal to the difference between the stabilization energy of the equilibrium structure of the species under consideration and the stabilization energy of the transition structure :

$$V = SE^o(M) - SE^o(M^{\neq}) \qquad [27]$$

Therefore it cannot strictly be used to determine only one of these terms. To illustrate this remark we shall successively consider the rotational barriers about the C-N bonds in aminoalkyl radicals and the barriers to internal rotation in 1,1-substituted allyl radicals. These data are collected in table XIV and XV respectively. The stabilization energies of the corresponding species and those of the radicals which most resemble the moieties containing the unpaired electron in the transition structures are also given in the same tables.

Table XIV. Rotational barriers in aminoalkyl radicals $H_2N\dot{C}HR$ (72)

$R\dot{C}HNH_2$	V_o [a]	$SE^o(H_2N\dot{C}HR)$ [a]	$SE^o(\neq)$ [a]	$SE^o(H_2\dot{C}R)$ [a]
$CH_3\dot{C}HNH_2$	7.6	7.89	0.29	-0.5
$(CH_3)_2CH\dot{C}HNH_2$	7.3	-	-	-
$(CH_3)_3C\dot{C}HNH_2$	7.5	-	-	-
$(CH_3)_3COCO\dot{C}HNH_2$	14.9	(19.95) [b]	(5.05)	(3.43) [b]
$HC{\equiv}\dot{C}HNH_2$	10.5	17.81	5.31	10.56
$NC\dot{C}HNH_2$	11	14.96	3.96	6.6

[a] In kcal mol^{-1}

[b] Values corresponding to R≡COOH

Table XV. Rotational barriers in allylic radicals (kcal mol^{-1})(73)

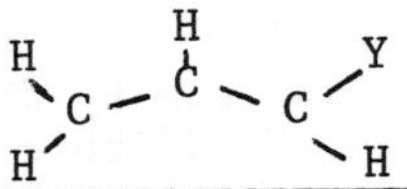

X	Y	V_o	SE^o(allylic)	$SE^o(\neq)$	$SE^o(H\dot{C}XY)$
H	H	15.7	13.21	-2.49	-2.65
CN	H	10.6	16.69	6.09	6.60
H	CN	9.8	17.19	7.39	6.60
MeO	H	14.7	-	-	3.44
H	MeO	14.3	-	-	3.44

Table XV. (Continued)

X	Y	V_o	SE^o(allylic)	$SE^o(\neq)$	$SE^o(H\dot{C}XY)$
OH	H	-	16.16	-	2.53
H	OH	-	16.10	-	2.53
CN	CN	9.6	12.84	3.24	3.34
MeO	MeO	12.0	-	-	-
OH	OH	-	20.07	-	8.73
MeO	CN	6.0	-	-	-
CN	MeO	6.1	-	-	-
OH	CN	-	16.81	-	5.83
CN	OH	-	15.93	-	5.83

It is seen that the empirical stabilization energies of the transition structures

$$SE^o(\neq) = SE^o(\text{radical}) - V_o \qquad [27]$$

are well correlated to those of the corresponding radicals :

$$SE^o(\neq) = 0.625 + 0.461\ SE^o(CH_2R) \qquad r=0.995 \qquad [28]$$

and

$$SE^o(\neq) = 0.091 + 0.998\ SE^o(\dot{C}HXY) \qquad r=0.992 \qquad [29]$$

On the whole, an energy change cannot generally be used to determine any property, particularly the stabilization energy, of one of the species involved in the corresponding elementary process. Other examples of this important remark will be given in the following parts of this paper.

2° The approach based on ESR data

As shown by the data collected in table XVI, there exists a correlation between the hyperfine splitting constants $a(H_\alpha)$ of a series of carbon centered radicals and the corresponding stabilization energies. It may be written as :

$$a(H_\alpha) = 22.123 - 0.523\ SE^o(R^\cdot) \qquad r=0.966 \qquad [30]$$

Thus, the terms $a(H_\alpha)$ can, at least qualitatively, provide information on the thermodynamic stabilization of carbon centered free radicals due to the delocalization of their unpaired electron.

4. A LIMITED APPROACH TO FREE RADICAL CHEMISTRY

In this part, we shall show that the concept of stabilization energy can be used to rationalize the thermochemistry and the kinetics of different elementary processes involving free radicals. We shall more particularly analyse the effects of substituents on the various parameters of the following processes :

- hydrogen transfer reactions;
- carbon-carbon bond homolyses of substituted ethanes;
- free radical recombination reactions;
- radical additions to substituted ethylenes.

For this purpose, we shall use either experimental results when they are available or data obtained by various theoretical approaches.

Table XVI. Comparison of $a(H_\alpha)$, G and SE^o, kcal mol^{-1}

Radical	$a(H_\alpha)$	SE^o
$\dot{C}H_2CH_3$	22.38 (24)	-0.50
$\dot{C}H(CN)_2$	19.18 (31)	3.34
$\dot{C}H_2COOH$	21.31 (33)	3.43
$\dot{C}H_2CHO$	19.00 (36)	4.66 (74)
$\dot{C}H(CN)(OH)$	18.10 (30)	5.83
$\dot{C}H(CH_3)(COCH_3)$	18.82 (34)	6.25 (44)
$\dot{C}H_2CN$	20.88 (30)	6.60
$\dot{C}H_2C_6H_5$	16.34 (76)	11.61
$\dot{C}H_2CHCH_2$	14.81 (77)	13.21
$\dot{C}H(C_6H_5)_2$	14.48 (76)	14.54 (75)
$\dot{C}H(NH_2)(CN)$	14.90 (32)	14.96
$CH_3\dot{C}HCHCH_2$	13.83 (77)	15.43
$CH_2CH\dot{C}HOH$	13.15 (77)	16.13
$\dot{C}H(NH_2)(COOH)$	11.77 (40)	19.95

4.1. Hydrogen transfer reactions.

The hydrogen transfer reactions :

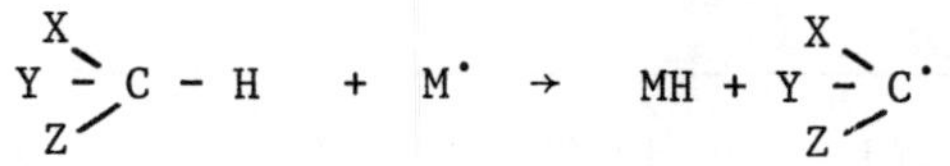

are tightly related to the carbon-hydrogen bond homolyses of the molecules involved in these processes. As a matter of fact, their heat of reaction is nothing else but a difference of bond dissociation energies :

$$\Delta H^o_o = BDE(C\text{-}H) - BDE(M\text{-}H) \qquad [31]$$

Thus, the thermochemistry of hydrogen transfer reactions can be rationalized in terms of stabilization energies. Indeed, using equation [19], one readily obtains the following approximate relation, for a given attacking radical :

$$\Delta H^o_o \simeq E^o(C\text{-}H) + SE^o(CHXYZ) - SE^o(\dot{C}XYZ) + C \qquad [32]$$

It is seen that the most exothermic hydrogen abstraction reactions correspond to the most destabilized substituted methanes and (or) the most stabilized carbon centered radicals. Some typical examples are given in table XVII.

We may wonder whether the kinetic parameters of hydrogen transfer reactions can be rationalized as the corresponding heats of reaction. To answer this question, we shall try to find an explicit expression of the activation barriers of these processes in terms of stabilization energies. Then, we intend to calculate the corresponding rate constants using the formalism of the transition state theory including a tunnelling correction. Indeed, reliable results can be obtained by this approach, provided that precise barrier heights are available.

Table XVII. Heats of reaction of hydrogen tranfer processes[a] $XYZ\dot{C}H + CH_3 \rightarrow CH_4 + \dot{C}XYZ$

CHXYZ	$SE^o(CHXYZ)$	$SE^o(CXYZ)$	ΔSE^o	ΔH^o_o [b]
$CH_2(OH)_2$	6.06	8.73	-2.67	-14.72
$CH_2(NH_2)_2$	2.15	10.21	-8.06	-18.01
$CH_2(CH{=}CH_2)_2$	3.54	20.31	-16.77	-22.11
$CH_2(OH)(CN)$	-7.19	5.83	-13.02	-22.65
$CH_2(CN)_2$	-13.40	3.34	-16.74	-23.95

Table XVII. (Continued).

CHXYZ	$SE^{o}(CHXYZ)$	$SE^{o}(\dot{C}XYZ)$	ΔSE^{o}	ΔH_{o}^{o} [b]
$CH_2(NH_2)(CN)$	-2.41	14.96	-17.37	-25.96
$CH_2(OH)(CH{=}CH_2)$	-1.91	16.13	-18.04	-27.50
$CH_2(NH_2)(COOH)$	-1.27	19.95	-21.22	-30.09
$CH(CN)_3$	-35.03	-5.16	-29.87	-32.87
$CH(NH_2)(CN)_2$	-20.24	12.08	-32.32	-38.90

[a] In kcal mol^{-1}, at 298.15 K.

[b] $\Delta H_{o}^{o} = -\ 12.053 + 0.774\ \Delta SE^{o}$ $r = 0.973$

It is well known that the activation barriers calculated by the ab initio methods of quantum chemistry are generally not very accurate due to the partial neglect of correlation effects in the reactants and the transition structure. It is the reason why we have developed a semi-empirical procedure for estimating very simply the activation barriers of atom transfer reactions (12,14,23).
The activation barrier of the exchange reaction :

$$AB + C \rightarrow A + BC$$

can be written :

$$\Delta E_1^{\neq}(ABC) = D_{AB} - E(A...B...C) \qquad [33]$$

where D_{AB} is the electronic bond dissociation of the bond A-B and E(A...B...C) is the energy required to dissociate the activated complex into the fragments A, B and C. This quantity may be written as a sum of bond contributions :

$$E(A...B...C) = E(A...B) + E(B...C) \qquad [34]$$

We represent the dissociation of the bonds A-B and B-C by the following Morse curves :

$$M_{AB}(R) = D_{AB}\{(1-\exp[a_{AB}(R_e^{AB}-R)])^2 - 1\} = D_{AB}\mu_{AB}(R) \qquad [35]$$

$$M_{BC}(R) = D_{BC}\{(1-\exp[a_{BC}(R_e^{BC}-R)])^2 - 1\} = D_{BC}\mu_{BC}(R) \qquad [36]$$

Assuming that A...B and B...C (at the transition state) are both one-electron bonds, equation [34] can be more explicitly written as :

$$E(A...B...C) = -\frac{1}{2}[M_{AB}(R^{\neq}_{AB}) + M_{BC}(R^{\neq}_{BC})] \quad [37]$$

where $R^{\neq}_{AB}$ and $R^{\neq}_{BC}$ are the corresponding bond lengths at the transition state. Thus, equation [33] becomes :

$$\Delta E^{\neq}_{1}(ABC) = D_{AB} + \frac{1}{2}[M_{AB}(R^{\neq}_{AB}) + M_{BC}(R^{\neq}_{BC})] \quad [38]$$

The activation barrier of the reverse reaction :

$$A + BC \rightarrow AB + C$$

is also easy to deduce. It is written :

$$\Delta E^{\neq}_{2}(ABC) = D_{BC} + \frac{1}{2}[M_{AB}(R^{\neq}_{AB}) + M_{BC}(R^{\neq}_{BC})] \quad [39]$$

Finally, the reaction energies are evaluated by the relation :

$$\Delta E^{1}_{o} = \Delta E^{\neq}_{1}(ABC) - \Delta E^{\neq}_{2}(ABC) = D_{AB} - D_{BC} = -\Delta E^{2}_{o} \quad [40]$$

Therefore, the activation barrier of any atom transfer reaction can be estimated from the empirical Morse functions of the bonds involved in the process and the theoretical structure of the activated complex. The activation barriers of the identity reactions :

$$AB + A \rightarrow A + BA$$

and

$$CB + C \rightarrow C + BC$$

can be respectively written as :

$$\Delta E^{\neq}_{1}(ABA) = D_{AB} + M_{AB}(R'^{\neq}_{AB}) \quad [41]$$

and

$$\Delta E^{\neq}_{1}(CBC) = D_{BC} + M_{BC}(R'^{\neq}_{BC}) \quad [42]$$

where $R'^{\neq}_{AB}$ and $R'^{\neq}_{BC}$ are the lengths of the bonds AB and BC respectively in the symmetrical activated complexes. Assuming they are equal to the corresponding distances in the dissymmetrical transition structures, we obtain a very simple expression of the activation barrier of exchange reactions :

$$\Delta E^{\neq}_{1}(ABC) = \frac{1}{2}[\Delta E^{\neq}_{1}(ABA) + \Delta E^{\neq}_{1}(CBC)] + \frac{1}{2}\Delta E^{1}_{o} \quad [43]$$

The above formalism can be used not only to predict but also to rationalize the latter quantities (12,14). We shall only consider here the hydrogen transfer reactions leading to carbon centered free radicals ($\dot{C}XYZ$). Using the data of tables IX and XII and the procedure described in reference (14), we may systematically calculate the activation barriers of a large variety of hydrogen transfer reactions leading to carbon centered radicals :

$$XYZCH + M^{\cdot} \rightarrow \dot{C}XYZ + MH$$

As shown elsewhere (14) these activation barriers obey the following equation, for any given attacking radical :

$$\Delta E_1^{\neq}(RHM) = \frac{1}{2}\,\Delta E_o^1 + C \qquad [44]$$

This is merely an Evans-Polanyi relationship at 0 K, with fixed nuclei. Moreover, in accordance with equation [32], the following relations have been derived (14)

$$\Delta E_1^{\neq}(XH_2CHM) = 53.68 + 0.694\ \Delta SE^o - 0.374\ D(MH) \quad r=0.964 \quad [45]$$

$$\Delta E_1^{\neq}(XYHCHM) = 51.85 + 0.494\ \Delta SE^o - 0.374\ D(MH) \quad r=0.979 \quad [46]$$

$$\Delta E_1^{\neq}(XYZCHM) = 51.98 + 0.426\ \Delta SE^o - 0.374\ D(MH) \quad r=0.987 \quad [47]$$

Due to the tunnelling effect, there is a priori no reason that such correlations (eqns [44] to [47]) should hold at 300 K. On the whole, it is seen that, whatever the attacking radical may be, the lowest barriers are always associated to the largest negative values of ΔSE^o which correspond to highly destabilized molecules and (or) highly stabilized radicals. This important conclusion is illustrated by some specific examples collected in table XVIII.

Table XVIII. The effects of substituents on the barrier heights of hydrogen abstraction reactions from methane derivatives[a].

RH	$SE^o(RH)$	$SE^o(R^{\cdot})$	$\Delta E_1^{\neq}(M^{\cdot}:\dot{O}H)$	$\Delta E_1^{\neq}(M^{\cdot}:\dot{C}H_3)$	$\Delta E_1^{\neq}(M^{\cdot}:\dot{N}H_2)$
CH_4	4.20	-2.65	9.78	14.30	16.03
C_2H_6	0	-0.50	7.78	12.30	14.03
CH_3F	0	-2.06	7.78	12.30	14.03
$CH_2(CH_3)_2$	-0.27	1.82	5.27	9.79	11.51

Table XVIII. (Continued)

RH	$SE^o(RH)$	$SE^o(R^\cdot)$	$\Delta E_1^{\neq}(M^\cdot:\dot{O}H)$	$\Delta E_1^{\neq}(M^\cdot:\dot{C}H_3)$	$\Delta E_1^{\neq}(M^\cdot:\dot{N}H_2)$
CH_3CHO	0	4.66	4.63	9.14	10.87
$CH(CH_3)_3$	0.83	3.70	4.37	8.89	10.62
CH_3OH	0	2.53	3.94	8.45	10.18
CH_3OCH_3	0	3.44	3.77	8.29	10.02
CH_3CN	0	6.60	3.72	8.23	9.96
$CH_2(OH)_2$	6.06	8.73	2.34	6.85	8.58
CH_3NH_2	0	7.53	2.16	6.68	8.41
$CH_2(CH_3)(CN)$	0	8.35	2.11	6.63	8.35
$CH_2(NH_2)_2$	2.15	10.21	0.61	5.12	6.85
$CH_2(OH)(CN)$	-7.19	5.83	none	2.67	4.39
$CH_2(CN)_2$	-13.40	3.34	none	2.09	3.82
$CH_2(NH_2)(COOH)$	-1.27	19.95	none	none	0.66[b]
$CH(NH_2)(CN)_2$	-20.24	12.08	none	none	none
$CH(CN)_3$	-35.03	-5.16	none	none	none
$CH_2(NH_2)(CN)$	-2.41	14.96	none	1.04	2.77

[a] All quantities in kcal mol^{-1}, at 298.15 K.

[b] Extrapolated value.

Although barrier heights and activation energies are not directly comparable, we have found (14) that our semi-empirical activation barriers are roughly correlated with the corresponding activation energies. Thus, presumably, the substituents shall have, at least qualitatively, the same effects on barrier heights and activation energies. At the present time, we have only obtained preliminary results concerning the substituent effects on the rate constants of hydrogen abstraction reactions from methane derivatives (20).

We have calculated the rate constants and the kinetical parameters of the exchange reactions :

$$(NH_2)(CN)HC{-}H + M^{\bullet} \rightarrow \dot{C}H(NH_2)(CN) + MH$$

$$[M \equiv OH(I) \text{ and } OCH_3(II)]$$

at various temperatures, using the transition state theory formalism. The main results obtained in this work are summarized in table XIX and the Arrhenius plots of these reactions are respectively given in figures 3 and 4. It is seen that the abstraction reaction is much faster with OH than using methoxy radical, the major part of the rate enhancement being due to the increase of the preexponential factor A. Moreover, these two reactions exhibit a non Arrhenius behavior.

Table XIX. Kinetic parameters of the exchange reactions $CH_2(NH_2)(CN){+}M^{\bullet} \rightarrow \dot{C}H(NH_2)(CN){+}MH$ [$M \equiv OH(I)$ and $OCH_3(II)$]

Quantity	Reaction	T = 300 K	T = 1000 K
$\Delta E_1^{\neq}$ (kcal mol^{-1})	I	none	none
	II	4.06	4.06
log k (cm^3 mol^{-1})	I	13.07	12.82
	II	8.85	9.77
log A (cm^3 mol^{-1})	I	12.04	13.23
	II	9.06	10.94
E_a (kcal mol^{-1})	I	-1.41	1.90
	II	0.29	5.34
$\Delta S_1^{\neq}$	I	-27.51	-26.85
	II	-37.70	-36.54

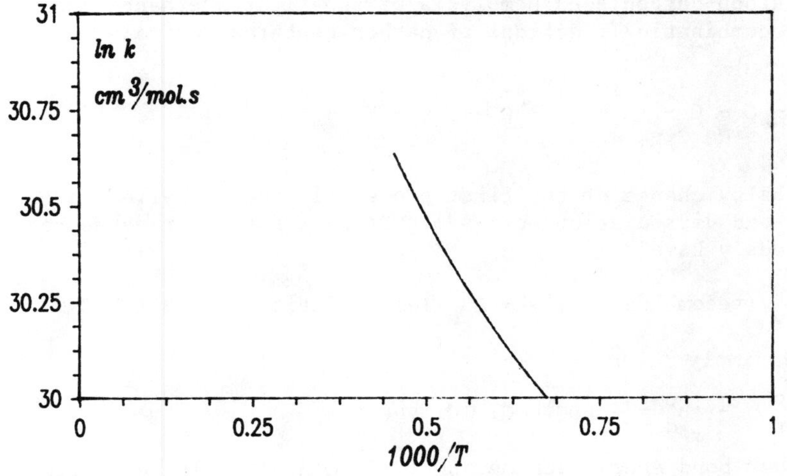

Figure 3. Arrhenius plots for the reaction : $CH_2(NH_2)(CN) + \dot{O}H \rightarrow \dot{C}H(NH_2)(CN) + H_2O$

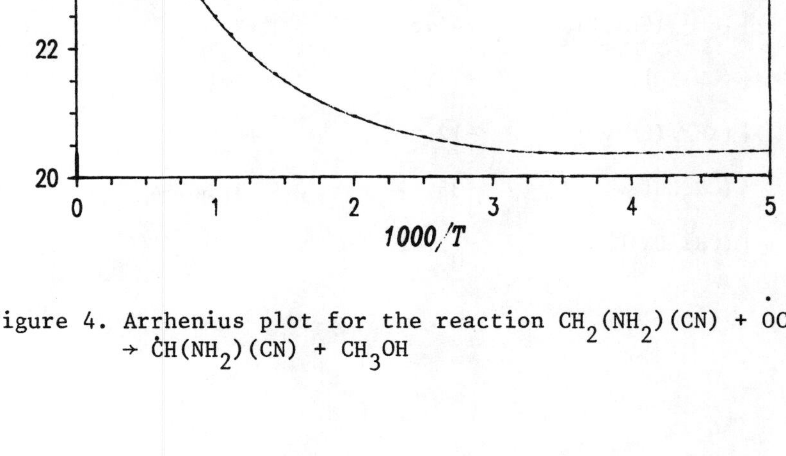

Figure 4. Arrhenius plot for the reaction $CH_2(NH_2)(CN) + \dot{O}CH_3 \rightarrow \dot{C}H(NH_2)(CN) + CH_3OH$

4.2. Carbon-carbon bond breaking and forming.

Let us now consider the two reverse reactions :
- the carbon-carbon bond homolysis of substituted ethanes and
- the recombination reactions of carbon centered radicals :

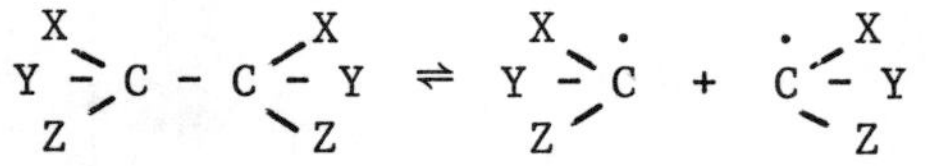

The enthalpy change of the first process is the so-called carbon-carbon bond dissociation energy : BDE(C-C). For the reverse reaction, we obviously have :

$$\Delta H^o_o \text{ (recombination)} = - \Delta H^o_o \text{ (dissociation)} = - \text{BDE(C-C)} \qquad [48]$$

or more simply :

$$\Delta H^o_r = - \Delta H^o_d = - \text{BDE(C-C)} \qquad [49]$$

Within our bond energy scheme, BDE(C-C) can strictly be written as :

$$\text{BDE(C-C)} = E^o\text{(C-C)} + SE^o\text{(ZYXCCXYZ)} - 2SE^o(\dot{C}\text{XYZ}) \qquad [50]$$

This equation allows us to rationalize the thermochemistry of the homolysis and recombination processes in terms of stabilization energies. Some typical values of carbon-carbon bond dissociation energies are given in table XX. It is seen, in accordance with relation [50], that the smallest BDE(C-C) correspond to the most destabilized substituted ethanes and (or) the most stabilized carbon centered radicals.

Table XX. Typical C-C bond dissociation energies (kcal mol^{-1}, 298.15 K)

Molecule (R-R)	SE^o(R-R)	SE^o(R$^\cdot$)	BDE (C-C)
$[(NH_2)(CN)_2C]_2$	-44.53[a]	12.08	16.36
$[(CN)_3C]_2$	-73.37[a]	-5.16	22.00
$[(NH_2)(CN)CH]_2$	-12.08[a]	14.96	43.05
$[(CN_2CH]_2$	-33.11[a]	3.34	45.26
$[(OH)(CN)CH]_2$	-13.31[a]	5.83	60.08

Table XX. (Continued).

Molecule (R-R)	SE^o(R-R)	SE^o(R•)	BDE(C-C)
$[(OH)_2CH]_2$	3.90[a]	8.73	71.49
$[Cl_3C]_2$	-49.89 (78)	-18.02 (44)	71.20
$[F_3C]_2$	33.81 (79)	10.68 (44)	97.50

[a] STO-3G/STO-3G results.

Now let us wonder whether the thermodynamic stabilization of a radical is related to its kinetic stabilization.

It is well known that most free radical are transient which means they are not kinetically stabilized. In other words, they recombine without activation energy and, correspondingly, their dimer dissociates also without activation energy. Thus,

$$\Delta H_d^{\neq} = \text{BDE(R-R)} \qquad [51]$$

On the other hand, some radicals corresponding to sterically hindered dimers are persistent and may even be stable. According to Griller and Ingold (41), persistence is principally a consequence of steric factors. Furthermore, according to Rüchardt (80-81) steric factors would also influence the rate of the carbon-carbon bond homolysis of substituted ethanes. At the present time, the existence of a barrier on the path of the recombination reaction of "sterically hindered" radicals remains an open question.

Although not numerous, there exist persistent and even stable "unhindered" radicals. Typical examples of such species are small nitroxides (82), hydrazyls (83), phenoxy radical (84) and nitric oxide. Besides, some persistent "unhindered" carbon centered radicals are known. We can mention the following species : aminodicyanomethyl radical (85) and tricyanomethyl radical which have a relatively low decay rate constant (31). Furthermore, as shown in table XXI, various types of free radicals recombine with a non negligible activation energy.

Table XXI. Activation energies of the recombination reaction of some free radicals (kcal mol^{-1}, 298.15 K).

$R^{\cdot}$	BDE(R-R)	$2SE^{o}(R^{\cdot})-SE^{o}(R-R)$	$(E_a)_r$
$(C_6H_5)_2\dot{C}OH$	32.5	52.55	1.0 (86)
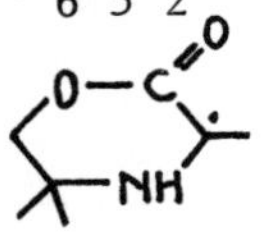	22.0	63.05	4.5 (87)
$(F_3CS)_3\dot{C}$	13.7	71.35	7.3 (88)
$(C_6H_5)(SCH_3)\dot{C}CN$	28.6	56.45	2.1 (89)
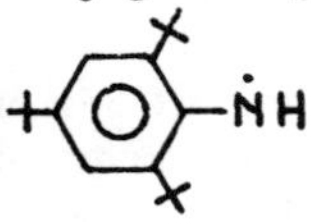	13.1	34.78	2.5 (90)
$(C_6H_5)_2N\dot{N}(COC_6H_5)$	9.2	38.68	7.4 (91)
$(F_3CS)_2\dot{N}$	7.7	40.18	10.7 (92)

It is seen that, for each series of radicals, the activation energies are correlated with the corresponding heats of reaction :

$$(E_a)_r = -\ 0.339\ BDE(C\text{-}C) + 11.938 \quad (r=0.999) \qquad [52]$$

for the recombination of carbon centered radicals and,

$$(E_a)_r = -\ 1.466\ BDE(N\text{-}N) + 21.526 \quad (r=0.990) \qquad [53]$$

for the recombinations of nitrogen centered species. These activation energies can also be written in terms of stabilization energies. The following relations are easily derived :

$$(E_a)_r = 0.339\ [2SE^{o}(R^{\cdot}) - SE^{o}(R\text{-}R)] - 16.926 \qquad [54]$$

and

$$(E_a)_r = 1.466\ [2SE^{o}(R^{\cdot}) - SE^{o}(R\text{-}R)] - 48.661 \qquad [55]$$

Thus, an electronic induced persistence would exist when the exothermicity of the recombination reaction becomes relatively small, namely :

$$\mathrm{BDE(C\text{-}C)} < 35 \text{ kcal mol}^{-1} \qquad [56]$$

and

$$\mathrm{BDE(N\text{-}N)} < 17 \text{ kcal mol}^{-1} \qquad [57]$$

or, in other words, when the radicals are relatively much more stabilized than their dimer :

$$2\mathrm{SE}(\dot{\mathrm{C}}\mathrm{XYZ}) - \mathrm{SE}^{o}(\mathrm{dimer}) > 50 \text{ kcal mol}^{-1} \qquad [58]$$

and

$$2\mathrm{SE}^{o}(\dot{\mathrm{N}}\mathrm{XY}) - \mathrm{SE}^{o}(\mathrm{dimer}) > 33 \text{ kcal mol}^{-1} \qquad [59]$$

This condition is most often verified when the radicals are stabilized and their dimer destabilized. On the whole, the thermodynamic stabilization of a free radical is rarely alone responsible for its kinetic stabilization i.e., its persistence.

The above formalism allows us also to rationalize the kinetics of the carbon-carbon bond homolysis of substituted ethanes. The activation energy of this process can be written as :

$$(E_a)_d = (E_a)_r + \mathrm{BDE(C\text{-}C)} \qquad [60]$$

or, more explicitly :

$$(E_a)_d = \Delta H_d^{\neq} = 68.124 - 0.661[2\mathrm{SE}^{o}(\mathrm{R}^\bullet) - \mathrm{SE}^{o}(\mathrm{R\text{-}R})] \qquad [61]$$

As shown below, this relation leads to a series of interesting applications. However, we must insist on the fact that it is only valid if BDE(C-C) is smaller than 35 kcal mol^{-1}.

i) Estimation of $\mathrm{SE}^{o}(\mathrm{R}^\bullet)$, $\Delta H_f(\mathrm{R}^\bullet)$, BDE(C-C) and $(E_a)_r$ from $\Delta H_d^{\neq}$ and $\Delta H_f^{o}(\mathrm{R\text{-}R})$ (kcal mol^{-1}, 298.15 K).

Examples :

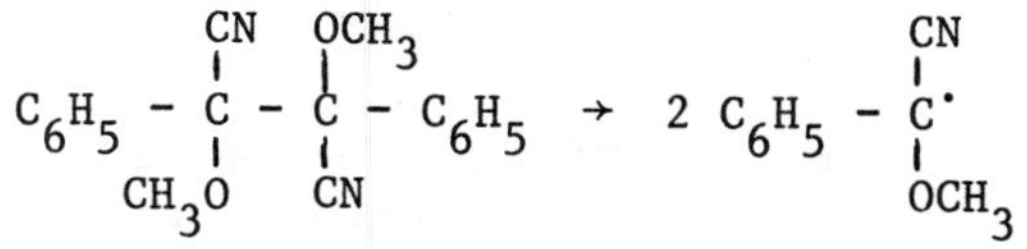

$\Delta H_d^{\neq} = 35.4$ (93) $\qquad 2\,\mathrm{SE}(\mathrm{R}^\bullet) - \mathrm{SE}^{o}(\mathrm{R\text{-}R}) = 49.51$

$\Delta H_f^{o}(\mathrm{R\text{-}R}) = 40.6$ (94) $\rightarrow \mathrm{SE}^{o}(\mathrm{R}^\bullet) = 14.14$

$SE^o(R\text{-}R) = -21.23$ $\quad$ $BDE(C\text{-}C) = 35.54$

$\Delta H_f^o(R^\cdot) = 37.83$

$(E_a)_r \sim 0$

The captodative radicals obtained in this reaction have a relatively large stabilization energy but no kinetic stabilization.

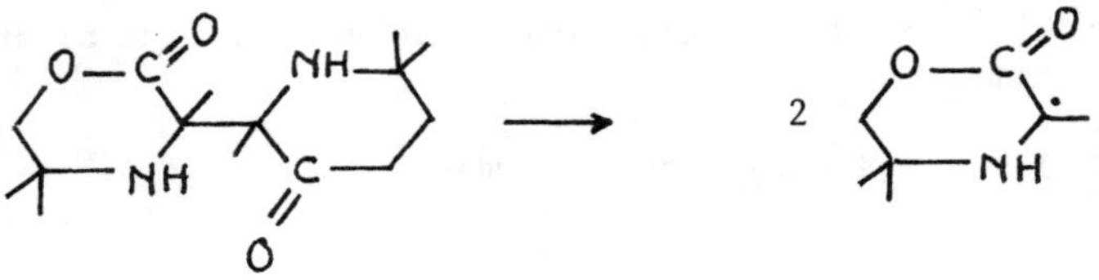

$\Delta H_d^{\neq} = 26.5$ (87) $\rightarrow$ $2\,SE^o(R^\cdot) - SE^o(R\text{-}R) = 62.97$

$(E_a)_r = 4.42$

The above captodative radical presumably has a large stabilization energy (estimated value $\sim$ 21 kcal mol^{-1}) but its kinetic stabilization is largely due to the important destabilization of its dimer ($\sim$ -21 kcal mol^{-1})

ii) Estimation of BDE(C-C), $(E_a)_r$ and $\Delta H_d^{\neq}$ from $\Delta H_f^o(R^\cdot)$ and $\Delta H_f^o(R\text{-}R)$ (kcal mol^{-1}, 298.15 K).

Examples :

$(CN)_3C - C(CN)_3 \rightarrow 2\,\dot{C}(CN)_3$

$\Delta H_f^o(R^\cdot) = 134.9$ $\quad$ $BDE(C\text{-}C) = 22.0$

$\Delta H_f^o(R\text{-}R) = 247.8$ $\rightarrow$ $(E_a)_r = 4.5$

$SE^o(R^\cdot) = -5.2$ $\quad$ $\Delta H_d^{\neq} = 26.5$

$SE^o(R\text{-}R) = -73.4$

The kinetic stabilization of the radical $C(CN)_3$ is essentially due to the destabilization of its dimer. This could be the explanation of the low decay rate constant of this species.

$(NH_2)(CN)_2C\text{-}C(CN)_2(NH_2) \rightarrow 2(NH_2)(CN)_2\dot{C}$

$\Delta H_f^o(R^\cdot)$	89.35		BDE(C-C)	= 16.36
$\Delta H_f^o(R\text{-}R)$	162.34	→	$(E_a)_r$	= 6.39
$SE^o(R^\cdot)$	12.08		$\Delta H_d^{\neq}$	= 22.75
$SE^o(R\text{-}R)$	-44.53			

The persistence of aminodicyanomethyl radical comes especially from the destabilization of its dimer even if its own thermodynamic stabilization is far from being negligible. On the whole, we have shown in this section that the concept of stabilization energy allows us to rationalize the effects of substituents on the self-reactivity of carbon centered radicals and on the C-C bond homolysis of the corresponding dimers.

In summary, the persistence of a free radical may be attributed either to the large destabilization of its dimer, due to steric or polar effects, - $(tBu)_3\dot{C}$ and $\dot{C}(CN)_3$ are typical examples of these two kind of species respectively - or to the large stabilization of the radical itself due to the delocalization of its unpaired electron or finally, most often, to the joint influence of all these factors. Pure electronic induced persistence must be very rare because it would required an exceptional stabilization energy of the radical and a negligible one of the corresponding dimer.

4.3. Radical additions to substituted alkenes.

A priori, the addition of a free radical to monosubstituted ethylene can lead to two other radicals :

$$M^\cdot + CH_2 = CH\text{-}X \rightarrow MCH_2\dot{C}H\text{-}X(I) \text{ or } \dot{C}H_2\text{-}CHMX(II)$$

The corresponding heats of reaction can be written as :

$$(\Delta H_o^o)_I = SE^o(M^\cdot) + \Sigma N_{AB}E_{AB}^o(M^\cdot) + SE^o(CH_2{=}CHX) + \Sigma N_{AB}E_{AB}^o(CH_2{=}CHX) - SE^o(I) - \Sigma N_{AB}E_{AB}^o(I) \quad [62]$$

and

$$(\Delta H_o^o)_{II} = SE^o(M^\cdot) + \Sigma N_{AB}E_{AB}^o(M^\cdot) + SE^o(CH_2{=}CHX) + \Sigma N_{AB}E_{AB}^o(CH_2{=}CHX) - SE^o(II) - \Sigma N_{AB}E_{AB}^o(II) \quad [63]$$

The difference between these two heats of reaction is practically equal to the difference between the stabilization energies of the species I and II. Then, in the absence of steric effects (if $M^\cdot \equiv H^\cdot$ for example), reaction I which obeys the so-called "anti-Markownikow" rule is thermochemically favored, the stabilization energy of radical I being generally larger than that of radical II. Thus, the orientation of radical additions to substituted alkenes could be, at least partly, interpreted in terms of the stabilization energies of radicals I and II possibly obtained at the first step of the reaction.

On the other hand, it is easy to verify that, in the series of monosubstituted ethylenes, the value of $(\Delta H^o_o)_I$ is mostly due to the term $\Delta\Sigma N_{AB}E^o_{AB}$. As this difference is not constant, the thermochemistry of radical additions to these compounds cannot be rationalized in terms of stabilization energies. In table XXII, we analyse the various contributions to the heats of reaction of some hydrogen atom additions to 1,1-disubstituted ethylenes. It is seen that the largest heats of reaction correspond either to a destabilized olefine [CH_2=C$(CN)_2$, for example] or a very stabilized radical [CH_3-C(NH_2)(CN), for example].

Table XXII. Heats of reaction of hydrogen atom additions to 1,1-disubstituted ethylenes (kcal mol^{-1}, 298.15 K)[a] $H^\bullet + CH_2=CXY \rightarrow CH_3-\dot{C}XY$

X=Y=CH_3	$CH_2=C(CH_3)_2$	$CH_3-\dot{C}(CH_3)_2$	Δ
$\Sigma N_{AB}E_{AB}$	1104.59	1139.40	-34.81
SE^o	0.07	3.7	-3.63
ΔH^o_a	1104.66	1143.10	-38.44 (ΔH^o_o)
X=Y=CN	$CH_2=C(CN)_2$	$CH_3-C(CN)_2$	Δ
$\Sigma N_{AB}E^o_{AB}$	940.20	975.04	-34.84
SE^o	-11.11	7.89	-19.00
ΔH^o_a	929.09	982.93	-53.84 (ΔH^o_o)
X=Y=OH	$CH_2=C(OH)_2$	$CH_3-C(OH)_2$	Δ
$\Sigma N_{AB}E^o_{AB}$	736.15	778.70	-42.55
SE^o	6.89	9.87	-2.98
ΔH^o_a	743.04	788.57	-45.53 (ΔH^o_o)

Table XXII. (Continued).

$X=NH_2$, Y=CN	$CH_2=C(NH_2)(CN)$	$CH_3-C(NH_2)(CN)$	Δ
$\Sigma N_{AB}E^o_{AB}$	897.97	936.65	-38.68
SE^o	2.71	16.08	-13.37
ΔH^o_a	900.68[b]	952.73[b]	-52.05 (ΔH^o_o)

[a] Except where otherwise specified, the data of this table are taken from reference (95).

[b] 6-31G/6-31G results (this work).

To end this section, we briefly describe the preliminary results of our theoretical calculations of the rate constants and the kinetic parameters of radical additions to substituted ethylenes.
We have actually chosen the following reactions :

$$H^{\cdot} + CH_2 = C\begin{matrix} X \\ Y \end{matrix} \rightarrow CH_3 - \dot{C}\begin{matrix} X \\ Y \end{matrix}$$

where X,Y represents the couples (H,H), (H,CN), (H,NH_2) and (CN,NH_2). The calculations were carried out using the methodology described elsewhere (12). On the other hand, the total energies and the vibrational frequencies were obtained using the GAUSSIAN 82 series of programs (96). The main results of this study are presented in table XXIII. Moreover, the electronic structures and the geometries of the reactants, activated complexes and products of the four chosen reactions are qualitatively described in figures 5 and 6.

In accordance with the large exothermicity of these reactions, the distance between the attacking hydrogen atom and the future β carbon of the product is very large in each transition structure (the transition state appears very early on the reaction path). Moreover it regularly increases with the exothermicity of the process. Accordingly the electronic structure of each activated complex is similar to that of the corresponding ethylene. However one observes a slight displacement of one of the β electrons of the double bond towards the $H-C_\beta$ bond in formation. On the other hand,an Evans-Polanyi relation, at OK and fixed nuclei, seems to exist for these reactions. An excellent correlation between the barrier heights and the corresponding activation energies is also observed, at any temperature.

Table XXIII. Calculated properties for the addition of hydrogen atom to substituted ethylenes.

Property	$CH_2=CH_2$	$CH_2=CHNH_2$	$CH_2=CHCN$	$CH_2=C(CN)(NH_2)$
$R(H...C_\beta)$ (Å)	2.019	1.987	2.190	2.232
$\Delta E^{\neq}$ (kcal mol^{-1})	2.33	3.29	(0)	(0)
ΔE^{o} (kcal mol^{-1})	-41.36	-39.64	-51.76	-53.65
log k (cm^3 mol^{-1} s^{-1})				
at 298.15 K	12.08	10.67	14.01	14.21
at 1000 K	13.62	10.82	14.09	14.22
log A (cm^3 mol^{-1} s^{-1})				
at 298.15 K	13.65	14.11	13.55	13.49
at 1000 K	14.66	14.11	13.53	14.70
E_a (kcal mol^{-1})				
at 298.15 K	2.14	3.33	-0.63	-0.99
at 1000 K	4.77	5.90	2.00	2.16

As shown in table XXIII and figure 7, the rate of H addition to vinylamine is decreased with respect to ethylene whereas the rates for addition to acrylonitrile and 1,1-aminocyanoethylene are increased. These differences essentially come from the variations of the activation energy from one reaction to another. It is worth noticing that acrylonitrile and 1,1-aminocyanoethylene have practically the same reactivity towards hydrogen atom which clearly behaves as a nucleophilic radical.

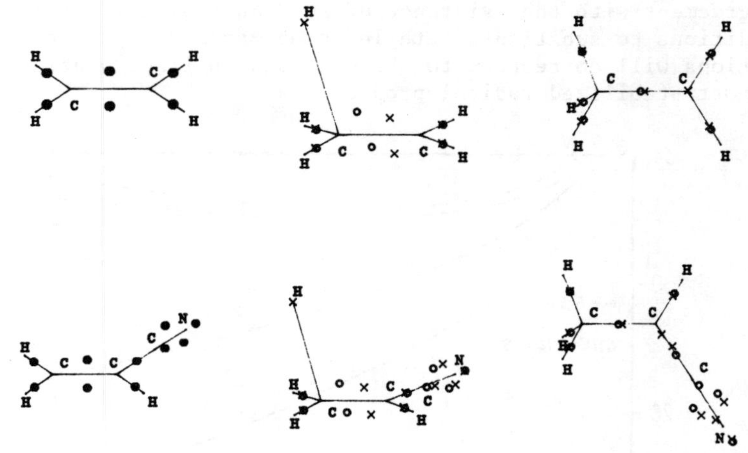

Figure 5. Electronic structures and geometries of the reactants, transition structures and products for H additions to ethylene and acrylonitrile (black dots represent electron pairs and the cross and open circles, α and β electrons respectively.

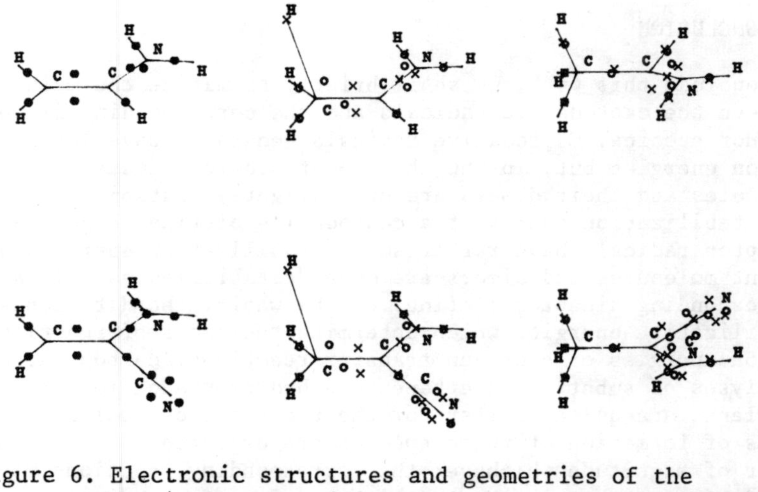

Figure 6. Electronic structures and geometries of the reactants, transition structures and products for H additions to vinylamine and 1,1-aminocyanoethylene (the symbols are the same as in figure 5).

In agreement with the existence of an Evans-Polanyi relation for the H additions to substituted ethylenes we may anticipate that the fastest reactions will correspond to the most destabilized olefines and (or) the most stabilized radical-product.

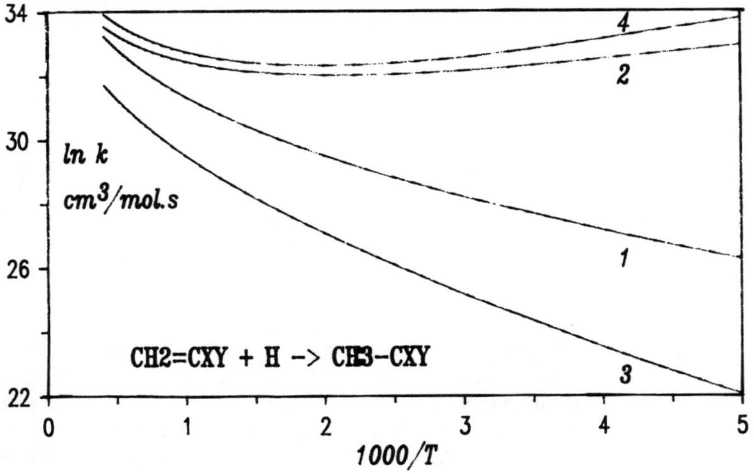

Figure 7. Arrhenius plots for the H additions to substituted ethylenes (1:X=Y=H; 2:X=CN,Y=H; 3:X=NH_2,Y=H; 4:X=CN,Y=NH_2).

5. CONCLUSION

To conclude this work, we shall briefly summarize the main differences between the captodative radicals and the corresponding dicaptor and didonor species. Captodative radicals generally have large stabilization energies but, in the absence of steric effects, their parent molecules and their dimers are only slightly destabilized. Further, the stabilization energy of a captodative olefine is generally small. Dicaptor radicals have rather small stabilization energies but their parent molecules and dimers are much destabilized as well as the corresponding dicaptor olefine. On the whole, the differences of stabilization energies which determine the heats of reaction and activation energies of hydrogen transfer reactions, carbon-carbon bond homolyses of substituted ethanes and hydrogen additions to substituted ethylenes are quite similar for the two types of radicals. Thus, the rates of formation of these species are expected to be of the same order of magnitude whichever the corresponding reactions may be. The difference between captodative and dicaptor radicals presumably lies in the relative rates of their recombination reactions with respect to other processes. The captodative radicals would generally prefer to dimerize and the dicaptor ones to undergo other reactions due to the destabilization of their dimer. Particularly in the presence of an olefine, dicaptor species would prefer the chain addition to the dimerization reaction. On the other hand, didonor radicals would always be obtained at lower rates than the other species but if they are formed, nothing but steric effects would hinder their dimerization.

6. ACKNOWLEDGEMENT

We wish to thank the Fonds National Belge de la Recherche Scientifique for a research grant (M.S.) and his continued financial support to our laboratory.

7. REFERENCES

1. G. Leroy, D. Peeters, A. Deplus and M. Tihange, Nouv. J. Chim., 3, 213 (1979).
2. D. Peeters and G. Leroy, Bull. Soc. Chim. Belg., 86, 833 (1977).
3. G. Leroy, Bull. Cl. Sci. Acad. R. Belg., LXIX, 352 (1983).
4. R. Daudel, G. Leroy, D. Peeters and M. Sana, Quantum Chemistry, Wiley, Chichester (1984).
5. G. Leroy in 'New Concepts of Free Radicals in Chemistry and Biology', Table Ronde Roussel Uclaf n° 41, 42 (1981).
6. G. Leroy, D. Peeters and C. Wilante, J. Mol. Struct. (Theochem) 120, 91 (1985).
7. G. Leroy, Internat. J. Quantum Chem., XXIII, 271 (1983).
8. G. Leroy, J. Mol. Struct. (Theochem), 93, 175 (1983).
9. G. Leroy in P.O. Löwdin (Ed.), Advances in Quantum Chemistry, Academic Press, New York, Vol. 17, pp 1-95 (1985).
10. G. Leroy, J. Mol. Struct. (Theochem), 120, 91 (1985).
11. G. Leroy, C. Wilante, D. Peeters and M.M. Uyewa, J. Mol. Struct. (Theochem), 124, 107 (1985).
12. G. Leroy, M. Sana and A. Tinant, Can. J. Chem., 63, 1447 (1985).
13. M. Sana and G. Leroy, J. Mol. Struct. (Theochem), in press.
14. G. Leroy and M. Sana, J. Mol. Struct. (Theochem), in press.
15. M. Sana, G. Leroy and J.L. Villaveces, Theoret. Chim. Acta (Berl.) 65, 109 (1984).
16. G. Leroy, D. Peeters, C. Wilante and M. Khalil, Nouv. J. Chim., 4, 403 (1980).
17. M. Sana and G. Leroy, J. Mol. Struct. (Theochem), 76, 259 (1981).
18. G. Leroy and D. Peeters, J. Mol. Struct. (Theochem), 85, 133 (1981).
19. D. Peeters, G. Leroy and C. Wilante, J. Mol. Struct. (Theochem), 107, 157 (1984).
20. G. Leroy, D. Peeters and M. Sana, Nouvelles de la Science et des Technologies, in press.
21. G. Leroy, in I.G. Csizmadia and R. Daudel (Eds), NATO Advanced Study Institute in Theoretical Organic Chemistry, Reidel, Dordrecht, pp 253-334 (1981).
22. G. Leroy, C. Wilante, D. Peeters and M. Khalil, Ann. Soc. Scient. Bruxelles, 95, 157 (1981).
23. G. Leroy, Bull. Soc. Chim. Belg., 94, 945 (1985).

24. R.W. Fessenden, J. Phys. Chem., 71, 74 (1967).
25. A.J. Dobbs, B.C. Gilbert and R.O.C. Norman, J. Chem. Soc., Perkin Trans II, 786 (1972).
26. G. Brunton, K.U. Ingold, B.P. Roberts, A.L.J. Beckwith and P.J. Krusic, J. Am. Chem. Soc., 99, 3177 (1977).
27. R.W. Fessenden and R.H. Schuler, J. Chem. Phys., 43, 2704 (1965).
28. H.M. Mc Connell, J. Chem. Phys., 24, 764 (1956).
29. H.G. Viehe, R. Merényi, L. Stella and Z. Janousek, Angew. Chem. Int. Edit. Engl., 18, 917 (1979).
30. P. Smith, R.A. Kaba, T.C. Smith, J.T. Pearson and P.B. Wood, J. Magn. Res., 18, 254 (1975).
31. R.A. Kaba and K.U. Ingold, J. Am. Chem. Soc., 98, 523 (1976).
32. D. Griller, D.C. Nonhebel and J.C. Walton, J. Chem. Soc., Perkin Trans II, 1373 (1983).
33. P. Smith, J.T. Pearson, P.B. Wood and T.C. Smith, J. Chem. Phys., 43, 1535 (1965).
34. H. Paul, H. Fischer, Helv. Chim. Acta, 56, 1575 (1973).
35. Landolt-Börnstein, Group II, Vol. 1 : 'Magnetic properties of free radicals', by H. Fischer, Springer-Verlag, Berlin-Heidelberg,New York (1965).
36. A.L. Buley, R.O.C. Norman and R.J. Pritchett, J. Chem. Soc. (B) 849 (1966).
37. J.C. Walton, J. Magn. Res., 52, 241 (1983).
38. W.T. Dixon, R.O.C. Norman and A.L. Buley, J. Chem. Soc., 3625 (1964).
39. L. Stella, R. Merényi, Z. Janousek, H.G. Viehe, P. Tordo and A. Munoz, J. Phys. Chem., 84, 304 (1980).
40. P. Neta and R.W. Fessenden, J. Phys. Chem., 75, 738 (1971).
41. D. Griller and K.U. Ingold, Acc. Chem. Res., 9, 13 (1976).
42. J.D. Pedley and J. Rylance, Computer Analysed Thermochemical data : Organic and Organometallic Compounds, University of Sussex (1977).
43. M.A. Flamm-Ter Meer, H.S. Beckhaus and C. Rüchardt, Thermochim. Acta, 80, 81 (1984).
44. D.F. Mc Millen and D.M. Golden, Ann. Rev. Phys. Chem., 33, pp. 493-532 (1982).
45. D. Griller and F.P. Lossing, J. Am. Chem. Soc., 103, 1586 (1981).
46. A.J. Colussi and S.W. Benson, Int. J. Chem. Kinetics, 9, 307 (1977).
47. M.A. Grela and A.J. Colussi, Int. J. Chem. Kinetics, 17, 257 (1985).
48. M.A. Grela and A.J. Colussi, J. Phys. Chem., 88, 5995 (1984).
49. J.D. Cox and D. Pilcher, 'Thermochemistry of Organic and Organometallic Compounds', Academic Press, New York (1970).
50. A.L. Castelhano and D. Griller, J. Am. Chem. Soc., 104, 3655 (1982).
51. A.S. Rodgers, J. Chao, R.C. Wilhoit and B.J. Zwolinski, J. Phys. Chem. Ref. Data, 3, 117 (1974).
52. X.W. An and M. Manson, J. Chem. Therm., 15, 287 (1983).
53. M. Rossi and D.M. Golden, Int. J. Chem. Kinetics, 11, 715 (1979).
54. K.D. King and T.T. Nguyen, J. Phys. Chem., 83, 1940 (1979).
55. T.T. Nguyen and K.D. King, J. Phys. Chem., 85, 3130 (1981).
56. D. Crans, T. Clark and P.v.R. Schleyer, Tetrahedron Lett., 21, 3681 (1980).
57. M. Szwarc, J. Chem. Phys., 16, 128 (1948).
58. S.W. Benson, J. Chem. Educ., 42, 502 (1965).

59. A.S. Rodgers, M.C.R. Wu and L. Kuitu, J. Phys. Chem., 76, 918 (1972).
60. J.Y. Chu, T.T. Nguyen and K.D. King, J. Phys. Chem., 86, 443 (1982).
61. S.W. Benson, Thermochemical Kinetics, 2nd ed., J. Wiley, New York (1976).
62. M.F. Dube and J.W. Timberlake, Tetrahedron, 36, 1753 (1980).
63. R.C. Zawalski, M. Lisiak, P. Kovacic, A. Luedtke and J.W. Timberlake, Tetrahedron Lett., 21, 425 (1980).
64. D. Barton, M. Hodgett, P. Skirving, M. Whelton, K. Winter and C. Vardy, Can. J. Chem., 61, 1712 (1983).
65. M. Rossi and D.M. Golden, Int. J. Chem. Kinetics, 11, 775 (1979).
66. P.S. Engel, Chem. Rev.,80, 99 (1980).
67. P.S. Engel, J.L. Wood, J.A. Sweet and J.L. Margrave, J. Am. Chem. Soc., 96, 2381 (1974).
68. Calculated value; ΔH_f^o (g) = 94.5 kcal mol^{-1}, from ΔH_f^o (c) = 76.6 kcal mol^{-1} (49) and an estimated value ΔH^o(sublimation) = 17.9 kcal mol^{-1}.
69. J. Kao and T-N Huang, J. Am. Chem. Soc., 101, 5546 (1979).
70. R.L. Montgomery and F.D. Rossini, J. Phys. Chem., 82, 575 (1978).
71. Calculated value; ΔH_f^o = 69.07 kcal mol^{-1}, from ΔH_f^o (c) = 54.7 kcal mol^{-1} (49) and an estimated value ΔH^o (sublimation) = 14.37 kcal mol^{-1}.
72. I. Macinnes, J-C. Walton and D.C. Nonhebel, J. Chem. Soc., Chem. Commun., 712 (1985).
73. H.G. Korth, P. Lommes and R. Sustmann, J. Am. Chem. Soc., 106, 663 (1984).
74. M. Rossi and D.M. Golden, Int. J. Chem. Kinetics, 11, 715 (1979).
75. M.J. Rossi, D.F. Mc Millen and D.M. Golden, J. Phys. Chem., 88, 5031 (1984).
76. R. Livingston, H. Zeldes and M.S. Conradi, J. Am. Chem. Soc., 101, 4312 (1979).
77. D. Griller, D.C. Nonhebel and J.C. Walton, J. Chem. Soc., Perkin Trans II, 1817 (1984).
78. J. Chao, A.S. Rodgers, R.C. Wilhoit and B.J. Zwolinski, J. Phys. Chem. Ref. Data, 3, 141 (1974).
79. S.S. Chen, A.S. Rodgers, J. Chao, R.C. Wilhoit and B.J. Zwolinski, J. Phys. Chem. Ref. Data, 4, 441 (1975).
80. C.Rüchardt and H.D. Beckhaus, Angew. Chem. Int. Ed. Engl., 19, 429 (1980).
81. C. Rüchardt, Top. Curr. Chem., 88, 1 (1980).
82. G. Chapelet-Letourneux, H. Lemaire and A. Rassat, Bull. Soc. Chim. France, 3283 (1965); K. Adamic, D.F. Bowman, T. Gillion and K.U. Ingold, J. Am. Chem. Soc., 93, 902 (1971); V. Malatesta and K.U. Ingold, J. Am. Chem. Soc., 103, 3094 (1981).
83. T. Balaban, Bul. Inst. Polithe "Gheorghe Gheorghiu-Deg" Ser. Chim. Metal, XLIII, 4 (1981).
84. D.C. Nonhebel, J.M. Tedder and J.C. Walton, Radicals, Cambridge University Press, London, New York (1979).
85. L. De Vries, J. Am. Chem. Soc., 99, 1482 (1977); N.E. Heimer, J. Org. Chem., 42, 3767 (1977).

86. S.A. Weiner, J. Am. Chem. Soc., 93, 6978 (1971).
87. R.W. Bennett, D.L. Wharry and T.G. Koch, J. Am. Chem. Soc., 102, 2345 (1980).
88. A. Haas, K. Schlosser, S. Steenken, J. Am. Chem. Soc., 301, 6282 (1979).
89. H. Birkhofer, H.-D. Beckhaus and C. Rüchardt, to appear in the NATO ASI Series on Substituent Effects in Radical Chemistry, Reidel, Dordrecht (1986). (this volume)
90. D. Griller, L.R.C. Barclay and K.U. Ingold, J. Am. Chem. Soc., 97, 6151 (1975).
91. W.K. Wilmarth and N. Schwartz, J. Am. Chem. Soc., 77, 4543 (1955).
92. K. Schlosser and S. Steenken, J. Am. Chem. Soc., 105, 1504 (1983).
93. C. Rüchardt and H.-D. Beckhaus in Synthetic Organic Chemistry, Springer Verlag, Berlin, Heidelberg, New York, Tokyo, pp. 1-23 (1986).
94. M. Zamkanei, J.H. Kaiser, H. Birhofer, H.-D. Beckhaus and C. Rüchardt, Chem. Ber., 116, 3216 (1983).
95. G. Leroy, M. Sana and C. Wilante, Compilation de données thermochimiques et théoriques, Rapport Technique n° 3, Laboratoire de Chimie Quantique, Louvain-la-Neuve (1985).
96. J.S. Binkley, M. Frish, K. Raghavachari, D.J. Defrees, H.B. Schlegel, R.N. Whiteside, E.M. Fludes, R. Seeger and J.A. Pople, Gaussian-82, Carnegie-Mellon Institute, Pittsburg University (1982).
97. W.J. Hehre, R. Ditchfield, H. Radom and J.A. Pople, J. Am. Chem. Soc., 92, 4796 (1970).
98. J.S. Binkley, R.A. Whiteside, R. Krishnan, R. Seeger, D.J. De Frees, H.B. Schlegel, S. Topiol, L.R. Kahn and J.A. Pople, Q.C.P.E., 13, 406 (1981).

SIGMA AND PI EFFECTS IN RADICALS

Timothy Clark
Institut für Organische Chemie der Friedrich-Alexander-Universität Erlangen-Nürnberg,
Henkestraße 42,
D-8520 Erlangen, Federal Republic of Germany

ABSTRACT. *Ab initio* molecular orbital calculations have been used to investigate the odd-electron sigma and pi interactions in radicals and radical ions. One- and three-electron sigma bond energies are found to be strongly dependent on the energy match between the interacting orbitals. Odd electron pi-bonds are strong for radical ions involving first row elements, but weak for neutral radicals and bonds between second row elements. The principles deduced from these studies are applied to some current problems in radical chemistry.

1. INTRODUCTION

Conventional Lewis acid - Lewis base chemistry has been extensively and successfully treated using *ab initio* molecular orbital calculations and simple perturbational molecular orbital (PMO) theory now forms the basis for the interpretation of organic reactions involving closed-shell species. The situation with radicals is, however, less satisfactory. Neither the unrestricted (UHF) nor the restricted (ROHF) Hartree-Fock formalisms are universally applicable. The former often suffers from severe spin contamination, especially for conjugated radicals, and the latter may predict spurious unsymmetrical structures to be the minimum for a symmetrical radical. In addition, both methods require more computer time than the equivalent calculations for closed-shell species, so that the theory of radicals and their reactions is less advanced than that of even-electron reactions. The recent advances in computer technology have, however made open-shell systems a more attractive subject for calculational studies, and recently considerable advances have been made. The qualitative MO theory of radicals and their reactions is also poorly developed. It has long been recognised that

H. G. Viehe et al. (eds.), Substituent Effects in Radical Chemistry, 49–68.
© 1986 by D. Reidel Publishing Co.

radicals can either react as nucleophiles, forming one-electron bonds as shown in Figure 1(a), or as electrophiles to give three-electron bonds (Figure 1(b)).

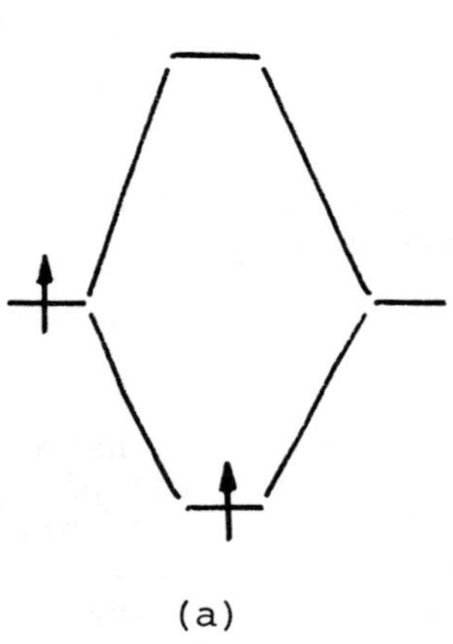

FIGURE 1. Orbital interaction diagrams for the nucleophilic (a) and electrophilic (b) modes of reaction of a radical singly occupied molecular orbital (SOMO) with a closed-shell reactant.

In both cases a stabilization about half as strong as a conventional two-electron bond is predicted, although this stabilization depends strongly on the overlap in the three-electron case.[2] This paper reports systematic investigations into these two types of interaction involving both sigma- and pi-type overlap and attempts to derive general principles for radical bonding and reactions.

2. METHOD

All calculations were performed with the GAUSSIAN 82 series of programs[3] adapted for our CYBER computer by A. Sawaryn and T. Kovář, or with the original VAX version of the program. Unless otherwise noted, optimizations were carried out using the UHF formalism and the 6-31G* basis set.[4] Energies quoted in the text include a second order Møller-Plesset (MP2) correction for electron correlation. Post-SCF calculations did not include the non-valence orbitals. Transition states were optimized using the standard routines included in GAUSSIAN 82. Unless otherwise noted, spin contamination was neglible.

3. SIGMA BONDS

3.1 Three-Electron Sigma Bonds

Earlier work[5-7] using the 4-31G basis set has shown that three-electron bonds in neutral systems are very weak, and that an energy match between the two partners in a three-electron bond is important. Consider a radical cation, $A^{+\cdot}$, and a Lewis base, B. The energy of a three-electron bond between $A^{+\cdot}$ and B is given by the energy released in reaction 1:

$$A^{+\cdot} + B \longrightarrow A\text{-}B^{+\cdot} \qquad (1)$$

A further energy variable, Δ(I.P.), is defined as the energy required to transfer an electron from B to $A^{+\cdot}$:

$$A^{+\cdot} + B \longrightarrow A + B^{+\cdot} \qquad (2)$$

This energy corresponds to the energy difference, E, between the two interacting orbitals in Figure 1(b), at least within the usual one-electron theoretical framework. A and B are chosen so that reaction 2 is always endothermic in the following discussion. This also means that $A^{+\cdot}$ and B are the most stable dissociation products of the radical complex $A\text{-}B^{+\cdot}$.
Table I shows the optimized A-B bondlengths, MP2/6-31G* three-electron bond energies, and Δ(I.P.) values for a series of $A\text{-}B^{+\cdot}$ radical cations, where A and B are argon and the hydrides of the elements phosphorus to chlorine. As noted in the earlier, more limited, 4-31G study,[7] the symmetrical complexes (A = B) are the most strongly bound. The larger 6-31G* basis set yields bond energies slightly lower than those obtained at 4-31G, which does not contain *d*-orbitals. The largest difference is found for $P_2H_6^{+\cdot}$, which is found to have a bond energy 5.1 kcal mol^{-1} lower than the MP2/4-31G value. In the earlier work a linear correlation between the logarithm of the three-electron bond energy and Δ(I.P.) was found.[7]
The present, more accurate, data and the additional argon-containing radical cation complexes reported here allow a more critical test of this hypothesis. Figure 2 shows a plot of the three-electron bond energy against Δ(I.P.) for the radical cations given in Table I. The dashed line is the least squares best fit exponential function obtained for the nine data points. The original proposition that the three-electron bond energies are roughly inversely logarithmically proportional to Δ(I.P.) is supported by the correlation, but a more detailed picture now emerges.

Table I. MP2/6-31G* three-electron bond energies (Equation 1) and Δ(I.P.) values (Equation 2, both in kcal mol^{-1}) and UHF/6-31G* three-electron bond lengths (r_{AB}, in Ångstrom) for radical cation complexes involving the elements phosphorus to argon.

A	B	r_{AB}	$-\Delta E$(Eq.1)	$-\Delta E(Eq.2)_{calc}$	$-\Delta E(Eq.2)_{exp.}$
Ar	Ar	2.515	24.0	0.0	0.0
HCl	Ar	3.280	2.3	71.5	69.8
H_2S	Ar	3.655	1.2	126.2	123.9
HCl	HCl	2.651	29.2	0.0	0.0
H_2S	HCl	3.177	9.5	54.7	54.1
H_3P	HCl	3.445	6.5	70.3	63.9
H_2S	H_2S	2.853	29.7	0.0	0.0
H_3P	H_2S	2.881	20.8	15.6	9.8
H_3P	H_3P	3.005	25.7	0.0	0.0

The elements sulfur and chlorine are found to form stronger three-electron bonds than phosphorus and argon. The deviations form the curve can be rationalized on this basis. The complex radical cations with the more strongly binding elements tend to lie above the line, those with phosphorus, and especially argon, lie below the curve. In general, however, a three-electron bond between elements of the second row can be expected to have a dissociation energy of 27.7 +/- 3 kcal mol^{-1}, falling off to about 11 kcal mol^{-1} at (I.P.) = 2 eV, and to about 3 kcal mol^{-1} at Δ(I.P.) = 5 eV. The energy of the reaction $Cl_{\cdot}$ + Cl^- $\rightarrow$ $Cl_2^{-\cdot}$, for instance , is about 29-30 kcal mol^{-1}, although the curve shown in Figure 2 was derived purely from radical cation calculations.

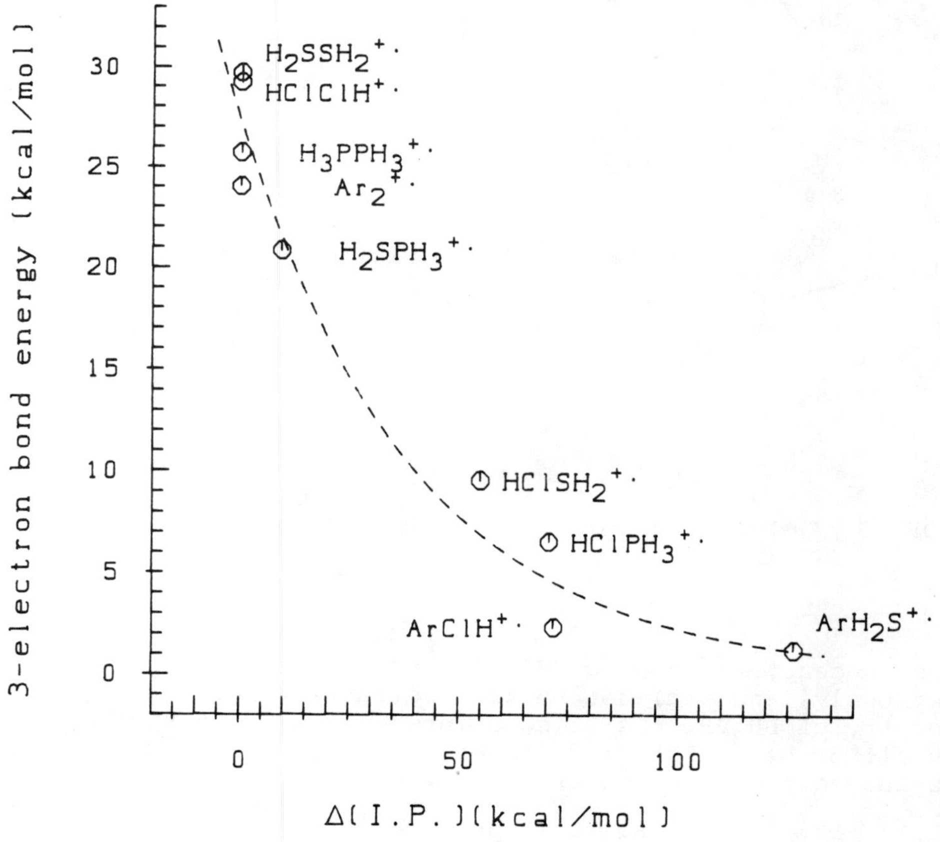

FIGURE 2. The dependence of the three electron bond energy on Δ(I.P.). All data are derived from MP2/6-31G* Calculations on the UHF/6-31G* optimized geometries.

The main features of the geometries of the radical cation complexes are similar to those found previously at 4-31G.[7] The $PH_3SH_2^{+\cdot}$ and $PH_3ClH^{+\cdot}$ radical ions are found to be most stable in the conformations shown in Figure 3. The coordination at phosphorus tends towards a trigonal bipyramidal (TBP) arrangement with the radical center occupying an equatorial position.

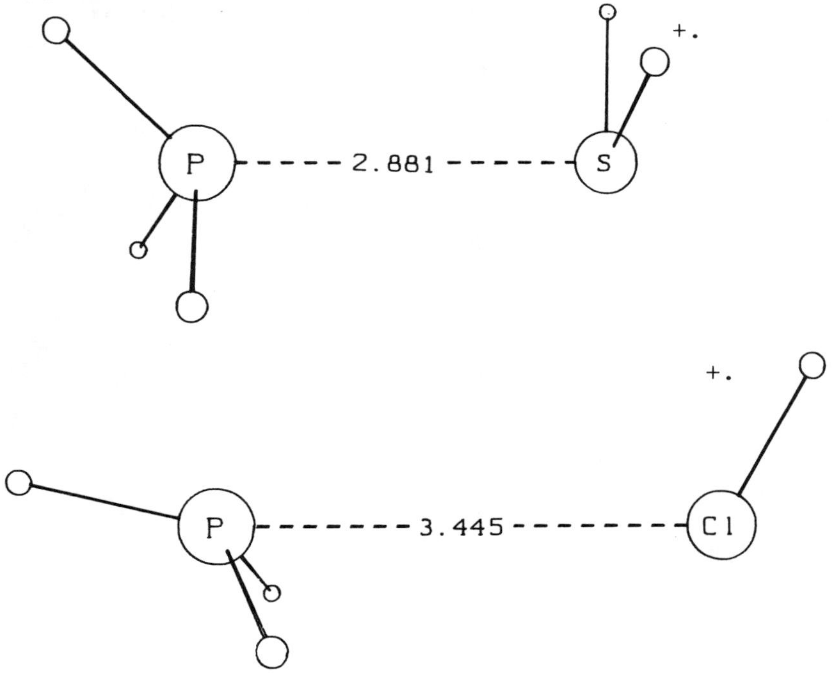

FIGURE 3. The most stable structures of $PH_3SH_2^{+\cdot}$ and $PH_3ClH^{+\cdot}$.

This tendency to adopt a TBP geometry has been discussed previously[7] and will not be treated further here. The geometry of $PH_3PH_3^{+\cdot}$ is of some interest as this species may either be an intermediate or the transition state in the degenerate electron transfer reaction

$$PH_3^{+\cdot} + PH_3 \longrightarrow PH_3 + PH_3^{+\cdot} \quad (3)$$

Although only C_{3v} symmetry was assumed, the only minimum (confirmed by diagonalization of the Hessian Matrix) found is the D_{3d} structure shown in Figure 4.

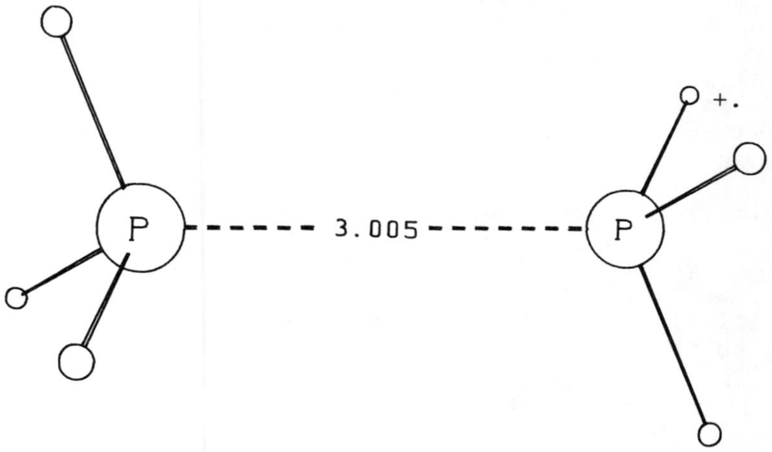

FIGURE 4. The 6-31G* optimized structure of $PH_3PH_3^{+\cdot}$.

The corresponding complex radical cations of the first-row elements nitrogen to neon are often not stable because they undergo extremely facile 1,2-hydrogen shifts, often without activation energy:

$$AH_n^{+\cdot} + BH_m \rightarrow \left[AH_n\text{-}BH_m^{+\cdot}\right] \rightarrow AH_{n-1}^{\cdot} + BH_{m-1}^{+} \qquad (4)$$

Table II, however, shows the three-electron bond energies, bond lengths, and (I.P.) values for some of the first-row radical cation complexes which can be calculated.

TABLE II. Three-electron bond energies, Δ(I.P.) values, and three electron-bond lengths for complex radical cations of the elements N-Ne (units and theoretical methods as in Table I).

A	B	r_{AB}	$-\Delta E$(Eq.1)
Ne	Ne	1.693	38.6
HF	HF	1.813	48.1
H_2O	H_2O	1.997	45.8
H_3N	H_3N	2.163	40.0

Once again, the elements of groups six and seven give the strongest three-electron bonds, followed by group five and the noble gas. The three-electron bond energies for the first-row elements are, however, considerably stronger than those for the equivalent second-row species. The average bond energy for the four symmetrical radical cation complexes reported here is 43 kcal mol^{-1}, compared with the 27.9 kcal mol^{-1} obtained from the best fit curve in Figure 2 for the symmetrical second-row radical cation complexes. Note that, especially in the case of $HFFH^{+\cdot}$, the three-electron bond energies exceed half the two electron bond energies for most of these complexes.

3.2 One electron sigma bonds

A series of symmetrical one-electron bonded radical complexes has been investigated. The results are summarized in Table III. The one-electron bond energy is defined according to the equation:

$$A + B^{\cdot} \longrightarrow AB^{\cdot} \qquad (5)$$

where A is usually a cation. The structure calculated for $CH_3CH_3^{+\cdot}$ corresponds to a one-electron bonded complex between the methyl cation and radical, not to the most stable form of the ethane radical cation.

TABLE III One-electron bond energies and lengths for some symmetrical radical complexes and for $BH_3CH_3^{\cdot}$. Units and theoretical methods are as given in Table I.

A	B	r_{AB}	$-\Delta E$(Eq.5)
Li^+	$Li^{\cdot}$	3.167	28.8
BeH^+	$BeH^{\cdot}$	2.479	49.4
BH_2^+	$BH_2^{\cdot}$	2.131	46.0
CH_3^+	$CH_3^{\cdot}$	1.976	51.0
AlH_2^+	$AlH_2^{\cdot}$	2.916	44.0
SiH_3^+	$SiH_3^{\cdot}$	2.699	21.8
BH_3	$CH_3^{\cdot}$	2.944	3.0

The radical cation complexes all show relatively strong one-electron bonds, the weakest being that found in $H_3SiSiH_3^{+\cdot}$ (21.8 kcal mol^{-1}) and the strongest in $H_3CCH_3^{+\cdot}$ (51.0 kcal mol^{-1}). The neutral radical complex $CH_3.BH_3$ gives a one-electron bond length almost 1 Ångstrom larger than that in the isoelectronic $H_3CCH_3^{+\cdot}$, and a bond energy of only 3.0 kcal mol^{-1}. This result is analogous to that found for neutral three-electron bonded complexes.[6] One electron bonds, however, show a larger variation in energy than the symmetrical three-electron bonds investigated. The large difference in bond energies between $H_2AlAlH_2^{+\cdot}$, which has a bond dissociation energy comparable to that of the first-row radical cation complexes, and $H_3SiSiH_3^{+\cdot}$, which is even more weakly bound than $Li_2^{+\cdot}$, is particularly unusual. Once again, howver, the charged species are considerably more strongly bound than the neutral complex, and the one-electron bond energies are larger than would be expected on the basis of the two-electron bond energies between the elements involved. Studies on the unsymmetrical one-electron bonded complexes are now under way.

4 ODD ELECTRON PI BONDS

4.1. Three-electron Pi bonds.

Three representative systems have been investigated to illustrate the principles involved in three-electron pi-bonding, the hydrazine radical cation,[8] the diphosphine radical cation[9], and the aminomethyl radical. The hydrazine radical cation (1) behaves typically for a charged species involving elements of the first row. The N-N bond is short and the nitrogen centers are almost planar. The transition state for rotation about the N-N bond (2) lies 37.9 kcal mol^{-1} in energy higher than (1), and has a significantly longer N-N bond:

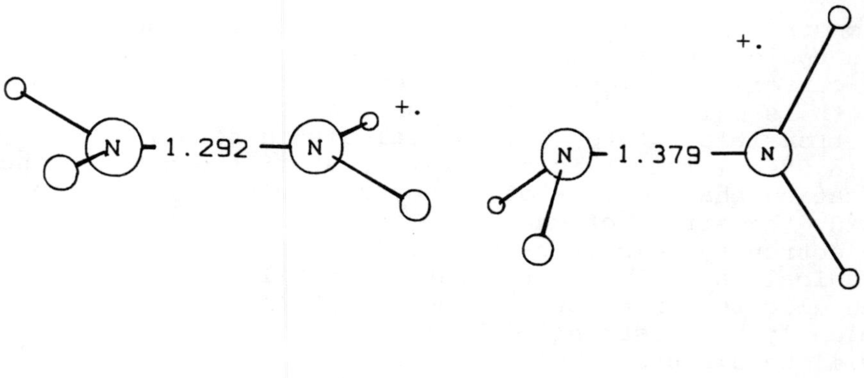

(1) (2)

This behavior indicates that the three-electron pi bond is about half as strong as a conventional two-electron pi-bond, and that it results in considerable shortening of the central N-N distance. The barrier to rotation of a substituted hydrazine radical cation has been measured in solution.[10] The diphosphine radical cation, on the other hand, adopts a C_{2h} anti conformation (3) in its most stable form. The bond length[9] is only slightly than that in the neutral diphosphine itself, and the rotation transition state (4) lies only 7.9 kcal mol^{-1} higher in energy than (3). Note, however, that this radical cation is stabilized by a sigma-pi interaction similar to that which was observed for the sigma and lone-pair orbitals in $H_2SSH_2^{+\cdot}$, although this interaction does not lead to a significant stabilization in the latter case.

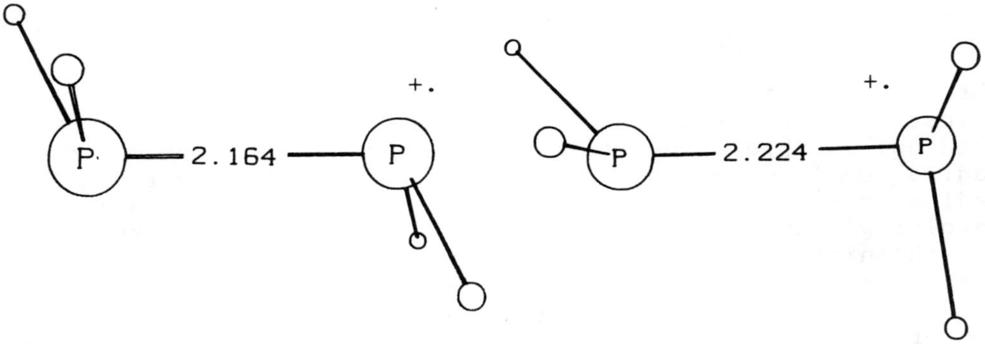

(3) (4)

The aminomethyl radical (5) is isoelectronic with the hydrazine radical cation (1). Nevertheless, the carbon and nitrogen centers are significantly non-planar, and the C-N bond length is only 0.018 Ångstrom shorter than in the rotation transition state (6). The calculated rotation barrier (8.6 kcal mol^{-1}) is closer to that of the diphosphine radical cation than to that of the hydrazine radical cation. Once again, the effect of charge on the bonding properties of odd-electron systems is particularly drastic. The aminomethyl radical has only a very weak pi-stabilization through the three-electron interaction, whereas the hydrazine radical cation clearly has a strong pi-bond. The origins of this effect will be discussed below, but it is clear that neutral radical systems can only be stabilized to a relatively moderate degree, compared with cations and anions, because the odd-electron interactions involved are far weaker than the predominantly two-electron interactions involved in the closed-shell species. On the other hand, charged radical ion systems should exhibit stabilization effects which are comparable with closed shell systems.

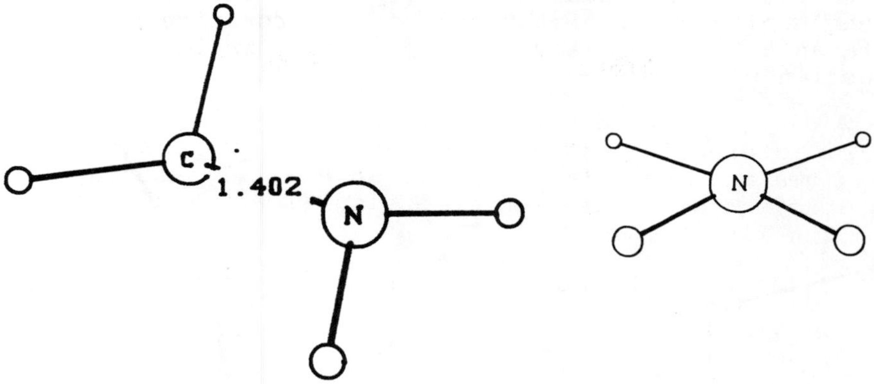

(5)

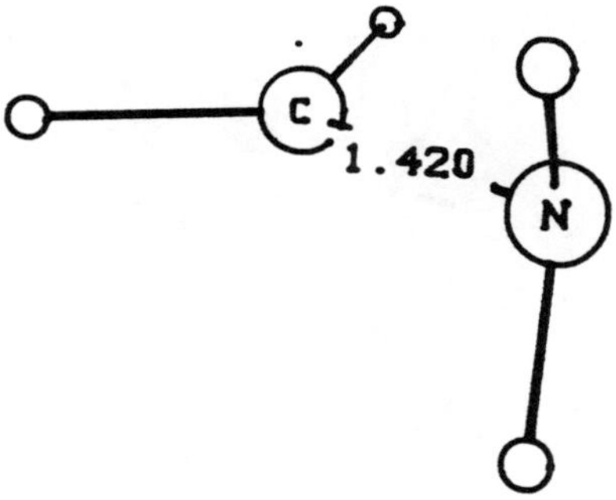

(6)

4.2 One-electron pi-bonds

The propene radical cation and the radical derived from a methyl-hydrogen abstraction from methyl borane, $BH_2CH_2^{\cdot}$, have been used as examples of systems with one-electron pi-bonds. The prototype olefin radical cation, ethylene$^{+\cdot}$, was not investigated as most ab initio methods predict a planar structure, whereas experimental measurements favor a twisted geometry. The propene radical cation is, however probably planar, although the experimental situation is not completely clear. UHF/6-31G* optimization of $CH_2CHCH_3^{+\cdot}$ without symmetry constraints leads to the planar structure (7). The calculated rotation barrier, via (8), is 25.5 kcal mol^{-1}, slightly less than half the barrier found in neutral olefins. As for the hydrazine radical cation, the is a significant lengthening of the pi-bond on rotation. This lengthening and the magnitude of the rotation barrier indicate an appreciable amount of pi stabilization in the one-electron pi-system.

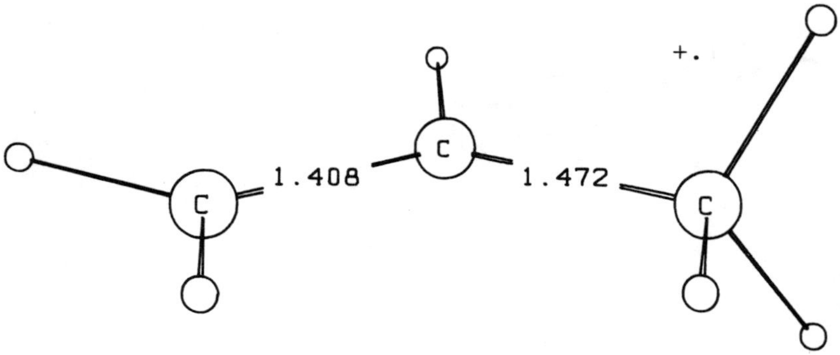

(7)

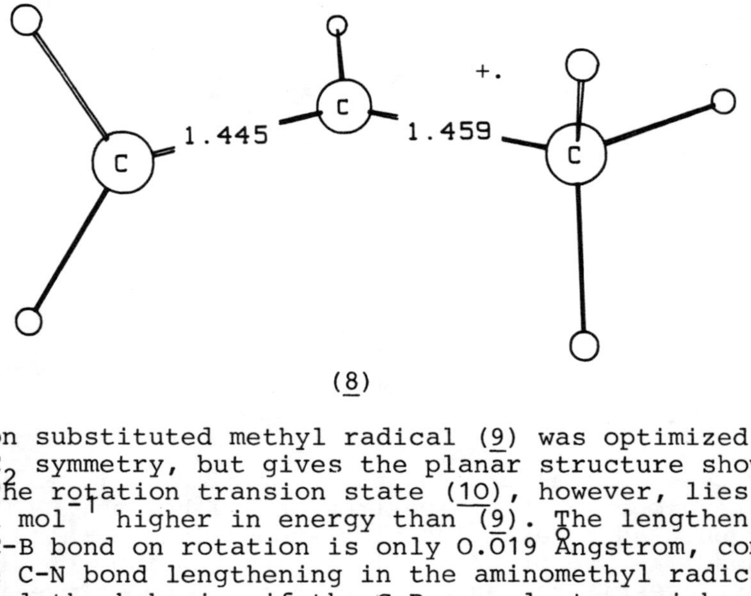

(8)

The boron substituted methyl radical (9) was optimized within C_2 symmetry, but gives the planar structure shown below. The rotation transion state (10), however, lies only 9.8 kcal mol^{-1} higher in energy than (9). The lengthening of the C-B bond on rotation is only 0.019 Ångstrom, comparable with the C-N bond lengthening in the aminomethyl radical. In general the behavior if the C-B one-electron pi-bond and the three-electron C-N pi-bond are very similar, with rotation barriers of about 9 kcal mol^{-1} and bond lengthenings on rotation of less than 0.02 Ångstrom. Although the difference between the charged propene radical cation and the uncharged $CH_2BH_2^{\cdot}$ radical is not quite as drastic as in the other examples, the effect of charge is still very evident in these one-electron bonded species.

(9)

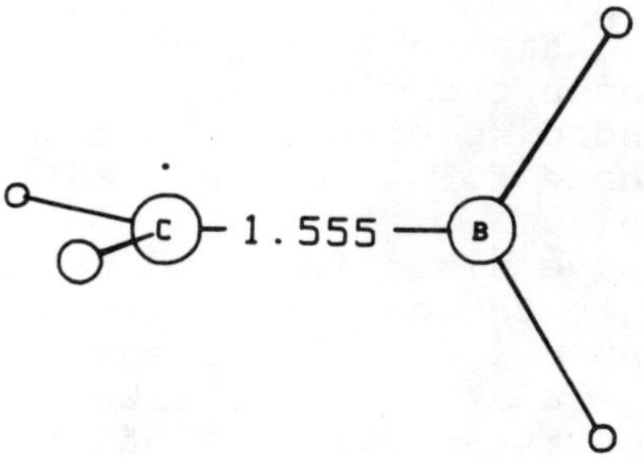

(10)

5. FACTORS AFFECTING ODD-ELECTRON BOND STRENGTHS

The odd-electron bonds investigated here, both sigma and pi and with one or three electrons, all show a strong dependence on the molecular charge. The radical cation systems are all strongly bonded, but the neutral radicals all show relatively small stabilizations. Further investigations are necessary in order to determine whether the exponential dependence of the odd-electron bond strength on Δ(I.P.) also applies to one-electron sigma bonds. This strong dependence on Δ(I.P.), or on the energy difference between the interacting orbitals in a single electron PMO picture, is best understood using valence bond theory with two "no bond resonance" structures

$$A^{+\cdot} \quad B \leftrightarrow A \quad B^{+\cdot}$$

for three-electron sigma bonded radical cations or

$$A^{+} \quad B^{\cdot} \leftrightarrow A^{\cdot} \quad B^{+}$$

for one-electron sigma bonded positively charged systems. The relative importance of these two resonance forms, and hence the strength of the odd-electron bond, depends on the relative energies of the A^{+} B/A B^{+} redox system. Energetically equivalent no-bond resonance forms yield strong bonds, which rapidly weaken as the energy difference between the two forms increases. Neutral radicals have weak interactions because one of the two resonance structures requires that charges be separated:

$$A^{\cdot} \quad B \leftrightarrow A^{-} \quad B^{+\cdot}$$

or

$$A \quad B^{\cdot} \leftrightarrow A^{-\cdot} \quad B^{+}$$

This charge separation is extremely unfavorable in the gas phase so that the right hand resonance structure always lies considerably higher in energy than the left hand one, and hence the odd electron bonds are weak. Similarly, the pi-bonded species can be described by energetically similar resonance forms for the radical ions, but not for the neutral radicals.

The above situation, and the calculations reported here, refer to the gas phase. In solution the charge-separated resonance forms should become energetically competitive, so that a strong stabilization of odd-electron bonded species in solution relative to the gas phase may be observable.

The bonding principles outlined above can, however be used to treat some common problems in radical and radical ion chemistry. The rest of this article will be concerned with some examples of radical substituent effects and related aspects of radical reactivity.

6. BRIDGED BETA-SUBSTITUTED ALKYL RADICALS

Although there has been extensive discussion of possible bridging in ß-halo alkyl radicals from an experimental viewpoint, theoretical studies have consistently favored open, classical structures. The MP2/6-31G* activation energies for a 1,2-halogen shift inthe ß-fluoro- and ß-chloro ethyl radicals, for instance, are 45.2 and 14.4 kcal mol^{-1}, respectively.[11] The symmetrical approach of a radical towards the center of a C-C double bond is exchange forbidden, but this type of structure involves a three-center-three-electron bond:

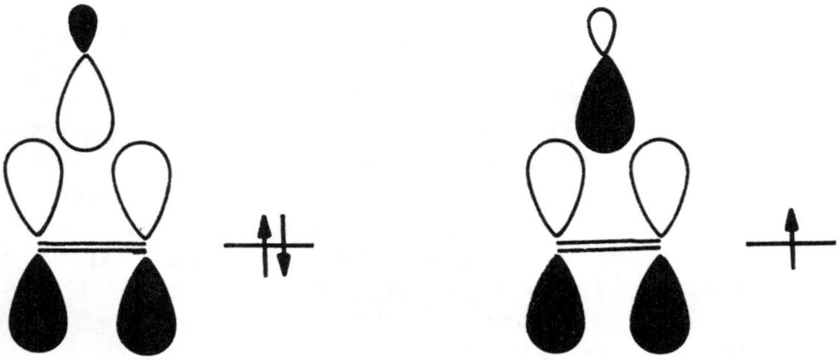

This suggests that bridging in radicals should be favored by a positive charge, as has been suggested by earlier theoretical work on ß-amino- and ß-hydroxy radicals.[12] Protonation of a ß-halo ethyl radical should give a strong stabilization because of the three-electron interaction, and therefore favor bridging. The protonated ß-fluoro- (11) and ß-chloro ethyl radicals (12) support this conclusion. The CCX bond angles in (11) and (12) are 91.9^{o} and 97.4^{o}, respectively, and the calculated barriers to a 1,2-halogen shift are only 0.05 and 1.57 kcal mol^{-1}, respectively.

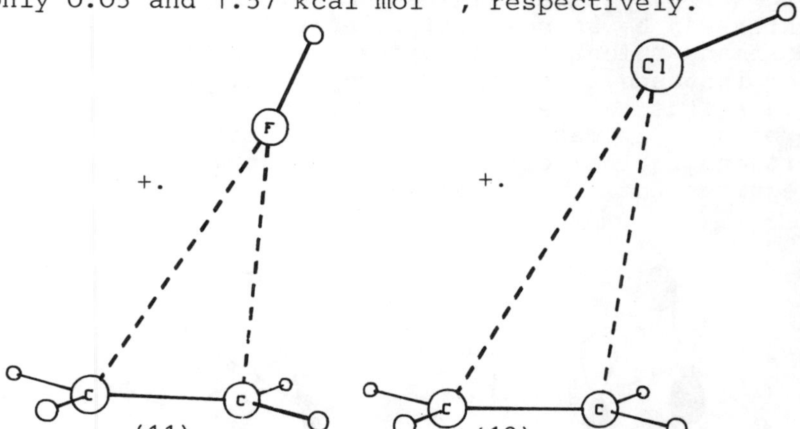

(11) (12)

A further, and possibly experimentally important, consequence of the extra stabilization in these protonated radicals is thatprotonation of the neutral ß-halo radicals is more favorable energetically than protonation of the corresponding alkyl halides. (11), for instance, has a calculated anergy for deprotonation (i.e. the proton affinity of the ß-fluoroethyl radical) of 171.8 kcal mol^{-1}, 17.3 kcal mol^{-1} higher than the calculated proton affinity of ethyl fluoride. Similarly, protonation of the ß-chloro ethyl radical to give (12) is 7.4 kcal mol^{-1} more favorable than protonation of ethyl chloride. This means that the neutral radicals are several $P_k a$ units more basic than the corresponding saturated compounds. The rotation barriers of (11) and (12) are substantial (22.6 and 10.5 kcal mol^{-1}, respectively). The high degree of double bond character in these protonated radicals, combined with the ease of protonation of the neutral radicals suggests that some degree of stereocontrol may be obtained in such radical reactions in acid media. Once again, however, it can be expected that solution conditions should favor bridged structures, and therefore lower the barriers to 1,2-shifts compared with the gas phase. The above examples also demonstrate the effect of the stronger three-electron bonds observed for elements of the first row. Although the barrier to a 1,2-fluorine shift in the neutral radical is much higher than the corresponding barrier for chlorine, the situation is reversed in the protonated radicals, the strong carbon-fluorine three-electron interaction favoring the bridged form. This trend is also reflected in the C-C rotation barriers of (11) and (12).

7. 1,3-SIGMATROPIC HYDROGEN SHIFT IN PROPENE$^{+\cdot}$

Sigmatropic reactions are some of the few radical reactions which can be successfully treated using the Woodward-Hoffmann rules. The degenerate 1,3-hydrogen shift in propene, for instance, is a high energy antarafacial process, but 1,n-shifts in olefin radical cations are well known facile processes in mass spectrometry. Considering once again the propene radical cation (7), the suprafacial 1,3-shift requires the orbital occupancy pattern shown in Figure 5(a):

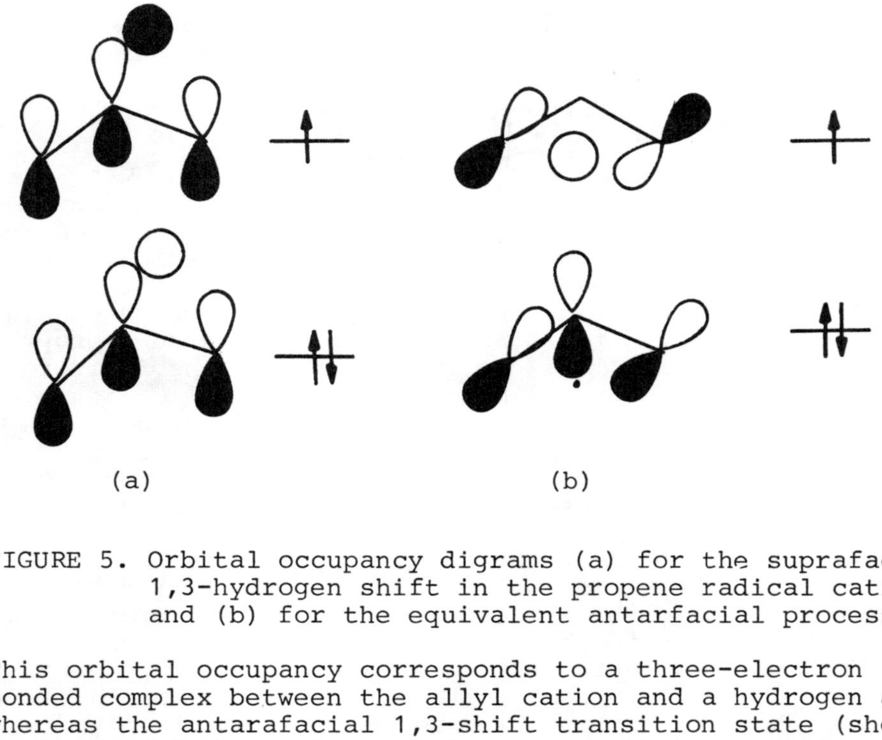

FIGURE 5. Orbital occupancy digrams (a) for the suprafacial 1,3-hydrogen shift in the propene radical cation, and (b) for the equivalent antarfacial process.

This orbital occupancy corresponds to a three-electron bonded complex between the allyl cation and a hydrogen atom, whereas the antarafacial 1,3-shift transition state (shown in Figure 5(b)) requires at least a partial occupation of the second pi-orbital in the allyl system (i.e. it contains a contribution from the allyl radical/proton redox form). The experimental heat of formation for a hydrogen atom plus an allyl cation is 278 kcal mol^{-1}, compared with 409 kcal mol^{-1} for a proton and an allyl radical. Therefore it is more favorable for tha propene radical cation to undergo a suprafacial shift and thus avoid the energetically unfavorable electron transfer from the hydrogen to the allyl fragment. The activation energy for this reaction can be expected to be the difference in energy between the propene eadical cation and an allyl cation plus a proton minus the three-electron binding energy in the transition state. This energy should, however, be substantial as the system is charged and because the atoms involved are carbon and hydrogen. The experimental heat of formation of the propene radical cation is about 229 kcal mol^{-1}, so that the maximum barrier to a suprafacial 1,3-hydrogen shift should be about 50 kcal mol^{-1}.

These predictions are confirmed by calculations. Despite an extensive search, it was not possible to find a transition state for the antarafacial shift, but the transition state for the suprafacial process (12) was easily located, and found to be only 29.6 kcal mol^{-1} less stable than the propene radical cation. This barrier, which is only slightly higher than the rotation barrier about the olefinic bond, is lowered by about 20 kcal mol^{-1} from the maximum value by the three-electron-four-center interaction shown in Figure 5(a).

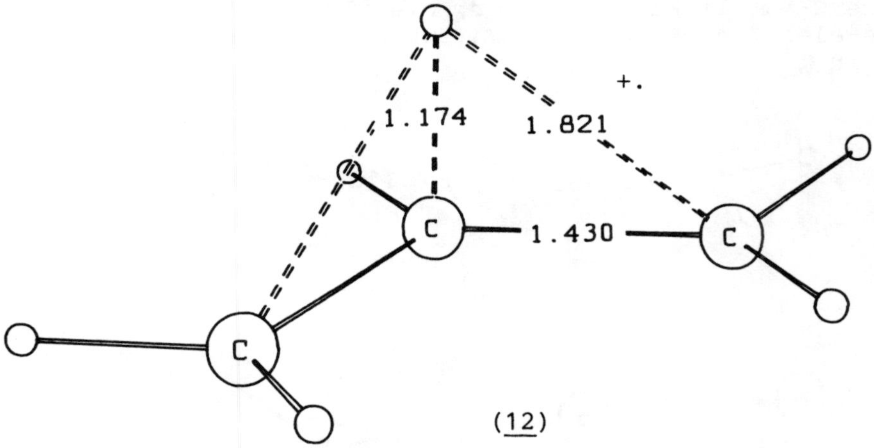

(12)

Note the startling change in behavior of propene towards a 1,3-sigmatropic hydrogen shift on removal of one electron. This process changes the stereochemistry of the 1,3-shift and also makes it a relatively facile reaction compared with the neutral parent molecule. The transition state (12) resembles a distorted edge-edge trimethylene radical cation, which has been suggested as a product of the cyclopropane radical cation, in which one of the central carbon hydrogen bonds has been weakened and distorted towards the center of the allyl system. As with most of the three-electron bonded species reported here, the energy of (12) is very sensitive to the inclusion of electron correlation (in this case via the MP2 correction).

8. CAPTO-DATIVE RADICALS

There has been much experimental and theoretical discussion of the concept of capto-dative radical stabilization. The cyano aminomethyl radical has been used as a typical organic example in this study, although cyanomethyl radicals suffer quite severe spin contamination in UHF calculations. This problem can, however, be largely removed by use of the MP2 correction, which decreases the spin contamination significantly. There are two significant aspects of the problem, the radical geometry and the stabilization energy. These will be discussed separately. The geometry of the aminomethyl radical (5) has been discussed above, and serves as a comparison with $NH_2CHCN^{\cdot}$. The cyanomethyl radical (13) is shown below with the cyano aminomethyl radical (14). (14) shows significant changes in the amino substituent geometry relative to (5). The $C-NH_2$ bond is shortened to 1.384 Ångstrom, compared with 1.402 in (5), and the amino group is almost planar. The geometry of the cyano group in (14) is essentially the same as that in (13). The changes in the $C-NH_2$ linkage, however, clearly indicate a stronger three-electron pi-interaction inthe capto-dative radical than in (5).

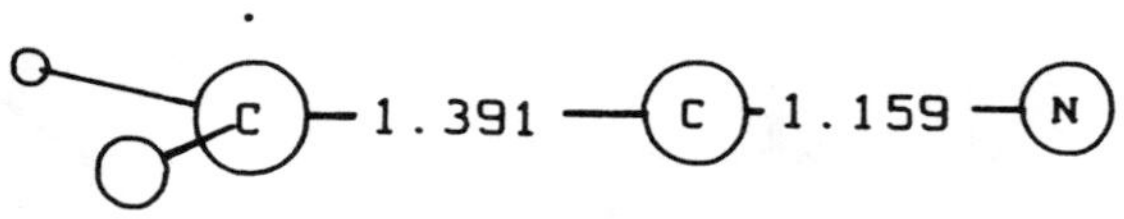

(13)

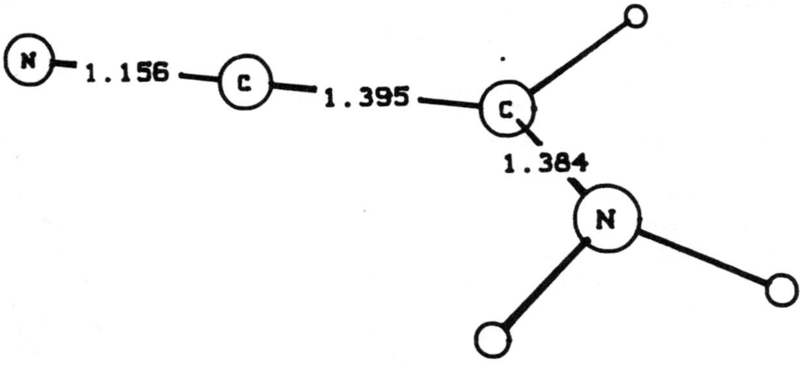

(14)

The calculated energies also reflect this extra stabilization at the MP2 level, but not at the uncorrected UHF level. This is an important point to consider in future calculations on potentially capto-dative systems, and helps to explain some of the conflicting conclusions reached in theoretical studies. The energy calculated for the group separation reaction

$$CH_3^{\bullet} + NH_2\dot{C}HCN \longrightarrow \dot{C}H_2NH_2 + \dot{C}H_2CN$$

is, for instance, only +0.1 kcal mol^{-1} at UHF/6-31G*, but +4.0 kcal mol^{-1} at MP2/6-31G*. At the uncorrelated level of theory, therefore, the stabilization found in the capto-dative radical is essentially the sum of the two individual group stabilizations, but in the more complete MP2 calculations there is a chemically significant extra stabilization.

These 4 kcal mol^{-1} extra stabilization are, however, not evident in bond fission reactions because a tetrahedral carbon substituted with amino and cyano groups also enjoys an extra anomeric effect stabilization. This energy, calculated using the bond separation reaction

$$CH_4 + NH_2CH_2CN \longrightarrow CH_3NH_2 + CH_3CN$$

is found to be 2.5 kcal mol^{-1}, giving a net bond weakening for a methylene hydrogen atom in NH_2CH_2CN of 1.5 kcal mol^{-1}. This is a small, but kinetically significant effect which is, however, much clearer when the radical geometry, which is only easily accessible through calculations, is considered.

ACKNOWLEDGMENTS

I am grateful to many colleagues, especially Profeesors P. v. R. Schleyer, S. F. Nelsen, K.-D. Asmus, and many others for stimulating discussions. This work would not have been possible without the cooperation of the staff of the Regionales Rechentrum Erlangen.

REFERENCES

1. See, for instance, W.J.Hehre, L.Radom, P.v.R.Schleyer, and J.A.Pople, "Ab Initio Molecular Orbital Theory", Wiley, New York, 1986.

2. N.C.Baird, J.Chem.Ed., 1977,54,291.

3. J.S.Binkley, R.A.Whiteside, K.Raghavachari, R.Seeger, D.J.DeFrees, H.B.Schlegel, M.J.Frisch, J.A.Pople, and L.A.Khan, Carnegie-Mellon University, 1982.

4. P.C.Hariharan and J.A.Pople, Theor.Chim.Acta, 1973,28, 213; M.M.Francl, W.J.Pietro, W.J.Hehre, J.S.Binkley, M.S.Gordon, D.J.DeFrees, and J.A.Pople, J.Chem.Phys., 1982,77,3654.

5. T.Clark, J.Comput.Chem., 1981,2,261.

6. T.Clark, J.Comput.Chem., 1982,3,112.

7. T.Clark, J.Comput.Chem., 1983,4,404.

8. T.Clark, unpublished results. See also reference 10.

9. T.Clark, J.Am.Chem.Soc., 1985,107,2597; D.Feller, E.R. Davidson, and W.T.Borden, ibid., 1985,107,2596.

10. S.F.Nelsen, G.T.Cunkle, D.H.Evans, and T.Clark, J.Am. Chem.Soc., 1983,105,5928.

11. T.Clark and M.C.R.Symons, J.C.S.Chem.Comm., 1986, in the press.

12. B.T.Golding and L.Radom, J.C.S.Chem.Comm., 1973,939; J.Fossey and J.-Y.Nedelec, Tetrahedron, 1981,37,2967.

THE CONTRASTING STRAIN ENERGIES OF SMALL RING CARBON AND SILICON RINGS. THE RELATIONSHIP WITH FREE RADICAL ENERGIES

Paul von Ragué Schleyer
Institut für Organische Chemie
Friedrich-Alexander-Universität Erlangen-Nürnberg,
Henkestrasse 42,
D-8520 Erlangen,
Federal Republic of Germany

Abstract. The similar strain energies of cyclopropane and of cyclobutane, despite the large differences in angle deformation, are due to 1,3-CC repulsions in the latter and rehybridization in the former. The weakness of the CC bonds in cyclopropane is compensated somewhat by CH bond strengthening. In silicon rings, angle deformation effects dominate, and trisilacyclopropane is much more strained than tetrasilacyclobutane and even more strained than cyclopropane. Silicon rehybridization on bonding is much less, and SiH bond stengthening does not compensate for SiSi bond weakening. The CH bond dissociation energies, which are uniquely high in cyclopropane, provide support for this interpretation.

1. Introduction

A century ago, Baeyer proposed his strain theory in a theoretical addendum to a preparative paper.[1] His idea that deviations from tetrahedral angles were chiefly responsible for the instability of 3- and 4-membered rings was accepted until the 1950's when measurements of heats of combustion revealed cyclopropane and cyclobutane to have nearly the same strain energy, 28 and 27 kcal/mol, respectively.[2,3] Since the angular deviations are much larger in cyclopropane than in cyclobutane, other factors clearly are important. But even today there is no generally accepted interpretation of the strain energies of these two most fundamental small ring hydrocarbons. This is the central problem addressed in this paper.

Quite recently, derivatives of the small ring silicon compounds, trisilacyclopropane[4] and tetrasilacyclobutane[5] have been synthesized, and their X-ray structures determined. However, no thermochemical data is available yet. This deficiency can be rectified theoretically as quite reliable ab initio calculations can be now carried out simply with generally available programs, like Gaussian 82.[6] The goals of this paper are to compare the strain energies of 3-and 4-membered carbon and silicon rings and to provide an explanation of their behavior.[7] The corresponding cycloalkyl and cyclosilyl

H. G. Viehe et al. (eds.), Substituent Effects in Radical Chemistry, 69–81.
© 1986 by D. Reidel Publishing Co.

radicals provide key information for this interpretation, and are included in this study as well.

2. Interpretations of the strain energies of cyclopropane and cyclobutane.

If Baeyer had been correct, the strain energy of cyclopropane would be much larger than that of cyclobutane. In fact, these energies are nearly the same. Either cyclobutane is higher in energy than it "ought to be", or some effects are reducing the energy of cyclopropane. Interpretations have generally tended to emphasize one possibility or the other, but both are operative.

2.1. Torsional Strain.

Other effects besides angle strain influence the energies of molecules. Torsional strain is one of these, and is exemplified by the 3 kcal/mol rotational barrier in ethane. This can be attributed to interactions in the degenerate ethane HOMO's, which have π^* symmetry.[8] These repulsive interactions are somewhat less in the staggered than in the eclipsed conformation. While eclipsed CH bonds are unfavorable, it is not clear to what extent this can be carried over to ring systems. Thus, planar cyclopentane has 5 ethane-like conformations, but only 10 eclipsed CH bonds, contributing about 1 kcal/mol each. Since planar cyclopentane has essentially no angle strain (the CCC bond angles are 108°), the entire strain can be attributed to torsional effects. Molecular mechanics calculations are no help, since they are parameterized arbitrarily in this regard.[9] However, 6-31G* ab initio calculations on planar cyclopentane show the strain energy to be 9.6 kcal/mol, i.e., just what one would expect from the 10 eclipsed CH bonds. The CC bonds are also eclipsed in planar cyclopentane, but these are involved in ring orbitals which evidently are no more unfavorable than those in acyclic counterparts.

On the basis of the number of eclipsed C-H bonds, cyclopropane and planar cyclobutane would be expected to have 6 and 8 kcal/mol torsional strain, respectively. In fact, cyclobutane is not planar, but has a rather low barrier (about 1.5 kcal/mol) to ring inversion.[3] Since some torsional strain is relieved on puckering, an estimate of about 7 kcal/mol of torsional strain in puckered cyclobutane seems reasonable. Due to the 60° ring angles, the cyclopropane methylenes are bent away from one another; this might decrease the torsional strain slightly, as well. Corrected for these torsional effects, the residual strain energies of cyclopropane and of cyclobutane still remain about the same, about 23 and 20 kcal/mol, respectively. We have not proceeded very much further in understanding our central problem.

2.2. Nonbonded Interactions.

Dunitz and Schomaker pointed out that cyclobutane suffers from transannular 1,3-CC nonbonded interactions.[10] These occur twice in cyclobutane, but are not present in cyclopropane because all carbons are bound together. Indeed, the 1,3-C-C distance in cyclobutane is only 2 Å, which is well within the van der Waals radius sum of carbon[9] and considerably below the 2.6 Å separations which are typical in acyclic and unstrained cyclic hydrocarbons. Wiberg recently provided evidence for the even greater importance of such 1,3-interactions in bicyclo[1.1.1]pentane.[11] But how does one evaluate the magnitude of these 1,3-interactions in cyclobutane? The standard parameters in molecular mechanics programs generally do not "work" for cyclobutane and "various adjustments have to be made".[9] The popular MM2 force field of Allinger indicates the nonbonded strain in cyclobutane to be 4 kcal/mol, but again this value depends on the arbitrary treatment. Using a molecular orbital method and comparing the results when overlap was included and excluded, Bauld estimated the 1,3-interaction energy to as high as 20-33 kcal/mol in cyclobutane.[12] Including the torsional correction, this would mean that the angle strain effect in cyclobutane would vanish fully! It is more reasonable to assume that angle strain amounts to about 10 kcal/mol, in contrast to the much larger angle strain in cyclopropane. This would account better for the cyclobutane vs cyclopropane thermochemical behavior, as the nonbonded contribution of Bauld probably is overestimated. We have come a definite step further in understanding our central problem.

2.3. "Sigma Aromaticity".

Quite recently, Dewar has proposed one of his typically audacious explanations. Noting that the six ring electrons in cyclopropane are "isoconjugate" with the six π electrons of benzene, he suggests that cyclopropane also should benefit from special "aromatic" stabilization.[13] Dewar argues that use of any reasonable bond angle bending force constant would lead to a much larger strain energy in cyclopropane, relative to cyclobutane, than is found experimentally. Ignoring the nonbonded and rehybridization contributions completely, he extrapolates a curve from cyclohexane through the strain energy points for cyclopentane and cyclobutane upwards. The large deviation from the resulting extrapolated energy of cyclopropane from the real value, some 55 kcal or so,[13] is to Dewar the "resonance stabilization". More recently, Cremer has espoused this argument, and also used it to interpret some of the chemical behavior and spectroscopic properties of cyclopropane.[14]

One should, in my opinion, regard "sigma aromaticity" with considerable scepticism. The effects which have been ascribed to this alleged phenomenon can easily be explained otherwise. Theoretically, I see no justification in such a proposal. There is no magic in six electrons; these can be quite unfavorable when the ring is small. Thus, planar triazacyclopropane would have 6 π electrons, but would be highly destabilized, since HOMO's are antibonding π* orbitals.

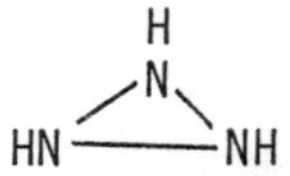

6 π antiaromatic

In cyclopropane, the highest pair of occupied MO's, first described by Walsh,[3,8] are quite high in energy and do not seem to exhibit any favorable characteristics. I think we must look elsewhere for the solution to our problem. There appears to be no need to invoke "sigma aromaticity" to explain the thermochemistry of cyclopropane.

2.4. Rehybridization.

Walsh, Coulson and Moffitt recognized that the extremely small bond angles in cyclopropane would lead to a marked carbon rehybridization.[3,8] The CC bonds would have more p and the CH bonds more s character. Indeed, an approximately sp^5-sp^2 description of the bonding in cyclopropane has been deduced theoretically,[3,8] and is supported remarkably well by the magnitude of the CH and CC coupling constants.[15]

Since C(sp^2)-H bonds are well known to be stronger than C(sp^3)-H bonds, it has been suggested several times that the strengthening of the CH bonds in cyclopropane would compensate to some extent for the weakening of the CC bonds. This argument, which was clearly stated in the popular textbook of Roberts and Caserio,[16] inter alia,-[12,17] also was overlooked by the "σ-aromaticity" proponents in their analyses.[13,14]

2.5. Free radical energies and bond dissociation energies

It is here that interest in free radical energies enter. Bond dissociation energies are the most direct way of comparing the strengths of different kinds of bonds. Table I summarizes representative energies pertinent to this problem.[18,19] Unfortunately, the experimental data are in a state of flux, and there is a definite tendency in more recent work towards larger values. Agreement has not been achieved. Hence, Table I lists the recommended values of McMillen and Golden,[18] but also provides the results of the newer determinations.[19]

Theory can help here. While the accurate calculation of bond dissociation energies directly is difficult, reliable relative energies are much easier to obtain. The reason for this is simple to understand. During a homolytic bond cleavage, e.g., of an alkane into an akyl radical and a hydrogen atom, the number of paired electrons decreases. This means that electron correlation effects change, and these differences are the hardest to describe quantum mechanically. However, relative energies can be compared by means of isogyric equations like eq. 1, in which spin is conserved on both sides.[6]

$$R\cdot + R'H \rightarrow RH + R'\cdot \qquad (1)$$

The evaluation of such equations is, for the most part, relatively

Table I. Energies of organic radicals, relative to methyl (kcal/mol)

Radical	McMillen, Golden[a]	Tsang[b]	MP4/6-31G* //6-31G*+ZPE[c]	IR[e]
CH_3	0.0	0.0	0.0	0.0
C_2H_5	-6.9	-4.6	-3.6	-3.6
1-C_3H_7	-7.2	-4.0	-3.5	-3.6
2-C_3H_7	-10.0	-5.8	-6.8	-6.2
t-C_4H_9	-11.9	-8.4	-7.5	-8.4
c-C_3H_5	+1.2		+3.2[d]	+5.2
c-C_4H_7	-8.6		–	-3.1
c-C_5H_9	-10.6		–	–
C_2H_3	+5±2		+10.4	+5.2
C_6H_5	+5.8±2		–	+6.3
C_2H	+27±5		+35.0	+29.6

[a] Data from ref. 18. Error amounts usually of ±1 kcal/mol are given, unless otherwise indicated, but these are optimistic.
[b] Data from ref. 19.
[c] Evaluated by means of equ. 1 from 6-31G*//6-31G* data, corrected for zero point energy differences. Much data has been taken from ref. 6. The t-C_4H_9 value is at 6-31G*.
[d] ZPE correction not available. Also see M.H. Lien and A.C. Hopkinson, J. Comput. Chem. **1985**, 6, 274.
[e] Values estimated by McKean[20] from infrared stretching frequencies.

insensitive to the size of the basis set and to the extent of inclusion of electron correlation. Methane and the methyl radical, whose energies appear to be well established experimentally,[18] can be used as standards. This is the basis for the theoretical values listed in Table I. The agreement with the newer experimental values for the larger alkyl radicals[19] is quite gratifying and helps to establish this set as the correct one.

The data in Table I show that the cyclopropane CH bond dissociation energy is 10 kcal/mol higher than typical secondary CH values whereas the difference in cyclobutane, about 1 kcal/mol, is an order of magnitude less. The value for cyclopropane is in line with other C(sp^2-H) values, e.g. those for ethylene and for benzene. It seems quite definite that the CH bonds in cyclopropane are significantly stronger than those in acyclic hydrocarbons whereas cyclobutane is more or less normal in this regard. Indeed, C-H stretching frequencies, which can be "isolated" by using deuterium-substituted substrates, also are significantly higher for cyclopropane than those in unstrained molecules.[20] This data can be used to estimate bond dissociation energies with impressive accuracy (Table I). Although this was done years ago,[20] the results agree better with the newer data in Table I.

The difficulty is that bond dissociation energies and bond energies are different. The latter are needed to solve our central problem and cannot be obtained unambiguously. We can measure or calculate the total energy of a molecule, but dividing this up into energies for bonds of different types is arbitrary. However, the two appear often to be directly related,[20] and interesting methods have been proposed.[21]

The CH bond dissociation energy in cyclopropane is not just due to the stronger CH bonds in the hydrocarbon, but also is enhanced by the increase in strain energy in going to the cyclopropyl radical. Molecular orbital explanations are equivalent, but simple strain arguments suffice to rationalize this behavior. Reduction of a 120° bond angle in the planar methyl radical to 60° costs more energy than reducing a 109.5° bond angle in methane to 60°. Translated to 3-membered ring systems, this means that the cyclopropyl radical prefers to be nonplanar, and has a rather large inversion barrier (for alkyl radicals) of 5 kcal/mol. The cyclobutyl radical is nonplanar, but both the increase in strain in going to approximately 90° angles and the inversion barrier are quite small.

Thus, only a portion of the larger C-H bond dissociation energy of cyclopropane can be ascribed to its enhanced CH bond energy. What portion? While we are still seeking ways to determine this quantitatively, it seems reasonable to estimate that the CH bond energy enhancement is 10 to 20% of the increase in bond dissociation energy (10 kcal/mol per CH bond).

Thus, if each of the six CH bonds in cyclopropane are stronger by 1-2 kcal/mol than normal (this estimate may be conservative),[16,21] considerable understanding of our central problem will result. The extreme bond angle deformation in cyclopropane weakens the CC bonds, but this is compensated by 10 kcal/mol of C-H bond strengthening.

This, rather than "σ-aromaticity" affords a much more satisfactory explanation for the apparently anomalous behavior of cyclopropane, particularly when torsional and nonbonded transannular effects are taken into consideration. Table II summarizes the energetic analysis presented above.

2.6. Conclusions. The origins of strain in small carbon rings.

Cyclopropane and cyclobutane have about the same total strain energy because angle deformation is only one of the factors involved. Torsional strain in both hydrocarbons is quite considerable, but not very different. Cyclobutane suffers from serious, 1,3-nonbonded repulsions, which raises its energy perhaps by as much as 10 kcal/mol. This effect is not present in cyclopropane, which, on the other hand benefits from rehybridization. The extreme weakening of the C-C bonds are compensated to some degree by the strengthening of C-H bonds. There is no need to invoke "σ-aromaticity".

3. Strain energies of small ring cyclosilanes

Small ring silicon compounds, having become available only relatively recently,[4,5] prompt an examination of their strain energies computationally. Indeed, the results are quite unexpected, and support the interpretations given above for the corresponding carbon compounds. Our calculations on strained ring systems (Table III) comprised of silicon have been carried out at various levels (3-21G, 3-21G*, 6-31G* and MP2/6-31G*), but the strain energies deduced from this data do not differ much from level to level. This in itself is revealing since d-type polarization functions are needed to describe the "bent" bonds in cyclopropane and in strained carbon ring systems. That d-orbitals have little effect on the relative energies of three membered ring silicon compounds suggests that "bent bonding" is less important for silicon than for carbon.

A few years ago, Gordon reported that the strain energy of silacyclopropane (**1**) was 10 kcal/mol higher than that of cyclopropane.[22] This seems surprising, since bond angle deformations at silicon require less energy than the corresponding deformations at carbon. This is shown by the experimental bending force constants[23] as well as model calculations on deformed structures. Gordon quite reasonably attributed his finding to the unusually small bond angle (47°) at silicon in silacyclopropane. (This angle is a geometrical consequence of the fact that Si-C bonds are longer than C-C bonds.) The strain energy increase at silicon, according to Gordon, is not compensated by the widening of the SiCC angles. In order to test this conclusion, we calculated trisilacyclopropane (**2**), which, obviously must have 60° ring angles. Much to our surprise, the strain energy at the four levels listed above were 37.5, 40.3, 49.6 and 38.2 kcal/mol, respectively. These values are 10 kcal/mol higher than the strain energy of cyclopropane, determined experimentally or by similar theoretical evaluations.[2,3]

Table II. Roughly estimated contributions to small ring strains (kcal/mol)

Contribution	c-$(CH_2)_3$	c-$(CH_2)_4$	c-$(SiH_2)_3$ (**2**)	c-$(SiH_2)_4$ (**3**)
XH Bond eclipsing	5	7	1	2
1,3-Repulsions	0	10	0	5
Angle strain	33	10	38	8
Rehybridization	-10	~0	-1	0
Total Strain	28	27	38[a]	18[a]

[a] Best values from Table III.

Table III. Strain energies of silicon ring systems, kcal/mol

Molecule	Point Group	3-21G//3-21G	3-21G*//3-21G*
c-$(SiH_2)_3$ (**2**)[a]	D_{3h}	37.5	40.3
c-$(SiH_2)_4$ (**3**)	D_{4h}	12.9	18.9
	D_{2d}	12.0	(18.4)[b]
c-$(SiH_2)_5$ (**4**)	D_{5h}	-2.2	-
$H_2Si(SiH)_2SiH_2$[c]	C_{2v}	74.8	70.6
$(SiH)_4$[d]	T_d	146.7	144.7

[a] Additional values: 39.6 at 6-31G*//6-31G*; 38.2 at MP2/6-31G*//6-31G*. [b] Incomplete optimization. [c] Tetrasilabicyclobutane.
[d] Tetrasilatetrahedrane.

A very recent paper by Schöller[7] on this topic claims that trisilacyclopropane (**2**) and cyclopropane have about the same strain energy. The reason for this discrepancy is that Schöller did not employ the proper method to evaluate his data, and his conclusion is incorrect. Strain energies should be based on "strain-free" group increments,[2,3] i.e., energy contributions of groups taken from similar, but strainless environments. Propane and trisilapropane are the simplest molecules containing strain-free CH_2 and SiH_2 groups, respectively. Hence, equations 2 and 3 should be used for strain energy evaluations, rather than the ones employed by Schöller (based on SiH_4 and Si_2H_6).

$$(SiH_2)_3\ (\mathbf{2}) + 3\ H_3SiSiH_3 \rightarrow 3\ H_3SiSiH_2SiH_3 \quad (2)$$

$$(SiH_2)_4\ (\mathbf{3}) + 4\ H_3SiSiH_3 \rightarrow 4\ H_3SiSiH_2SiH_3 \quad (3)$$

The strain energies of the 3-, 4-, and 5-membered silicon rings, properly evaluated (Table III), contrast remarkably with those for the carbon rings. Pentasilacyclopentane (**4**) is practically strain--free, whereas tetrasilacyclobutane (**3**) has only half the strain energy of trisilacyclopropane (**2**). Baeyer would have been pleased; expectations based on his strain theory are fulfilled very nicely. Perhaps Baeyer should have been a silicon chemist!

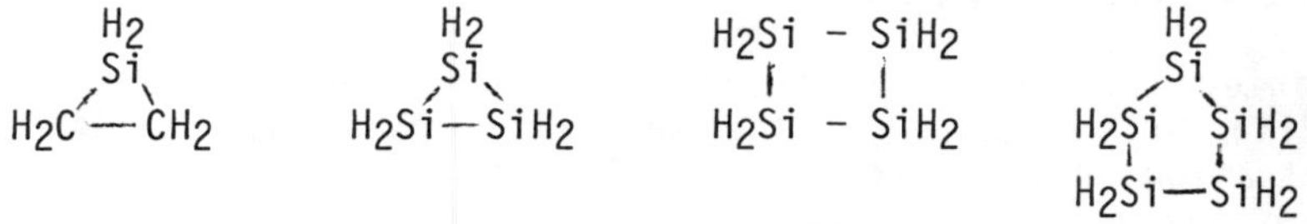

1 **2** **3** **4**

3.1. Interpretations. Strain in small silicon rings

Why do the strain energies of the silicon rings differ so much from their carbon counterparts? In first place, torsional effects are much less important in silanes than in hydrocarbons. The rotational barrier in disilane, Si_2H_6, is only about 1 kcal/mol (slightly less at the levels of theory we have employed). For this reason, **2-4** have been calculated by assuming planar geometries. Experimentally, tetrasilacyclobutane derivatives are known to be slightly nonplanar;[5] we find D_{2d} conformation calculationally to be more stable as well (but only by 1 kcal/mol). 1,3-SiSi nonbonded repulsions in **3** are indicated by the negative overlap populations, and perhaps are half as large as those in cyclobutane. Angle strain thus is only a part of the total strain energy of **3**. The lower strain energy in **3** than in cyclobutane thus agrees with expectations based on the slightly smaller bending force constants of silicon.

But what about trisilacyclopropane (**2**), why is its strain energy so much higher than that of cyclopropane if silicon bending is easier? To answer this question, we have calculated the hybridization at silicon bonds using natural population analysis[24] of natural localized molecular orbitals.[25] In **2**, the rehybridization is substantially less than that in cyclopropane. The Si-Si bonds are $sp^{2.9}$ and the SiH bonds are $sp^{1.9}$ hybridized, compared to $sp^{2.1}$ for SiH bonds in SiH_4. The corresponding values in cyclopropane, obtained by the same method are $sp^{4.5}$ for the CC bonds, and $sp^{2.7}$ for the CH bonds, compared to $sp^{3.2}$ for CH bonds in CH_4. The traditional sp^3 hybridization for CH_4 and SiH_4 is based only on the directionality of the bonds, but our values reflect the s- and p-orbital energies as well.[26] The valence electrons in silicon tend to populate s-orbitals more strongly. Hence, it is harder for silicon to form the bent bonds required in three membered rings. The consequences of this are clear. In trisilycyclopropane, the deformation of the angles to 60° weakens the Si-Si bonds considerably, but this is not compensated by any significant strengthening of the Si-H bonds. Thus, angle strain effects dominate in the silicon rings, although this is not the case in the carbon cycles. Trisilacyclopropane (**2**) has a high strain energy because of the large bond angle deformation which does not induce nearly as much rehybridization as in the carbon case. This also means that strained silicon-silicon bonds are not "bent" to the extent of their carbon counterparts, and the need for d-orbitals for an accurate description is much less.

3.2. Si-H bond dissociation energies

How can this interpretation be supported? Again, the bond dissociation energies leading to the corresponding radicals provide the answer. Unlike the situation in cyclopropane, we find the Si-H bond dissociation energy of trisilacyclopropane to be essentially the same as that of the secondary SiH_2 group in trisilapropane, Si_3H_8! Thus, there is no evidence for enhancement of the SiH bond energies in the small silicon rings, as a consequence of the large angle deformations. In other respects, the Si-H bond dissociation energies (our calculated data is given in Table IV) resemble those of their carbon counterparts (Table II). Primary silyl radicals are more stable than SiH_3, and secondary silyl radicals show an additional improvement. These radicals are all strongly nonplanar; this reflects the tendency of tri-substituted second row molecules to adopt pyramidal configurations with higher inversion barriers than those of their carbon counterparts. Second row elements prefer to place their nonbonding electrons, e.g., those in silyl radicals, in orbitals of higher s-character, since these are lower in energy.[27] Electrons in bonding MO's are farther from the nucleus; their hybridization is less crucial, and also change less with angle deformation.

3.3. Conclusion. Silicon Rings

Strain energies of small ring silicon compounds are dominated by angle deformation effects. Baeyer's ideas have proven to be largely correct for silicon compounds, although other factors are important

Table IV. Relative Si-H bond dissociation energies, kcal/mol.

Radical	3-21G*//3-21G*	6-31G*//6-31G*
SiH_3	0.0	0.0
SiH_3SiH_2	-2.9	-2.9[b]
$SiH_3SiH_2SiH_2$	-3.2	-3.2
$(SiH_3)_2SiH$	-	-5.6
c-$(SiH_2)_2SiH$	-6.1	-6.1

[a] Evaluated via: $SiH_3 + RH \rightarrow SiH_4 + R\cdot$ at the levels shown.
[b] The experimental difference is -4.0 kcal/mol. The MP3/6-31G* value is -2.9 kcal/mol.[6]

for the small carbon rings. While bending force constants for silicon are lower than their carbon counterparts, large bendings at silicon do not result in nearly as much rehybridization as is found in carbon. The weakening of Si-Si bonds is thus not compensated by appreciable strengthening of the SiH bonds as is the case for the corresponding carbon linkages. Table II compares the strain energy contributions (crudely estimated!), of the small carbon and silicon rings.

Acknowledgments

This work was supported by the Fonds der Chemischen Industrie and the Deutsche Forschungsgemeinschaft. This project is a collaboration with Alex J. Kos, who carried out a number of the calculations, and Alan E. Reed, who provided the population analyses, their interpretation, and valuable discussion. Some of the computations were done by Bernd Reindl, at ZMI, Tennenlohe; we are grateful for the use of facilities and the donation of computer time. We thank Peter Budzelaar for his interest and comments.

References

(1) A. Baeyer, Chem. Ber. **1885**, 18, 2269.
(2) P.v.R. Schleyer, J.E. Williams, and K.R. Blanchard, J. Am. Chem. Soc. **1970**, 92, 2377.
(3) A. Greenberg and J. Liebman, "Strained Organic Molecules", Academic Press, N.Y., 1978 and references cited.

(4) J.C. Dewan, S. Murakama, J.T. Snow, S. Collins, and S. Masamune, J. Chem. Soc., Chem. Commun. **1985**, 829 and earlier references cited.
(5) H. Matsumoto, K. Takasuna, M. Minemura, Y. Nagai, and M. Goto, J. Chem. Soc., Chem. Comm. **1985**, 1366.
(6) The Gaussian 82 program was employed for the calculations reported here: J.S. Binkley, M.J. Frisch, D.J. DeFrees, K. Raghavachari, R.A. Whiteside, H.B. Schlegel, E.M. Fluder, and J.A. Pople, Carnegie-Mellon University, Pittsburgh, P.A., Release H, 1984. Much data has been taken from the Carnegie-Mellon Quantum Chemistry Archive," R.A. Whiteside, M.J. Frisch, and J.A. Pople, Eds., 3rd ed., Carnegie-Mellon University, 1983. Also see, W.J. Hehre, L.Radom, P.v.R. Schleyer, and J.A. Pople, "Ab Initio Molecular Orbital Theory", Wiley, 1986.
(7) W. Schöller and T. Dabisch, J. Chem. Soc., Chem. Commun. **1985,** 1706 consider the strain energies of the silicon rings, but their values are not evaluated properly. Also see J. Rubio and F. Illas, J. Molec. Struct. **1984,** 110, 131.
(8) T.A. Albright, J.K. Burdett, and M.H. Whangbo, "Orbital Interaction in Chemistry," Wiley, New York, 1985.
(9) U. Burkert and N.L. Allinger, "Molecular Mechanics," ACS Monograph 177, American Chemical Society, Washington, 1982.
(10) J.D. Dunitz and V. Schomaker, J. Chem. Phys., **1952,** 20, 1703.
(11) K.N. Wiberg, Tetrahedron Lett., **1985,** 26, 599.
(12) N.L. Bauld, J. Cessak, and R.L. Holloway, J. Am. Chem. Soc. **1977,** 99, 8140.
(13) M.J.S. Dewar, J. Am. Chem. Soc. **1984,** 106, 669; Bull. Soc. Chim. Belg. **1979,** 88, 957. M.J.S. Dewar and M.L. McKee, Pure Appl. Chem. **1980,** 52, 1431. Also see I.Lee, K. Yang, and H.S. Kim, Tetrahedron **1985,** 41, 5007.
(14) D. Cremer and E. Kraka, J. Am. Chem. Soc. **1985,** 107, 3800.
(15) J. Wardeiner, W. Lüttke, R. Bergholz, and R. Machmek, Angew. Chem., Int. Ed. Emgl. **1982,** 21, 872. J.R. Wesener, D. Moskau, and H. Günther, J. Am. Chem. Soc. **1985,** 107, 7307.
(16) J.D. Roberts and M.C. Caserio, "Basic Principles of Organic Chemistry", Benjamin, New York, 1964, p. 113.
(17) K.U. Ingold, B. Maillard, and J.C. Walton, J. Chem. Soc., Perkin II, **1981**, 970.
(18) D.F. McMillen and D.M. Golden, Ann. Rev. Phys. Chem. **1982,** 33, 493.
(19) W. Tsang, J. Am. Chem. Soc. **1985,** 107, 2872 and literature cited.
(20) D.C. McKean, Chem. Soc. Rev. **1978,** 7, 399.
(21) The "in situ" bond energies estimated by C. Zhixing, Theoret. Chim. Acta (Berl.) **1985,** 66, 365, show a difference of 5.5 kcal/mol for comparably substituted $C(sp^3)$-H and $C(sp^2)$-H (olefinic) bonds, but no value for cyclopropane is given. The isolated CH stretching frequencies for ethylene and cyclopropane are essentially identical.[20]
(22) M.S. Gordon, J. Am. Chem. Soc. **1980,** 102, 7419.
(23) "JANAF Thermochemical Tables", D.R. Stull and H. Prophet, Eds., National Bureau of Standards, Washington, 2nd Ed., 1971.

(24) A.E. Reed, R.B. Weinstock, and F. Weinhold, J. Chem. Phys. **1985**, 83, 735.
(25) A.E. Reed and F. Weinhold, J. Chem. Phys. **1985**, 83, 1736.
(26) For a clear explanation, see E. Swithes, W.N. Lipscomb, and M.D. Newton, J. Am. Chem. Soc. **1970**, 92, 3847. Also see ref. 27.
(27) W. Kutzelnigg, Angew. Chem., Int. Ed. Engl. **1984**, 23, 272.

ALLYLIC AND ALLENIC CAPTO-DATIVE STABILIZATION IN NITROGEN CENTERED FREE RADICALS

Daniel Kost,* Morton Raban*[1] and Kalman Aviram
Department of Chemistry
Ben Gurion University of the Negev
Beer-Sheva 84105
Israel

ABSTRACT: The ability of both the single electron as well as the nonbonding electron lone-pair on central nitrogen in the title radicals to interact with acceptor substituents creates two distinct capto-dative type aminyls: allylic and allenic. The discussion is supported by ab-initio UHF-SCF-MO calculations of formylhydrazyl and boranyl-hydrazyl as model radicals.

The extraordinary stabilization of radicals substituted by both electron donor and acceptor groups has been termed the "capto-dative" effect.[2] The phenomenon has been well established experimentally[3] and also demonstrated and analyzed theoretically.[4] Most of the discussion on capto-dative stabilization focused on carbon centered radicals, and no particular distinction has been made between the latter and radicals centered about other elements. Among the most extensively studied free radicals, and indeed among the most stable ones, are those in which the central atom is nitrogen. Thus N,N-diphenyl-N'-picrylhydrazyl(**1**) and its derivatives have been known for years, and have recently been studied by Balaban and coworkers.[5] In the present article we wish to discuss nitrogen centered capto-dative radicals, and to point out the differences in electronic structure between them and their carbon centered analogues.

Capto-dative stabilization in carbon centered radicals involves the simultaneous conjugation of the donor and acceptor π-orbitals with the singly occupied molecular orbital (SOMO) on carbon. The nodal planes of the π-orbitals of all three centers are coplanar, in analogy to the allyl system, and the extra stability arises from a favorable three orbital mixing, which is more effective than the sum of both of the two-orbital interactions.[2,4] In an analogous way simultaneous (allylic) conjugation can also operate in nitrogen centered radicals (aminyls). However, in the case of aminyls, an additional nonbonding orbital is present, the lone pair of electrons which normally would lie in the nodal plane of the π-system. Potentially both the lone-pair orbital and the SOMO can conjugate with the acceptor substituent orbital, and the

H. G. Viehe et al. (eds.), Substituent Effects in Radical Chemistry, 83–86.
© 1986 by D. Reidel Publishing Co.

preferred structure will be the one in which the gain in energy due to conjugation is greatest. Thus, the combined stabilizing effect in nitrogen radicals may, in principle, be more effective than in carbon radicals, in the event that conjugation of the acceptor substituent with the lone pair leads to better stabilization than with the SOMO.

We may define two basic structures of nitrogen centered radicals, based on the analogy with hydrocarbon skeletons: The allylic type (**2**), in which all three components (donor, acceptor and radical) are conjugated in a single π-system; and the allenic type (**3**), in which the donor conjugates with SOMO and the acceptor is conjugated with the nitrogen lone pair in two orthogonal π-planes. The term "capto-dative" has been generally applied for both types of aminyls, without distinction. It may be noted, however, that only in the allylic type can there be a true synergistic effect, in the sense that the combined conjugation with SOMO is more effective than the sum of individual contributions of the substituents.

To demonstrate these two types of radicals we have employed non-empirical SCF-MO theory at the 4-31G level to calculate push-pull aminyls.[6] We chose as model aminyls the boranylhydrazyl, H_2N-N-BH_2(**4**) and formylhydrazyl, H_2N-N-CHO(**5**) radicals. The results of the calculations are presented in the Table. It is immediately evident from the Table that both types of radicals, allylic and allenic, are possible, and that depending on the substituents each one may be the more stable conformation of the radical. With the relatively weak acceptor group in **5** the allylic structure **5a** is preferred by 11.7 kcal/mole relative to the more stable of the two allenics, **5b**. Hence we may conclude that in presence of weak acceptor substituents aminyls behave in a manner similar to normal, carbon centered, capto-dative radicals. In the presence of the powerful acceptor group BH_2 in **4**, however, the stable conformation is found to be **4c**, of the allenic type: conjugation of the acceptor with the p-orbital on central nitrogen is so strong that the resulting π-bonding orbital is lower in energy than the hybrid (roughly sp^2) lone pair orbital. As a result the lone pair now occupies this π-orbital to form a two-electron bond, and the SOMO becomes the in-plane hybrid orbital. The donor substituent follows this shift of electron occupancy: it can never conjugate with a lone pair or any two-elecron orbital, since such conjugation would be repulsive. Therefore in **4c** the donor substituent is turned by 90° out of the molecular plane, and is conjugated, to some extent, with the in-plane SOMO.

An interesting manifestation of the capto-dative effect is found when the linear allylic and allenic forms, **5d** and **5e**, are compared. In these structures both the single electron and the lone pair on nitrogen occupy pure atomic p-orbitals. Torsion about the C-N bond will lead from one structure to the other. The Table shows that both structures are essentially equal in energy, indicating a negligible barrier for topomerization. In terms of electronic interactions the main difference between the two conformers is that in **5d** the carbonyl group is conjugated with the single electron, whereas in **5e** it is conjugated with the lone pair. In both structures the NH_2-donor interacts with SOMO. It is thus evident that the stabilization gained by interaction of the

TABLE: Calculated Energies and Optimized Geometries of Push Pull Aminyls.

Cmpd.	Relative Energy Kcal/mol	Bond Lengths, A[a] N-N	N-X	C=O	C-H	Bond Angles, deg.[a] N-N-X	N-C-O	N-C-H
4a	4.98	1.32	1.44			117.8		
4b[b]	3.75	1.30	1.35			180.0		
4c	0.00	1.31	1.37			144.8		
4d	29.42	1.27	1.41			180.0		
5a	0.00	1.33	1.35	1.26	1.08	116.4	120.7	119.6
5b	11.70	1.34	1.41	1.20	1.08	115.6	124.4	113.8
5c	20.58	1.30	1.34	1.21	1.09	134.1	126.1	111.8
5d	32.97	1.28	1.29	1.29	1.08	180.0	123.3	120.0
5e	32.02	1.30	1.33	1.22	1.08	180.0	126.7	112.0

a. The following values were kept constant: N-H, 0.99A; B-H, 1.18A; Angles: N-N-H, 121.0°; N-B-H, 122.0°.
b. Since **4b** optimized to a linear structure, **4b** and **4e** are identical

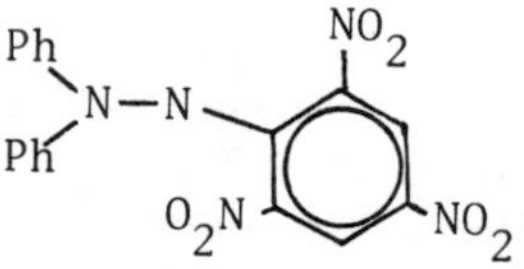

1

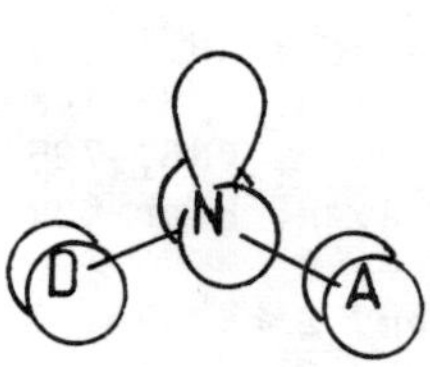

2 (Allylic)

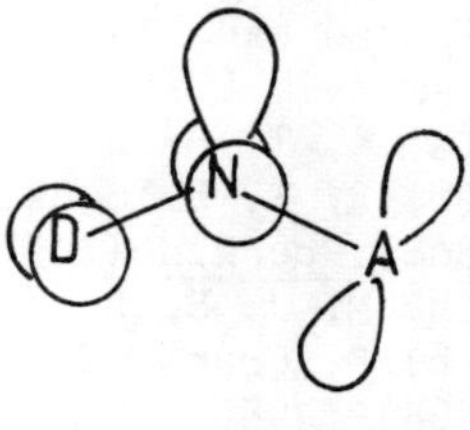

3 (Allenic)

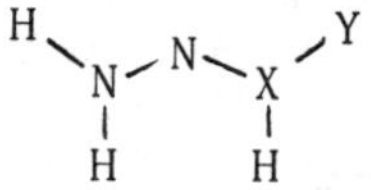

a (allylic)

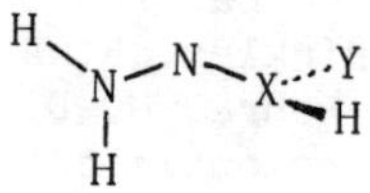

b (allenic)

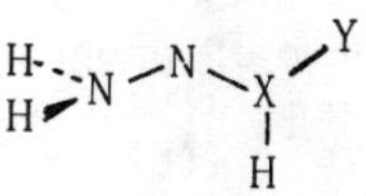

c (allenic)

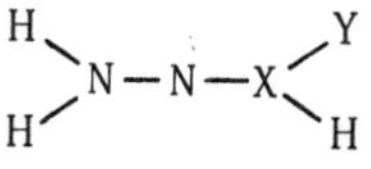

d (allylic)

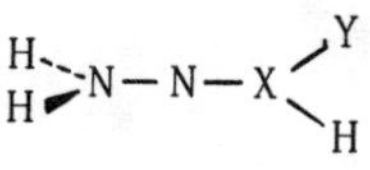

e (allenic)

4, X = B; Y = H
5, X = C; Y = O

carbonyl with a doubly occupied p-orbital, in the absence of assistance from the donor, approximately equals that gained from interaction with the singly occupied orbital, when it is accompanied by simultaneous interaction with the donor. The capto-dative effect, the interaction of the donor with SOMO in addition to the interaction of the acceptor, is thus roughly worth one half of the unassisted interaction of the carbonyl acceptor with a lone pair in an equivalent (atomic p) orbital.

The maximum stabilization of an aminyl would naturally be obtained if both the single electron and the lone pair could be conjugated with acceptor groups, in addition to the donor-SOMO interaction. This can be achieved by groups which have π-acceptor properties in two perpendicular planes, such as nitrile. Indeed 1-cyano-2,2-diphenylhydrazyl showed remarkable stability in benzene solution as well as in the crystalline state. Such a radical retains the stabilizing effects of both the allylic and the allenic type structures.

ACKNOWLEDGEMENT: We thank the Fund for Basic Research, administered by the Israel Academy of Sciences and Humanities (D.K.), and the National Institutes of Health M.B.R.S. Program (M.R.), for financial support.

REFERENCES

1. Visiting Professor, on sabbatical leave from Wayne State University, Detroit, MI 48202, U.S.A.
2. For an excellent recent review see: H. G. Viehe, Z. Janousek, R. Merenyi, L. Stella, Acc. Chem. Res., **18**, 148 (1985).
3. (a) A. T. Balaban, Rev. Roum. Chim., **16**, 725 (1971).
   (b) R. W. Baldock, P. Hudson, A. R. Katritzky, F. Soti, J. Chem. Soc., perkin I, 1422 (1974).
4. (a) M. J. S. Dewar, J. Am. Chem. Soc., **74**, 3353 (1952).
   (b) D. Crans, T. Clark, P. v. R. Schleyer, Tetrahedron Lett., **21**, 3681 (1980).
   (c) G. Leroy, D. Peeters, Theochem., **2**, 133 (1981).
5. M. T. Caproiu, M. Elian, N. Grecu, N. Negoita, A. T. Balaban, J. Chem. Soc., Perkin II, 591 (1983).
6. (a) UHF-SCF Calculations were carried out using the GAUSSIAN-76 system of programs: J. S. Binkley, R. A. Whiteside, P. Hariharan, R. Seeger, J. A. Pople, W. J. Hehre, M. D. Newton, QCPE, **11**, 368 (1979). (b) Partial geometry optimization was performed as noted on the Table.
7. N. Negoita, R. Baican, A. T. Balaban, Tetrahedron, **30**, 73 (1974).

CONFORMATIONAL PREFERENCES IN β-SUBSTITUTED ETHYL RADICALS: A QUANTITATIVE ANALYSIS OF HYPERCONJUGATION, HOMOCONJUGATION, d-ORBITAL AND STERIC EFFECTS.

Jacques FOSSEY
L.E.C.S.O. - C.N.R.S.
B.P. 28
94320 Thiais
France

ABSTRACT. The conformational preference in XCH_2CH_2 was analysed by a quantitative PMO treatment[1]. Due to a balance between the SOMO effect in the O rotamer and the π effect in the E rotamer, the rotational barriers are generally small. The increase of the barriers when X is a second-row element is mainly due to the hyperconjugation with the C-X orbitals. The principal contribution to the hyperconjugation is the delocalisation of the single electron from the SOMO into the σ^{x}C-Cl when X=Cl, and the delocalisation from the σC-X into the SOMO when X=SH, PH_2, SiH_3. The other effects are small or negligible.

1. INTRODUCTION

β-substituted alkyl free-radicals are transient species often implied in free radical chemistry. They exhibit an interesting diversity of conformational preferences as deduced from ESR measurements[2]. Recent *ab-initio* UHF-MO studies found almost a free rotation around the C-C bond if X is a first-row element. But with second-row substituents a notable barrier is found, the O rotamer being favoured[3].

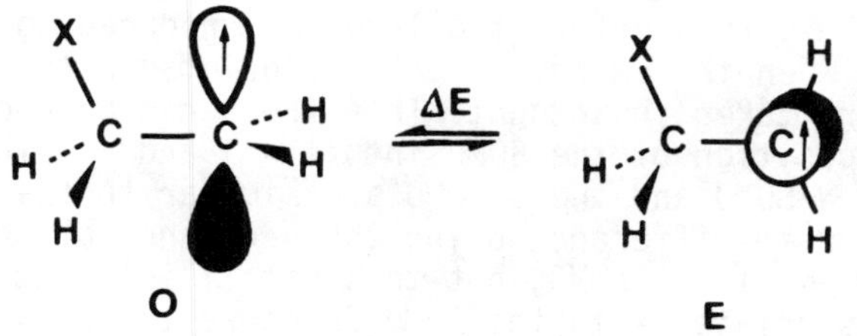

Figure 1. Equilibrium between O and E rotamers of XCH_2CH_2

This second-row effect has been ascribed to three different types of conjugations involving the SOMO: the hyperconjugation with the C-X orbitals, the homoconjugation with p lone pairs or d-orbitals on X. However the relative importance of these effects has not been clearly established yet. In this paper a complete detailed analysis is presented.

H. G. Viehe et al. (eds.), Substituent Effects in Radical Chemistry, 87–90.
© 1986 by D. Reidel Publishing Co.

2. RESULTS AND DISCUSSION

For each radical, XCH_2- and $-CH_2$ fragment MO's (FMO's) are computed by a procedure described elsewhere[5]. The energy effects associated with the various types of orbital interactions, given in Table I, are calculated according to the PMO formulas[1].

Table I $\Delta E = E(E) - E(0)$. $\Delta E(PMO/\sigma)$, $\Delta E(PMO/SOMO)$, $\Delta E(PMO/\pi)$, $\Delta E(PMO/a')$, $\Delta E(PMO/a'')$ denote the partial PMO energy differences associated with the σ, SOMO, π, a' or a" orbitals of the $-CH_2$ fragment. ΔE(PMO/TOTAL) is the sum of the first three terms. ΔE(SCF) refers to the SCF (STO-3G//3-21G for X=CH_3, NH_2, OH, F and STO-3G*//3-21G* for X=SiH_3, PH_2, SH, Cl) energy differences. ΔE are given in kcal/mol.

X	CH_3	NH_2	OH	F	SiH_3	PH_2	SH	Cl
ΔE(SCF)	0.09	0.30	-0.13	0.58	1.36	1.66	1.46	2.81
ΔE(PMO/TOTAL)	0.06	-0.57	-0.66	0.21	1.06	1.55	1.26	4.34
$\Delta E(PMO/\sigma)$	0.27	0.07	-0.09	0.13	0.22	0.24	0.12	0.12
ΔE(PMO/SOMO)	0.64	2.69	4.02	6.60	-0.83	0.95	1.94	7.67
$\Delta E(PMO/\pi)$	-0.85	-3.33	-4.58	-6.26	1.67	0.36	-0.79	-3.45
$\Delta E(PMO/a')$	4.28	3.94	4.11	4.90	5.42	6.13	5.99	9.30
$\Delta E(PMO/a'')$	-4.49	-4.58	-4.67	-4.56	-4.58	-4.82	-4.85	-5.08

ΔE(SCF) values are very small when X is a first-row substituent and they are notably larger when X is a second-row substituent. Their comparisons with ΔE(PMO/TOTAL) are satisfactory: therefore the orbital interactions reproduce well the origin of the barriers around the C-C bond in XCH_2CH_2.

2.1. Effect of the SOMO on ΔE

The usual approximation of considering only the effect of the SOMO on ΔE should lead to conformational preference, except for X=SiH_3, in favour of the 0 rotamer. Furthermore that preference should increase with the X electronegativity. But, this effect too large, is reduced to the correct order of magnitude when the energy effects associated with the π FMO's of $-CH_2$ ($\Delta E(PMO/\pi)$) are taken in account. This trend can be understood considering the decomposition of the SOMO into its α and β spin components (cf Figure 2). The α-SOMO and the β-SOMO are similar to the π and π^{X} FMO's of $-CH_2$. The main difference being the occupancy of the π orbital which is twice that of the α-SOMO, but this effect is counterbalanced by a larger overlap associated with the SOMO. Because these two types of orbitals lie on orthogonal planes, the XCH_2 orbitals which interact with the SOMO's in one rotamer, interact with the π orbitals in the other one, and vice versa. For this reason, ΔE(PMO/SOMO) and $\Delta E(PMO/\pi)$ have opposite signs. Consequently, ΔE(PMO/TOTAL) which includes both interactions are small in magnitude. That fact is more specially true when X is a first-row substituent than when it is a second-row one.

2.2. Effect of a'-type orbitals on ΔE

Because both rotamers have a plane of symmetry, orbitals are classified either symmetric (a'-type) or antisymmetric (a"-type). Interactions between a' FMO's favour the O rotamer, while interactions between a" FMO's favour the E rotamer (cf Table I). Furthermore ΔE(PMO/a") values are not affected by the nature of X, since X belongs to the nodal plane of the a" orbitals. For first-row X there is an almost perfect balance between a' and a" interactions. The origin of the second-row effect comes from the increase of the ΔE(PMO/a') values when X is a second-row element.

ΔE(PMO/a') was decomposed into contributions due to four subgroups of XCH_2: dX, X, C-X and $-CH_2-$.

Table II Contributions to ΔE(PMO/a') of FMO's of XCH_2- subgroups: dX, X, C-X, $-CH_2-$.

	CH_3	SiH_3	NH_2	PH_2	OH	SH	F	Cl
ΔE(PMO/a':dX)	-	0.35	-	0.31	-	0.47	-	0.59
ΔE(PMO/a':X)	0.99	0.64	0.65	1.15	0.64	0.56	0.67	1.24
ΔE(PMO/a':C-X)	2.34	3.28	2.62	3.53	2.45	3.76	3.22	6.03
ΔE(PMO/a':$-CH_2-$)	0.95	1.15	0.67	1.14	1.02	1.20	1.01	1.44

It is interesting to compare in Table II the values for a X first-row substituent to their corresponding ones for a X second-row substitutent i.e. the second-row effect. Interactions with a' d-orbitals on X are small but nevertheless reinforce the second-row effect. Interactions with a':X, i.e. all a' FMO's belonging to X except the d-orbitals, do not show a clear tendency. They reinforce the second-row effect when X is PH_2 and Cl, but they reduce it when X is SiH_3 and SH. These effects are mainly steric in nature[5]. Interactions with a' FMO's of $-CH_2-$, i.e. σ-CH_2- and σ^x-CH_2-, contribue slightly to the second-row effect. The main contribution to ΔE(PMO/a'), and the larger second-row effect, come from interactions with the a' FMO's of C-X, i.e. σC-X and σ^xC-X.

2.3. Detailed analysis of the hyperconjugation with C-X

All interactions given in Table III and plotted on Figure 2 imply the C-X orbitals. The steric parts, due to repulsions between occupied orbitals, are the strongest ones, but they have a small effect on ΔE (O favoured) and they slightly contribue to the second-row effect since their differences are slightly larger for the corresponding second-row radicals. Hyperconjugaison was split into two contributions D and A. These two contributions favour the O rotamer and they are mainly responsible of the second-row effect. When comparing X=Cl to X=F, the SOMO(D) interaction, which imply an electron delocalization from the α-SOMO into the σ^xC-Cl, is, by a strong stabilization of the O rotamer, the main contribution to the second-row effect. For the other radicals, the difference between first-row and second-row comes from the SOMO(A) interaction which imply an electron delocalization from the σC-X into the β-SOMO.

Table III Calculated PMO interactions (in kcal/mol) between C-X FMO's and a' FMO's of $-CH_2$ (SOMO's in the O rotamer or π's in the E rotamer). S (Steric), D (Donor) and A (Acceptor) refer to interactions shown on Figure 2.

		CH_3	SiH_3	NH_2	PH_2	OH	SH	F	Cl
	(S)	7.66	10.55	6.11	9.22	6.49	9.81	5.24	8.50
SOMO	(D)	-1.75	-1.04	-2.46	-2.06	-2.87	-2.48	-3.73	-6.28
	(A)	-0.80	-2.43	-0.58	-2.02	-0.35	-1.65	-0.19	-0.88
	(S)	8.27	11.54	6.59	9.90	6.62	10.66	5.56	9.15
	(D)	-0.52	-0.33	-0.68	-0.59	-0.76	-0.66	-0.95	-1.53
	(A)	-0.30	-0.85	-0.22	-0.64	-0.14	-0.56	-0.06	-0.25

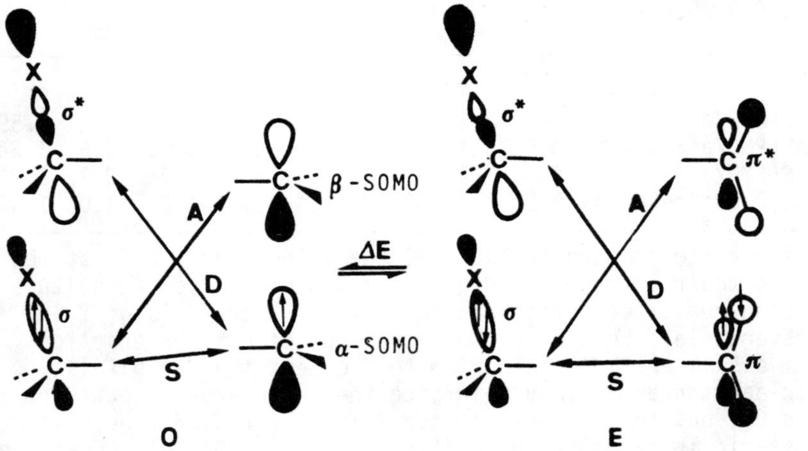

Figure 2. Schematic representation of interactions with C-X FMO's

3. CONCLUSIONS

Low rotational barriers in XCH_2CH_2 are due to the opposite effect of the SOMO's and π FMO's of $-CH_2$. The second-row effect, i.e. the enhanced preference for the O rotamer, comes from the α-SOMO↔σ^{*}C-Cl interaction when X=Cl, and from the β-SOMO↔σC-X one when X=SH, PH_2, SiH_3.

REFERENCES

1. Bernardi F., Bottoni A. and Fossey J., Theoret.Chim.Acta (Berl), **61**, 252 (1982).
2. Kochi J.K., *Advances in Free-Radical Chemistry*, Williams G.H. Editor, Vol. V, Chapter 4, Elek P. London (1975).
3. Fossey J. and Bernardi F. unpublished results.
4. Rossi A.R. and Wood D.E., J.Amer.Chem.Soc., **98**, 3452 (1976).
5. Bernardi F., Bottoni A., Fossey J. and Sorba J., Tetrahedron in press.

SUBSTITUENT EFFECTS ON THE THERMOCHEMISTRY AND ELECTROCHEMISTRY OF FREE RADICALS[1]

J.A. Hawari, J.M. Kanabus-Kaminska, D.D.M. Wayner and D. Griller*
Division of Chemistry, National Research Council of Canada
Ottawa, Ontario, Canada K1A 0R6

ABSTRACT

The role of substituent effects in free radical and carbene chemistry was investigated using a variety of techniques including e.s.r. spectroscopy, measurement of appearance energies, electrochemistry and kinetics. The thermodynamic data obtained in some of these experiments were used to probe the magnitude of substituent effects and their impact on group additivity methods. Electrochemical experiments showed that the removal of an electron from a substituent had a profound effect on the strengths of the neighbouring C-H bonds. Finally, examples drawn from carbene chemistry showed that sterically crowded substituents could be used to induce major electronic changes in these molecules.

1. INTRODUCTION

In organic chemistry, the concept of the substituent effect is frequently associated with Hammett's work on linear free energy relationships and the effect of electron withdrawing and donating groups on heterolytic reactions. Such effects are substantially less important in free radical processes and, while they play a role, the free radical chemist is far more concerned with the function of substituents which can stabilize the unpaired electron through delocalization.

Electron delocalization manifests itself in the rates of homolytic reactions. However, it is more precisely evaluated through the measuement of spin densities using electron spin resonance (e.s.r.) spectroscopy and through measurements of the thermochemical properties of free radicals. These thermodynamic characteristics - bond strengths and oxidation or reduction potentials of free radicals - form the subject of this paper.

H. G. Viehe et al. (eds.), Substituent Effects in Radical Chemistry, 91–105.
© 1986 by D. Reidel Publishing Co.

2. NO SUBSTITUENT EFFECT AT ALL!

2.1. Background

While it is tempting to focus attention on dramatic substituent effects, it is worth remembering that a major subject in free radical chemistry is based on the hypothesis that many substituent effects are trivially small. The concept of group additivity,[2] which allows the calculation of heats of formation of radicals and molecules, depends on this principle although it has been seldom verified.

As an example, consider the case of secondary alkyl radicals $RR'\dot{C}H$. The group additivity concept requires that the thermodynamic properties of this set of radicals are independent of the structure of the alkyl groups R and R'. In other words, the strengths of chemical bonds between secondary alkyl radicals and a specific group or atom X are constant for all R and R'. Only when this is true, can we compute heats of formation by adding contributions from the various groups which make up the radical.[2]

Most free radical chemists would take this hypothesis for granted. However, if we are to apply the group additivity method, we require that it be true over a very narrow range i.e. that the strengths of bonds between X and the secondary alkyl radicals do not vary by more than ca. 500 cal mol^{-1}. The hypothesis can therefore only be verified to the required precision by measuring bond strengths relative to one another, since the experimental errors involved in absolute measurements generally exceed this limit.

2.2. The Radical Buffer

One method of measuring relative bond strengths with high precision makes use of the "radical buffer"[3-4] which can be established between alkyl radicals and their parent iodides, eq. 1. We have studied

$$R\cdot + R'I \rightleftharpoons RI + R'\cdot \qquad (1)$$

buffers of this kind by monitoring the radicals in solution using e.s.r. spectroscopy. For alkyl radicals, rate constants for the forward and back reactions are normally $>10^6\ M^{-1}s^{-1}$. Under the conditions of very low radical concentrations which apply in the e.s.r. experiment, the rates of these reactions are very much greater than those for the bimolecular self-reactions of the radicals so that their relative concentrations accurately reflect the equilibrium constant $K = k_1/k_{-1}$.[4]

It was found that the values of K were, within error, independent of the method of radical generation and of the radical concentrations, implying that equilibria were indeed established under the experimental conditions.[4] Moreover, the data formed a self-consistent set when different combinations of alkyl iodides were used to set up the buffer. The methyl radical was taken as the standard for comparison and the equilibrium constants thus obtained are reported in Table I.

Table I. Equilibrium Constants, K, as Measured by the Radical Buffer Technique and Related Thermodynamic Data in Kcal mol^{-1}

R'•	K_{300}	$\Delta H_f(R'\bullet)$	BDE(R'-I)	BDE(R'-H)
Me•	1.0	34.4[a]	56.5	104.4
Et•	20.1	28.0	55.5	100.3
n-Pr•	14.6	22.8	55.7	99.7
i-Pr•	468	19.2	54.1	96.1
s-Bu•	245	13.9	54.2	96.4
c-C_5H_9•	394	25.1	53.6	95.6
t-Bu•	1.7×10^4	9.4	52.1	93.9

[a] Reference value

The data fall into distinct classes. Taking methyl as the standard with K = 1, the primary alkyl radicals give values of ca. 20 for K whereas values of ca. 300 obtain for the secondary alkyls. Tert-butyl is again quite different with $K = 1.7 \times 10^4$. However, within any class of radicals e.g. secondary alkyls, the overall variation in K is less than a factor of 2 which, in free energy, corresponds to 410 cal mol^{-1} at 25°. Since the entropies of the components in the radical buffers are known,[4] it is easily verified that the effective constancy of K within a class of radicals is not due to compensating entropic and enthalpic effects. The data therefore lend strong support to the assumptions made about the additivity of enthalpies in the group approach.

The equilibrium constants can be combined with entropic and enthalpic data to calculate the heats of formation of the radicals and hence bond dissociation energies, BDE, eq. 2. As expected, the latter

$$BDE(R-X) = \Delta H_f(R\bullet) + \Delta H_f(X\bullet) - \Delta H_f(RX) \quad (2)$$

again fall into distinct classes which reflect the substitution patterns of the radicals themselves further supporting the assumptions about enthalpy which are implicit in the group method.

2.3. Bond Strengths in Disulfides and Tetrasulfides

Measurements of the strengths of carbon-sulfur bonds in mixed disulfides,[5] BDE(RSS-R'), are consistent with the principles described above, since the data were independent of the nature of R but showed an appropriate dependence on R'.

The experimental data on a series of mixed disulfides RSSBu-t were obtained by an appearance energy method.[6] The technique makes use of an energy resolved electron beam to cleave molecules in the gas phase, to an ion and a radical, eq. 3. The ions are detected using a

$$R_1\text{-}R_2 + e \xrightarrow{AE} R_1^+ + R_2\cdot + 2e \qquad (3)$$

quadrupole mass spectrometer. In a typical experiment, the electron energy is slowly increased until a current due to the desired ion is detected. This "appearance energy", AE, is related to the thermodynamic parameters of interest via eq. 4.

$$AE(3) = \Delta H_f(R_1^+) + \Delta H_f(R_2\cdot) - \Delta H_f(R_1\text{-}R_2) \qquad (4)$$

The appearance energies for the formation of t-Bu^+ were measured for the series of disulfides RSSBu-t and were found to be the same, within experimental error, eq. 5, Table II. The results immediately demonstrate

$$RSSBu\text{-}\underline{t} + e \xrightarrow{AE} RSS\cdot + \underline{t}\text{-}Bu^+ + 2e \qquad (5)$$

that the strength of the S-Bu-t bond is independent of the nature of R. The strength of the bond was calculated by taking the average value of AE and by applying values for the heats of formation of the tert-butyl radical[4] (9.4 kcal mol^{-1}) and its corresponding carbonium ion (166.2 kcal mol^{-1})[7] eq. 6,7. The data led to BDE(RSS-Bu-t) = 52.6 ± 1.0 kcal mol^{-1}.

$$BDE(RSS\text{-}Bu\text{-}\underline{t}) = \Delta H_f(RSS\cdot) + \Delta H_f(\underline{t}\text{-}Bu\cdot) - \Delta H_f(RSS\text{-}Bu\text{-}\underline{t}) \qquad (6)$$

$$= AE(5) - \Delta H_f(\underline{t}\text{-}Bu^+) + \Delta H_f(\underline{t}\text{-}Bu\cdot) \qquad (7)$$

Given this result and a proof of the applicability of group additivities to this system, it was a simple matter to extend the calculations to give the heats of formation of the perthiyl radicals by substituting the heats of formation of the mixed disulfides[5] into eq. 6 (Table II). With this data in hand, the calculations were further extended to calculate the effect of changing R' on BDE(MeSS-R'), eq. 8. As expected,

$$BDE(MeSS\text{-}R') = \Delta H_f(MeSS\cdot) + \Delta H_f(R'\cdot) - \Delta H_f(MeSSR') \qquad (8)$$

changing the structure of the group directly attached to the bond being broken had a discernable effect, despite the mounting experimental errors. That is, the bond was strongest for R' = Me and weakest for R' = t-Bu, Table III. Finally, the heats of formation of the perthiyl radicals were used to calculate the strengths of the central S-S bonds in dialkyl tetrasulfides, eq. 9, Table IV.

$$BDE(RSS\text{-}SSR) = 2.\Delta H_f(RSS\cdot) - \Delta H_f(RS_4R) \qquad (9)$$

Again the strengths of these S-S bonds were, within experimental error, independent of the nature of the alkyl groups since these were one bond removed from that being broken.

Table II. Energy for Reaction (5). Heats of Formation of Mixed Disulfides and of Perthiyl Radicals.

RSSBu-*t*	AE(5)[a] (eV)	ΔH_f(RSSBu-*t*) (kcal mol^{-1})	ΔH_f(RSS•)[b] (kcal mol^{-1})
MeSSBu-*t*	9.08	-26.8	16.4
EtSSBu-*t*	9.05	-32.8	10.4
i-PrSSBu-*t*	9.08	-39.9	3.3
t-BuSSBu-*t*	9.12	-47.8	-4.6

[a]Error ±0.08 eV; 1 eV = 23.069 kcal mol^{-1}.
[b]Based on the average value of AE(5) = 9.08 ± 0.04 eV. Relative error ±1.0 kcal mol^{-1}; absolute error ±2.0 kcal mol^{-1}.

Table III. Bond Dissociation Energies for MeSS-R'[a]

R'	ΔH_f(MeSSR') (kcal mol^{-1})	ΔH_f(R'•)[b] (kcal mol^{-1})	BDE(MeSS-R')[c] (kcal mol^{-1})
Me	-5.8	34.4	56.6
Et	-11.8	28.0	56.2
i-Pr	-18.9	19.2	54.5
t-Bu	-26.8	9.4	52.6

[a]ΔH_f(MeSS•) = 16.4 kcal mol^{-1}, Table II.
[b]Data from ref. (4).
[c]Relative error ±1 kcal mol^{-1}; absolute error ±2 kcal mol^{-1}.

Table IV. Bond Dissociation Energies for RSS-SSR[a]

R•'	$\Delta H_f(RS_4R)$[b] (kcal mol^{-1})	BDE(RSS-SSR)[c] (kcal mol^{-1})
Me	0.10	32.9
Et	-11.0	31.8
i-Pr	-25.4	32.0
t-Bu	-41.6	32.4

[a]Values for ΔH_f(RSS•) taken from Table II.
[b]From ref. 5.
[c]Relative error ±1 kcal mol^{-1}; absolute error ±2 kcal mol^{-1}.

2.4. Generality of the Assumption

The group additivity method requires that, under specified conditions, the strengths of chemical bonds are insensitive to the nature of the substituents. The above experiments provide data which support the concept. However, it is important to recognize that this assumption is fundamental to the method and that it is not restricted to group additivity contributions which relate directly to heats of formation of radicals.

Some of the extended calculations which were used in the disulfide case illustrate the generality of this requirement. For example, the heats of formation of mixed disulfides MeSSR' were required to calculate the effect of R' on BDE(MeSS-R'), Table III. As is often the case, these heats of formation have not been determined experimentally and were obtained by applying the group additivity approach. Thus, the heat of formation of MeSSR' was assumed to be equal to the average of the known heats of formation of the symmetric disulfides MeSSMe and R'SSR'. This is exactly equivalent to assuming that the strengths of the central S-S bonds are the same in all of the disulfides, eq. 10-13.

If $$\mathrm{BDE(MeS{-}SMe)} = 2.\Delta H_f(\mathrm{MeS\bullet}) - \Delta H_f(\mathrm{MeSSMe}) \qquad (10)$$

$$= \mathrm{BDE(R'S{-}SR')} = 2.\Delta H_f(\mathrm{R'S\bullet}) - \Delta H_f(\mathrm{R'SSR'}) \qquad (11)$$

$$= \mathrm{BDE(MeS{-}SR')} = \Delta H_f(\mathrm{MeS\bullet}) + \Delta H_f(\mathrm{R'S\bullet}) - \Delta H_f(\mathrm{MeSSR'}) \qquad (12)$$

Then $$\Delta H_f(\mathrm{MeSSR'}) = \{\Delta H_f(\mathrm{MeSSMe}) + \Delta H_f(\mathrm{R'SSR'})\}/2 \qquad (13)$$

If we are to apply the group additivity method to extend experimental measurements of thermochemical data to unknown systems, we require that the strengths of chemical bonds be constant for well defined classes of substituents. While the above cases support the concept, the example described below shows that there is at least one rather unexpected exception to the principle.

3. THE AMINO GROUP - A POWERFUL STABILIZING SUBSTITUENT

3.1. Background

In view of the results described above for alkyl and perthiyl radicals, it would be reasonable to suppose that the strengths of bonds between alpha-aminoalkyl radicals, $R_2N\dot{C}R_2$, and other atoms or groups would be independent of the substituents on nitrogen, since they are remote from the radical centre. In fact, this does not appear to be the case and the thermodynamic properties of these radicals show a rather large dependence on the pattern of substitution at the amino group.

3.2. The Experimental Results

The appearance energy method was used to study the fragmentation of a series of ethylenediamines.[8] These undergo fragmentation to give aminoalkyl radicals and their corresponding ions, eq. 3, where $R_1 = R_2 = R_2NCR_2$. The heats of formation of the latter are reasonably well defined, as are those of the ethylenediamines themselves. As a result, the heats of formation of the aminoalkyl radicals can be calculated from the experimental data.[8] These results lead to the bond dissociation energies in the parent amines, eq. 2, Table V.

Table V. Heats of Formation and Ionization Energies for α-Aminoalkyl Radicals.

R_1-R_2	$\Delta H_f(R_1-R_2)$ kcal mol^{-1}	$\Delta H_f(R_2\cdot)$ kcal mol^{-1}	BDE(R_2-H) kcal mol^{-1}	IE($R_2\cdot$) eV
$Me_2NCH_2-CH_2NMe_2$	-4.7	26	84	5.7
$Me_2NCH_2-CH_2NHMe$	-3.6	30	87	5.9
$H_2NCH_2-CH_2NH_2$	-4.3	36	94	6.1
$H_2N(Me)CH-CH(Me)NH_2$	-20.9	26	90	5.7
$H_2N(Me)_2C-C(Me_2)NH_2$	-36.0	17	89	5.4

3.3. Discussion

Some of the results described above conform to conventional expectations for the system. Thus, the C-H bond strength in CH_3NH_2 is approximately 2 kcal mol^{-1} less than that in methanol implying that the $-NH_2$ substituent has a slightly greater stabilizing effect than -OH.[9] In addition, successive substitutions of hydrogen by methyl at the alpha-carbon lead to small reductions in the C-H bond strengths.[8] However, the most surprising changes are brought about by methyl substitution at nitrogen. This has a dramatic stabilizing effect to the extent that the C-H bond dissociation energy in trimethylamine is only 84 kcal mol^{-1}. The dimethylamino substituent therefore has a greater effect on the C-H bond dissociation energy than does phenyl since the C-H bond strength in trimethylamine is some 4 kcal mol^{-1} less than that in toluene.[8-9]

The dramatic substituent effect due to the dimethylamino group is reflected in the high degree of spin-delocalization in the amino radical itself[10] and in the ease of hydrogen abstraction its parent amine. For example, the rate constant for H abstraction from trimethylamine by tert-butoxyl[11] is 1.1×10^8 mol^{-1}s^{-1} and is therefore about two to three orders of magnitude greater than rate constants for the analogous reaction at hydrocarbons.[8]

The phenomenon can be rationalized in terms of the theory of the three electron bond. In simple terms, this predicts that an interaction between the unpaired electron and the nitrogen lone pair would lead to a strong stabilization of two of the electrons, with an accompanying destabilization of the third. The destabilized electron should therefore be easily removed from the radical and, as a consequence, the radicals should have low ionization energies.

The ionization energies, IE, for the alpha-aminoalkyl radicals are given in Table V and were calculated using eq. 14. They are more

$$IE(R\cdot) = \Delta H_f(R^+) - \Delta H_f(R\cdot) \tag{14}$$

than 2 eV lower than those for simple alkyl radicals which fall in the range 7-10 eV.[12] In fact, the ionization energies for these radicals are similar to those for lithium (5.39 eV) and sodium (5.14 eV) which explains why these radicals behave as potent reducing agents in many chemical systems.

3.4. Conclusion

The amino group exerts a powerful substituent effect on an adjacent radical centre which is manifest in reduced bond dissociation energies for the parent amines. The stabilizing effect is dramatically enhanced by successive methyl substitution at the amino nitrogen. All aminoalkyl radicals have low ionization potentials which explains their role as strong reducing agents. This property should be reflected in the electrochemistry of alpha-amino alkyl radicals which by inference should have low oxidation potentials.

4. SUBSTITUENT EFFECTS IN THE ELECTROCHEMISTRY OF RADICALS

4.1. Background

Oxidation and reduction potentials of free radicals are thermodynamic properties which can reflect the effect of substituents on the radical centre. However, these data have proved to be particularly difficult to measure. This is because free radicals, by their very nature, are short lived intermediates and cannot therefore be used as substrates in conventional electrochemical experiments. As a consequence, carbonium and carbanions have been used as starting materials with all of the obvious attendant difficulties.

In an attempt to overcome this problem, we have devised a technique which allows transient radicals to be used as the starting materials for electrochemical measurements.[13] As a result, we have beer able to obtain some information on substituent effects in electrochemical oxidations and reductions.

4.2. The Technique

The apparatus used in this work is shown schematically in Fig. I. It relies on the modulation principle for signal-to-noise enhancement. Radicals were generated by uv photolysis of appropriate precursors contained in the sample cell. The photolysis light was modulated by a mechanical chopper fitted with irises so that the light intensity at the sample cell rose and fell in intensity as a sine wave. Since the radicals were short lived, their concentrations rose and fell with the light intensity.

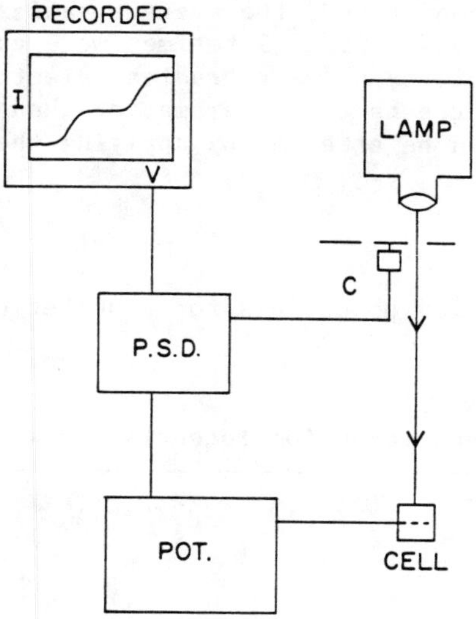

Fig. 1. Diagram of electrochemical apparatus. C-light chopper; POT - potentiostat. PSD - phase sensitive detector.

The sample cell itself was a simple flow cell fitted with a gold working electrode, a counter electrode and a reference electrode all of which were controlled by a conventional potentiostat. Flowing the sample was simply designed to prevent difficulties associated with sample depletion or product accumulation. The output of the potentiostat was fed to a phase sensitive detector which was driven by a reference signal taken from the light chopper. In simple terms, the phase sensitive detector searched the output of the potentiostat for a signal which oscillated at the same frequency as the light was modulated, and the signal which had this characteristic was that due to the radical since its concentration was modulated by the potolysis.

When the voltage at the working electrode was such that the radical was neither oxidized or reduced, no signal was detected at the phase sensitive detector. However, when the voltage was slowly scanned through the oxidation or reduction potentials of the radicals small currents were detected so that a polarogram due to the radical was obtained. Radicals were generated by the reactions shown in eqs. 15-17 and the measured oxidation and reduction potentials were independent of the method of radical generation.

$$RC(O)R \xrightarrow{h\nu} 2\ R\bullet + CO \quad (15)$$

$$\underline{t}\text{-BuO-OBu-}\underline{t} \xrightarrow{h\nu} 2\ \underline{t}\text{-BuO}\bullet \quad (16)$$

$$\underline{t}\text{-BuO}\bullet + R\text{-}H \longrightarrow \underline{t}\text{-BuOH} + R\bullet \quad (17)$$

Under the experimental conditions, the average radical concentrations were ca. 10^{-7} to 10^{-8} M and radical lifetimes were ca. 10^{-3} s. However, currents of only ca. 1 microamp due to radical electrochemistry were easily detected, which attests to the dramatic enhancements in signal-to-noise levels which can be attained by applying the modulation technique.

4.3. Results

The oxidation and reduction potentials for a number of radicals are given in Table VI.

Table VI. Oxidation and Reduction Potentials of Some Free Radicals

R	$E_{\frac{1}{2}}^{OX}(R\bullet)^a$ V	$E_{\frac{1}{2}}^{RED}(R\bullet)^a$ V	$E_{\frac{1}{2}}^{OX}(RH)^a$ V
$PhCH_2$	0.73	-1.45	2.28
Ph_2CH	0.35	-1.14	2.14
$PhCMe_2$	0.13	-1.77	2.20
Me_2NCH_2	-1.03		1.12
Et_2NCHMe	-1.12		0.96
$(PhCH_2)_2NCHPh$	-0.92		1.32

[a]Versus the saturated calomel electrode.

It is immediately obvious that the alpha-aminoalkyl radicals have oxidation potentials which are far lower than those for the simple hydrocarbon radicals. The results is consistent with the observation of very low ionization potentials for the aminolkyl radicals in the gas phase and with their general chemical properties.

The electochemical data contain a good deal of thermodynamic information. For example, the free energy difference in solution between a carbonium ion and its carbanion is equal to the difference between the oxidation and reduction potentials of the corresponding radical. The gas phase equivalent is the difference between the ionization potential and the electron affinity of the radical. For benzyl in the gas phase, this difference is 186 kcal mol^{-1}, while in solution it is only ca. 50 kcal mol^{-1} which points to a dramatic and preferential stabilization of one of the ions in going from the gas phase to solution.[13]

The data can be further extended by the use of thermochemical cycles. An example is given in eq. 18,19. Using this we derived the

$$\begin{array}{ccc} R\text{-}H & \xrightarrow{\text{BDE}} & R\bullet + H\bullet \\ E_{\frac{1}{2}}^{OX}\downarrow & & \downarrow E_{\frac{1}{2}}^{OX} \\ RH^{\bullet +} & \xrightarrow{\text{BDE}} & R^{+} + H\bullet \end{array} \qquad \begin{array}{c} (18) \\ \\ (19) \end{array}$$

difference between the C-H bond dissociation energies in hydrocarbons or amines and their corresponding radical cations Table VII. These differences are large - between 35 and 55 kcal mol^{-1} - and show a significant substituent effect for the aromatic hydrocarbon radicals, which reflect variations in their oxidation potentials. In instances where BDE(R-H) were known, the absolute values of BDE($RH^{\bullet +}$) were obtained.

Table VII. Bond Dissociation Energies for Radical Cations, $RH^{\bullet +}$, in kcal mol^{-1} at 300 K.

R	ΔH(18)-ΔH(19)[a]	BDE(R-H)	ΔH(19)[b]	ΔH(20)[b]
$PhCH_2$	36	88	52	-17
Ph_2CH	41	84	43	-18
$PhC(Me)_2$	48	84	36	-19
Me_2NCH_2	50	84	34	6
Et_2NCHMe	48			
$(PhCH_2)_2CHPh$	52			

[a]Error ±2 kcal mol^{-1}. [b]Error ±3 kcal mol^{-1}.

Interestingly, for the few tertiary amines which have been investigated so far, the difference in bond dissociation energies seems to be insensitive to the substitution pattern. However, the dominant effect is that a change in the substituent at the radical centre from $-NH_2$ to $-^{+}NR_2$ causes a reduction in the C-H bond dissociation energy of ca. 50 kcal mol^{-1}.

More complex cycles can be used to estimate the enthalpy change for an alternative pathway for the decomposition of the radical cation, eq. 20, Table VII. In general, proton loss seems to be the preferred mode of decomposition.

$$RH^{\bullet +} \longrightarrow R\bullet + H^{+} \qquad (20)$$

5. CARBENES - USING BULKY SUBSTITUENTS TO CHANGE THEIR ELECTRONIC PROPERTIES

5.1. Background

The chemistry of a carbene is generally thought to be dictated by the nature of its spin state.[14] Since carbenes have two unpaired electrons, the readily accessible states are singlet and triplet. Typical reactions of the singlet state are insertion into the O-H bonds of alcohols and stereospecific olefin addition. By contrast, the triplet state carries out radical-like chemistry. Reactions of the triplet state include atom abstractions, reactions with other triplet molecules such as oxygen and non-stereospecific olefin additions.

These broad characterizations hold fairly well for carbenes which have singlet groundstates.[14] However, they do not appear to apply with the same rigour to many triplet carbenes.[14-15] For example, laser flash photolysis studies of triplet diphenylcarbene show that its optical absorption spectrum is quenched by methanol, a typical scavenger of the singlet carbene.[14-15] To account for this apparent inconsistency, it was proposed[16] that the triplet and singlet states of the carbene were in equilibrium so that at room temperature the triplet ground state as a reservoir for the reactive singlet which, in turn, reacted with the methanol.[14]

Unfortunately, this appealing explanation has not stood the test of kinetic scrutiny[14-18] and direct reaction between the triplet state and molecules like methanol seems to be a real possibility.[17] Of course, a direct reaction requires that the carbene-methanol complex accesses a singlet state in order to produce the final product in its groundstate, eq. 21. While the singlet state of the complex is quite

$$Ph_2C:^T + MeOH \longrightarrow [Ph_2C:\cdots\cdot MeOH]^S \longrightarrow Ph_2CHOMe \qquad (21)$$

different to that of the free carbene, the ease with which it is formed ought to be indirectly related to the energy gap between the triplet and singlet states of the free carbene.

The laser flash photolysis results are disquieting because they undermine the notion that triplet carbenes have a distinct chemistry of their own. This raises the question - does the chemistry of any triplet diarylcarbene conform to the dictates of spin selection?

5.2. The Answer

The obvious means of addressing this problem was to search for carbene which have very large triplet - singlet energy separations.[14] Fortunately, the literature contains some hints on how to set about this.[19] Theory indicates that the triplet state of a diarylcarbene is slightly destabilized by an expansion of the central C-C-C bond angle.[20] However, such an expansion is predicted to have a dramatic destabilizing effect on the singlet state of the carbene thus increasing the singlet - triplet energy gap.

The e.s.r. spectrum of triplet dimesitylcarbene[21] suggests that it is an appropriate candidate. The spectroscopic E value for a triplet carbene is a crude measure of the extent of bending at the central carbon. For example E = 570 MHz for diphenyl carbene in which the central C-C-C angle is 148° and E would be zero for a completely linear carbene. The value observed for dimesitylcarbene was 240 MHz, indicating that the central angle in the carbene is far greater than that in diphenylcarbene and pointed to an enhancement in the triplet-singlet energy separation. In effect, the ortho-methyl groups of dimesitylcarbene push against the faces of the phenyl ring opposite and the compression is relieved by angle expansion.

Laser flash photolysis experiments[22] on triplet dimesitylcarbene at room temperature showed that did not react with methanol and indeed all aspects of its chemistry conformed to the behaviour which has been anticipated, but so seldom found, for a triplet carbene.

The steric influence of the ortho-methyl groups in dimesitylcarbene primarily caused a geometrical change. However, this simple alteration was accompanied by an electronic change which had a profound effect on the chemistry of the carbene. If the interpretation is correct, it should be possible to achieve a gradation in the electronic properties of carbenes simply by changing the steric bulk of the substituents. Table VIII shows that this is indeed the case. Rate constants for the reactions of the carbene with methanol are inversely related to the E values for the carbenes. That is, the more bent the carbene, the faster its reaction with methanol and hence the smaller the energy separation between the triplet and singlet states.

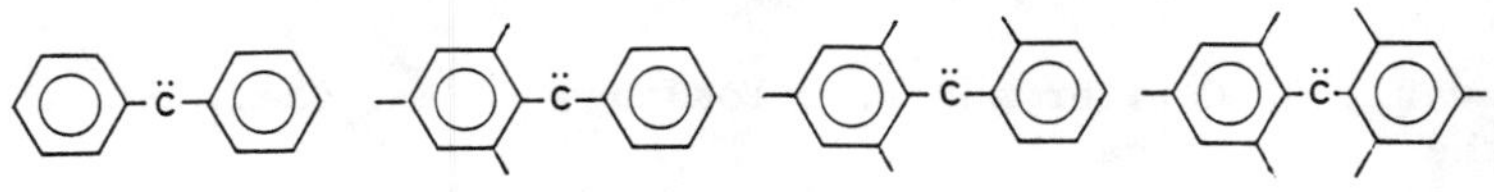

I II III IV

Table VIII. Carbenes I-IV: Spectroscopic E Values (77K); Rate Constants for their Reactions with Methanol (300K).

Carbene	k^a $M^{-1}s^{-1}$	E MHz
I	$\sim 10^7$	580
II	8×10^4	465
III	2×10^4	400
IV	$<10^3$	250

[a]In isooctane.

The results for carbenes illustrate that substituent effects may be more subtle than is at first perceived. In this case, a phenomenon which was fundamentally an electronic change was achieved by manipulating a steric substituent effect.

6. SUMMARY

Studies of the thermodynamic properties of free radicals provided some insights into the role of substituent effects. In some instances the complete absence of such effects was a distinct advantage since it allowed the extension of thermochemical measurements to unknown systems by group additivity methods. However, this was clearly not a general property of free radicals as was illustrated or alpha-aminoalkyls. Radical electrochemistry appeared to be quite sensitive to small substituent changes and was grossly affected by the removal of an electron from a ligand. Substantial electronic effects on carbene chemistry were brought about by changing the steric bulk of ligands attached to the carbene centre. While these studies were, to some extent, outside conventional approaches to the investigation of substituent effects, we nevertheless hope that they represent a useful contribution to the field.

REFERENCES

(1) Issued as NRCC publication No. 25214.
(2) S.W. Benson "Thermochemical Kinetics", 2nd Ed. John Wiley, New York, N.Y., 1976.
(3) R. Hiatt and S.W. Benson, J. Am. Chem. Soc. 94 (1973) 385; Int. J. Chem. Kinetics, 4 (1972) 151; ibid, 5 (1973) 385.
(4) A.L. Castelhano and D. Griller, J. Am. Chem. Soc., 104 (1982) 3655.
(5) J.A. Hawari, D. Griller and F.P. Lossing, J. Am. Chem. Soc., submitted for publication.
(6) F.P. Lossing and J.C. Traeger, Int. J. Mass Spectrosc. Ion Proc., 19 (1976) 9.
(7) J.C. Traeger and R.G. McLoughlin, J. Am. Chem. Soc., 103 (1981) 3647.
(8) T.J. Burkey, A.L. Castelhano, D. Griller and F.P. Lossing, J. Am. Chem. Soc., 105 (1983) 4701.
(9) D.F. McMillen and D.M. Golden, Ann. Rev. Phys. Chem., 33 (1982) 493.
(10) P.R. Marriott, A.L. Castelhano and D. Griller, Can. J. Chem., 60 (1982) 274.
(11) D. Griller, J.A. Howard, P.R. Marriott and J.C. Scaiano, J. Am. Chem. Soc., 103 (1981) 619.
(12) H.M. Rosenstock, K. Draxl, B.W. Steiner and J.T. Herron, J. Phys. Chem. Ref. Data, Suppl. b (1977) 1.
(13) D.D.M. Wayner and D. Griller, J. Am. Chem. Soc., in press.

(14) D. Griller, A.S. Nazran and J.C. Scaiano, Acc. Chem. Res., 17 (1984) 283.
(15) D. Griller, A.S. Nazran and J.C. Scaiano, Tetrahedron, 41 (1984) 1525.
(16) K.B. Eisenthal, N.J. Turro, M. Aikawa, J.A. Butcher, Jr., C. Dupuy, G. Hefferson, W. Hetherington, G.M. Korenowski, M.J. McAuliffe, J. Am. Chem. Soc., 102 (1980) 6563.
(17) D. Griller, A.S. Nazran and J.C. Scaiano, J. Am. Chem. Soc., 106 (1984) 198.
(18) A.S. Nazran and D. Griller, J. Am. Chem. Soc., 107 (1985) 4613.
(19) H.E. Zimmerman and D.H. Paskovich, J. Am. Chem. Soc., 86 (1964) 2149.
(20) J. Metcalfe and E.A. Halevi, J. Chem. Soc., Perkin Trans. 2, (1977) 634; R. Hoffmann, G.D. Zeiss and G.W. Van Dine, J. Am. Chem. Soc., 90 (1968) 1485.
(21) A.S. Nazran, E.J. Gabe, Y. LePage, D.J. Northcott, J.M. Park and D. Griller, J. Am. Chem. Soc., 105 (1983) 2912.
(22) A.S. Nazran and D. Griller, J. Am. Chem. Soc., 106 (1984) 543.

SOME APPLICATIONS OF ESR SPECTROSCOPY TO THE STUDY OF SUBSTITUENT EFFECTS[1]

G. W. Burton, L. J. Johnston,[2] J. C. Walton[3] and K. U. Ingold
Division of Chemistry
National Research Council of Canada
Ottawa, Ontario K1A 0R6
Canada

ABSTRACT. ESR Spectroscopy has been applied to help answer the following questions: What makes a good phenolic antioxidant? What controls the inversion rates of non-planar radicals? How "large" is the $CH_2^{\bullet}$ group and does the unpaired electron confer a unique "substituent character" upon this group?

WHAT MAKES A GOOD PHENOLIC ANTIOXIDANT?

Autoxidation, which is the more or less spontaneous reaction which many organic compounds (RH) undergo with atmospheric oxygen at ambient temperatures, is a free radical chain reaction:

Initiation: Production of $R^{\bullet}$ (1)

Propagation: $$R^{\bullet} + O_2 \longrightarrow ROO^{\bullet} \quad (2)$$

$$ROO^{\bullet} + RH \longrightarrow R^{\bullet} + ROOH \quad (3)$$

Termination: $$ROO^{\bullet} + ROO^{\bullet} \longrightarrow \text{Molecular Products} \quad (4)$$

The rate of an autoxidation can generally be reduced by the addition of certain compounds known as antioxidants. For convenience, antioxidants are divided into two classes. Preventive, or primary, antioxidants reduce the rate of initiation of free radical chains, most commonly by converting the free-radical producing hydroperoxidic products of prior autoxidation (ROOH) to innocuous products, such as the corresponding alcohol. Chain-breaking, or secondary, antioxidants trap the chain-propagating peroxyl radicals ($ROO^{\bullet}$) and thereby reduce the length of the autoxidation chain. Phenols (ArOH) constitute the major class of chain-breaking antioxidants. Their mechanism of action can be described by reactions 5 and 6:[4]

H. G. Viehe et al. (eds.), Substituent Effects in Radical Chemistry, 107–122.
© 1986 by D. Reidel Publishing Co.

$$ROO^{\bullet} + ArOH \longrightarrow ROOH + ArO^{\bullet} \quad (5)$$

$$ROO^{\bullet} + ArO^{\bullet} \longrightarrow \text{Molecular Products} \quad (6)$$

Each molecule of ArOH stops two oxidation chains because the resonance stabilized phenoxyl radical ($ArO^{\bullet}$) is too unreactive to continue the chain.

In living organisms and in food the autoxidation process is known as lipid peroxidation. It has long been recognized that Vitamin E, i.e., α-tocopherol (1, R = $(CH_2CH_2CH_2CHCH_3)_3CH_3$), is a hydrophobic, peroxyl radical-trapping, chain-breaking antioxidant which is present in the lipid fraction of biological materials.

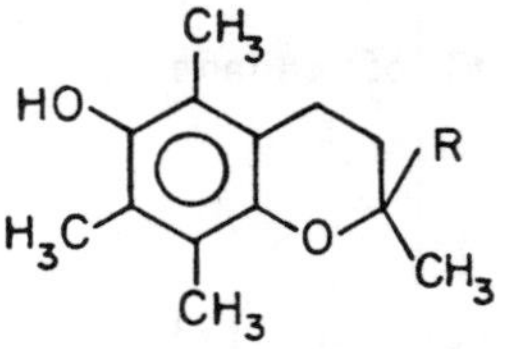

1

In the early 1980's,[5,6] we showed by means of _in vitro_ kinetic studies that α-tocopherol was the best chain-breaking phenolic antioxidant known at that time. That is, the rate constant ($\underline{k}$) for the reaction of peroxyl radicals with α-tocopherol was much higher than the values found for any of the synthetic phenolic antioxidants used in commerce. (For α-tocopherol $\underline{k} = 3.2 \times 10^6\ M^{-1}\ s^{-1}$).[7,8] We have also employed a different experimental technique in which we extracted the lipid fraction from biological fluids and tissues and compared the vitamin E concentration (measured by HPLC) with the total concentration of all chain-breaking antioxidants (measured by "titration" with peroxyl radicals).[9-12] Our results showed that vitamin E is certainly the major (and probably the only) lipid soluble chain-breaking antioxidant in living organisms.[9-12]

The facts outlined above stimulated us to undertake a detailed examination of those structural features of α-tocopherol that serve to make it such an outstanding antioxidant.[7,8,13] One result of these studies has been the discovery that 2-substituted-5-hydroxy-2,4,6,7-tetramethyldihydrobenzofurans (2) are better antioxidants (by ca. 50 - 80%) than α-tocopherol and related 2-substituted-6-hydroxy-2,5,7,8-tetramethylchromans.[7,8] This work has led to the synthesis of 2 (R = $(CH_2CH_2CH_2CHCH_3)_3CH_3$) which actually has more vitamin E activity than vitamin E itself![14]

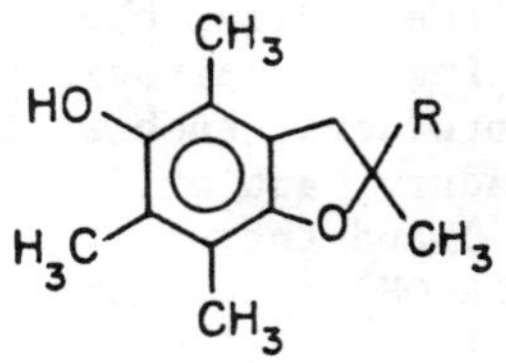

2

During these studies two things became clear. First, that the high antioxidant activity of **1**, and the even greater reactivity of **2**, could be ascribed to stabilization of the phenoxyl radical formed in reaction 5 by conjugative delocalization of the p-type lone pair on the oxygen para to the OH group, i.e.,

The necessary orbital overlap for such conjugative electron delocalization is enforced in **1** (and even more strongly enforced in **2**) by the heterocyclic ring. Second, that the magnitude of k could be related to the thermodynamic stabilization of the phenoxyl radical (i.e., to the strength of the phenolic O-H bond). Thermodynamic stabilization of the phenoxyl should depend on the extent of conjugative electron delocalization from the para oxygen and this, in turn, should depend on the nature of the R group. Thus, an inductively electron-withdrawing R group should reduce conjugative electron delocalization because it will make the development of positive charge on the para oxygen more difficult. For any particular class of 4-alkoxyphenols we would therefore expect that k will decrease (i.e., antioxidant activity will decrease) as the inductive electron-withdrawing capacity of the substituent on the para oxygen increases. This is exactly what is observed with **1** as the R group is changed from CH_3 to CO_2H; the inductive effect of these R groups being given by their σ_I substituent constants (see Table I and Figure 1).

Table I.

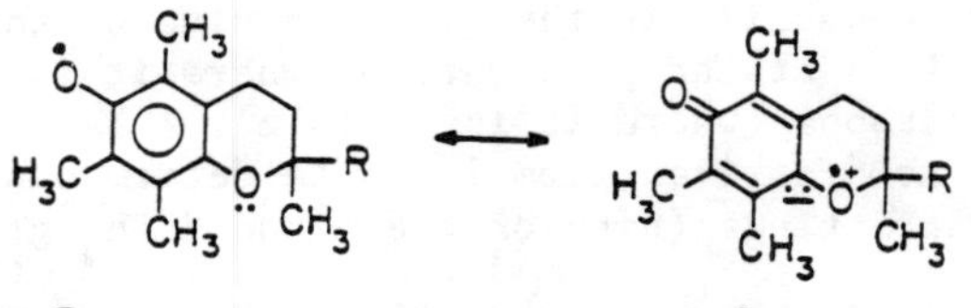

R	σ_I	$10^{-5}k(M^{-1}s^{-1})$
CH_3	-0.01	38
$CH_2CH_2CO_2H$	0.07	37
$CH_2CH_2CO_2CH_3$	0.07	33
CH_2OH	0.11	27
$CH_2CO_2CH_3$	0.19	27
CH_2CO_2H	0.19	19
OCH_2	0.30	15
CO_2H	0.30	11
CO_2CH_3	0.32	18

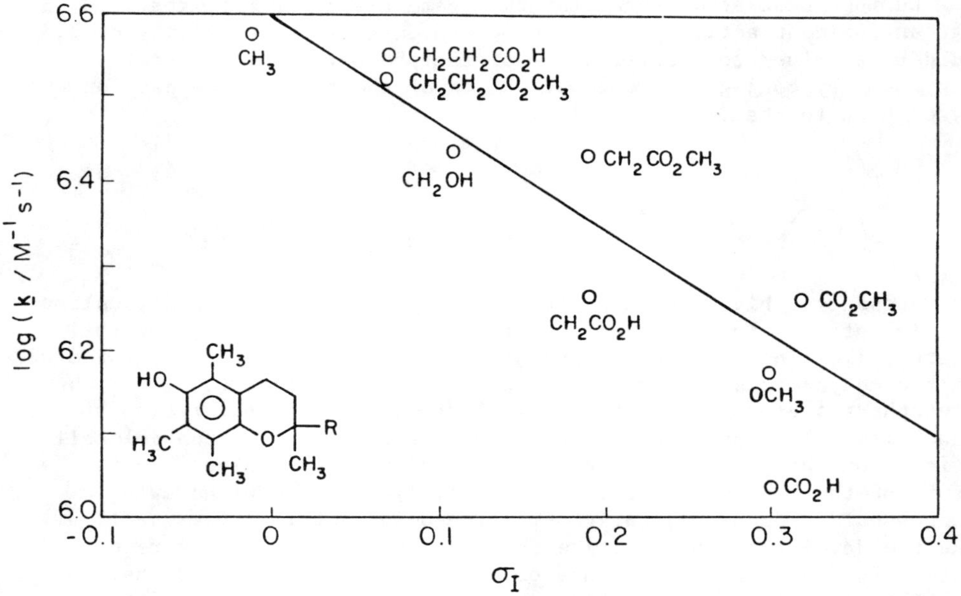

Figure 1. Log($\underline{k}$/M^{-1} s^{-1}) vs. σ_I for **1**.

Of course, these kinetic results do not in themselves "prove" that electron-withdrawing R groups in **1** actually do reduce conjugative electron delocalization in the corresponding aryloxyl radicals. Fortunately, we have a probe which will allow us to check whether the postulated effect of R is or is not a real effect. This probe is ESR spectroscopy. The spin-density in the aromatic ring of the aryloxyl radicals will be localized at the para carbon (where it is "invisible") and at the two ortho carbons (where it is visible). That is, relative spin-densities in the ArO$^{\bullet}$ derived from **1** can be determined by measuring the hyperfine splittings (hfs) of the 5 and 7 CH_3 groups.

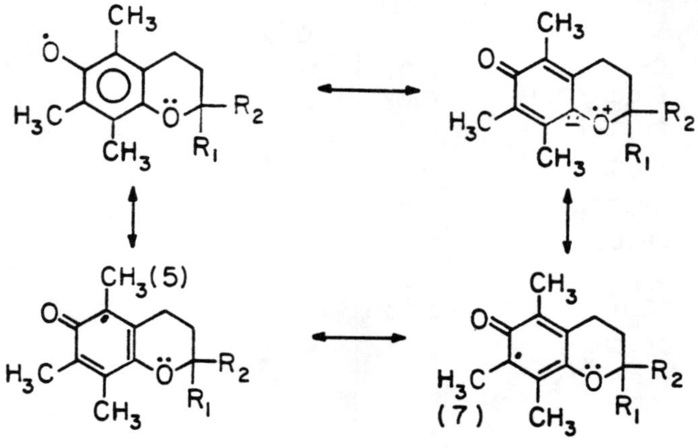

We would expect that as conjugative electron delocalization from the para oxygen is decreased by placing more and more strongly electron-withdrawing groups on the carbon atom at position 2, that the unpaired electron would become more and more localized in the aromatic ring. As a consequence, $\underline{a}^H$(5 CH_3) and $\underline{a}^H$(7 CH_3) should increase and the antioxidant activity of the parent hydroxychroman should decrease. This is exactly what is observed (see Table II and Figure 2).

Table II.

R_1	R_2	$a^H(5CH_3)$	$a^H(7CH_3)$	$10^{-5}k\,(M^{-1}s^{-1})$
CH_3	CH_3	5.94	4.51	38
CH_3	$C_{16}H_{33}$	5.98	4.57	32
CH_3	$CH_2CH_2CO_2H$	6.00	4.56	37
H	H	6.00	4.62	27
CH_3	$CH_2CO_2CH_3$	6.15	4.60	27
CH_3	OCH_3	6.05	4.95	15
CH_3	CO_2CH_3	6.35	4.72	18

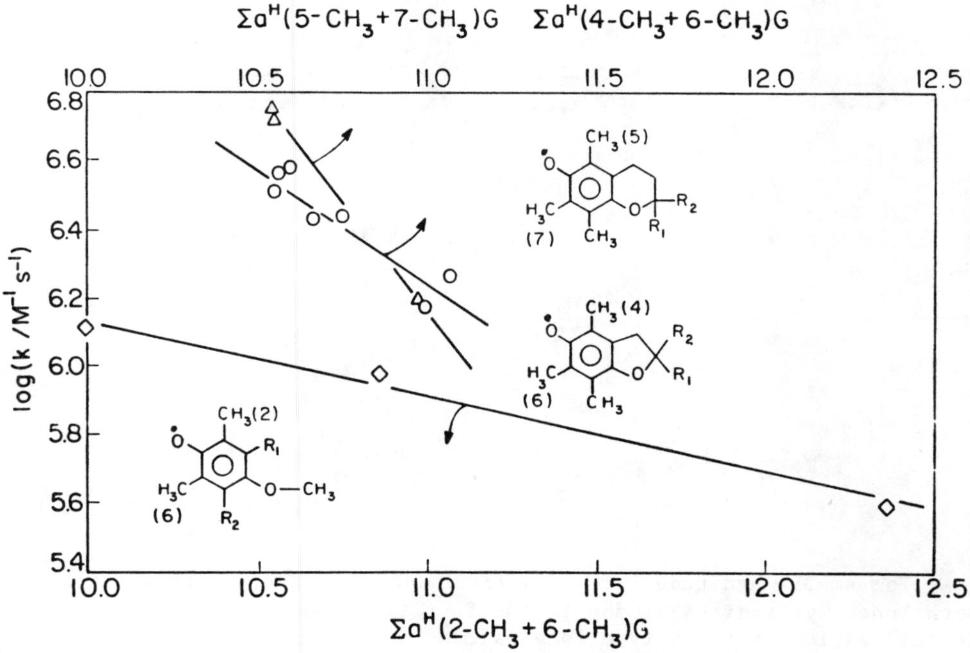

Figure 2. Log($\underline{k}$/M^{-1} s^{-1}) vs. Σa^H for ortho methyl groups in the aroxyl radicals derived from 1 (O), from 2 (Δ), and from 4-methoxy-2,6-dimethylphenols ( ).

Similar effects are observed in 4-methoxy-2,6-dimethylphenols and their radicals and in 2 and their radicals (see Fig. 2).

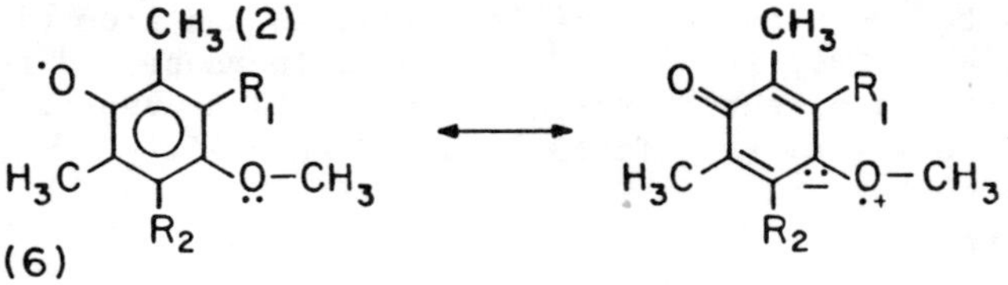

(6)

R_1	R_2	$a^H(2CH_3)$	$a^H(6CH_3)$	$10^{-5}k(M^{-1}s^{-1})$
H	CH_3	5.5	4.5	13
H	H	5.43	5.43	9.4
CH_3	CH_3	6.18	6.18	3.9

For the acyclic 4-methoxy phenols, many of which make good antioxidants (see *k* values in Fig. 2), we have an additional probe for spin-density on the para oxygen atom, the magnitude of $a^H(OCH_3)$.

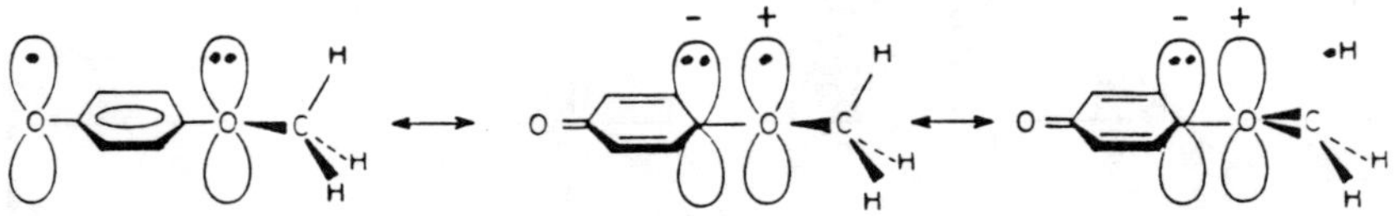

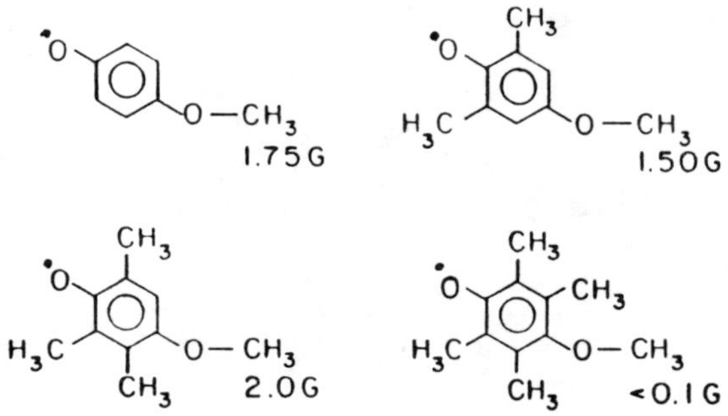

The small magnitude (< 0.1 G) of $a^H(OCH_3)$ in the 4-methoxytetramethylphenoxyl radical is due to the fact that conjugative electron delocalization of the p type lone pair of electrons of the methoxyl oxygen cannot occur. This is because the steric effect of the two meta methyl groups forces the $O\text{-}CH_3$ bond into a position where it is perpendicular to the aromatic plane (as has been shown by X-ray analysis of the parent phenol).[5] The p type lone pair of the methoxyl oxygen therefore lies in the plane of the aromatic ring and hence is orthogonal to the SOMO. A natural consequence is that 4-methoxytetramethylphenol ($\underline{k}$ = 3.9 x 10^5 M^{-1} s^{-1}) is a less active antioxidant than 4-methoxy-2,3,6-trimethylphenol (k = 13 x 10^5 M^{-1} s^{-1}) and 4-methoxy-2,6-dimethylphenol ($\underline{k}$ = 9.4 x 10^5 M^{-1} s^{-1}).[8] What is rather more surprising in view of the known electron-withdrawing inductive effect of the methoxy group (σ_I = 0.30) is that the "perpendicular" methoxy in 4-methoxytetramethylphenol is more "activating" than a 4-methyl group (i.e., pentamethylphenol) or a hydrogen atom (i.e., 2,3,5,6-tetramethylphenol).

	10^4 k ($M^{-1}s^{-1}$)	σ_I
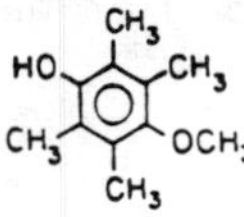	39	0.30
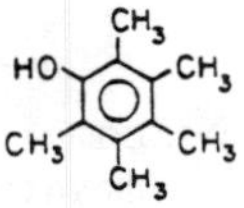	36	-0.01
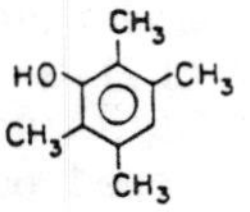	7.5	0.00

This completely unexpected substituent effect contradicts the general belief[15] that a "perpendicular" alkoxy group will be "deactivating" (relative to CH_3 and to H) because the +M activating effect of its p type lone pair will be inoperative, leaving only the -I effect. Our results show that this is not the case. The -I effect of a perpendicular alkoxy group is more than outweighed by a residual +M effect which we attribute to a resonance contribution from the other lone pair on the oxygen.[16]

WHAT CONTROLS THE INVERSION RATES OF NON-PLANAR RADICALS?

We know that trivalent radicals, $a(b)\dot{A}X$, are generally non-planar for A = Si, Ge and Sn, irrespective of the nature of a, b and X. For A = carbon, significant non-planarity occurs when 2 (or 3) of the substituents are strongly electron withdrawing (e.g., a,b = F or OR), and when a and b form a small ring (e.g., $H_2C{=}\dot{C}{-}X$, $c{-}C_3H_4^{\bullet}{-}X$). Perhaps the "last" major kinetic problem relating to free radicals in homogeneous solution has to do with the effect of a, b, A and X on the rates of inversion. For example, within a particular family of radicals what effect does a change in X have on the out of plane angle (θ) and on the barrier height (B)?

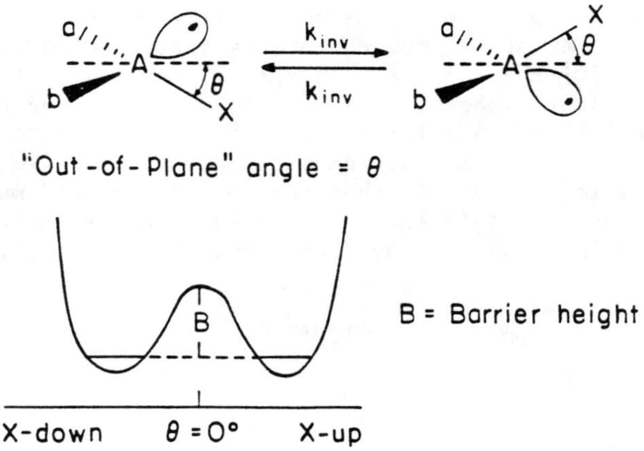

We began to look at this general problem by combining our own laser flash kinetic studies on silicon-[17] and germanium-[18] centered radicals with product studies by other workers and have been able to estimate rate constants for the inversion of methyl-α-naphthyl-phenylsilyl and -germyl radicals of ca. 6.8 x 10^9 M^{-1} s^{-1} and ca. 9.6 x 10^8 M^{-1} s^{-1}, respectively, at 80°C. We have now turned our attention to 1-substituted cyclopropyl radicals and have estimated two inversion rates by combining laser flash kinetic studies[19,20] with product studies.[20]

The ESR spectrum of c-$C_3H_5^{\bullet}$ shows a^{H_α} (1H) ≈ -6.7 G and a^{H_β}(4H) ≈ + 23.5 G.[21] The small magnitude of the α-H hfs provides evidence that in its equilibrium configuration the cyclopropyl radical is non-planar.

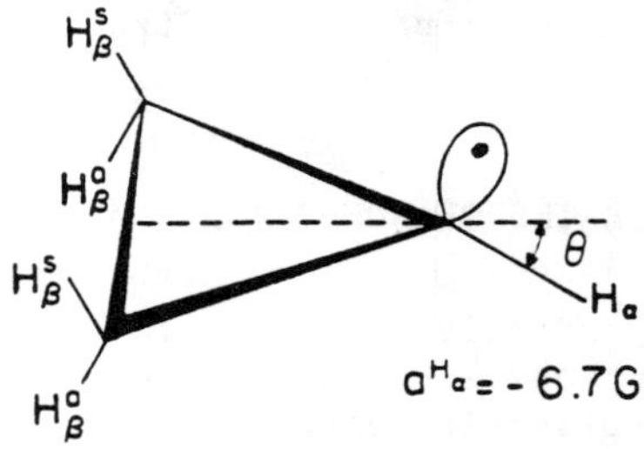

INDO calculations suggest that Θ ~ 30-35° and B = 3.2 kcal/mol,[22] while *ab initio* calculations give Θ = 41°, B = 3.8 kcal/mol[23] and Θ = 39.3°, B = 3.0 kcal/mol.[24] The INDO calculated hfs for the *syn* and anti β-H's were nearly equal to each other for all values of Θ indicating that rapid inversion at the radical center is not necessarily required to explain the observed magnetic equivalence of the four β-H's at temperatures as low as 77 K.[22] Nevertheless, it did seem worth while to examine the ESR spectrum of $c\text{-}C_3H_5^{\bullet}$ at still lower temperatures because ab initio calculations[23] indicate that the β-H's that are syn to the unpaired electron should have significantly larger hfs than the anti β-H's. More importantly, Kawamura et al[25] have found that in various "configurationally locked" substituted cyclopropyl radicals the hfs of syn β-H's are about three times larger than the hfs of the anti β-H's.

Cyclopropyl radicals were prepared on the surface of a rotating cryostat at 77 K by co-condensing molecular beams of $c\text{-}C_3H_5Br$, Na atoms and $c\text{-}C_3H_6$ (as an inert matrix).

$$c\text{-}C_3H_5Br + Na \longrightarrow c\text{-}C_3H_5^{\bullet} + NaBr$$

The ESR spectrum of these radicals was essentially unchanged when they were cooled from 77 to 4 K. It does not therefore appear possible to produce an ESR observable "freezing-out" of the inversion either because it is too fast or because the syn and anti β-H's really do have nearly identical hfs (or both).

We therefore decided to combine laser flash kinetic measurements of the trapping of $c\text{-}C_3H_5^{\bullet}$ with product studies on appropriately labelled cyclopropyl radicals. The kinetic measurements were made using the "probe" technique as indicated below:[19]

LASER FLASH PHOTOLYSIS

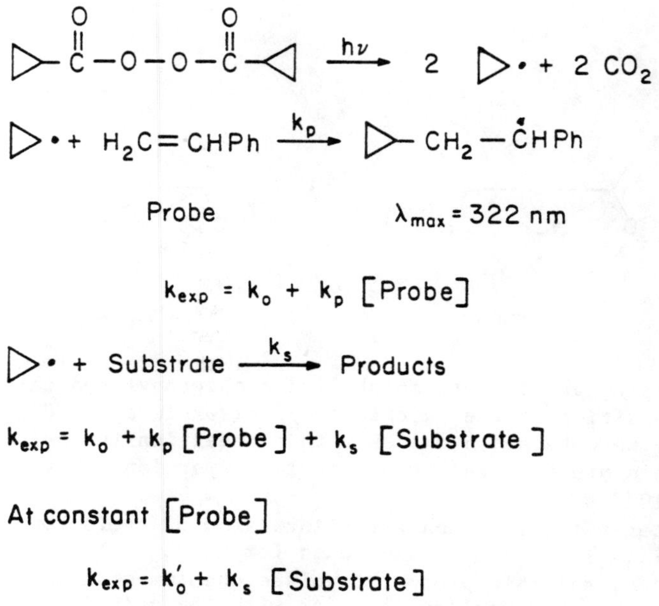

We chose CCl_3Br as our trap both because the cyclopropyl radical reacts rapidly with CCl_3Br ($k = 2.8 \times 10^9\ M^{-1}\ s^{-1}$ at ambient temperatures)[19] and because appropriate deuterium labelling of our cyclopropyl source (following the procedure of Kobayashi and Lambert[26]) would yield cyclopropyl bromides in which the extent of retention of configuration could be determined by NMR.

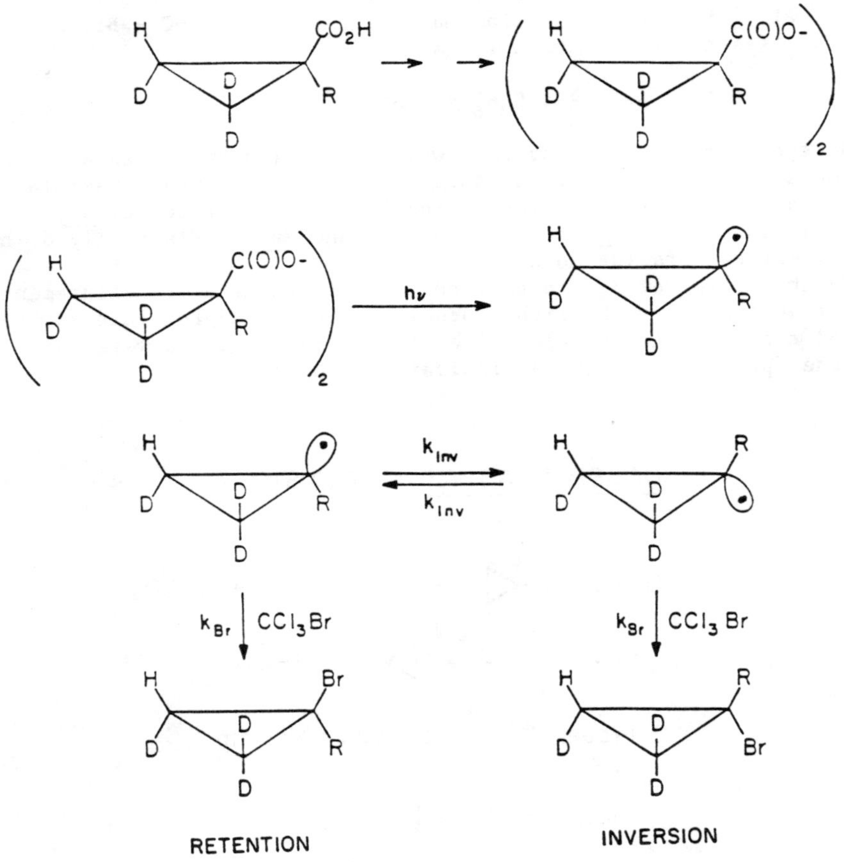

On generating the deuterium-labelled cyclopropyl radical by the thermal decomposition of the parent diacyl peroxide at 71°C in neat CCl_3Br (9.8 M) we obtained bromides with a retention/inversion ratio of ca. 1.07 . This yields a rate constant for inversion of cyclopropyl, k_{inv}, of ca. $10^{12}\ s^{-1}$.

It is interesting to compare the inversion of cyclopropyl with that of ammonia. The average time taken for an NH_3 molecule to invert is 2.5×10^{-11} s, and this process involves quantum-mechanical tunneling through a 5.9 kcal/mol barrier with the principal vibrational mode for inversion having a frequency of 950 cm^{-1}.[27] Thermally activated processes leading to an increase in the rate of inversion are relatively unimportant except at very high temperatures.

Cyclopropyl inversion through a (calculated)[22-24] barrier of 3.0 - 3.8 kcal/mol[28] must also occur largely by tunneling and hence it cannot be "frozen-out" by going to low temperatures. Our failure to observe any difference between the <u>syn</u> and <u>anti</u> β-H hfs at 4K therefore sheds no light on the difference between Kochi et al's[22] conclusion that <u>syn</u> and <u>anti</u> β-H hfs are approximately equal, and Kawamura et al's[25] conclusion that the former will be about three times as large as the latter. The question as to whether the <u>syn</u> and <u>anti</u> β-H's <u>of the parent cyclopropyl radical</u> do or do not have intrinsically "equal" hfs has, therefore, no real meaning because the <u>syn</u> and <u>anti</u> β-H's can never be distinguished by ESR specroscopy.

We have also carried out trapping experiments with a stereospecifically deuterium-ring-labelled 1-methylcyclopropyl radical.[20] In this case we found a retention/inversion ratio of 1.47 in neat $CC\ell_3Br$ at 71°C. Combining this ratio with the rate constant for bromine abstraction from $CC\ell_3Br$ measured by laser flash photolysis (viz.[20] $3.7 \times 10^9\ M^{-1}\ s^{-1}$ at room temperature) yields $\underline{k}_{inv} = 2.1 \times 10^{11}\ s^{-1}$. Inversion by tunneling must be unimportant because of the methyl group's mass. We can therefore assume that the Arrhenius pre-exponential factor for inversion will be ca. $2 \times 10^{13}\ s^{-1}$, which yields a barrier of ca. 3.1 kcal/mol. This barrier height is in excellent agreement with the barrier heights calculated for cyclopropyl[22-24] which implies that the potential functions for the inversion of 1-methylcyclopropyl and cyclopropyl must be rather similar. This seems intuitively reasonable and if it is true it implies that the two radicals should, in their equilibrium configurations, be approximately equally pyramidal. This in turn implies that the α-^{13}C hfs should be approximately equal, and such we have found is the case.[20] The relatively large magnitude of these α-^{13}C hfs (~ 96-98 G, see below) indicates that these two radicals are distinctly non-planar. In contrast, the <u>1</u>-(trimethylsilyl)cyclopropyl radical has an α-^{13}C hfs of only 40.9 G.[29] This radical must therefore be nearly planar, a consequence, presumably, of dπ-pπ overlap or its equivalent.[29] Chemical trapping of $c\text{-}C_3H_4^{\bullet}SiMe_3$ before inversion would seem to be out of the question.

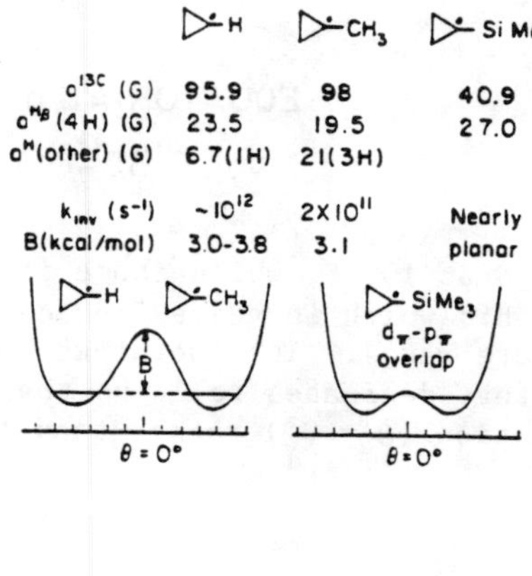

HOW "LARGE" IS THE $CH_2^{\bullet}$ GROUP AND DOES THE UNPAIRED ELECTRON CONFER A UNIQUE "SUBSTITUENT CHARACTER" UPON THIS GROUP?

One of the most popular methods for measuring relative effective "sizes" of substituent groups, R, is to examine R substituted cyclohexanes. The ratio of c-$C_6H_{11}R$ conformers having R in the axial position to those having R in the equatorial position is measured (generally by NMR) and the effective "size" of R is related to the difference in free energy of the two conformers, $-\Delta G°$, at 300 K.[30,31] Thus, for methylcyclohexane the axial/equatorial conformational ratio is 0.054 at 300 K, yielding $-\Delta G°(CH_3)$ = 1.74 kcal/mol. What is the situation for the cyclohexylmethyl radical? That is, can the axial and equatorial conformers be distinguished by their ESR spectra and does the CH_2 group have a conformational preference?

Axial Equatorial

We have found that the axial and equatorial conformers of cyclohexylmethyl and of 4-alkylcyclohexylmethyl radicals can be readily distinguished by the magnitude of their β-H hfs.[32,33] Those radicals in which the $CH_2^{\bullet}$ groups is axial have βH hfs which increase from

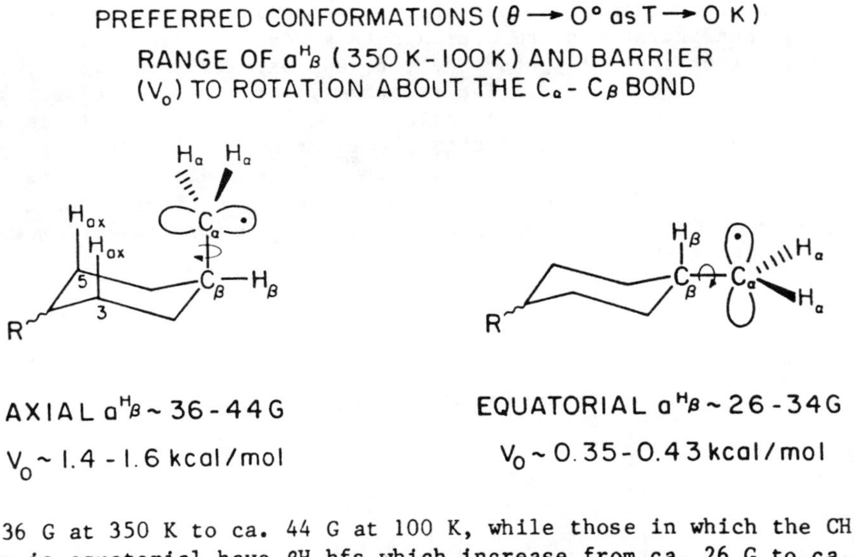

ca. 36 G at 350 K to ca. 44 G at 100 K, while those in which the $CH_2^{\bullet}$ group is equatorial have βH hfs which increase from ca. 26 G to ca. 34 G over the same temperature range. The fact that both of these βH hfs increase as the temperature decreases tells us that both conformers prefer the eclipsed conformation (Θ = 0°) rather than the staggered conformation (Θ = 90°).

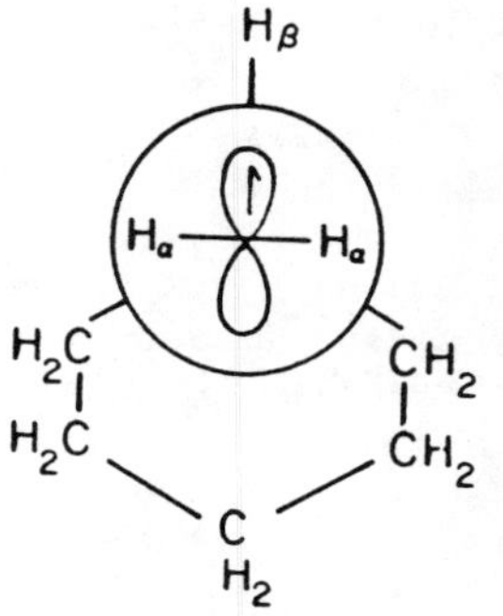

Eclipsed, Θ = 0°

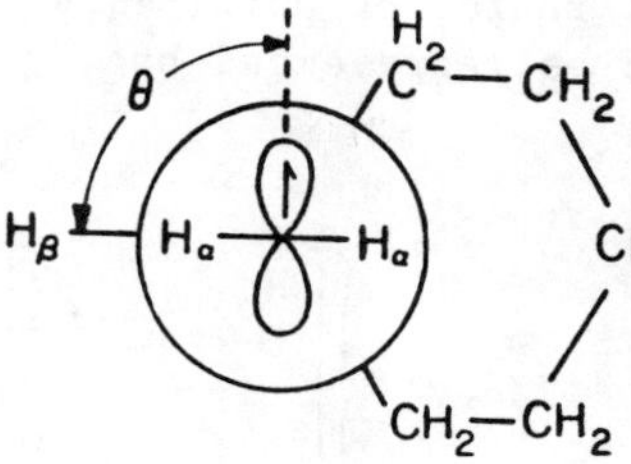

Staggered, Θ = 90°

The β-H hfs for axial conformers are larger than for equatorial conformers because the eclipsed conformation is more strongly favored for an axial $CH_2^{\bullet}$ group. This is due to the fact that rotation of the axial $CH_2^{\bullet}$ group is sterically hindered by the axial hydrogens at positions 3 and 5. The $CH_2^{\bullet}$ rotation barriers, V_o, are calculated to be about 1 kcal/mol greater for the axial than for the equatorial conformer.[33]

The radicals generated by bromine atom abstraction from cis-4-methylcyclohexylmethyl bromide at temperatures <200 K reflect the relative concentrations of the two conformers of the starting bromide as measured by NMR. This is because cyclohexyl ring inversion is slow on the ESR time scale at these low temperatures. At temperatures above 200 K ring inversion is rapid and our results lead to an approximate value of 0.96 kcal/mol for $-\Delta G°(CH_2^{\bullet})$ at 300 K. A more precise measurement obtained by measuring axial/equatorial conformer ratios for the unsubstituted cyclohexylmethyl radical yielded $-\Delta G°(CH_2^{\bullet})$ = 0.71 kcal/mol at this temperature. Comparison with the $-\Delta G°$ values for some other small groups[30,31] shows that the effective size of $CH_2^{\bullet}$ is greater than OH or OCH_3 (which might be expected) but smaller than NH_2 (which is surprising).

	$(-\Delta G°)_{300K}$ kcal/mol
$CH_2^{\bullet}$	0.71
CH_3	1.74
OH	0.52*
OCH_3	0.60
NH_2	1.20*

*In aprotic solvents

The actual ratios of axial/equatorial conformers as a function of temperature can be represented by:

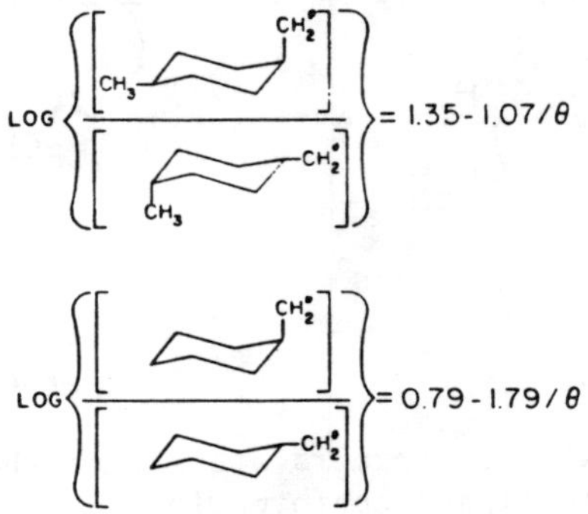

where Θ = 2.3 RT kcal/mol. Our results imply that the $CH_2^{\bullet}$ group has a unique and totally unexpected effect as a substituent on a cyclohexane ring. That is, the axial conformer is entropically favored and hence it becomes the predominant conformer at elevated temperatures: specifically at T > 173 K for cis-4-methylcyclohexylmethyl (which is within the temperature range of our measurements) and at T > 495 K for cyclohexylmethyl (which is outside our experimental temperature range).

In this meeting, we are all agreed that substituents can and do have effects on radical chemistry. If the results described above are substantiated it would appear that radicals can have unexpected substituent effects of their own.

ACKNOWLEDGEMENT

K.U.I. and J.C.W. thank NATO for the award of a research grant without which the present work would not have been undertaken.

REFERENCES AND NOTES

(1) Issued as NRCC No. 25149.
(2) NRCC Research Associate.
(3) NRCC Visiting Scientist. Permanent address: Department of Chemistry, University of St. Andrews, St. Andrews, Scotland, KY16 9ST.
(4) Howard, J.A.; Ingold, K.U. Nature 1962, 195, 280-281.
(5) Burton, G.W.; LePage, Y.; Gabe, E.J.; Ingold, K.U. J. Am. Chem. Soc. 1980, 102, 7791-7792.
(6) Burton, G.W.; Ingold, K.U. J. Am. Chem. Soc. 1981, 103, 6472-6477.
(7) Burton, G.W.; Hughes, L.; Ingold, K.U. J. Am. Chem. Soc. 1983, 105, 5950-5951.
(8) Burton, G.W.; Doba, T.; Gabe, E.; Hughes, L.; Lee, F.L.; Prasad, L.; Ingold, K.U. J. Am. Chem. Soc. 1985, 107, 0000-0000.
(9) Burton, G.W.; Joyce, A.; Ingold, K.U. The Lancet August 1982, 327.
(10) Burton, G.W.; Joyce, A.; Ingold, K.U. Arch. Biochem. Biophys. 1983, 221, 281-290.
(11) Cheeseman, K.H.; Collins, M.; Proudfoot, K.; Slater, T.F.; Burton, G.W.; Webb, A.C.; Ingold, K.U. Biochem. J. 1986, 000, 000-000.
(12) Slater, T.F.; Cheeseman, K.H.; Collins, M.; Maddix, S.; Milia, A.; Proudfoot, K.; Burton, G.W.; Webb, A.; Ingold, K.U. Phil. Trans. R. Soc. London 1985, B311, 000-000.
(13) Doba, T.; Burton, G.W.; Ingold, K.U. J. Am. Chem. Soc. 1983, 105, 6505-6506.
(14) Ingold, K.U.; Burton, G.W.; Foster, D.O.; Zuker, M.; Hughes, L.; Lacelle, S.; Lusztyk, E.; Slaby, M. Science submitted.
(15) Taft, R.W., Jr.; Evans, H.D. J. Chem. Phys. 1957, 27, 1427-1428.
(16) Of course, this raises the question as to why a(OCH_3) should be so small for a "perpendicular" OCH_3 group. We tentatively suggest that the answer to this question is that α and β spin reach the methoxyl's hydrogens in approximately equal amounts and "cancel out"; the α-spin by C-H hyperconjugation to the α-spin on oxygen and the β-spin via C-O hyperconjugation to the α-spin on the 4-carbon of the aromatic ring, followed by C-H spin polarization.
(17) Chatgilialoglu, C.; Ingold, K.U.; Scaiano, J.C. J. Am. Chem. Soc. 1982, 104, 5123-5127.
(18) Ingold, K.U.; Lusztyk, J.; Scaiano, J.C. J. Am. Chem. Soc. 1984, 106, 343-348.
(19) Johnston, L.J.; Scaiano, J.C.; Ingold, K.U. J. Am. Chem. Soc. 1984, 106, 4877-4881.
(20) Johnston, L.J.; Ingold, K.U. J. Am. Chem. Soc. 1986, 108, 0000-0000.
(21) Kaptein, R. In: Chemically Induced Magneic Polarization; Lepley, A.R.; Closs, G.L., Eds.; Wiley-Interscience: New York, 1973; pp. 137-196.
(22) Kochi, J.K.; Bakuzis, P.; Krusic, P.J. J. Am. Chem. Soc. 1973, 95, 1516-1526.

(23) Ellinger, Y.; Subra, R.; Levy, B.; Millie, P.; Berthier, G. J. Chem. Phys. 1975, 62, 10-29.
(24) Dupuis, M.; Pacansky, J. J. Chem. Phys. 1982, 76, 2511-2515.
(25) Kawamura, T.; Tsumura, M.; Yokomichi, Y.; Yonezawa, T. J. Am. Chem. Soc. 1977, 99, 8251-8256.
(26) Kobayashi, K.; Lambert, J.B. J. Org. Chem. 1977, 42, 1254-1256.
(27) Herzberg, G. Molecular Spectra and Molecular Structure. II. Infrared and Raman Spectra of Polyatomic Molecules; Van Nostrand: New York, 1945; pp. 221-224.
(28) Calculated[24] pyramidal bending frequency for $c\text{-}C_3H_5^{\bullet} = 713\ cm^{-1}$.
(29) Paquette, L.A.; Hoppe, M.; Johnston, L.J.; Ingold, K.U. Tetrahedron Lett. 1985, 26, 0000-0000.
(30) Eliel, E.L.; Allinger, N.L.; Angyal, S.J.; Morrison, G.A. Conformational Analysis; Interscience: New York, 1967.
(31) Hirsch, J.A. Topics in Stereochemistry 1967, 1, 199-222.
(32) Ingold, K.U.; Walton, J.C. J. Am. Chem. Soc. 1985, 107, 6315-6317.
(33) Ingold, K.U.; Walton, J.C. J. Chem. Soc., Perkin Trans. 2 1986, 0000-0000.

SUBSTITUENT EFFECTS ON ABSOLUTE RATE CONSTANTS AND ARRHENIUS PARAMETERS FOR THE ADDITION OF TERT-BUTYL RADICALS TO ALKENES

Hanns Fischer
Physikalisch-Chemisches Institut der Universität
Winterthurerstrasse 190
CH-8057 Zürich, Switzerland

ABSTRACT. Absolute rate constants and their temperature dependencies are reported for the addition of tert-butyl radicals to 29 substituted ethenes, mostly in 2-propanol solution. They were measured by time resolved and modulation electron spin resonance and by muon spin rotation. At 300 K the rate constants cover the range from 60 $M^{-1}s^{-1}$ (cis-butene-2) to 10^6 $M^{-1}s^{-1}$ (acrylonitrile). With few exceptions, the frequency factors obey log $A/M^{-1}s^{-1}$ = 7.5±.5. For mono- and 1,1-disubstituted ethenes log k_{300} increases and the activation energy decreases linearly with increasing electron affinity of the olefins. For the addition of tert-butyl to olefins, which bear chloro- and methyl- substituents at the site of addition, steric effects are separated from the polar effects caused by electron affinity changes by comparison with corresponding unsubstituted compounds. The strong effects of electron affinity point to a dominance of SOMO-LUMO interactions in the transition state for the addition of the highly nucleophilic tert-butyl radical.

1. INTRODUCTION

Recently, Tedder and Walton [1] and Giese [2] have reviewed the rates and orientations of free radical addition to olefins which, in general, represent a "complex interplay of polar, steric, and bond-strength terms", heavily influenced by the nature and the positions of substituents. Very important rules and conclusions were established which in the formulation of Giese [2] are as follows: (1) Substituents at the not attacked carbon atom of the olefin (β-substituents) exert essentially only polar effects on the rates of addition. (2) Substituents at the attacked carbon atom (α-substituents) exert both polar and steric effects on the addition rates. (3) Radical substituents induce polar and steric effects. (4) The addition of alkyl radicals to alkenes is characterized by an early transition state, and the polar substituent effects can be described in terms of frontier molecular orbital theory. (5) The regioselectivity is determined by steric substituent effects. Furthermore [1], the polar and steric substituent effects, at

H. G. Viehe et al. (eds.), Substituent Effects in Radical Chemistry, 123–142.
© 1986 by D. Reidel Publishing Co.

least in gross terms, manifest themselves mainly in the activation energy . The frequency factors for alkyl radical addition are usually about log $A/M^{-1}s^{-1}$ = 7.5±1.0 except for very strong polar contributions which may enhance the frequency factors above the normal values [3].

These conclusions were reached mainly by comparisons of relative rate constants of additions to various substituted olefins and their temperature dependencies with the help of absolute Arrhenius parameters known for rather few reactions. Though they allow the predictions of qualitative trends in terms of radical nucleo-or electrophilicities and on polar Hammett-σ-and steric Taft-E_s-parameters of the substituents [1,2] further corroborations by large series of absolute rate data on many reactions are clearly desirable. Moreover, it is felt that such large series should allow the quantification of the "effects" in terms of molecular properties such as ionization potentials, electron affinities and deformation energies. In fact, Bayes et al. [4] have recently shown that log k of the reactions of methyl, ethyl, 2-propyl and tert-butyl radicals with oxygen and ozone increases nearly linearly with decreasing difference of the radical ionization potentials and the substrate electron affinities.

Guided by these views we have carried out a series of experiments on absolute rate constants for the addition of tert-butyl radicals to a large number of substituted alkenes in solution [5], part of which have been published earlier [6]. Various reasons suggested the selection of the radical and of the alkene substrates: The chemical sources for tert-butyl, i.e. photolysis of di-tert-butyl ketone or addition of muonium to isobutane, are well characterized [7,8], and many reactions of this radical have been studied in our group. Then, the tert-butyl radical has a very low ionization potential, IP = 6.92 eV [9], and is highly nucleophilic , therefore [10,11]. For this case, frontier molecular orbital theory [12] predicts particularly strong influences of interactions between the radical's singly occupied orbital (SOMO) and the olefin's lowest unoccupied orbital (LUMO) on the addition rates. They reflect a partial electron transfer in the transition state from radical to olefin depending on the substituents "polar" effect, which should correlate with the electron affinities (EA) of the olefins. Correspondingly, these were chosen to span a wide range. Further, to separate steric from polar effects, pairs of α-unsubstituted and substituted compounds were selected. This paper reviews the methods employed and the results and supplements our previous work [6] by the inclusion of effects of α- and β-methylation and of additional data on reaction mechanisms.

2. EXPERIMENTAL PROCEDURES

The experimental arrangement for steady-state and time-resolved electron spin resonance (ESR) with photochemical radical generation and many of the experimental procedures have been described previously [6,7,10]. These techniques were used for all olefins mentioned below except acrylonitrile. In principle, tert-butyl radicals are generated by photolysis of di-tert-butyl ketone. This involves a Type I cleavage followed by an activated decarbonylation of the pivaloyl radical. Our most recent estimate for the rate constant of this process is [5] $\log k_D/s^{-1}$ 13.5 - 44.8/θ, where θ = 2.3 RT in kJ°mole^{-1}. Thus, for T $\geqq$ 250 K it is complete within less then 100 µs, the usual experimental time resolution, so that the formation of two tert-butyl radicals can be considered as instantaneous. In liquids, tert-butyl self-terminates by disproportionation and combination with diffusion controlled rates, describable by a second-order lifetime $\tau_2 = (2k_t[R]_o)^{-1}$, where $[R]_o$ is the steady-state concentration. If intermittent photolysis is applied and if $[R]_o$ is reached in the on-period the decay of [R] in the off-period is solely determined by τ_2. Pseudo-first order reactions of tert-butyl with the solvent or added substrates manifest themselves by a deviation of the decay part of the kinetic trace from a pure second-order behaviour. We have shown previously [7,10], that for not too large perturbations of the self-termination by these additional processes, the part of the kinetic trace decaying from the stationary concentration $[R]_o$ can be analyzed via

$$[R] = [R]_o \cdot \exp(-t/\tau_1) \cdot (1+\frac{1}{2}(1+(1+4\tau_2/\tau_1)^{1/2} \cdot t/\tau_2)^{-1} \quad (1)$$

to yield τ_2 and the pseudo-first order life-time

$$\tau_1^{-1} = k_3[S] + k[E] \quad (2)$$

where the first term characterizes the reaction with the solvent and the second reaction with substrate, here the alkene. The conditions for the validity of eq.(1) have been discussed [6,7,10] and involve (1) [S], [E] >> $[R]_o \cong 10^{-6}$ M, (2) negligible depletion of substrate [E] during the dwell time of the solution in the reaction zone, and (3) $\tau_1 \geqq \tau_2$. Whereas condition (1) was always fulfilled, the two other conditions required special precautions [6] for many cases. They restricted, together with the condition for fast decarbonylation of pivaloyl mentioned above, the range of temperatures and substrate concentrations which could be applied. The solvent used in nearly all time resolving experiments was 2-propanol for which $\log k_s/M^{-1}s^{-1} = 5.5 - 29.5/\theta$ with θ = 2.3 RT in kJ/mole has been determined [6], and was applied in the extraction of k from τ_1 via eq.(2).

All solutions (typically .17 - .46 M in ketone and 10^{-2} - 1 M in substrate) were freed from oxygen by purging with helium prior to use. All chemicals were purchased in the purest available form and, if not otherwise mentioned below, used without further purification. In individual runs 50'000 to 500'000 individual time profiles [R](t) were sampled. τ_1 and τ_2 resulted from fits of eq.(1) to the decay part. For several substrates different substrate concentrations were used and the pseudo-first order behaviour of τ_1 (eq.(2)) was ensured. The Arrhenius parameters of k were obtained by direct fits of k(T) to the Arrhenius law. Errors given below represent one standard deviation.

Acrylonitrile reacts so fast with tert-butyl that even at rather low olefin concentrations the addition could not be monitored by time resolved ESR. Two other methods were applied to determine the rate constant for this olefin near room temperature. The first was modulation ESR spectroscopy, where one measures the response of spectrum amplitude and phase of the adduct radical $(CH_3)_3CCH_2\dot{C}HCN$ as function of the frequency of modulation of tert-butyl generation. Analysis of the phase lag caused by the addition reaction analogously as in [13] yielded the lifetime for addition $\tau_a^{-1} = k\cdot[E]$ from which k resulted ([E] = .2 M in 2-propanol). Unfortunately, the temperature range for these experiments was very restricted. Below 273 K addition of pivaloyl to acrylonitrile interfered and above 310 K the phase lag became unmeasurably small. Consequently, the Arrhenius parameters could not be determined. The second method was muon spin rotation. Implantation of positive muons in liquid isobutene yields muonated tert-butyl radicals [8] $CH_2Mu\dot{C}(CH_3)_2$ in which one proton is replaced by a positive muon. Addition of acrylonitrile caused a broadening of the μSR-lines from which τ_a^{-1} was extracted following the procedures of ref.[14].

Several acrylonitrile concentrations were applied and the results showed that τ_a^{-1} increased linearly with concentration as expected for an addition reaction. The details on the experiments involving acrylonitrile and other fast reacting monomers will be published elsewhere.

First vertical gas phase electron affinities of the olefins as needed for the discussion were taken from the literature or kindly measured for this work by Dr. M. Allan, University of Fribourg, by electron energy loss spectroscopy [15].

3. RESULTS

3.1. Reaction Mechanisms

Tert-butyl may react with substituted olefins by addition to both carbon atoms of the double bonds, by H-abstraction from substituents, by addition to unsaturated side groups (phenyl) or even by complete electron transfer. To establish the dominant mechanisms steady-state ESR experiments were carried out to detect the secondary radicals resulting from the reactions. In these studies higher substrate concentrations and/or higher temperatures than those used in the kinetic runs were employed. Furthermore, for methyl substituted alkenes, where H-abstraction to give allyl radicals appeared very likely, these allyl radicals were produced in separate experiments by reacting tert-butoxy radicals from di-tert-butyl peroxide or triplet acetone with the neat olefins. The results of these studies are as follows: For nearly all olefins investigated tert-butyl radicals react highly predominantly by addition to the less substituted carbon of the double bond. The spectral features of the resulting alkyl radicals are given in Table I and confirm their structures. Addition to the CH_2-group and not to the ring is also

TABLE I ESR-Parameters of tert-Butyl Adduct Radicals in 2-Propanol

Olefin	Radical [a)]	T[K]	g	Coupling Constants [G]
$CH_2 = CHOC_2H_5$	$RCH_2\dot{C}HOC_2H_5$	295	2.0031	H_α : 15.2, $2H_\beta$: 15.9
$CH_2 = CHSi(OEt)_3$	$RCH_2\dot{C}HSi(OEt)_3$	300	2.0026	H_α : 20.2, $2H_\beta$: 25.5
$CH_2 = CHSiCl_3$	$RCH_2\dot{C}HSiCl_3$	293	2.0026	H_α : 20.1, $2H_\beta$: 25.1
$CH_2 = CHCN$	$RCH_2\dot{C}HCN$	263	2.0028	H_α : 20.0, $2H_\beta$: 17.9, ^{14}N : 3.5
$CH_2 = CHCN$	$RCH_2\dot{C}HCN$	295	2.0028	H_α : 20.0, $2H_\beta$: 18.3, ^{14}N : 3.5
$CH_3CH = CHCN$ (c:t = 9:1)	$R(CH_3)CH\dot{C}HCN$	242	2.0028	H_α : 19.9, $2H_\beta$: 17.4, ^{14}N : 3.4
$CH_2 = CCl_2$	$RCH_2\dot{C}Cl_2$	283	2.0070	$2H_\beta$: 11.4, ^{35}Cl : 3.4, ^{37}Cl : 2.8
$CH_3CH = CCl_2$	$R(CH_3)CH\dot{C}Cl_2$	311	2.0068	H_β : 7.2, 2Cl : 3.5
$CH_2 = C(CH_3)CN$	$RCH_2\dot{C}(CH_3)CN$	337	2.0028	$2H_\beta$: 13.2, $3H_\beta$: 21.0, ^{14}N : 3.3
$CH_2 = C(CH_3)OAc$	$RCH_2\dot{C}(CH_3)OAc$	335	–	$2H_\beta$: 14.1, $3H_\beta$: 22.7
$CH_2 = C(CH_3)OCH_3$	$RCH_2\dot{C}(CH_3)OCH_3$	294	–	$2H_\beta$: 12.3, $3H_\beta$: 21.0, $3H_\gamma$: 1.5
$CH_2 = C(CH_3)Cl$	$RCH_2\dot{C}(CH_3)Cl$	286	–	$2H_\beta$: 12.8, $3H_\beta$: 22.6, Cl : 1.9
$CH_2 = C(CH_3)CH(CH_3)_2$	$RCH_2\dot{C}(CH_3)CH(CH_3)_2$	309	–	$2H_\beta$: 13.1, $3H_\beta$: 22.9, H_β : 8.2
$CH_2 = C(CH_3)C(CH_3)_3$	$RCH_2\dot{C}(CH_3)C(CH_3)_3$	322	–	$2H_\beta$: 14.5, $3H_\beta$: 21.7

a) $R = C(CH_3)_3$

dominant for styrene, which showed the appearance of a benzyl type radical with many sharp lines. In a few cases the olefins of Table I gave rise to weak lines of other radicals which, however, were not of allyl type as found by comparison with the spectra of the separately produced allyl radicals. These weak lines disappeared at high flow rates so that they presumably belong to radicals derived from some reaction products. Since their yields never exceeded about 5% of the yields of the main radicals they are kinetically unimportant. One important exception was di-methyl-butene-2 (tetramethylethene) for which tert-butyl was found to react to about 85% by H-abstraction and by about 15% by addition. We had assumed previously [6] addition as the principal pathway and have to revise this now. A further remarkable finding was obtained for 1-chloro-propene-1 (90% cis/10% trans) which gave the same radical as 2-chloro-propene-1, namely $(CH_3)_3CCH_2\dot{C}(CH_3)Cl$. NMR-studies showed that the 1-chloro-propene contained traces (≅ 3%) of 2-chloro-propene which could not be removed by purification. The exclusive observation of the 2-chloro-propene adduct is then obviously due to a much more rapid addition of tert-butyl to 2-chloro-propene than to 1-chloro-propene, and the ratio of rate constants must be larger than about 300.

In total, the results of these studies confirm the well known [1,2] regioselectivity of the addition of branched alkyl radicals to olefins: For mono- and 1,1-disubstituted alkenes addition occurs nearly exclusively at the unsubstituted carbon. For trisubstituted alkenes addition occurs at the less substituted carbon and for 1,2-disubstituted alkenes at the carbon bearing the less bulky substituent (CH_3 vs. CN). Four substituents seem to be required for reactions at side groups to become important (tetramethylethene).

3.2. Reaction Kinetics

Figure 1 shows two typical kinetic traces obtained for two different concentrations of vinyltriethoxysilane together with results of fitting eq.(1) to the tert-butyl decay profiles in the time window denoted by arrows. With increasing substrate concentration τ_2 increases slightly because of a lowering of $[R]_0$ by partial reaction in the on-period. τ_1 decreases by a factor of 5 which corresponds to the increase of substrate concentration (eq.(2)). Figure 2 shows τ_1^{-1} corrected for the reaction of tert-butyl with the solvent 2-propanol (less than 3% for all concentrations) versus substrate concentration. The linear dependence confirms pseudo-first order kinetics, and the slope of the straight line yields k = 11'600 $M^{-1}s^{-1}$ at 267 K. Similar checks for the pseudo-first order of the addition reactions were carried out for acrylonitrile and 2,3,3-trimethyl-butene-1 ($CH_2 = C(CH_3)C(CH_3)_3$). For the other 26 olefins investigated such systematic and time-consuming variations of substrate were not undertaken, partially because of limited solubilities and too high reaction rates, which lead to severe depletion of substrate for low concentrations and to a too rapid decay at high concentrations. However, in most cases at least a few different concentrations were applied, and the consistency of rate constants obtained with them ensured the pseudo-first order behaviour.

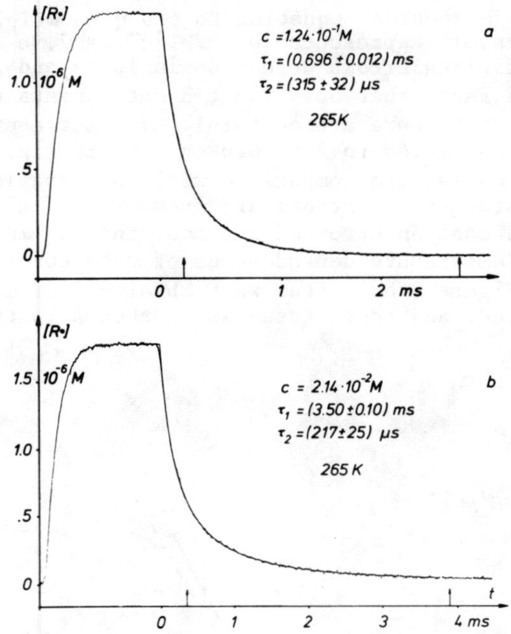

Figure 1. Tert-butyl concentration vs. time in 2-propanol containing vinyltriethoxysilane.

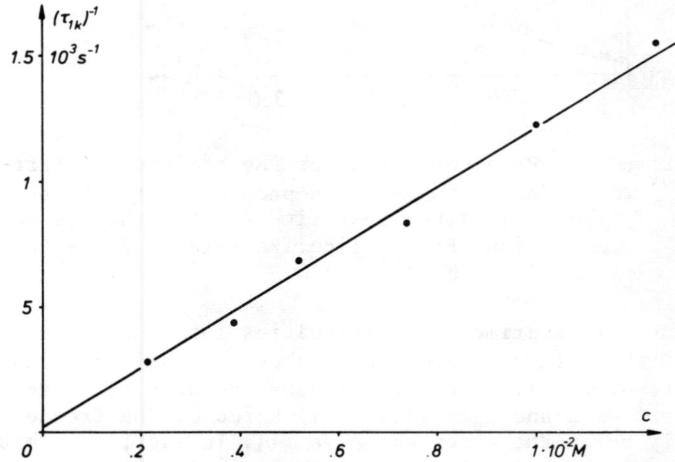

Figure 2. Pseudo-first order plot for the reaction of tert-butyl with vinyltriethoxysilane at 267 K.

Figure 3 shows the rate constants of addition of tert-butyl to 2-chloro-propene-1 plotted versus temperature. The full line is the best SIMPLEX fit of the Arrhenius equation to the data weighted with their experimental errors. It represents log $A/M^{-1}s^{-1}$ = 7.96 and E_a = 22.4 kJ/mole, and the fitting errors were ±.06 in log A and ±.3 kJ/mole in E_a. In view of the facts that only about 3 data points were taken for each temperature, that there are certainly systematic errors and that a fit with log A restricted to 7.5 (broken line in fig.3, E_a = (19.88±.02) kJ/mole) is also compatible with the experimental errors we believe that the statistical errors are much too low. A safe guess is certainly for this case an error of not more than .5 in log A and 3 kJ/mole in E_a. Temperature dependencies of rate constants of the quality shown in Figure 3 or better were obtained for most of the 29 alkenes investigated, and these cases are marked A in the tables of results (II,III).

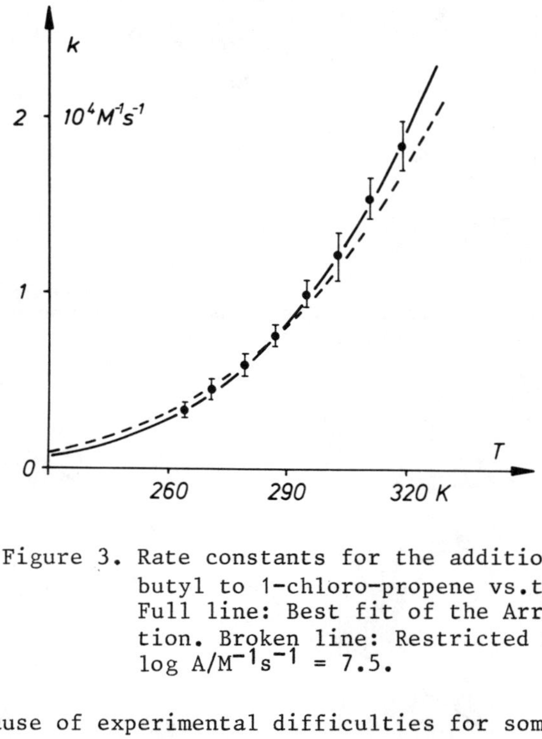

Figure 3. Rate constants for the addition of tert-butyl to 1-chloro-propene vs.temperature. Full line: Best fit of the Arrhenius equation. Broken line: Restricted fit with log $A/M^{-1}s^{-1}$ = 7.5.

However, because of experimental difficulties for some compounds a larger scatter of the data occurred, and one particularly bad example is given in figure 4. For 1,1-dichloro-propene the best fit gave log $A/M^{-1}s^{-1}$ = 6.0±.2 and E_a = (16.3±1.1) kJ/mole. The frequency factor is considerably below the accepted range [1]. In fact, the result of the restricted fit with log A = 7.5 (broken line) also accommodates the data, yet gives a much higher activation energy of E_a = (24.6±0.1) kJ/mole. Cases where the data had larger scatter than those shown in figure 3 are labelled B in the tables of results (II,III), and for these we guess the true errors in log A as at most ±1.8 and of E_a as at most ±12 kJ/mole.

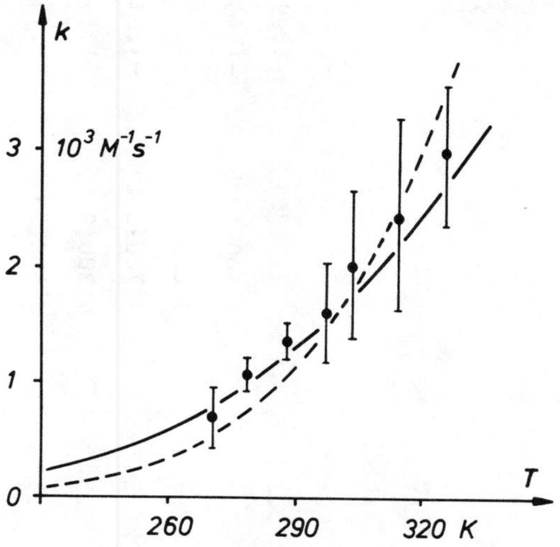

Figure 4. Rate constants for the addition of tert.- butyl to 1,1-dichloro-propene vs. temperature. Full line. Best fit of the Arrhenius equation. Broken line: Restricted fit with log $A/M^{-1}s^{-1}$ = 7.5.

Table II and III summarize the measured rate constants and their Arrhenius parameters for the reaction of tert-butyl with substituted alkenes. It seems very unlikely that the reactions occur otherwise than by addition to the least substituted end of the double bonds since we have presented proof for that fact for 13 of the compounds in chapter 3.1 and since the others are of very similar structure. In tables II and III we display the temperature ranges employed, the ranges of substrate concentrations and the number n of individual decay profiles of the type of figure 1 which were evaluated. A and B classify the experimental scatter of data and indicate the estimated error of the Arrhenius parameters. The errors resulting from the fits were much smaller for each case but seem unrealistic. Experimental errors are given for the rate constants at 300 K only. The last column gives the electron affinities of the compounds. Finally, numbers given in brackets for activation energies resulted from fits restricted with log $A/M^{-1}s^{-1}$ = 7.5.

Experiments on possible solvent effects on the rate constants were carried out for t-1,2-dichloroethene and for 2-methoxypropene. For the former compound 2-propanol, methylcyclopentane (258 K - 331 K) and benzene (296 K - 323 K) were used, while for the latter 2-propanol and n-heptane (271 K - 318 K) were applied. At equal temperatures the rate constants observed with different solvents agreed within the error limits (tables II and III), i.e. within about 20%. However, the data presented in table II for acrylonitrile in two different solvents

TABLE II Rate data for the addition of tert-butyl radicals to alkenes $CH_2 = CR_1R_2$ in 2-propanol

R_1	R_2	T[K]	c[M]	n	class[a)]	$\log A/M^{-1}s^{-1}$	E_a[kJ/mole]	$k_{300}[M^{-1}s^{-1}]$	EA[eV]
H	$C(CH_3)_3$	267-307	.76	15	B	6.1	22.5(34.5)[b)]	150(30)	-1.73 [16]
CH_3	$C(CH_3)_3$	271-318	.4-.8	25	B	7.2	28.7	180(20)	-1.77
CH_3	OCH_3	271-312	.8	24	B	5.6	19.0(29.6)[b)]	220(30)	-2.48
CH_3	$CH(CH_3)_2$	271-329	.4-.8	33	A	7.6	29.2	310(50)	-1.87
H	OCH_2CH_3	253-322	.2-.4	16	A	7.7	29.0	390(50)	-2.24
H	$CH(CH_3)_2$	265-293	.36	12	B	6.8	23.6(27.4)[b)]	530(50)	
CH_3	CH_3	265-315	.5-1.9	18	B	7.5	26.2	740(80)	-2.19 [17]
H	CH_3	252-309	.16-.4	24	A	7.4	25.3	920(70)	-1.99 [17]
H	CH_2CH_3	254-271	.71	6	A	7.5	25.5	1'100(100)	-1.90 [16]
H	H	264-313	.07-.17	30	A	7.9[c)]	27.6	1'250[c)] (50)	-1.78 [17]
CH_3	$OCOCH_3$	271-314	.05-.1	24	A	7.6	25.2	1'700(300)	-1.51
H	$OCOCH_3$	267-323	.12-.23	18	A	7.2	20.5	4'200(200)	-1.19
H	$Si(CH_3)_3$	266-322	.1	17	A	7.0	17.3	9'600(200)	-1.14
CH_3	Cl	264-318	.05	24	A	7.9	22.2	11'400(10^3)	-1.44
H	Cl	265-322	.013	16	A	7.2	16.6	16'500($1.5\cdot10^3$)	-1.28 [18]
H	$SiCl_3$	265-312	.028	11	A	7.2	16.7	17'000($2\cdot10^3$)	- .52 [16]

TABLE II Rate data for the addition of tert-butyl radicals to alkenes $CH_2 = CR_1R_2$ in 2-propanol continuation

R_1	R_2	T[K]	c[M]	n	class[a)]	$\log A/M^{-1}s^{-1}$	E_a[kJ/mole]	$k_{300}[M^{-1}s^{-1}]$	EA[eV]
H	$Si(OEt)_3$	267-305	.021-.12	24	A	7.3	16.0	$27'500(2\cdot10^3)$	-1.11
H	C_6H_5	265-322	.008-.01	11	B	7.6	14.3	$132'000(8\cdot10^3)$	- .25[17]
Cl	Cl	265-315	.01	9	A	8.0	13.8	$350'000(2\cdot10^4)$	- .76[18]
H	C_5NH_4	266-286	.009	7	B	7.4	9.7	$460'000(10^5)$	- .41
H	CN	298	1-2	4[d)]			(6.5)[b)]	$2.4\cdot10^6(2\cdot10^5)$	- .21
H	CN	298	.2	5[e)]			(8.6)[b)]	$1\cdot10^6(4\cdot10^5)$	- .21

a) A : estimated error for log A : ≦ .5, for E_a ≦ 3 kJ/mole.
B : estimated error for log A : ≦ 1.8, for E_a ≦ 12 kJ/mole.

b) From rate constants with the restriction of $\log A/M^{-1}s^{-1} = 7.5$.

c) Statistically corrected for 2 equivalent sites of addition.

d) μSR-measurement, solvent isobutene.

e) modulation ESR-spectroscopy.

TABLE III Rate data for the addition of tert-butyl radicals to 1,2- di- and polysubstituted alkenes.

Alkene	T[K]	c[M]	n	class[a)]	log $A/M^{-1}s^{-1}$	E_a[kJ/mole]	$k_{300}[M^{-1}s^{-1}]$	EA[eV]
c-$CH_3CH=CHCH_3$	255-315	.71	15	B	6.1[c)]	26.3(34.5)[b)]	31[c)](7)	-2.22 [17]
$CH_3CH=C(CH_3)_2$	250-320	.28	14	B	7.1	29.4	80(20)	-2.24 [17]
$CH_3CH=CCl_2$	271-314	.016-.1	22	B	6.0	16.3(24.6)[b)]	1'600(200)	-1.13
$CH_3CH=CHCN$ (c:t =9:1)	271-326	.01-.02	24	A	7.7	18.4	30'000(10^3)	≧-.27
c-CHCl =CHCl	261-296	.26	7	B	7.4[c)]	24.2	1'350[c)](200)	-1.11 [18]
t-CHCl =CHCl	254-331	.052-.058	36	A	7.0[c)]	19.5	3'650[c)](150)	-.80 [18]
CHCl =CCl_2	266-323	.046	17	A	7.3	17.6	17'000($2\cdot10^3$)	-.59 [18]
CCl_2 =CCl_2	245-300	.29	10	A	7.3[c)]	24.1	1'100[c)](100)	-.30 [18]

a)-c) see Table II.

differ by a factor of 2.4 which by large exceeds the experimental error. Finally, we note that attempts to measure the rates of reaction of tert-butyl with butadiene, methacrylonitrile and acrylic acid esters by time-resolved ESR failed because these compounds reacted to fast even at temperatures as low as 260 K. We conclude that for these alkenes k_{300} must be larger than $10^6 M^{-1}s^{-1}$.

4. DISCUSSION

4.1. General remarks

The rate constants compiled in tables II and III span a range of 5 orders of magnitude. With a few exceptions the frequency factors are in the narrow range of $7.0 \leq \log A/M^{-1}s^{-1} \leq 8.0$ so that the large variation of rate constants is mainly due to substituent effects on activation energies. This has been formulated previously by Tedder et al [1]. Moreover, the few exceptions all belong to experiments of class B which have large errors. The average of frequency factors from experiments of class A is $\log A/M^{-1}s^{-1} = 7.5 \pm .3$. Within this class the frequency factors do not show an evident regular variation with substituent size or character and deserve no detailed discussion, therefore. The fairly high, but accurately measured, value for ethene parallels similar findings [1] for the addition of $\cdot CH_3$ and $\cdot CF_3$ radicals to this compound, which have been rationalized in terms of reaction entropies. The activation energies in tables II and III range from about 30 kJ/mole to about 7 kJ/mole.

4.2. Alkenes of type $CH_2 = CR_1R_2$, the "polar" effect.

As mentioned in the introduction the substituents of these alkenes should exert essentially only polar effects on the rates for addition at the unsubstituted carbon. For the highly nucleophilic tert-butyl radical an early transition state is expected, and the polar effects should be describable in terms of frontier molecular orbital theory. If we take only SOMO-LUMO interactions into account the activation energy contains one additive term which characterizes the partial electron transfer from radical to olefin in the transition state [12].

$$\Delta = - \frac{c_\alpha^2}{\varepsilon_L - \varepsilon_S} \beta^2 \qquad (3)$$

c_α^2 is the LUMO density at the attacked carbon, β is an exchange integral, and $\varepsilon_L - \varepsilon_S$ is the SOMO-LUMO energy gap. If the transition state has a similar geometry for all reactions considered, as likely from the similarity of frequency factors, then β should be approximately constant. Further, c_α^2 varies little with substitution and is $c_\alpha^2 \cong .5$ for most olefins except styrene and 2-vinyl pyridine, where $c_\alpha^2 \cong .23$ [16]. The energy gap can be approximated by the difference of the radical's ionization potential and the olefin's electron affinity plus an attractive Coulomb term [16], i.e. $\varepsilon_L - \varepsilon_S = IP - EA - C$. From the data of table II and IP = 6.92 eV one notes that IP-EA varies with substitution in the rather narrow range $7.1 \leqq IP - EA \leqq 9.4$ eV. For a reasonable mean distance between radical and olefin in the transition state of about 2.8 Å the Coulomb term becomes $C \cong 5$ eV as commonly suggested [16], and $\varepsilon_L - \varepsilon_S$ varies from 2.1 to 4.4 eV. Since the variation is smaller than the mean value (3.25 eV) an expansion of eq.(3) about the mean value suggests itself and leads to

$$E_a = \text{const.} - \frac{c_\alpha^2 \beta^2}{(IP - \overline{EA} - C)^2} \, EA \qquad (4)$$

where higher order terms are neglected. Eq.(4) predicts a linear dependence of the activation energy on EA. The corresponding plot is shown in figure 5, and the expected trend is clearly observed. The best fit to a straight line is $E_a = (7.9 \pm 1.2) - (9.5 \pm .8) \cdot EA$ (r = .94). Since more data are now included the parameters differ slightly from those presented earlier [6]. Since the frequency factors are essentially equal a linear dependence on EA should also hold for $\log k_{300}$. This is shown in figure 6 where the straight line is $\log k_{300}/M^{-1}s^{-1} = (6.0 \pm .3) + (1.6 \pm .2) \cdot EA$ (r = .90). In view of the crudeness of the approach the deviations of the data from the straight lines in figures 5 and 6 is not surprising and in part expected since c_α, β and C are not truely constant and since steric effects of the substituents and bond-strength terms are completely ignored. Moreover, it should be noted also that for

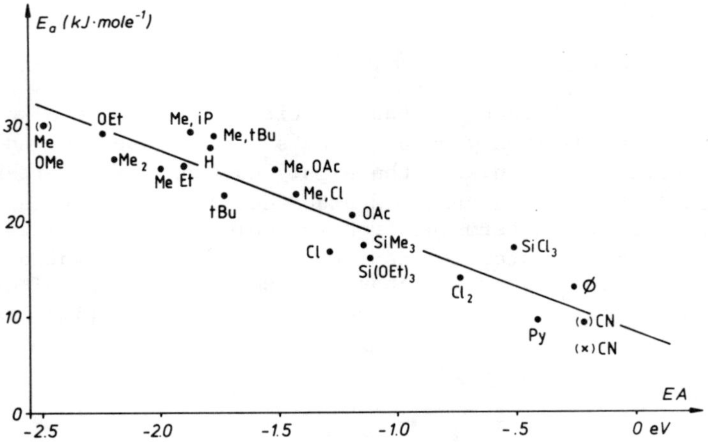

Figure 5. Activation energies for the addition of tert-butyl to alkenes $CH_2 = CR_1R_2$ vs. their electron affinities. Data in brackets are derived from fits with $\log A/M^{-1}s^{-1} = 7.5$. x: Solvent isobutene.

most cases the deviations are smaller than the experimental errors. Furthermore, evaluation of the slope of the line in figure 5 in terms of eq.(4) leads with $c_\alpha^2 = .5$ to $\beta = 1.4$ eV which seems reasonable in view of the spectroscopic value of 3 eV for benzene [12].

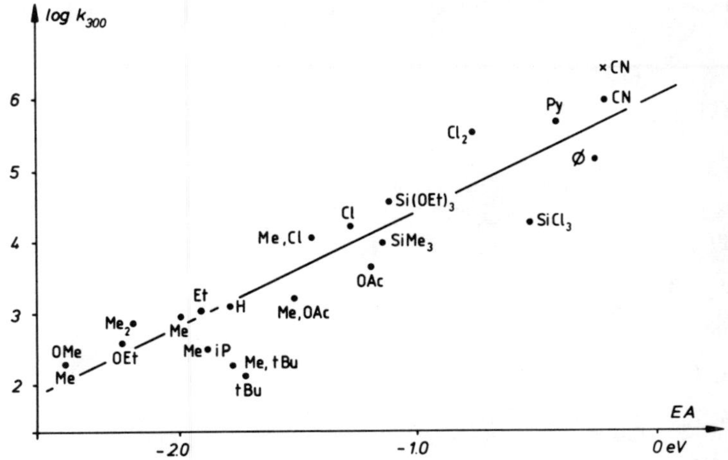

Figure 6. Rate constants ($M^{-1}s^{-1}$) for the addition of tert-butyl to alkenes $CH_2 = CR_1R_2$ vs. their electron affinities. x: Solvent isobutene.

It is expected that correlations of activation energies or rate constants with electron affinities of olefins as found here should also hold for other nucleophilic alkyl radicals. In fact, this has shown earlier [6] for cyclohexyl and methyl additions, though for the less nucleophilic methyl radical (IP = 9.84 eV [9]) the effect is less pronounced. For strongly electrophilic radicals SOMO-HOMO interactions should be dominant, and one expects a linear increase of E_a and a decrease of log k with increasing ionization potential of the alkene. For reactions of the dicyanomethyl radical $\dot{C}H(CN)_2$ Riemenschneider and Boldt have shown this to be the case [17]. For tert-butyl, however, there is no correlation of E_a with the IP of the olefins, as expected.

In the past, polar substituent effect on E_a or log k have often been treated in terms of Hammett type σ constants [2]. For the monosubstituted alkenes of Table II except styrene we also find a good linear correlation, $\log k_{300} = (3.4\pm.1)+(4.2\pm.4)\cdot\sigma$, (r = .98), with the point for styrene falling off by more than 1.5 units. This seems due to the fact that most σ-constants vary as EA, except for styrene. For disubstituted alkenes the polar effects of the substituents are not simply additive. This is shown in table IV where the effects of introducing one methyl group as the second substituent are compared for several cases. Obviously, the changes of the electron affinities are not constant and are reflected in the rate constants.

TABLE IV. Effects of β-Methylation

Olefins	$EA(CH_3)-EA(H)$ [eV]	$k_{300}(H)/k_{300}(CH_3)$
$CH_2 = CHC(CH_3)_3$ / $CH_2 = C(CH_3)C(CH_3)_3$	-.04	.83±.23
$CH_2 = CHCl$/ $CH_2 = C(CH_3)Cl$	-.16	1.45±.30
$CH_2 = CHCH_3$/ $CH_2 = C(CH_3)_2$	-.20	1.24±.30
$CH_2 = CH_2$/ $CH_2 = CHCH_3$	-.21	1.36±.20
$CH_2 = CHOAlk$/ $CH_2 = C(CH_3)OAlk$	-.32	2.41±.10

4.3. Polysubstituted alkenes. The steric effect of α-chloro and methyl groups.

For the alkenes of tables III addition of tert-butyl occurs at a carbon atom which bears methyl- or chloro-substituents. A comparison of the data of tables II and III demonstrates that for similar electron affinities the rate constants for the α-substituted alkenes are by more than one order of magnitude lower while the activation energies are higher than those observed for the mono- and 1,1-disubstituted alkenes. We attribute this to the steric effect of the α-substituents [1,2]. Figures 7 and 8 show that this effect can be separated from the polar effects of α-substitution. For this analysis we use log k_{300} since for many of the α-substituted alkenes the activation energies have rather large errors. In figure 7 log k_{300} for ethene and the chloroethenes are plotted versus their electron affinities. The data for ethene, vinylchloride and vinylidene chloride fall on a straight line with similar parameters as that of figure 6 ($\log k_{300}/M^{-1}s^{-1} = 6.8+2.0 \cdot EA$). The two 1,2-dichloroethenes and trichloroethene follow a linear correlation

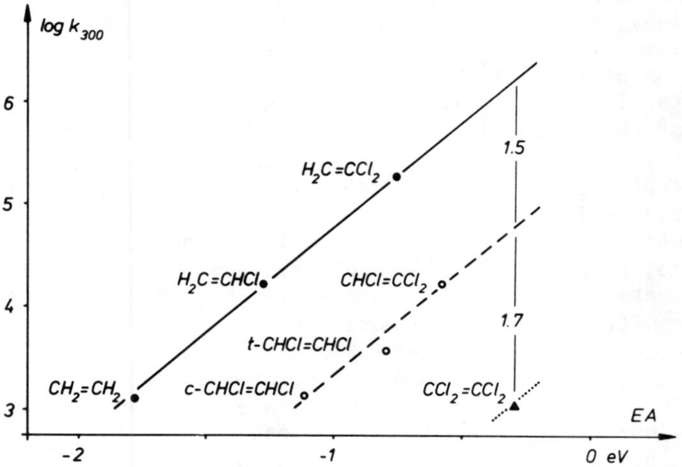

Figure 7. Rate constants ($M^{-1}s^{-1}$) for the addition of tert-butyl to ethene and chloroethenes in 2-propanol versus their electron affinities. Data for ethene, 1,2-dichloroethene and tetrachloroethene statistically corrected.

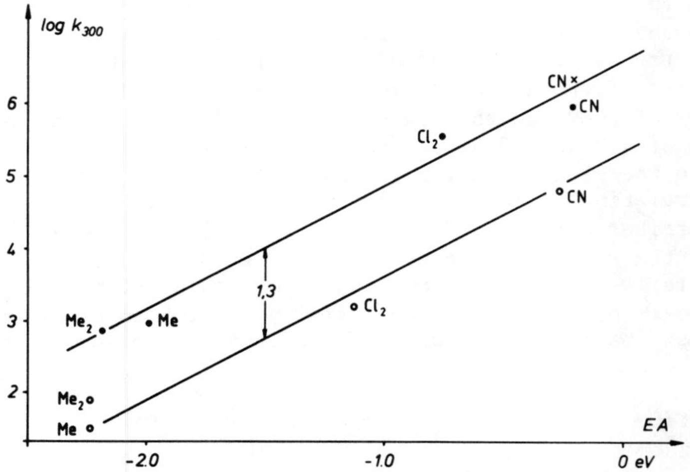

Figure 8. Rate constants ($M^{-1}s^{-1}$) for the addition of 1-mono- and 1,1-disubstituted alkenes (•) and to the corresponding 2-methylsubstituted alkenes (o) versus their electron affinities. x: Solvent isobutene.

with the same slope but shifted by 1.5 units (a factor of 32 in k_{300} to lower values. Further, the point for tetrachloroethene is again lowered by 1.7 units. The interpretation is straightforeward: For equal electron affinities the polar effects are equal and the vertical displacements of -1.5 and -1.7 units measure the steric effects of introduction of the first and the second chlorine atom at the site of addition. This further implies that the steric effect is independent of the β-substituents, and that the α-steric effects are approximately additive for chlorine atoms. Figure 8 shows the same type of plot for the steric effect of α-methylation. The line drawn through the points for isobutene, propene, vinylidene chloride and acrylonitrile again has similar parameters as that of figure 6 ($\log k_{300}/M^{-1}s^{-1} = 6.7+1.7 \cdot EA$), and the line with the same slope drawn through the points for the α-methylated species is displaced by 1.3 units to lower values. Obviously, an α-methyl group causes a steric effect of a factor of 20 on the rate constants. It is smaller than that of a chlorine atom but larger than estimated in our earlier paper. There, less data were considered and tetramethylethylene was erroneously taken into account [6]. Expressed in terms of activation energies the α-steric effects amount to decreases of about 9 kJ/mole for chlorine and about 7.5 kJ/mole for methyl substitution.

Steric effects of α-substituents will of course be different for different attacking radicals. However, it is interesting to note that in their study of the addition of the electrophilic radical $\cdot CH(CN)_2$ to alkyl substituted alkenes Riemenschneider and Boldt [17] presented a plot of ln k vs. IP which appears very similar to our figure 8, and from which an α-steric effect of a methyl group of -1.2 in log k-units can be deduced which is similar to our findings. Further, Giese [2] has shown that the introduction of an α-chlorine atom lowers the addition rate constant of the nucleophilic cyclohexyl radical to methylacrylate by a factor of .067 and that this is composed of polar and steric contributions of 4.3 and .016, respectively. From the average increase of +0.24 eV in the electron affinity on α-chlorination (tables II,III) our linear correlations predict a polar contribution of about 3.0 and a steric contribution of .05 for tert-butyl addition. The same author attributed the rate decrease by a factor of .011 found for α-methylation alone to steric effects. For tert-butyl we would estimate a decrease by about a factor of .5 to electron affinity and .05 to steric effects. In gross terms, the magnitudes of the effects are reasonably comparable.

5. CONCLUSION

The experimental data presented in this paper are in general agreement with the well established rules of radical addition to olefins [1,2] summarized in the introduction. Going beyond these we show that for highly nucleophilic radicals, such as tert-butyl, polar effects of β-substituents on rates manifest themselves in correlations of activation energy and rate constants with the olefin electron affinities, and that the strong steric effects of α-substituents can well be separated from their polar effects. We have indicated that similar data treatments are possible for highly electrophilic radicals where the ionization potential of the olefin should determine the polar effects. Furthermore, we point out that the use of substituent constants for the analysis of polar effects appears to us little attractive since introduction of a given substituent in different olefins does not change the electron affinity always by the same amount. We believe that the treatment of data presented here should be applicable with similar success to other radical additions, as well. As far as literature data on rate constants for other systems were available the limited knowledge of electron affinities prevented this application, so far. Finally, we admit that our treatment completely ignores steric effects of β-substituents or bond strength effects which, though of minor importance, do affect the data [1,2]. We believe that many more large series of rate constants and Arrhenius parameters on radical additions must become available before others than the gross features can be put in quantitative terms.

6. Acknowledgement

I owe all the experimental data to I. Beranek, M. Heming, F. Jent, K. Münger, H. Paul and E. Roduner of this institute. Communication of results prior to publication by Professor P. Boldt, Braunschweig, and financial support of this work by the Swiss National Foundation for Scientific Research is also gratefully acknowledged.

7. REFERENCES

1. J.M. Tedder, and J.C. Walton, Acc. Chem. Res., 9, 183 (1976), Am. Chem. Soc. Symp., 66, 107 (1978), Adv. Phys. Org. Chem., 18, 51 (1978), Tetrahedron, 36, 701 (1980), Angew. Chemie, 94, 433 (1982).
2. B. Giese, Angew. Chemie, 95, 771 (1983).
3. C. Chatgilialoglu, K.U. Ingold and J.C. Scaiano, J. Am. Chem. Soc., 104, 5123 (1982).
4. R. Paltenghi, E.A. Ogryzlo and K.D. Bayes, J. Phys. Chem., 88, 2595 (1984), and references therein.
5. K. Münger, Ph. D. Thesis, University of Zürich (1985), I. Beranek, Diploma Thesis, Unviversity of Zürich (1985), F. Jent, Diploma Thesis, University of Zürich (1985), M. Heming and E. Roduner, unpublished.
6. K. Münger and H. Fischer, Int. J. Chem. Kinetics, 17, 809 (1985).
7. K. Münger and H. Fischer, Int. J. Chem. Kinetics, 16, 1213 (1984) and references given therein.
8. E. Roduner, W. Strub, P. Burkhard, J. Hochmann, P.W. Percival, H. Fischer, M. Ramos and B.C. Webster, Chem. Phys., 67, 275 (1982).
9. F.A. Houle and J.L. Beauchamp, J. Am. Chem. Soc., 101, 4067 (1979).
10. H.-R. Dütsch and H. Fischer, Int. J. Chem. Kinetics, 13, 527 (1981); 14, 195 (1982).

11. W.A. Pryor, F.Y. Tang, R.H. Tang and D.F. Church, J. Am. Chem. Soc., 104, 2885 (1982).
12. K. Fukui, Fortschr.Chem. Forsch., 15, 1 (1970); L. Salem, J. Am. Chem. Soc., 90, 543 (1968); I. Fleming, 'Frontier Orbitals and Organic Chemical Reactions', Wiley, London, 1976.
13. H. Paul, Int. J. Chem. Kinetics, 11, 495 (1979).
14. P. Burkhard, E. Roduner, J. Hochmann and H. Fischer, J. Phys. Chem., 88, 773 (1984).
15. M. Allan, Helv. Chim. Acta, 65, 2008 (1982).
16. J.C. Giordan, J. Am. Chem. Soc., 105, 6544 (1983).
17. K.D. Jordan and P.D. Burrow, Acc. Chem. Res., 11, 341 (1978).
18. P.D. Burrow, A. Modelli, N.S. Chiu and K.D. Jordan, Chem. Phys. Lett., 82, 270 (1981).
19. N.D. Epiotis, 'Theory of Organic Reactions', Springer, Berlin, 1978.
20. K. Riemenschneider, Ph. D. Thesis, Technical University Braunschweig (1977). Private communication by P. Boldt.

ALLYLIC RADICALS - AN ESR SPECTROSCOPICAL STUDY

Reiner Sustmann
Institut für Organische Chemie
Universität Essen
D-4300 Essen 1
West Germany

ABSTRACT. ESR-spectroscopical results on allylic radicals are used to discuss such topics as electron delocalization, allyl resonance energy, syn-anti-isomerization of allylic radicals and substituent effects in free radicals. The effect of thermodynamical stabilization of free radicals on reaction rates and reaction paths is analysed.

1. INTRODUCTION

The study of organic free radicals by electron spin resonance permits direct insight into structure and spin density distribution. Furthermore conformations, conformational changes and isomerizations may be studied in favourable cases. Last but not least kinetic ESR spectroscopy allows the determination of reaction rates. ESR spectroscopy is therefore a powerful tool for the elucidation of the behaviour of free radicals.

The allyl radical has attracted physical organic chemists over a long period of time. It constitutes the simplest conjugated radical and therefore is basic with respect to phenomena like delocalization or resonance. It also can serve as a model to study such areas as
- substituent effects in free radical chemistry -
- isomerization about partial double bonds -
- equilibria between configurational isomers -
- influence of radical stabilization on reaction paths -
- influence of radical stabilization on reaction rates -

These topics will be discussed on the basis of our own work in this area and with the aid of literature data.

H. G. Viehe et al. (eds.), Substituent Effects in Radical Chemistry, 143–166.
© 1986 by D. Reidel Publishing Co.

2. STRUCTURE OF THE ALLYL RADICAL

Since its first detection /1/ the ESR spectrum of the allyl radical has been recorded under various conditions. Fig. 1 reports values for the isotropic coupling constants and g-value of the undeuterated and some deuterated derivatives /2,3/. Anisotropic parameters, which were obtained in a matrix isolation study /4/, are displayed in Fig. 2.

ESR-DATA FOR ALLYL RADICALS IN SOLUTION

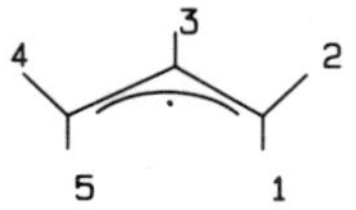	LIT.	T C	G-FACTOR (+/-0.00003)	A1	A2	A3 (MTESLA)	A4	A5
	A	-130	2.00254	1.390	1.481	0.406	1.481	1.390
	B	+230	—	1.400	1.482	0.453	1.482	1.400
	C	-60	2.00257	0.212	1.484	0.405	1.484	1.392
	C	-60	2.00257	1.391	0.225	0.404	1.483	1.391
	C	-60	2.00256	1.392	1.482	0.063	1.483	1.392

A) J.K.KOCHI AND P.J.KRUSIC, J.AM.CHEM.SOC.90.7157(1968)
B) P.J.KRUSIC, P.MEAKIN AND B.E.SMART, J.AM.CHEM.SOC.96.6211(1974)
C) H.-G.KORTH, H.TRILL AND R.SUSTMANN, J.AM.CHEM.SOC.103.4483(1981)

Figure 1. ESR-hyperfine coupling parameters for allylic radicals

Deuteration unambiguously assigns the coupling constants to specific hydrogen atoms proving a greater value for the anti than for the syn protons at the terminal carbon atom. The temperature dependence of +1.3 mG/^{o}C for the coupling constant of the central hydrogen atom confirms the negative spin density at carbon atom 2.

A semiempirical MNDO-calculation /5/ and an advanced ab-initio calculation /6/ give almost identical results for bond lengths and angles. The central CCC-angle is greater than 120° and a CC-bond length of 1.38 **Å** supports our picture of a 1 and 1/2 CC-bond in this radical.

A number of theoretical investigations have treated the spin density distribution in the allyl radical. INDO /5/ calculations and the McConnell treatment /7/ show negative spin-density at C-2 and positive values at C-1. INDO hyperfine coupling constants, obtained with the MNDO optimised structure, confirm the experimental assignment.

ANISOTROPIC COUPLING PARAMETERS FOR ALLYL RADICAL
(MT)

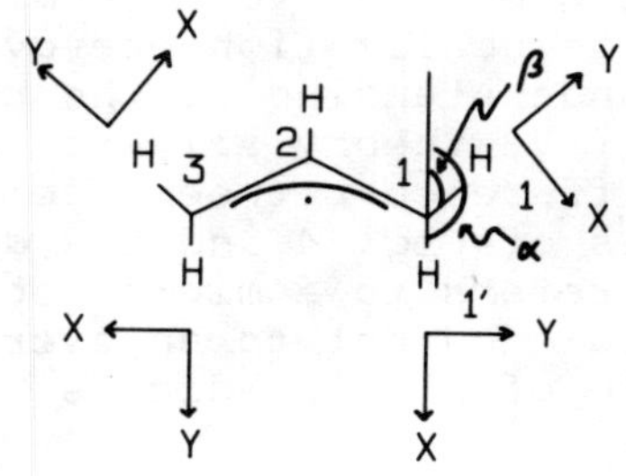

	X	Y	Z	ISOTROPIC
G	2.0026	2.0031	2.0023	2.0027
$A_{H1'}$	2.11	0.68	1.41	1.40
A_{H1}	2.32	0.76	1.46	1.51
A_{H2}	0.61	0.38	0.26	0.42

$\alpha = 184°$, $\beta = 59°$, $\alpha - \beta = \measuredangle$ H1C1H1' $= 125°$

Figure 2. Anisotropic ESR parameter for the allyl radical /4/

3-21G-MCHF-CALCULATION

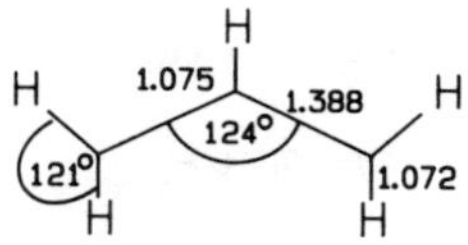

SPINDENSITYDISTR.
EXPTL.

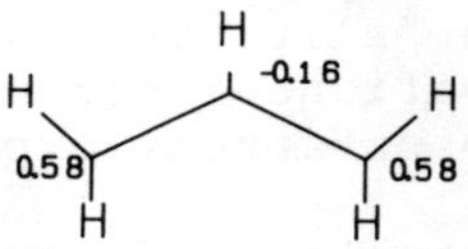

T. TANAKA AND M. DUPUIS, J.AM.CHEM.SOC.105.1713.(1983)
R. W. FESSENDEN AND R. H. SCHULER, J. CHEM. PHYS. 39,2147(1963)

MNDO-CALCULATION

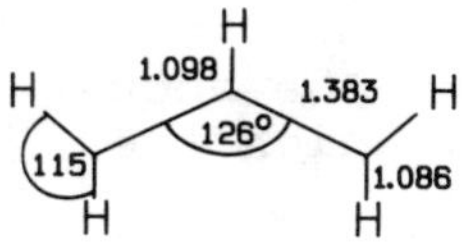

SPINDENSITYDISTR
INDO

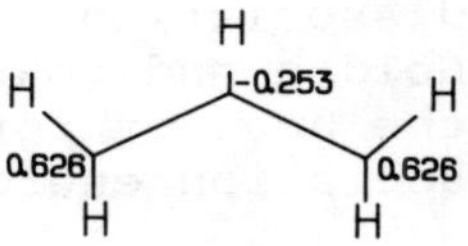

R.SUSTMANN AND W.MUELLER 1983

Figure 3. Spin density distribution in the allyl radical

3. ALLYL DELOCALIZATION ENERGY

Many attempts have been made to determine experimentally values for allyl delocalization energy, a quantity which is basic for an understanding of the phenomenon of conjugation. In terms of VB-theory allyl delocalization energy is defined as difference between the energy of the two localized structures of Fig. 4 and the delocalized radical. This quantity refers to a microscopic property of a single allyl radical and not to an averaged quantity measured for an ensemble of allyl radicals.

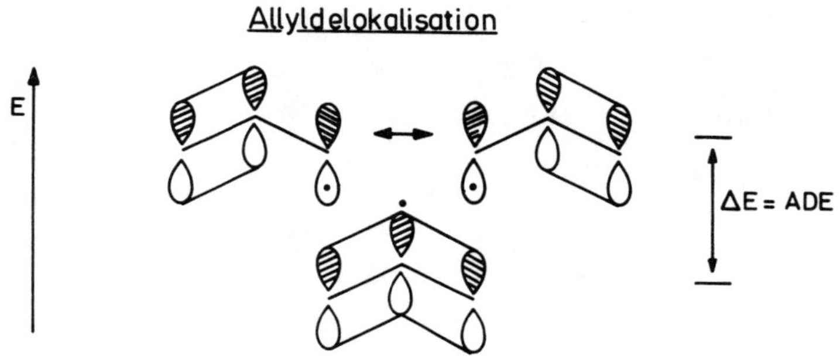

Figure 4. Definition of allyl delocalization energy

Because it is not possible to verify and investigate experimentally the localized structure, all attempts to determine allyl delocalization energy will suffer from approximations inherent in the model chosen.

Two different experimental approaches will be mentioned:

- determination of the stabilization energy as difference between the C-H-bond dissociation energies in propane respectively methane and propene.
- determination of the stabilization energy from activation parameters of isomerization processes in which allylic radicals are involved as intermediates.

In Fig. 5 I have displayed the procedure which relies on bond dissociation energies principally worked out by Benson, Golden and coworkers /8/. In this scheme the accuracy of the heat of formations is crucial, from which the bond dissociation energies are calculated.

ALLYL - DELOCALIZATION ENERGY ACCORDING TO GOLDEN AND BENSON

$$ADE = DH^{O}_{298}\ (n\text{-}C_3H_7) - DH^{O}_{298}\ (\text{Allyl-H})$$

$$DH^{O}_{298}\ (A\text{-}B) = \Delta H^{O}_{f,298}(A) + \Delta H^{O}_{f,298}(B) - \Delta H^{O}_{f,298}(A\text{-}B)$$

$$\Delta H^{O}_{f,298}(\text{Allyl}) = 39.1 +/- 1.5 \text{ kcal/mol}$$

$$DH^{O}_{298}(\text{Allyl-H}) = 86.3 +/- 1.5 \text{ kcal/mol}$$

$$DH^{O}_{298}(n\text{-}C_3H_7\text{-}H) = 98 \text{ kcal/mol}$$

$$ADE = 11.7 +/- 2.0 \text{ KCAL/MOL}$$

M.ROSSI, K.D. KING AND D.M. GOLDEN, J.AM.CHEM.SOC.101,1223,(1979)

Figure 5. Determination of allyl delocalization energy

For instance the experimentally determined $H^{O}_{f,300}$ for allyl radical has been revised several times. The most recent determination from equilibrium measurements between allyl and hexadiene-1,5 by VLP-pyrolysis is 39.1 $\pm$ 1.5 kcal/mol /9/. Inherent in the determination of H_f(Allyl) by this procedure is the assumption that allyl dimerization occurs without activation. Another basic quantity is BDE(C_3H_7-H) of propane. The value of 98 kcal/mol /10/ used by Rossi and Golden /9/ leads to an Allyl Decolalization Energy (ADE) of 11.7 kcal/mol /9/. Improvements in the determination of experimental data, therefore has caused changes in ADE over the years. It can not be excluded that further adjustments will be required in the future. I want to mention only the recent discussion of bond dissociation energies (BDE) by Tsang /11/.

Fig. 6 explains the determination of ARE from isomerization reactions. A differentiation is made between ADE and ARE in order to characterize the experimental procedures for the evaluation of the interaction of an unpaired electron with an adjacent double bond. In an elegant study Roth and his collaborators /12/ obtained a value of 14.0 $\pm$ 2 kcal/mol from the activation energies of isomerization of optically active 1,3-disubstituted allenes. Doering /13/ approached the problem from an investigation of the isomerization of hexatrienes. Depending on the reference system, ethylene or butene-2, he proposes values of 13.1 $\pm$ 2.0 resp. 12.2 $\pm$ 2.0 kcal/mol for ARE. In both procedures ARE is calculated in a thermochemical cycle in order to take account of model inherent approximations.

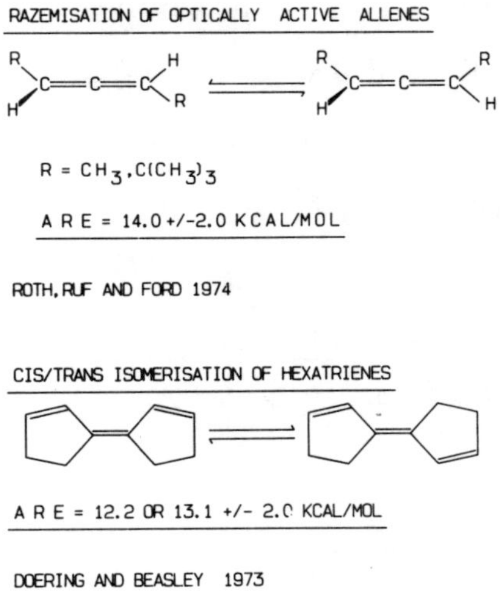

Figure 6. Determination of Allyl Resonance Energy (ARE) from isomerization processes.

The use of rotational barriers for the determination of the allyl resonance energy in these experiments suggest an even better path to this quantity (Fig.7).

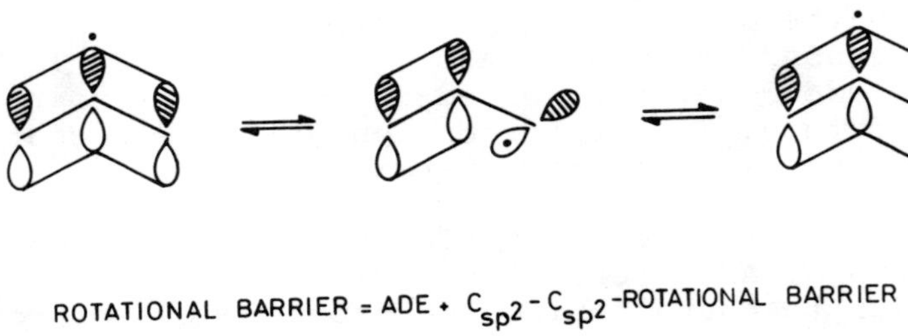

Figure 7. Rotational barrier in the allyl radical as measure of Allyl Resonance Energy.

The transition state of rotation about a CC-bond in allyl corresponds to a decoupling of double bond and unpaired electron. It should therefore be a direct measure of the resonance energy, except that a correction for a C_{sp2}-C_{sp2}-single bond rotational barrier might modify the quantity slightly.

Krusic /3/ made an attempt to determine the rotational barrier in the allyl radical by ESR line shape analysis. As it proved impossible to observe line shape effects in the allyl radical ESR spectrum up to presumably the highest experimentally accessible temperature we analyzed kinetically the syn/anti-isomerization of 1- ^{2}H -allyl radical /2/. The barrier to rotation is 15.7 ± 1.0 kcal/mol. This value, if corrected by ca. 1.0 kcal/mol for a CC-single bond rotation leads to about 14.5 kcal/mol for ARE. This is at the upper limit of the mentioned values for ADE and ARE.

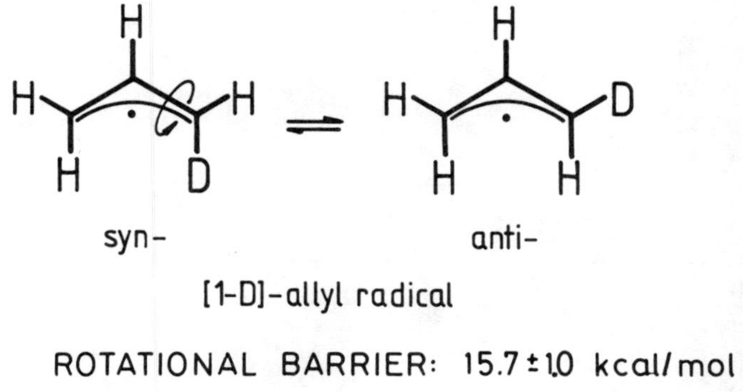

Figure 8. Syn/anti-isomerization of 1-[^{2}H]-allyl radical

However, it is not surprising that the values, obtained in different experimental setups, are not identical; the methods of determination vary. Principally, we can measure only differences between states and as these are not identical in the above experiments we should not expect complete coincidence.

It seems to us that the direct determination of ARE from the rotational barrier in the allyl radical is the method which depends the least on additional approximations. Therefore it should be a good measure of Allyl-Resonance-Energy.

The only correction necessary is the barrier to rotation about a CC-single bond between two sp^2-hybridized carbon atoms. However, from an operational point of view one might even arrive at the conclusion that the value of the barrier itself constitutes the best measure of Allyl-Resonance-Energy.

4. SUBSTITUENT EFFECTS IN FREE RADICAL CHEMISTRY

4.1. Spin density distribution and structure

Substitution in the allyl radical changes the isotropic hyperfine coupling parameters of the remaining hydrogen atoms. Two aspects will be discussed in this section: change in hyperfine coupling of the hydrogen atoms at C-3 and differences in hyperfine couplings between hydrogen atoms at C-1 and C-3 if one substituent X is introduced. In Fig. 9 hyperfine coupling constants are compared /14-17/. The difference to the unsubstituted allyl radical is very small if methyl replaces a hydrogen atom.

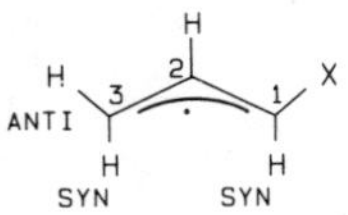

	AH1(SYN)	AH3(SYN)	AH3(ANTI)	ρ_1	ρ_2	ρ_3	$\sum_{i=1}^{3} \rho_i$
X = H	1.390	1.390	1.481	0.627	-0.253	0.627	1.000
X = CH3	1.383	1.383	1.478	—	—	—	—
X = CH=CH2	1.16	0.96	1.04	—	—	—	—
X = OCH3	1.307	1.311	1.417	0.600	-0.240	0.520	0.880
X = NH2	1.16	1.21	1.25	—	—	—	—
X = SC(CH3)3	1.150	1.193	1.231	—	—	—	—
X = CN	1.418	1.305	1.220	0.589	-0.242	0.516	0.863

Figure 9. Hyperfine coupling constants and spin densities in monosubstituted allylic radicals.

Alkyl substitution does not influence the distribution of the unpaired spin. The other substituents may be classified in three groups: 1. Vinyl as a conjugating group consisting only of carbon atoms, 2. amino-, alkoxy- and thioalkyl-substituent as heteroatom substituents with a lone pair, 3. cyano- and methoxycarbonyl- as π-conjugating groups with electron withdrawing character. Vinyl as a substituent leads to a remarkable decrease in hyperfine coupling constants at C-3. It is the highest withdrawal of unpaired electron density induced by one of these substituents. Even though methoxy, amino and thioalkoxy groups contribute two extra electrons to the conjugated π-system, there is a noteworthy difference between methoxy and amino or thioalkoxy on the other side. The latter two substituents lead to a decrease in coupling at C-3 which is similar to the effect of the electron withdrawing cyano group. In contrast to cyano, which has a high coupling constant at C-1, both groups decrease also this hyperfine coupling constant markedly. This indicates that both amino and thioalkoxy are able to take over considerable unpaired electron density. INDO-calculated spin densities are also recorded in Fig. 9.

Fig. 10 lists some hyperfine coupling constants and INDO calculated spin densities for 1,1-disubstituted allylic radicals.

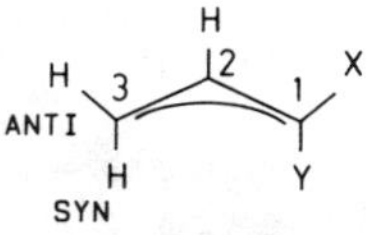	AH(SYN) (mT)	AH(ANTI)	ρ_1	ρ_2	ρ_3	$\sum_{i=1}^{3} \rho_i$
X=Y=CN	1.140	1.216	0.541	-0.223	0.443	0.761
X=Y=OCH3	1.435	1.491	0.554	-0.221	0.492	0.825
X=CN, Y=OCH3	1.006	1.093	0.568	-0.217	0.419	0.770

Figure 10. Hyperfine coupling parameters and spin densities for 1,1-disubstituted allylic radicals.

A second cyano group leads to a further decrease of the averaged couplings at C-3 whereas a second methoxy group causes an increase. The combined action of methoxy and cyano is similar to the effect of one vinyl group on the distribution of the unpaired spin.

In Fig. 11 some values of CC-bond lenghts obtained by MNDO-geometry optimisations are compared/5/. Noteworthy seems to be that the higher the withdrawal of unpaired spin density from the α-position the more unsymmetrical the carbon frame work becomes. In other words, the allylic character of the radical changes to that of a vinyl substituted methyl radical.

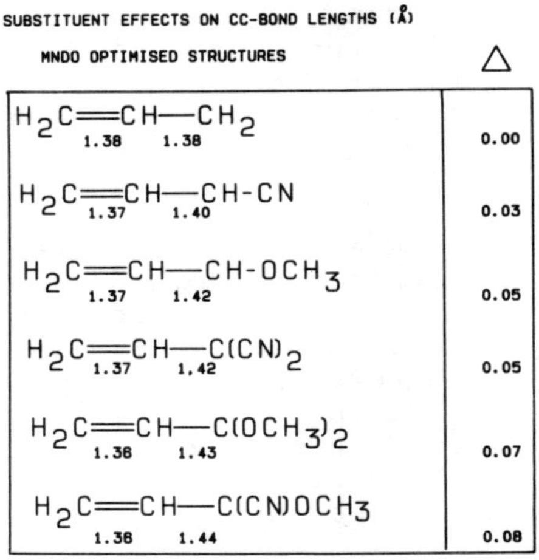

Figure 11. CC-bond lengths in allylic radicals from MNDO optimisations.

4.2. Delocalization and stability of allylic radicals

Delocalization of an odd electron provides stabilization. This is the concept of resonance. From the rotational barrier of 15.7 kcal/mol in the unsubstituted allyl radical we determined a resonance energy of 14.5 kcal/mol. How does replacement of a hydrogen atom by other substituents influence delocalization and at the same time the stability of radicals?

The variation in the magnitude of hyperfine coupling constants can be taken as an indication of delocalization. Even if this in itself does not provide a measure of stabilization in energetic terms, it can serve to establish a relative scale for the influence of substituents. The delocalizing power of a substituent was cast in a quantitative scheme by Fischer/18/ in 1964. He proposed a formula by which the spin density at a sp^2-hybridised carbon atom could be calculated from increments for each substituent at that atom. The variation in α-hyperfine couplings as a function of substituents in the aromatic part of benzylic radicals was used by Arnold /19/ to develop a $\delta_{\alpha}^{\cdot}$-scale. The implication is that the coupling constants reflect a property which correlates with changes in linear free energy. In the context of the action of several substituents on a radical center, Stella /20/ developed a formula (Fig. 11) which relies on the earlier developments.

Substituent parameters

based on spin density distribution

$$S_x^{exp} = 1 - \frac{a_x}{a_o}$$

$$S_{xy}^{exp} = 1 - \frac{a_{xy}}{a_o}$$

$$S_{xy}^{calc} = 1 - \left[(1 - S_x^{exp})(1 - S_y^{exp})\right]$$

$$\Delta S_{xy} = \frac{S_{xy}^{exp} - S_{xy}^{calc}}{S_{xy}^{calc}} \cdot 100$$

$\Delta S_{xy} > 0$ synergetic action of substituents

$\Delta S_{xy} < 0$ antagonistic action of substituents

L. Stella a. L. Sylvander

Figure 12. Definition of delocalization parameters for substituents according to Stella.

A substituent parameter S_x^{exp} is defined as difference between 1.0 and the ratio of the coupling constant in the substituted to the unsubstituted reference system. Following the formalism of Fischer /18/ the calculated delocalization power of several substituents S_{xy}^{calc} is obtained as product of the individual values. From this and the experimental S_{xy}^{exp} a ΔS_{xy} is calculated. This quantity allows to recognize, whether the action of two substituents on the spin density distribution is additive, less than additive or surpasses additivity. In Fig. 13 we show such an analysis for allylic radicals, generated and measured in our laboratory.

What is the significance of these values with respect to the problem of thermodynamic stabilization of free radicals by substituents? From the qualitative statement that delocalization of unpaired electron density causes stabilization, one has progressed to a quantitative model. This allows a classification of different substituents, certainly an important advancement. It does not, however, provide values for stabilization energies.

Stabilization of Allylic Radicals by Substituents

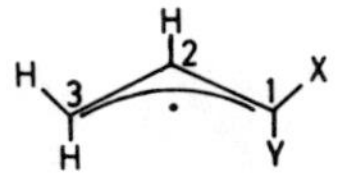

X	Y	$(a_{H_3}+a_{H_{3'}})/2$ [G]	ΔS_{xy}(%)
H	H	14.37	
CO_2CH_3	H	13.22	
H	CO_2CH_3	13.64	
CN	H	12.79	
H	CN	12.63	
CH_3O	H	13.41	
H	CH_3O	13.09	
$CH{=}CH_2$	H	10.01	
H	$CH{=}CH_2$	10.32	
$CO_2C_2H_5$	$CO_2C_2H_5$	13.10	-30
CN	CN	11.73	-16
CH_3O	CH_3O	14.63	-112
$CH{=}CH_2$	$CH{=}CH_2$	7.2	0
CH_3O	CN	10.53	+48
CN	CH_3O	10.53	+41

Figure 13. Delocalization parameters ΔS_{xy} for allylic radicals

5. EQUILIBRIA BETWEEN CONFIGURATIONAL ISOMERS

Monosubstituted allylic radicals can exist in a syn- or anti-conformation (Fig. 14). Fig. 14 displays for a few cases the position of the equilibrium which follows from ESR-spectroscopical studies. A bulky substituent, like a t-butyl group, prefers the anti-position. There is no syn-tert-butyl substituted allyl radical recognizable in equilibrium at room temperature. A model shows that there is strong 1,3-interaction between the substituent and one allylic hydrogen atom, a situation not present for the anti configuration. A different explanation has to be invoked for cyano- and methoxy substituted allyl radical. Cyano prefers by ΔG = 0.14 kcal/mol a syn arrangement, the same is true for methoxy (ΔG = 0.48 kcal/mol). In the 1-cyano-1-methoxy allyl radical the methoxy group occupies the syn position in the preferred isomer. The polar character of these substituents, leading to dipolar interactions in the molecule, might be made responsible for these observations.

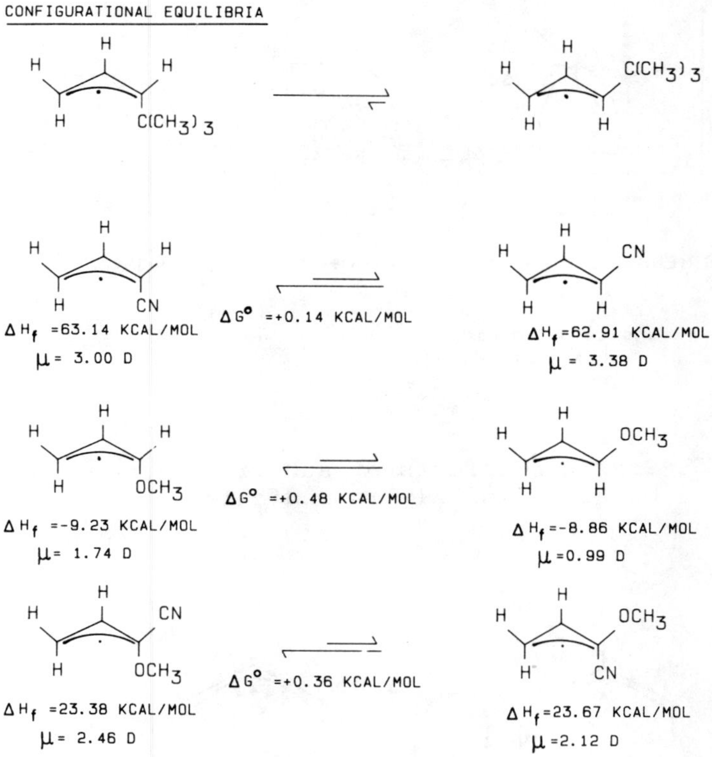

Figure 14. Configurational equilibria of allylic radicals

6. ISOMERIZATION ABOUT PARTIAL DOUBLE BONDS

In order to assess the effect of substituents on the barrier height for isomerization of allylic radicals one has to remember that a substituent will influence both the ground state and the transition state of isomerization. Fig. 15 explains the problems involved in such an undertakement. Both states might be stabilized or destabilized to equal or unequal extent, or one will be stabilized and the other destabilized. A priori, we cannot predict the situation. More important, however, such experiments will not provide absolute values for substituent effects!
In Fig. 16 I want to demonstrate a case where destabilization of the ground state of an allylic radical dominates the rotational barrier.

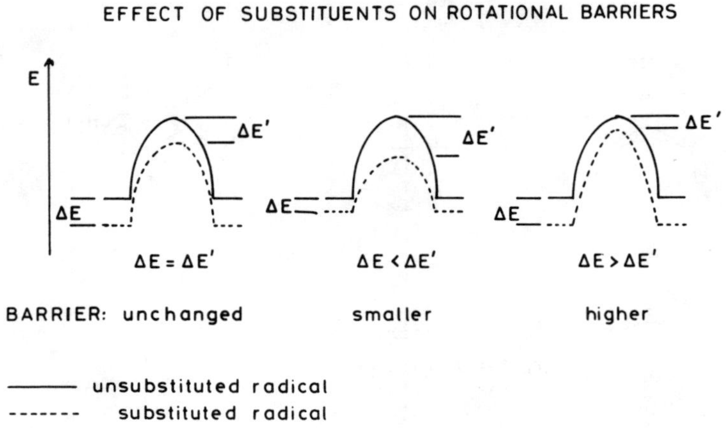

Figure 15. Possible effects of substituents on rotational barriers in allylic radicals.

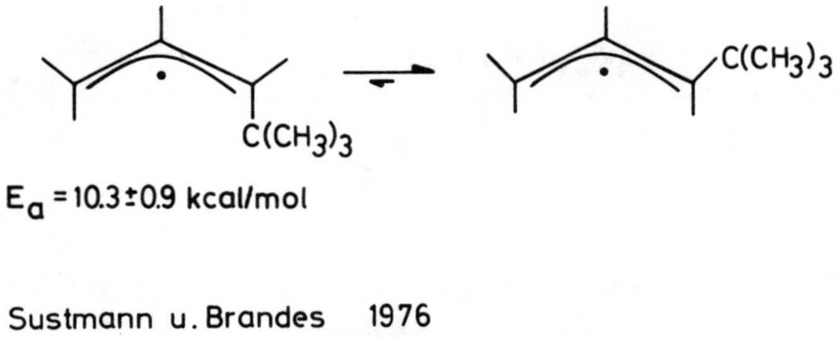

Figure 16. Rotational barrier in syn 1-t-butyl-allyl radical

The barrier to rotation syn anti in syn-1-t-butyl allyl radical, as determined ESR-spectroscopically some years ago /21/, is 10.7 kcal/mol. It is by 5 kcal/mol lower than the barrier in the unsubstituted allyl radical. Even though there will be an electronic substituent effect, nobody will question the predominance of a steric effect. Thus, the low barrier is mainly due to a raise of the ground state energy of syn-tert-butyl allyl radical as compared to the transition state for rotation.

A different situation will be encountered when electronic effects will govern the picture. Fig. 17 shows schematically the expectation.

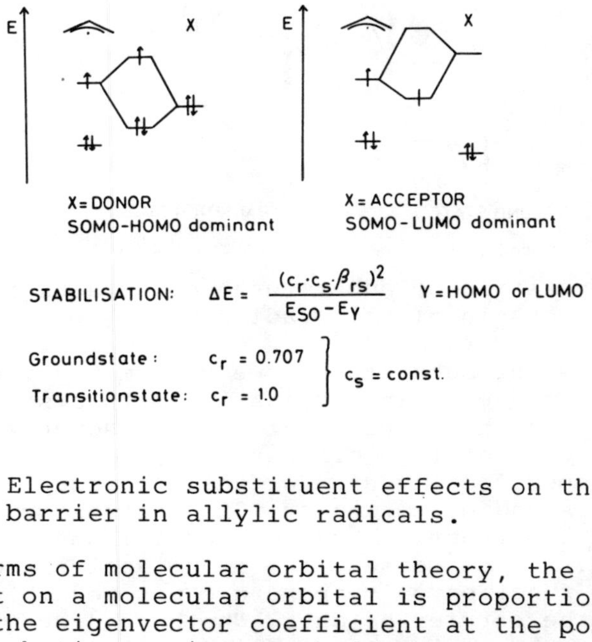

Figure 17. Electronic substituent effects on the rotational barrier in allylic radicals.

In terms of molecular orbital theory, the action of a substituent on a molecular orbital is proportional to the square of the eigenvector coefficient at the position where the substituent is attached. For our problem, the singly occupied MO in ground and transition state will have the deciding effect for the rotational barrier.

In the transition state the square of the eigenvector coefficient at the radical center equals 1.0, the highest value it can assume. In the ground state it is 0.5. If a substituent effect exists it should always be higher in the transition state. The situation as far as rotational barriers in allylic radicals are concerned is illustrated in Fig. 18. In total a lowering of the rotational barrier is expected.

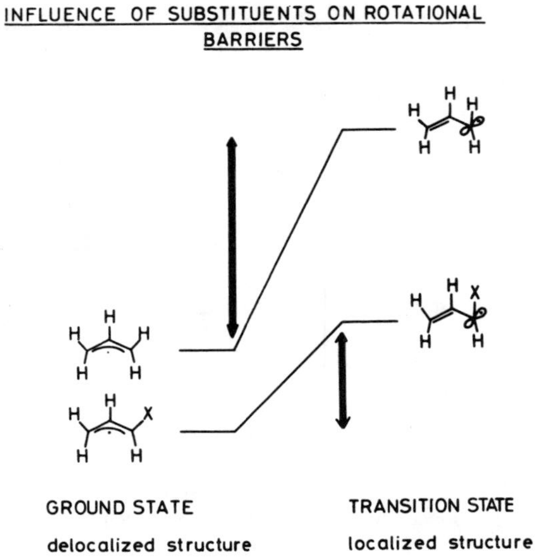

Figure 18. Expectation of substituent effects on rotational barriers in allylic radicals.

For a number of substituents we determined their influence, either alone or in combination with a second like or unlike group on the barrier height for isomerization /15,22,23/. Experimentally this was achieved in most cases by generating specifically syn- or anti-configurations of allylic radicals and measuring by ESR spectroscopy the appearance and the fraction of the second isomer as function of temperature. In some cases line shape analysis of the ESR spectra as function of temperature was possible /22/. Fig. 19 lists the barrier heights. Whether a substituent is considered an electron acceptor or an electron donor it lowers the barrier to rotation of the deuterated allylic radical. Cyano does this to a higher extent than methoxy, this parallels results of other studies on bond dissociation energies /24/. Two more points should be mentioned. Two like substituents do not lower the barrier additively, and two unlike groups do show in this experiment at least an additive interaction with the unpaired electron. This latter point is explained in more detail in Fig. 20.

RADICAL	E_a(kcal/mol)
syn-CH_2=CH-ĊHD ⇌ anti-CH_2=CH-ĊHD	15.7 ±1.0
anti-CH_2=CH-ĊHCN → syn-CH_2=CH-ĊHCN	9.8 ±1.1
syn-CH_2=CH-ĊHCN → anti-CH_2=CH-ĊHCN	10.6 ±1.3
anti-CH_2=CH-ĊH(OCH_3) → syn-CH_2=CH-ĊH(OCH_3)	14.3 ±1.4
syn-CH_2=CH-ĊH(OCH_3) → anti-CH_2=CH-ĊH(OCH_3)	14.7 ±1.5
CH_2=C(C(CH_3)$_3$)-Ċ(CN)$_2$	9.6 ±0.3
CH_2=C(CH_3)-Ċ(CN)$_2$	> 12
CH_2=CH-Ċ(OCH_3)$_2$	> 12
CH_2=CH-Ċ(OCH_2)$_2$	> 13.5

Figure 19. Substituent effects on rotational barriers in allylic radicals.

		E_A (kcal/mol)	ΔE_A
(1)	H_2C=CH–ĊHD	15.7	—
(2)	H_2C=CH–ĊHCN	10.2	5.5
(3)	H_2C= CH–ĊHOCH$_3$	14.5	1.2
(4)	H_2C= CH–Ċ(CN)OCH$_3$	6.0	9.7

In case of additivity of substituent effects the barrier should be 9.0 kcal/mol

A barrier of 6.0 kcal/mol shows a synergetic effect of ca 3 kcal/mol

Figure 20. Analysis of cd-substitution on the rotational barrier of 1-cyano-1-methoxy allyl radical.

A calculation shows that the additive expectation value for the rotational barrier in 1-cyano-1-methoxy allyl radical should be 9.6 kcal/mol. The measured value exceeds the "additive" quantity by 2.9 kcal/mol. A thorough error propagation analysis, however, proves that there is an uncertainty of $\pm$ 2.0 kcal/mol. Therefore, two unlike substituents do act at least additively, probably more than that on the rotational barrier in an allyl radical. As the change in barrier height is an expression of differences in influence on ground state and transition state, respectively, one may reason that the "true" or "absolute" stabilizing power of individual substituents should be bigger. If therefore values for stabilization, deduced here, compare well with results from other measurements this has to be considered rather fortuitous. A comparison should be possible only as far as relative sequences are concerned.

Can a connection be made between the ΔS_{xy} values, obtained from spin density distributions and the energy values for the effect of a substituent on the rotational barrier? The ΔS_{xy} values were deduced from experimental coupling constants. The underlying idea was that delocalization expresses itself in a decrease of the observed hyperfine splitting constants.

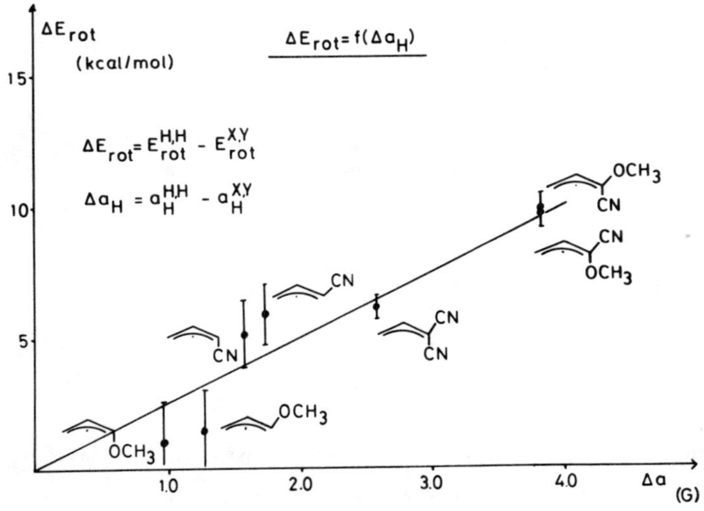

Figure 21. Relationship between rotational barriers and hyperfine coupling constants in allylic radicals.

If one plots the difference in rotational barrier between the almost unperturbed 1-d-allyl radical and the substituted ones against Δa_H, the difference between the averaged coupling constants in 3-position of the allylic radicals, one immediately realizes that there must exist a relationship between spin delocalization and stabilization. That this is only roughly linear, should be no surprise. It is even questionable whether it should be a linear relationship.

A different graphical representation, which resembles a proposal by Walton /25/, plots a_H (methyl) and the averaged hyperfine coupling constant at C-3 in substituted allylic radicals against the total stabilization of a methyl radical by all substituents. The values are calculated in such a way that the lowering of the rotational barrier of substituted allylic radicals is added as stabilisation to the stabilization of the unsubstituted allyl radical.

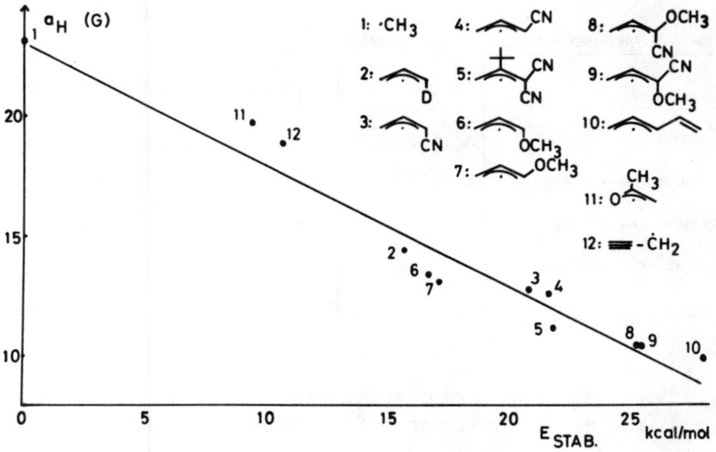

Figure 22. Relationship between coupling constants in allyl radicals and stabilization energies.

The highest stabilization is found for pentadienyl radical. A bis-vinyl substituted methyl radical should be stabilized by at least 28 kcal/mol.

What is the maximum of thermodynamical stabilization which can be given to a methyl radical? Obviously, it has to be a trisubstituted radical. What kind of substituent does contribute the greatest amount?
In the cases considered here, this is the vinyl group. Thus trivinyl-methyl radical should be the "winner" in our collection. A second vinyl group however has a smaller effect than the first. For the third group we anticipate even less, i.e. we might expect a total stabilization of 30-35 kcal/mol. This could be exceeded, if dienyl or polyenyl groups would be attached to methyl. The difficulty to generate such radicals will be rather prohibitive for experimental realisation.

7. STABILIZATION AND REACTION RATES

Does additional thermodynamical stability confer kinetic stability to substituted methyl radicals? It seems that Ingold and Griller /26/ addressed this problem for the first time in their definition of persistency. The termination rates for allyl radical and different substituted allylic radicals were determined by kinetic ESR-spectroscopy /2,23,27,28/ These radicals lead in high yields to dimers, the expected and obeyed rate law is of pure second order. Fig. 23 lists some rate data, together with activation parameters for recombination.

RATE CONSTANTS FOR DIMERISATION OF ALLYLIC RADICALS

	T K	$2K_T$ $10^9\ M^{-1}s^{-1}$	E_A (KCAL/MOL)
$H_2C{=}CH{-}CH_2$	275.5	0.96 ± 0.01	2.76 ± 0.06
SYN-$H_2C{=}CH{-}CH{-}CN$	277	4.01 ± 0.15	2.59 ± 0.15
ANTI-$H_2C{=}CH{-}CH{-}CN$	278	4.59 ± 0.12	3.07 ± 0.11
$H_2C{=}CH{-}C(CN)OCH_3$	233	0.8 ± 0.05	---

Figure 23. Termination rate constants and activation energies for recombination of allylic radicals.

Typical are rate constants of a magnitude which is characteristic for diffusion-controlled reactions. This is supported by calculated rate constants from the von Smoluchowski equation. We, therefore, must conclude that the amount of thermodynamical stability conferred to methyl by substitution with vinyl and substituted vinyl groups does not provide life times which are longer than those of methyl radicals under comparable conditions. Does this observation fit our expectation?

In the process of dimerization a CC-bond is formed. The bond energy involved is roughly 85-90 kcal/mol for an unperturbed CC-single bond. Substitution reduces this, the stabilization energy of the underlying radical giving a good estimate of the amount of lowering. In the most stabilized cases we expect a CC-bond energy which is still of the order of 45-50 kcal/mol, i.e. the reactions have to be considered as highly exothermic. As a consequence radical dimerization possesses an early transition state, i.e. a low barrier of activation and a high reaction rate. It remains to be seen whether a further decrease in free enthalpy of reaction will cause an activation energy for dimerization.

In an empirical attempt /29/ to find a correlation between reaction rates and free enthalpies of reaction we collected a number of such data from the literature. In Fig. 24 a plot of log k versus ΔG_r^o

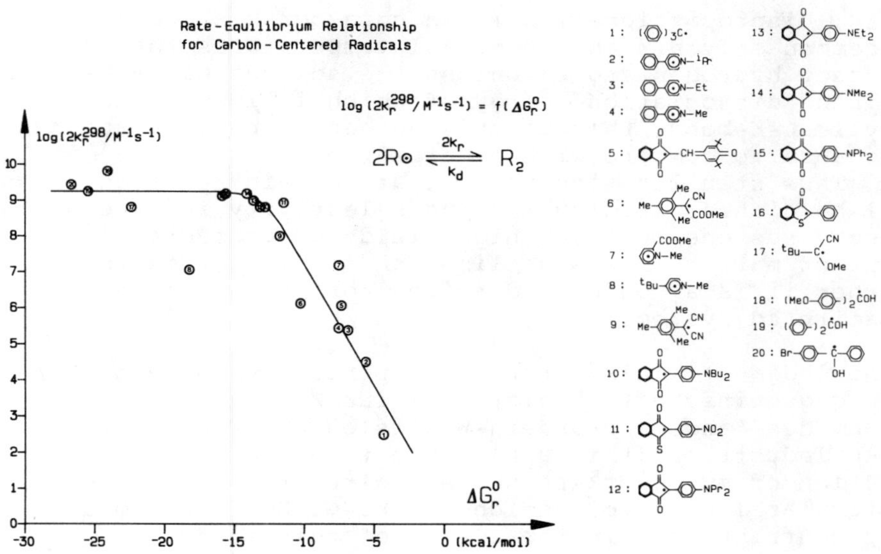

Figure 24. log k vs. ΔG_r^O for dimerization of carbon centered radicals

for the dimerization of carbon centered radicals is shown. Obviously there exists a relationship and interestingly a decrease of reaction rate is observed at low negative free enthalpies of reaction. It seems to support the idea that the reaction rate for radical dimerization depends on the exothermicity of the reaction if ΔG_r^O is smaller than a certain threshold. This kind of graphical representation is reminiscent of those for electron transfer reactions. Marcus theory /30/ and its variation by Rehm and Weller /31/ predicts such a behaviour for this type of reaction.

8. STABILITY AND REACTION PATHS

Free radicals can undergo a variety of reactions. Termination by bimolecular self reaction and disproportionation are reactions with low or no activation. Abstraction reactions also belong to the ubiquitous events. Rearrangements do occur, but often can not compete with other processes. How does stabilization of a free radical influence the possibilities for reactions? Allyl radicals are a good case for discussion. If they are not involved in chain reactions they undergo bimolecular selfreaction, giving dimeric products. Disproportionation is an unfavourable process./32/ In common solvents and at normal temperatures they do not abstract hydrogen atoms. Unless the solvent has a C-H-bond, the bond dissociation energy of which is lower than an allylic C-H-bond, this is an endothermic process and has no chance to compete with dimerization.

Thus stabilization means that certain reaction channels will be closed, leading to high selectivity in others. In those cases one observes high yields of products. It does not matter, how stabilization is brought to the radical, whether it is a conjugated radical or perhaps a capto-dative substituted system.

Acknowledgment: This work was supported by Deutsche Forschungsgemeinschaft, Ministerium für Wissenschaft und Forschung des Landes Nordrhein-Westfalen and Fonds der Chemischen Industrie. It is a pleasure to acknowledge the collaboration of my coworkers who are cited the references. In particular I want to mention Dr. H.-G. Korth who made important contributions in the course of this study.

9. REFERENCES

1) B. Smaller and M.S. Matheson, J. Chem. Phys. **28**, 1169 (1958)

2) H.-G. Korth, H. Trill and R. Sustmann, J. Am. Chem. Soc. **103**, 4483 (1981)

3) P.J. Krusic, P. Meakin and B.E. Smart, J. Am. Chem. Soc. **96**, 6211 (1974)

4) G. Maier, H.P. Reisenauer, B. Rohde and K. Dehnicke, Chem. Ber. **116**, 732 (1983)

5) R. Sustmann and W. Müller, unpublished

6) T. Tanaka and M. Dupuis, J. Am. Chem. Soc. **105**, 1713 (1983)

7) R.W. Fessenden and R.H. Schuler, J. Chem. Phys. **39**, 2147 (1963)

8) D.M. Golden and S.W. Benson, Chem. Rev. **69**, 125 (1969)

9) M. Rossi, K.D. King and D.M. Golden, J. Am. Chem. Soc. **101**, 1223 (1979)

10) K.W. Egger and A.T. Cocks, Helv. Chim. acta **56**, 1516 (1973); J.A. Kerr, Chem. Rev. **68**, 465 (1968)

11) W. Tsang, J. Am. Chem. Soc. **107**, 2872 (1985)

12) W.R. Roth, G. Ruf and P.W. Ford, Chem. Ber. **107**, 48 (1974)

13) W. v. E. Doering and G.H. Beasley, Tetrahedron **29**, 2231 (1973)

14) R. Sustmann, H. Trill and D. Brandes, Chem. Ber. **110**, 245 (1977)

15) R. Sustmann, H. Trill, F. Vahrenholt and D. Brandes Chem. Ber. **110**, 255 (1977)

16) D. Griller, D.C. Nonhebel and J.C. Walton, J. Chem. Soc. Perkin Trans II, **1984**, 1817

17) D. Griller, D.C. Nonhebel and J.C. Walton, J. Chem. Soc. Perkin Trans II, **1983**, 1373

18) H. Fischer, Z. Naturforsch. **19a**, 866 (1964)
H. Fischer, Z. Naturforsch. **20a**, 428 (1965)

19) J.M. Dust and D.R. Arnold, J. Am. Chem. Soc. **105**, 1221 (1983)

20) L. Sylvander, L. Stella, H.-G. Korth and R. Sustmann, Tetrahedron Lett. **26**, 749 (1985)

21) R. Sustmann and D. Brandes, Chem. Ber. **109**, 354 (1976)

22) R. Sustmann and H. Trill, J. Am. Chem. Soc. **96**, 4343 (1974)

23) H.-G. Korth, P. Lommes and R. Sustmann, J. Am. Chem. Soc. **106**, 663 (1984)

24) M. Zamkanei, J.H. Kaiser, H. Birkofer, H.D. Beckhaus and Ch. Rüchardt, Chem. Ber. **116**, 3216 (1983)

25) J.C. Walton, J. Chem. Soc. Perkin Trans. II, **1983**, 1043

26) D. Griller and K.U. Ingold, Acc. Chem. Res. **9**, 13 (1976)

27) H.J. Hefter, C.H.S. Wu and G.S. Hammond, J. Am. Chem. Soc. **95**, 851 (1973)

28) H.-G. Korth, R. Sustmann, P. Lommes and W. Sicking, Int. J. Chem. Kinet. **15**, 267 (1983)

29) R. Sustmann and H.-G. Korth, unpublished

30) L. Eberson, Adv. Phys. Org. Chem. **18**, 79 (1982)

31) D. Rehm and A. Weller, Ber. Bunsenges. Phys. Chem. **73**, 834 (1969)

32) R. Klein and R.D. Kelley, J. Phys. Chem. **79**, 1780 (1975)

ESR-SPECTROSCOPICAL INVESTIGATION OF CAPTO-DATIVE SUBSTITUENT EFFECTS IN BENZYL RADICALS

H.-G. Korth, P. Lommes, and R. Sustmann
Institut für Organische Chemie
Universität - GHS Essen
D-4300 Essen, West Germany

L. Sylvander and L. Stella
Laboratoire de Chimie Organique B
Université d'Aix-Marseille III
F-13397 Marseille, France

ABSTRACT. ESR spectra of three series of mono- and disubstituted benzyl radicals have been investigated and their hyperfine splitting constants have been analyzed in terms of substituent-induced spin delocalization. On the basis of the dependence of radical stabilization on spin delocalization it is concluded that capto-dative substitution implies a synergetic substituent effect on radical stabilization, whereas antagonistic effects are observed in the case of two identical substituents.

1. INTRODUCTION

In view of the synthetic potential of 'capto-dative' substituted radical systems /1/, a relative substituent scale should be of value. Desirable would be the evaluation of substituent effects in the ground state of the radicals. For delocalized radicals, the measurement of the spin density distribution by ESR spectroscopy should be an appropriate tool. This approach is based on the interrelation of spin delocalization and thermodynamical stabilization /2,3/.

Here we report on the investigation of the ESR hyperfine coupling (hfc) constants of mono- and disubstituted benzyl radicals **1**, p-methylbenzyl radicals **2**, and α-methyl-benzyl radicals **3**. The p-hydrogen hfc constants in **1**, and the hydrogen hfc values of the methyl groups in **2** and **3**, respectively, serve as probes for the substituent-induced spin density variations, using the unsubstituted radicals (X = Y = H) as reference systems.

We define experimental delocalization parameters S^{exp} for each of the series /3-5/:

H. G. Viehe et al. (eds.), Substituent Effects in Radical Chemistry, 167–170.
© 1986 by D. Reidel Publishing Co.

$$S^{exp}_{X,Y} = 1 - (a_{X,Y}/a_{HH})$$

By definition, S^{exp} is zero for the unsubstituted reference radicals and the value increases with the stabilizing power of the substituent(s).

For disubstituted radicals we calculate S^{calc}_{XY} values, which represent the superimposed effects S_{XH}, S_{YH} of two independent substituents X, Y:

$$S^{calc}_{XY} = 1 - (1 - S^{exp}_{XH}) (1 - S^{exp}_{YH})$$

The relative difference ΔS between S^{exp}_{XY} and S^{calc}_{XY} is regarded as a measure for the synergetic (more than additive) or antagonistic (less than additive) behaviour of the XY pair:

$$\Delta S\ (\%) = 100 (S^{exp}_{XY} - S^{calc}_{XY}) / S^{calc}_{XY}$$

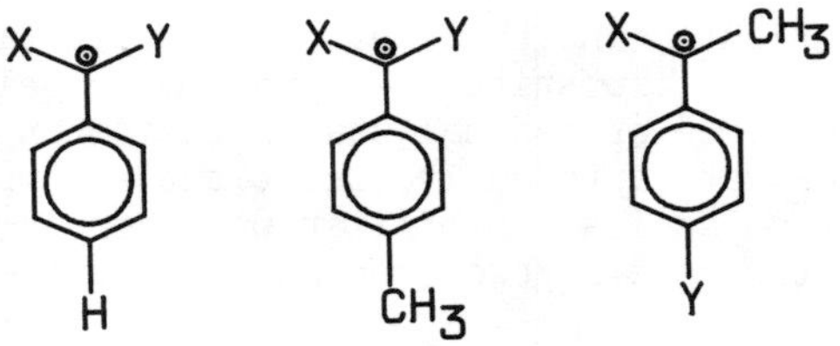

1 **2** **3**

2. EXPERIMENTAL

Radicals **1** - **3** were generated in solution (mostly chlorobenzene) either by hydrogen abstraction from the parent hydrogen compounds with photolytically generated tert-butoxyl radicals, or by thermal dissociation of the radical dimers. The hyperfine coupling constants were evaluated from the ESR spectra via computer simulation. The preservation of planarity in the benzylic systems was deduced from the magnitudes of the coupling constants and supported by MNDO/INDO calculations.

3. RESULTS

The appropriate hfc constants, experimental and calculated delocalization parameters are collected in Tables 1 and 2:

TABLE 1 Delocalization parameters for α-substituted benzyl radicals **1** and p-methyl-benzyl radicals **2**.

radical	X	Y	a_H (mT)[a]	S^{exp}	S^{calc}	ΔS (%)
1	H	H	0.619	0		
	Me	H	0.598	0.034		
	OMe	H	0.573	0.074		
	COOH	H	0.545	0.120		
	COOEt	H	0.543	0.123		
	COOMe	H	0.541	0.126		
	CN	H	0.526	0.150		
	NH_2	H	0.503	0.187		
	SEt	H	0.498	0.195		
	SMe	H	0.496	0.199		
	Me	OMe	0.625	-0.010	0.105	-109
	OMe	OMe	0.603	0.026	0.143	-82
	COOMe	COOMe	0.548	0.115	0.236	-51
	COOEt	COOEt	0.543	0.123	0.231	-47
	CN	COOMe	0.515	0.168	0.257	-35
	CN	CN	0.503	0.187	0.278	-33
	SMe	SMe	0.419	0.323	0.358	-10
	SEt	SEt	0.406	0.344	0.353	-3
	CN	Me	0.470	0.241	0.179	+35
	COOMe	OMe	0.425	0.313	0.191	+64
	COOMe	SMe	0.42	0.32	0.30	+7
	COOMe	SEt	0.413	0.333	0.296	+12
	CN	OMe	0.409	0.339	0.213	+59
	CN	NH_2	0.387	0.375	0.309	+21
	COOMe	NH_2	0.385	0.378	0.289	+31
	CN	SEt	0.368	0.405	0.316	+28
	CN	NMe_2	0.302	0.512		
2	H	H	0.653	0		
	CN	H	0.592	0.093		
	OMe	H	0.590	0.096		
	SEt	H	0.529	0.190		
	OMe	OMe	0.640	0.020	0.183	-89
	CN	CN	0.590	0.096	0.177	-46
	COOMe	COOMe	0.580	0.112		
	CN	COOMe	0.560	0.142		
	SEt	SEt	0.437	0.331	0.344	-4
	CN	OMe	0.444	0.320	0.180	+78
	CN	SEt	0.405	0.380	0.265	+43
	CN	NMe_2	0.323	0.505		

a) hfc of p-hydrogen in **1**, p-methyl group in **2**.

TABLE 2 Delocalization parameters for α,p-substituted α-methyl-benzyl radicals **3**.

X	Y	a_H (mT)[a)]	S^{exp}	S^{calc}	ΔS (%)
H	H	1.770	0		
H	OMe	1.725	0.025		
H	CN	1.690	0.045		
COOEt	H	1.660	0.062		
CN	H	1.600	0.096		
OMe	H	1.550	0.124		
CN	CN	1.590	0.102	0.137	-25
OMe	OMe	1.575	0.110	0.146	-25
CN	SMe	1.51	0.147		
CN	NH_2	1.50	0.153		
CN	OMe	1.461	0.175	0.119	+48
OMe	CN	1.395	0.212	0.163	+30

a) hfc of α-methyl group.

4. CONCLUSIONS

The introduction of a single substituent into the benzylic systems leads to a reduction of the spin density in the aromatic ring, regardless of the electron-donor or -acceptor ability of the substituent.

The delocalization effects on the methyl groups in **2** or **3** are essentially the same as on the p-hydrogen atom in **1**. Thus, configurational spin polarization effects or polar effects seem to play an unimportant role in these systems (almost planar radical centre, freely rotating methyl groups).

The combination of two dative (donor) or two capto (acceptor) substituents produces delocalization (stabilization) effects which are less than additive (antagonistic; negative sign of ΔS).

The combination of dative and capto substituents produces delocalization (stabilization) effects which are more than additive (synergetic; positive sign of ΔS).

5. REFERENCES

1) H.G. Viehe, Z. Janousek, R. Merényi, and L. Stella, Acc. Chem. Res. **18**, 148 (1985).
2) J.C. Walton, Rev. Chem. Intermed. **5**, 249 (1985).
3) J.H. Dust and D.R. Arnold, J. Am. Chem. Soc. **105**, 1221 (1983).
4) L. Sylvander, L. Stella, H.G. Korth, and R. Sustmann, Tetrahedron Lett **26**, 749 (1985).
5) H. Fischer, Z. Naturforsch. **19a**, 866 (1964).

THE EFFECT OF SUBSTITUENTS ON BENZYLIC RADICAL ESR HYPERFINE COUPLING CONSTANTS. THE $\sigma^{\bullet}_{\alpha}$ SCALE BASED UPON SPIN DELOCALIZATION.[1]

Donald R. Arnold
Department of Chemistry
Dalhousie University
Halifax, Nova Scotia
B3H 4J3
Canada

ABSTRACT. The effects of substituents in the meta and para positions of benzylic radicals on the α- and β-hydrogen hyperfine coupling constants (hfc's) are discussed. The relationship between hfc and radical stabilization energy is also described. The development of the $\sigma^{\bullet}_{\alpha}$ substituent parameter scale is reviewed; and, examples of the application of the $\sigma^{\bullet}_{\alpha}$ parameters in Hammett type linear free-energy correlations are given.

INTRODUCTION

The generally accepted definition of radical stability is that which was proposed by Benson, based upon relative carbon-hydrogen homolytic bond dissociation energy (2). For example, the carbon-hydrogen bond dissociation energy of ethane is 98.2 ± 1 kcal mol^{-1} the bond dissociation energy of the benzylic bond of toluene is 88.0 ± 1 kcal mol^{-1}, the difference (10.2 kcal mol^{-1}) is attributed to the resonance stabilization energy of the benzyl radical (2,3).

This is of course an exceedingly useful concept. Much of what is known about relative reactivity in free radical reactions can be explained simply on the basis of relative carbon-hydrogen bond dissociation energy. For example, the selectivity observed upon free radical substitution (eg. bromination) of a hydrocarbon, where the product determining step is hydrogen abstraction from a primary, secondary, or tertiary position, is explained nicely in terms of the relative carbon-hydrogen bond dissociation energy (4).

[1] This is Part 7 of this series. (Part 6 is reference 1.)

H. G. Viehe et al. (eds.), Substituent Effects in Radical Chemistry, 171–188.
© 1986 by D. Reidel Publishing Co.

Nevertheless, homolytic bond dissociation energy is fundamentally the **difference** in energy between the molecule with the intact bond and the radicals. It is, therefore, an over simplification to attribute the effect of variation in structure on the bond dissociation energy to the effect on the radical only; there must also be a substituent effect on the bond in the intact molecule. As a result, relative bond dissociation energies are not independent of the bond being broken (5). Bond dissociation energies are, in fact, known to be dependent on steric effects and on polar substituent effects on the bond, not just on the radical (6).

This understanding is already well developed when considering the stability of ionic intermediates formed upon **heterolytic** bond dissociation. For example, the stability of anions, including carbanions, is usefully described in terms of the relative Bronsted acidity (7). It is recognized, however, that this order cannot be applied more generally to other Lewis acids. And, the theory of "Hard and Soft Acids and Bases" has been developed to account for the variation observed in the general case (8). The same must be true for homolytic cleavage. We therefore must consider additional approaches for defining radical stability.

DISCUSSION

We have suggested that a substituent parameter scale ($\sigma^{\bullet}_{a}$) based upon the hyperfine coupling constant (hfc) due to the a-hydrogens in the electron spin resonance (ESR) spectra of substituted benzylic radicals could accurately reflect spin delocalization (9). The a-hydrogen hyperfine coupling constant is related to the spin density at the benzylic position by the McConnell equation (Equation 1) where Q is a proportionality constant and ρ is the spin density (10). Variation in the a-hfc should, therefore, represent the variation in spin density at the benzylic position of substituted benzylic radicals, relative to the benzyl radical.

Equation 1 $\quad a\text{-hfc} = Q\rho$

The $\sigma^{\bullet}_{a}$ values (Table 1) were determined from the analysis of the ESR spectra (Table 2) and equation 2 (9).The form of this expression was chosen so that the $\sigma^{\bullet}_{a}$ for the unsubstituted benzyl radical would be zero, as is the case for all the other substituent parameter scales. Substituents that lead to less spin density at the benzylic position have positive $\sigma^{\bullet}_{a}$ values. The ratio of the hyperfine coupling constants, and not the logarithm of this ratio, was used

Table 1. $\sigma^{\bullet}_{a}$ values [a] (9).

Substituent	$\sigma^{\bullet}_{a}$	Substituent	$\sigma^{\bullet}_{a}$
4-SMe	0.063	4-Cl	0.011
4-COMe	0.060	4-i-Pr	0.009
4-SPh	0.058	4-t-Bu	0.008
4-COPh	0.055	4-S(O_2)Me	0.005
4-COOMe	0.043	3-Me	0.002
4-CN	0.040	4-OCOPh	0.000
4-SCOMe	0.029	H	0.000
4-S(O)Ph	0.026	3-OMe	-0.001
4-OMe	0.018	3-OPh	-0.002
4-OPh	0.018	4-OCOMe	-0.005
4-S(O)Me	0.018	2-F	-0.007
4-S(O_2)Ph	0.018	3-Cl	-0.007
4-SiMe_3	0.017	3-F	-0.009
4-S(O)OMe	0.016	4-CF_3	-0.009
4-Me	0.015	4-F	-0.011
4-SO_3Me	0.013	3-COOMe	-0.014
4-Et	0.012	3-CF_3	-0.017
		3-CN	-0.026

[a] Defined by equation 2.

initially because it was assumed that the variation in a-hfc would be inversely proportional to the variation in stabilization energy associated with the delocalization of the spin away from the benzylic position. This assumption has subsequently been shown to be correct (11). The $\sigma^{\bullet}_{a}$ scale measures the component of energy that can be attributed to spin delocalization in the substituted benzyl radical in comparison to the unsubstituted case.

Equation 2 $$\sigma^{\bullet}_{a} = 1 - (\text{hfc } a\text{-H}_{X} \ / \ \text{hfc } a\text{-H}_{O})$$

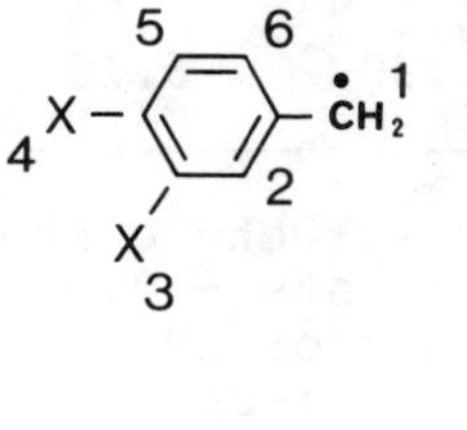

Table 2. The hfc constants of substituted benzyl radicals (9).[a]

Sub.	a_1	a_2	a_3	a_4	a_5	a_6
4-SMe	15.23	5.08	1.73	0.90[b]	1.73	5.08
4-COMe	15.28	5.00	1.75	0.50[b]	1.75	5.00
4-SPh	15.30	5.03	1.85	-	1.85	5.03
4-COPh	15.35	4.98	1.76	-	1.76	4.98
4-COOMe	15.55	5.05	1.75	0.37[b]	1.75	5.05
4-CN	15.60	5.00	1.78	0.96[c]	1.78	5.00
4-SCOMe	15.77	5.03	1.81	-	1.81	5.03
4-S(O)Ph	15.83	5.10	1.80	-	1.80	5.10
4-OMe	15.95	5.02	1.60	0.75[b]	1.60	5.02
4-OPh	15.95	5.11	1.70	-	1.70	5.11
4-S(O)Me	15.95	5.03	1.75	0.35[b]	1.75	5.03
4-S(O_2)Ph	15.95	5.05	1.80	-	1.80	5.05
4-SiMe_3	15.97	5.03	1.71	-	1.71	5.03
4-S(O)OMe	15.99	5.07	1.73	-	1.73	5.07
4-Me	16.00	5.05	1.60	6.50[b]	1.60	5.05
4-S(O_2)OMe	16.04	5.03	1.71	-	1.71	5.03
4-Et	16.05	5.00	1.85	3.25[d]	1.85	5.00
4-Cl	16.07	5.24	1.75	0.50[e]	1.75	5.24
				0.60[e]		
4-i-Pr	16.10	5.10	1.80	2.85[f]	1.80	5.10
4-t-Bu	16.12	5.10	1.75	-	1.75	5.10
4-S(O_2)Me	16.17	5.09	1.78	1.05[b]	1.78	5.09
3,5-di-Me	16.19	5.20	3.35	6.08	3.35	5.20
3-Me	16.22	5.15	3.38[b]	6.15	1.75	5.00
4-OCOPh	16.25	5.22	1.80	-	1.80	5.22
H	16.25	5.10	1.70	6.13	1.70	5.10
3-OMe	16.27	5.30	-	6.30	1.65	4.50
3-OPh	16.29	5.15	-	6.15	1.75	5.00
4-OCOMe	16.33	5.27	1.80	-	1.80	5.27
3-Cl	16.37	5.15	-	6.33	1.83	5.05
2-F	16.42	5.28	1.78	6.32	1.72	8.18[g]
	16.32					
3-F	16.39	5.15	4.72[g]	6.19	1.80	4.95
4-CF_3	16.39	5.19	1.76	6.88[h]	1.76	5.19
4-F	16.42	5.30	1.75	14.43[g]	1.75	5.30
3-COOMe	16.48	5.22	-	6.15	1.75	4.98
3-CF_3	16.53	5.18	3.30[h]	6.13	1.80	4.38
3-CN	16.68	5.25	0.30[c]	6.18	1.80	4.95

[a] Positions indicated in the structure above. Values are believed to be accurate ± 0.03 G. [b] Hydrogens of CH_3. [c] Nitrogen of CN. [d] Hydrogens of ethyl. [e] Cl. [f] Hydrogen of isopropyl. [g] F. [h] Fluorine of CF_3.

The $\sigma^{\bullet}_{a}$ scale is particularly useful because it describes the effect of the substituent on the radical only; $\sigma^{\bullet}_{a}$ is an intrinsic property of the radical. Hammett linear free-energy correlations can be expected when using $\sigma^{\bullet}_{a}$ to describe the characteristics of free radicals or free radical like species, such as, triplet excited states, radical ions, and methylenes. Relative rates of reactions involving these intermediates should also correlate with $\sigma^{\bullet}_{a}$. Such varied behaviour as the reduction potential of substituted benzophenone derivatives to the ketyl radical (12), the n,π^* triplet energy of benzophenone derivatives (12), the zero-field splitting parameter (D) of diphenylmethylenes (12,13), the relative rates of the thermal isomerization of methylenearylcyclopropanes (9,14), and the relative rates of thermal decomposition of substituted dibenzylmercury compounds (15) correlate with $\sigma^{\bullet}_{a}$. Some examples will be discussed below.

When $\sigma^{\bullet}_{a}$ values are applied in a Hammett free-energy relationship to explain or predict free radical reactivity in general, it must be kept in mind that:

1.) Correlation is to be expected only if the transition state is "late" along the reaction coordinate. That is, when the transition state resembles the intermediate radical.

2.) Other effects, both steric and electronic, which influence the transition state are not accounted for by $\sigma^{\bullet}_{a}$ alone, and they must be considered separately. An extended Hammett treatment can be used to account for additional polar effects.

It seemed intuitively obvious that for benzylic radicals, remote (meta or para) substituents which lead to a decrease in spin density at the α-position would also cause that benzylic radical to be more stable at that position. These substituents would cause an increase in the contribution of those valence bond structures which increase π-bonding with the methylene (eg.**1b** and **c**); and a corresponding increase in resonance stabilization energy. This effect should be manifest in an increase in the barrier associated with rotation of the methylene as valence bond structures like **1b** and **c** increase the double bond character to the benzylic methylene.

This increase in stabilization energy associated with increased delocalization of the spin away from the benzylic position is not to be confused with the total stabilization energy or total resonance energy of the radical. It is only that incremental stabilization energy that is associated with the delocalization of spin away from a specific position, in this case the benzylic methylene (11).

Of all the valence bond structures that contribute to the overall structure of the benzyl radical, only **1a** will lead directly to α-hfc. If substituents increase the relative importance of one or more of the other structures (**1b-g**) we can expect that that benzylic radical will be easier to form (by bond homolysis) and less reactive **at the benzylic position.**

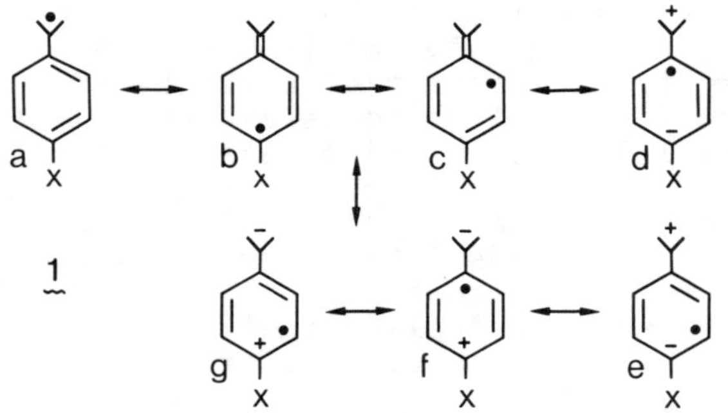

Consider, in the first approximation, that the McConnell equation with the same proportionality constant (Q) applies to both the methyl and benzyl radicals. Since the α-hfc for the methyl radical (planar) is 24.65 G (16), and the α-hfc for the benzyl radical is 16.25 G, structure **1a** contributes about 66 % to the overall structure of the benzyl radical.

Equation 2 defines $\sigma^{\bullet}_{\alpha}$ for substituents in terms of the α-hfc constants. Considering equation 2 in the usual form of a Hammett linear free-energy relationship sets $\rho^{\bullet}$ equal to one. While this may give a qualitative indication of the effect of substituents on the stability of the radical at the benzylic position, it is fundamentally important to establish the quantitative energy equivalence of this variation.

By extending the reasoning just described, the π-stabilization energy of a π-radical, SE($R^{\bullet}$), can be represented by equation 3,

Equation 3 $$SE(R^{\bullet}) = V_2(CH_2^{\bullet}) - V_2^{*}$$

where $V_2(CH_2^{\bullet})$ and V_2^{*} are the rotational barriers of the $CH_2^{\bullet}$ group of the radical $R^{\bullet} = ZCH_2^{\bullet}$, in the presence and absence of spin delocalization respectively (17). In a complex radical this π-stabilization energy is just the additional stabilization that results from delocalization of spin away from a specific position ($CH_2^{\bullet}$) (11). For example, the stabilization energy of the benzylic radical is defined by equation 3, taking $V_2(CH_2^{\bullet})$ as the rotational barrier around the benzylic bond of the radical, and assuming the rotational barrier of the methylene in the absence of spin delocalization (V_2^*) to be 1 kcal mol^{-1} (18).

The barrier for rotation of the benzylic methylene group has not been measured; but, data are available for the α-phenethyl radical ($V_2(CH_2^{\cdot})$ = 13.4 kcal mol^{-1}) (19) and since these radicals are so similar this value can be used (see below). From equation 3, the π-stabilization energy of the benzylic radical is 12.4 kcal mol^{-1}. Since we know from the α-hfc that the valence bond structure **1a** contributes about two-thirds to the overall structure of the benzylic radical this barrier can be attributed to the involvement of those other structures (**1b-g**) which delocalize the spin away from the benzylic position.

The ESR spectrum and the barrier for rotation of the terminal methylene of several π-stabilized radicals have been measured. From these data, plotted in Figure 1, a quantitative assessment of the relationship between α-hfc and the stabilization energy of the radical has been made (equation 4) (11).

Equation 4 $SE(R^{\cdot})(\text{kcal mol}^{-1}) = 34.7 \pm 0.7 - (1.37 \pm 0.05)\alpha\text{-hfc (G)}$

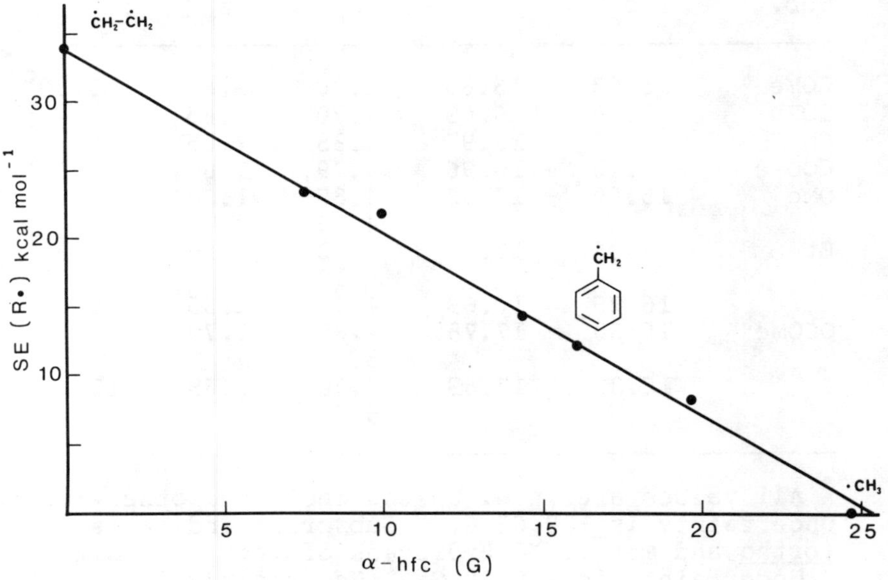

Figure 1. A plot of the π-stabilization energy ($SE(R^{\cdot})$) versus the ESR hfc of the α-H of the CH_2 group (α-hfc) in π-radicals of the form $ZCH_2^{\cdot}$. The correlation coefficient is r = 0.997 (11).

A reduction in α-hfc of a benzylic radical by 1.0 G corresponds to an increase in stabilization energy of 1.4 kcal mol^{-1}. The range of known $\sigma^{\bullet}_{\alpha}$ values (9), from *para*-thiomethyl (0.063) to *meta*-cyano (-0.026) represents a variation in benzylic radical stability of 1.9 kcal mol^{-1}.

While the $\sigma^{\bullet}_{\alpha}$ scale can certainly be applied to benzylic radicals; it is important to know if these same parameters can be applied, in general, to more complex substituted benzylic radicals. To answer this question we have studied the effect of substituents on the hyperfine coupling of α-phenethyl (1) and cumyl radicals (20). These hyperfine coupling constants are listed in Tables 3 and 4.

Figure 2 is a plot of the α-hfc of the α-phenethyl radicals versus the α-hfc of the benzylic radicals. The good correlation (α-hfc(phen.) = (1.20 $\pm$ 0.21)α-hfc(benzyl) - (3.4 $\pm$ 3.3); r = 0.979) indicates that the substituents affect the spin density in nearly the same way for both series (1).

Table 3. Hyperfine coupling constants for *para*-substituted phenethyl radicals [a] (1).

Sub.	a_{α} [b]	a_{β} [b]	a_o [b]	a_m [b]	a_x [b]
COMe	15.03	16.65	4.70	1.65	0.30[c]
COPh	15.08	16.69	4.70	1.65	-
CN	15.35	16.97	4.88	1.65	0.80[d]
COOMe	15.43	16.96	4.79	1.60	0.30[c]
OMe	15.60[e]	17.25	4.85[f] 4.95	1.50[f]	0.70[c,f]
Et	16.06	17.51	4.72[f] 5.12	1.60[f]	2.95[f,g]
H	16.25	17.69	4.95	1.65	5.90
OCOMe	16.30	17.78	4.85 5.15	1.70	-
F	16.33	17.89	5.00 5.30	1.69	13.46[h]

[a] All values are in G. Unless indicated otherwise the uncertainty is $\pm$ 0.03 G. [b] Subscript indicates position (*ortho* and *meta*). [c] Hydrogens of methyl. [d] Nitrogen of CN. [e] Uncertainty is $\pm$ 0.10 G. [f] Uncertainty is $\pm$ 0.06 G. [g] α-Hydrogens of ethyl. [h] ^{19}F.

Table 4. Hyperfine coupling constants for *para*-substituted cumyl radicals [a] (19).

Sub.	a_{β} [b]	a_o [b]	a_m [b]	a_x [b]
COMe	15.07	4.55	1.55	<0.50[c]
COPh	15.07	4.52	1.53	–
CN	15.37	4.60[d]	1.60[d]	0.80[d,e]
COOMe	15.40	4.57	1.55	<0.35[c]
SPh	15.60	4.72	1.65	–
$S(O)_2Ph$	15.76	4.61	1.55	–
Cl	16.03	4.85	1.65	0.36[f], 0.30[g]
OMe	16.10	4.75[d]	1.50[d]	0.60[c,d]
OPh	16.10	4.82	1.62	–
Me	16.10	4.70[d]	1.60[d]	5.80[c,d]
Et	16.13	4.60[d]	1.60[d]	2.90[d,h]
CF_3	16.15	4.70	1.58	8.40[i]
i-Pr	16.17	4.75	1.55	2.35[j]
t-Bu	16.17	4.75	1.55	–
OCOPh	16.24	4.83	1.65	–
H	16.28	4.70	1.65	5.55
OCOMe	16.30	4.85	1.65	–
F	16.35	4.75	1.55	12.65[k]

[a] All values are in G. Unless indicated otherwise the uncertainty is ± 0.03 G. [b] Subscript indicates position. [c] Hydrogens of methyl. [d] Uncertainty is ± 0.06 G. [e] Nitrogen of CN. [f] ^{35}Cl. [g] ^{37}Cl. [h] α-Hydrogens of ethyl. [i] Fluorine of CF_3. [j] α-Hydrogen of isopropyl. [k] ^{19}F.

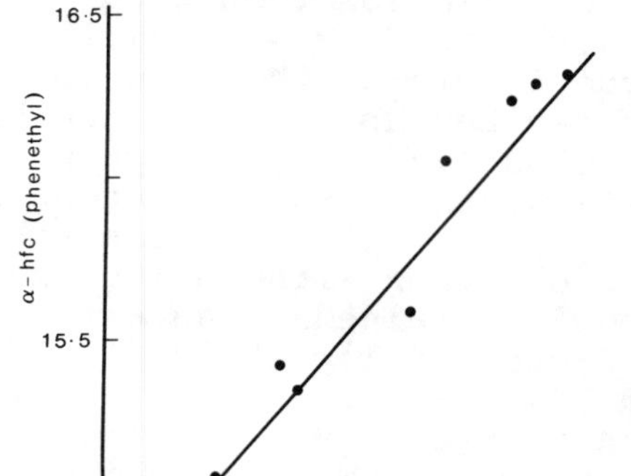

Figure 2. A plot of the α-hfc of the α-phenethyl radicals versus the α-hfc of the benzyl radicals (1).

The α-hfc for the benzyl radical is usually only slightly greater than the α-hfc for the analogous phenethyl radical. For the nine compounds studied the difference is only 0.15 ± 0.15 G. There is, in fact, no difference for the unsubstituted benzyl and phenethyl radicals. The difference in α-hfc of the para-hydrogens of the benzyl and phenethyl radicals is also small (0.23 G). The average decrease of 0.15 G in the benzylic α-hfc is equivalent to an increase in stabilization of only 0.21 kcal mol^{-1}; yet, the difference in the benzylic carbon-hydrogen bond dissociation energy between toluene and ethylbenzene is 2.6 kcal mol^{-1} (2a). This discrepancy can be taken as evidence that most of the difference in bond dissociation energy is, in this case, due to differences in the carbon-hydrogen bonds of the intact molecules.

Another important fact made apparent by analyses of the data in Table 3, is that there is a linear correlation between the α-hfc and the β-hfc of the substituted phenethyl radicals. The mechanism of hyperfine coupling is different for the α- and β-positions; the main cause of coupling being due to spin polarization for hydrogens in the α-position and hyperconjugation in the case of β-hydrogens. Theory predicts that both the α- and β-hydrogen hyperfine coupling constants are proportional to the spin density; and, the observed linear correlation provides good experimental evidence to support this prediction.

We can now discuss the effect of substituents on the cumyl radicals (Table 4) confident in the fact that the β-hfc gives an accurate indication of variation in spin density which can be compared to the α-hfc of the analogous benzyl radical.

The plot (Figure 3) of the β-hfc of substituted cumyl radicals versus the α-hfc of the analogous benzyl radicals does **not** give as good a correlation (β-hfc(cumyl) = 1.06 ± 0.23 (α-hfc(benzyl)) - 0.93 ± 3.7; r = 0.92)(19). Furthermore, it is evident, from figure 3, that the deviations from the linear relationship can be explained on the basis of charge effects on spin delocalization. The points above the line are for those substituents generally recognized as being electron donating ($\sigma_p < 0$) while the points below the line are associated with electron withdrawing groups. In fact, a good correlation **is** obtained by an extended Hammett treatment of the data (β-hfc(cumyl) = 0.92 ± 0.13(α-hfc(benzyl)) - 0.45 ± 0.15 σ_p + 1.3 ± 2.1; r = 0.98) (19). We take this as evidence that charge separated valence bond structures such as (**1d** and **e**) are more significant in the cumyl radical series, relative to the benzyl radicals. This is in keeping with the increased stability of the carbocation in the case of cumyl, relative to benzyl.

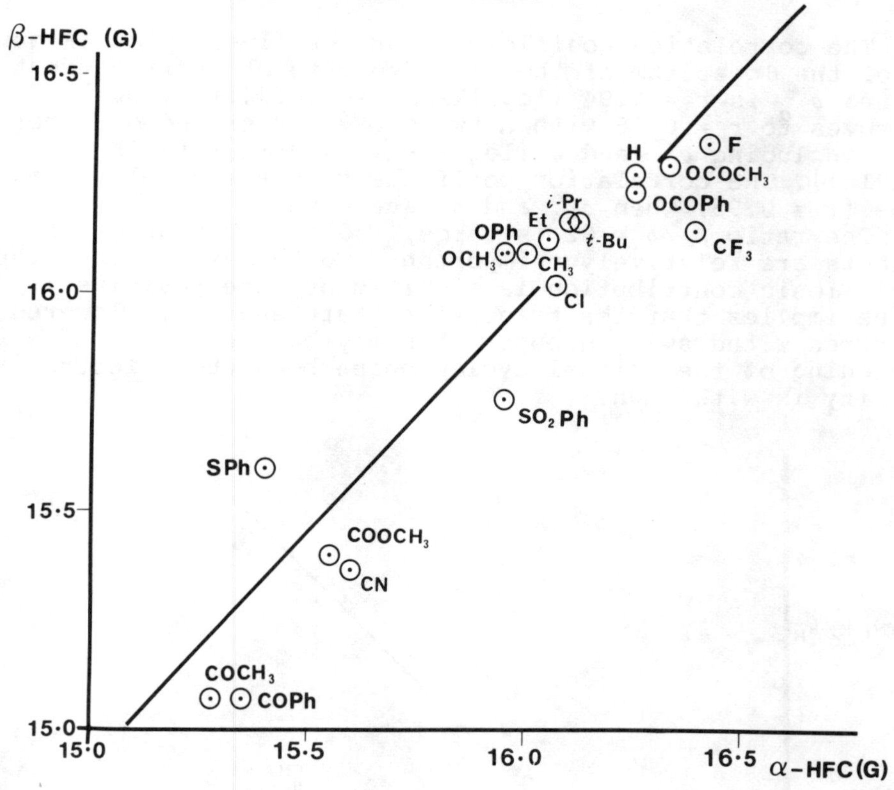

Figure 3. A plot of the β-hfc of the cumyl radical versus the α-hfc of the benzyl radical (19).

It should be clear from this discussion that charge density has a direct influence on spin density. Therefore, simple attempts to assess a contribution due to polar effects on the transition state of a free radical reaction using, for example, an extended Hammett treatment, will include this intrinsic polar effect on radical stability along with any polar effect on the reaction.

We return now to a discussion of applications of the $\sigma_a^\bullet$ parameters in linear free-energy relationships.

The effect of substituents on the thermal rearrangement of methylenearylcyclopropanes has been studied by Creary and his co-workers (14). The transition state for this reaction should have radical (biradical) character and polar factors resulting from variation in charge separation at the transition state should be relatively unimportant. Data are now available for eighteen derivatives.

The correlation coefficient for the linear plot (Figure 4) of the logarithm of the relative rate of rearrangement versus $\sigma^{\bullet}_{a}$ is r = 0.94 ($\log (k/K_H) = 0.0490 + 7.26\sigma^{\bullet}_{a}$).[2] This improves to r = 0.96 with a two parameter extended Hammett plot including $\sigma^{\bullet}_{a}$ and σ ($\log k/k_H = 0.0131 + (7.48\sigma^{\bullet}_{a} + 0.121\sigma)$). The correlation coefficient is essentially the same (r = 0.95) when $\sigma^{\bullet}_{a}$ and σ^{+} are used.

The ratio $\rho^{\bullet}/\rho = 62^{a}$ is large, indicating that polar effects are relatively unimportant. To the extent that the small ionic contribution is significant, the positive ρ value implies that the transition state energy is lowered by electron withdrawing groups. This may be the result of a weakening of the initial cyclopropane bond when electron density is withdrawn.

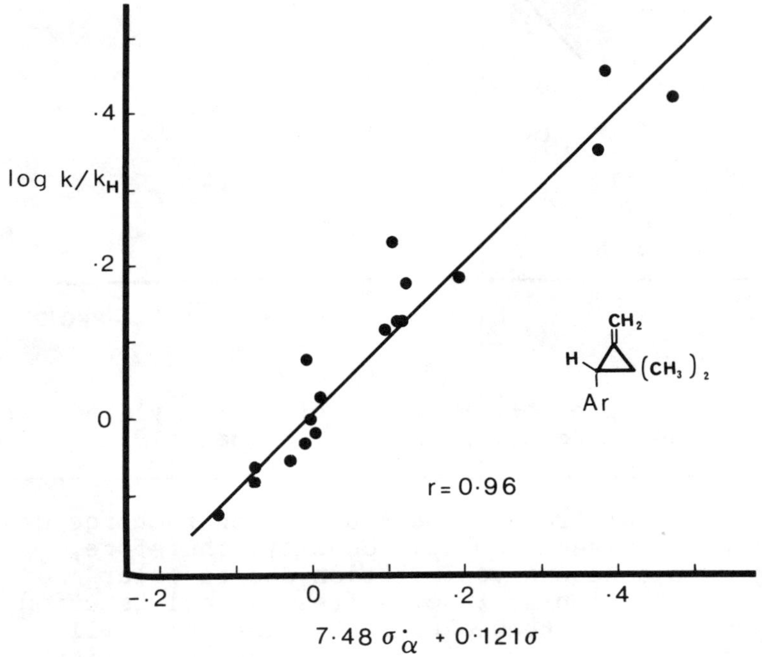

Figure 4. The relative rate of the methylenearylcyclopropane rearrangement versus $\sigma^{\bullet}_{a}$ and σ.

---

[2] The ionic substituent parameters were taken from reference 20. The correlation coefficient was corrected for degrees of freedom.

Creary and his co-workers have very recently extended their study to the methylenearylcyclopropane esters (14c). This series is of particular interest because the relative rates should reflect any radical stabilization resulting from an increased contribution of the charge separated valence bond structures, such as **1f** and **g**, when the aryl ring is substituted with electron donating substituents.

Data are available for only eight derivatives. The correlation coefficient (r = 0.87) for the linear plot (Figure 5) of the relative rate of rearrangement versus $\sigma^{\bullet}_{a}$ and σ^{+} is not as good as for the parent series ($\log k/k_H = 0.170 + 4.15\sigma^{\bullet}_{a} - 0.315\sigma^{+}$); however, the observed trend is interesting. The ratio of $\rho^{\bullet}/\rho^{+} = 13$ has decreased significantly; which is consistent with a greater involvement of polar factors. The ρ^{+} value is now negative. And, there is essentially no correlation when σ is used instead of σ^{+}. All of these facts support the suggestion (14c) that charge separated valence bond structures (**1f** and **g**) are providing additional stabilization of the radical.

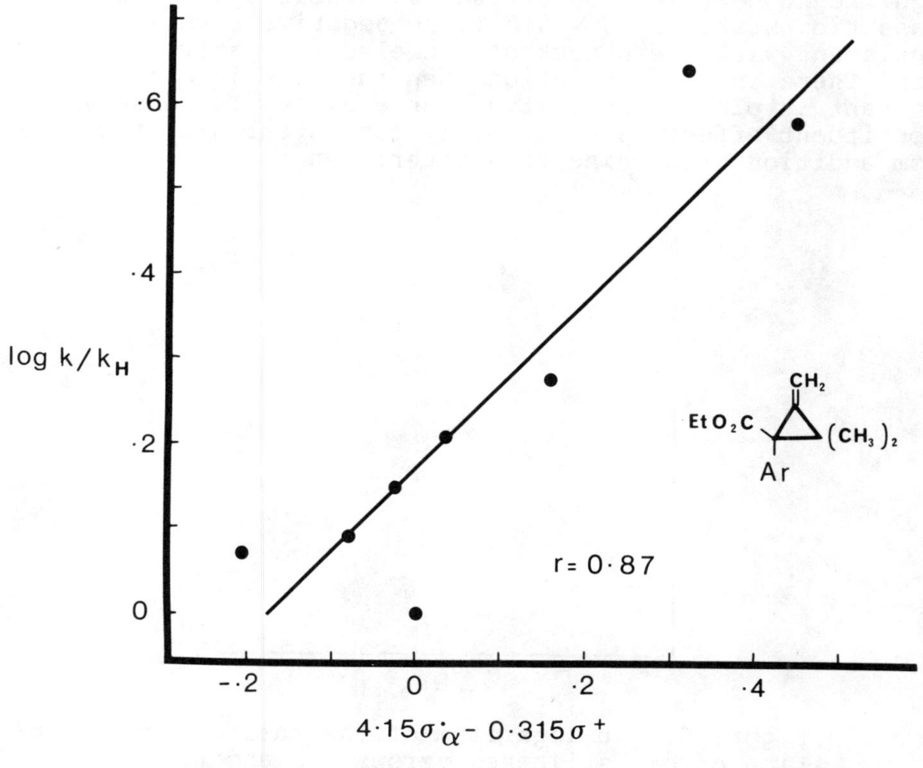

Figure 5. The relative rate of the methylenearyl-cyclopropane-ester rearrangement versus $\sigma^{\bullet}_{a}$ and σ^{+}.

The iodine atom-catalyzed isomerization of symmetrically substituted cis-stilbenes has been studied (21). The mechanism for this reaction undoubtedly involves the reversible addition of the iodine atom to the double bond, accompanied by rotation around the single bond of the intermediate benzylic radical. There has been some debate concerning the identity of the rate determining step: is it the addition of the iodine atom (22); or, is it the rotation around the single bond of the benzylic radical (21)?

The authors observed a good correlation between the logarithm of the relative rate of the isomerization and the Hammett σ parameters for the substituents; however, those derivatives bearing substituents which effectively stabilize the benzylic radical (eg. para-cyano and para-nitro) were excluded from the correlation analysis (21).

Data are available for only eight derivatives; nevertheless, it is clear that an extended Hammett equation incorporating both $\sigma^{\bullet}_{\alpha}$ and σ can account for the observed relative rates. The linear equation ($\log k = 0.0384 + 9.45\sigma^{\bullet}_{\alpha} - 0.963\sigma$; $r = 0.99$) plotted in figure 6, indicates a significant contribution of radical stability at the transition state ($\rho^{\bullet}/\rho = 9.8$). The negative ρ value is consistent with the attack of the electronegative iodine atom. There is no correlation when the logarithm of the rate constant is plotted against $\sigma^{\bullet}_{\alpha}$ or σ alone. The observed substituent effects are certainly consistent with the iodine atom addition step being rate determining.

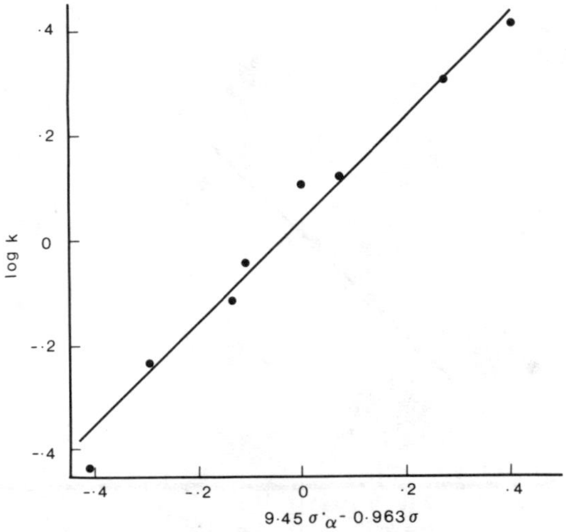

Figure 6. The logarithm of the rate of isomerization of cis-stilbenes versus $\sigma^{\bullet}_{\alpha}$ and σ.

Professor R.A. Jackson and his co-workers have studied the thermal decomposition of a series of dibenzylmercury compounds (15). These compounds decompose by a first order homolytic cleavage of the carbon-mercury bond. This bond must have some ionic character, and σ_0 (Taft's inductive substituent parameter) was chosen to account for this polar effect. Deviations of the logarithm of the relative rate, from a correlation established with three meta derivatives, was attributed to variation in benzylic radical stability (15). This approach assumes that there is no effect of substituents in the meta position on benzylic radical stability, which is clearly unjustified in view of the variation in the a-hfc of the meta substituted benzyl radicals.

Data are available for nine derivatives. There is no correlation between the logarithm of the relative rate and $\sigma^{\bullet}_{a}$ or σ alone; but, the correlation improves considerably when both $\sigma^{\bullet}_{a}$ and σ are used ($\log k = -3.94 + 7.03\sigma^{\bullet}_{a} - 0.757\sigma$; $r = 0.89$). The ratio $\rho^{\bullet}/\rho = 9.3$ indicates a significant contribution of polar effects at the transition state.

Previous publications in this series have discussed the general features of the $\sigma^{\bullet}_{a}$ scale and the mode of interaction of each substituent with the benzylic radical center (9,10). Some of these conclusions will be summarized here.

Most of the para substituents delocalize spin density more effectively than the unsubstituted radical ($\sigma^{\bullet}_{a}>0$). Those with non-bonding lone pairs can interact by donating an electron; they are β-spin donors. Unsaturated substituents can accept an electron, that is, they are a-spin acceptors. In the terminology of simple perturbation molecular orbital theory, the orbital interaction involved with the a-spin acceptors is three-electron, three-orbital which leads to a net two-electron stabilization; while, with the β-spin donors the orbital interactions are two-orbital, three-electron and lead to a net one-electron stabilization (23). For the a-spin acceptors, spin delocalization increases in the order $-CF_3 < -CH_3 < -CN < -COR$ which is consistent with increased interaction as the energy of the π- (or σ)-orbital of the a-spin acceptor decreases. For the β-spin donors, spin delocalization increases in the order $-F < -Cl < -OR < -SR$ which is consistent with increased interaction as the energy of the lone pair of the β-spin donor decreases.

These substituent effects involve direct interaction with the radical through the π-system; there are also effects that are transmitted through the σ-framework. As usual, the effect of substituents in the meta position can be used to assess these inductive effects, since direct interaction through the π-system is prevented.

There is a correlation (r = 0.93) of the α-hfc of the meta substituted benzyl radicals with σ_m. Meta substituents that inductively withdraw electrons lead to a spin density in the benzylic position which is greater than for the unsubstituted benzyl radical.

The mechanism by which the inductive withdrawal of electron density leads to a decrease in spin delocalization at the benzylic position is not as easily visualized (11). One explanation is based upon the recognition that the inductive withdrawal of electron density will increase localization of α-spin density at the benzylic position either by decreasing the overlap between the interacting centers, or, by increasing the energy separation between the interacting systems. The electron-withdrawing groups lower the energy of the aromatic π-system, which,from simple perturbation molecular orbital theory, results in a decreased interaction between the benzylic carbon and the aromatic system (23,24).

CONCLUSIONS

Analysis of the hydrogen-hfc of substituted benzylic radicals gives information regarding the effect of substituents on the distribution of spin throughout the radical. Since the stability of benzylic radicals (that is, the ease of formation and relative reactivity) is related to the spin density at the benzylic position, the α-hfc constants can be used to define benzylic radical stability. This hypothesis has been couched in terms of a Hammett type substituent parameter, $\sigma^{\bullet}_{\alpha}$, which can be used in linear free-energy relationships to correlate properties and reactivity of benzylic radicals.

ACKNOWLEDGMENTS

This work was supported by a grant from the Natural Sciences and Engineering Research Council of Canada. I am particularly grateful to my co-workers whose names appear as coauthors of the publications in this series. I am also pleased to acknowledge the receipt of information from Professor Creary prior to publication.

REFERENCES

1. D.R. Arnold, A.M.de P. Nicholas, and K.M. Young. Can.J.Chem., **64**, in press (April, 1986).
2. (a) D.F. McMillen and D.M. Golden. Ann.Rev.Phys.Chem., **33**, 493 (1982); (b) S.W. Benson. J.Chem.Educ., **42**, 502 (1965).
3. M.J. Rossi, D.M. Golden. J.Am.Chem.Soc., **101**, 1230 (1979); (b) D.J. DeFrees, R.T.McIver, Jr., and W.J. Here. J.Am.Chem.Soc.,**102**, 3334 (1980).
4. M.L. Poutsma. Free Radicals., Vol.II. Edited by J.K. Kochi. Wiley, New York. 1973. Chapt. 15.
5. (a) A.M.de P. Nicholas and D.R. Arnold. Can.J.Chem., **62**, 1850 (1984); (b) A.M.de P. Nicholas and D.R. Arnold. Can.J.Chem., **62**, 1860 (1984).
6. (a) A.A. Zavitsas and J.A. Pinto. J.Am.Chem.Soc., **94**, 7390 (1972); (b) C. Ruchardt. Angew.Chem.,Int.Ed.Engl., **9**, 830 (1970).
7. D.J. Cram. Fundamendals of Carbanion Chemistry, Academic Press. New York. 1965. p.19.
8. (a) R.G. Pearson. J.Am.Chem.Soc., **85**, 3533 (1963); (b) R.G. Parr and R.G. Pearson. J.Am.Chem.Soc., **105**, 7512 (1983).
9. (a) J.M. Dust and D.R. Arnold. J.Am.Chem.Soc., **105**, 1221, 6531 (1983); (b) D.D.M. Wayner and D.R. Arnold. Can.J.Chem., **62**, 1164 (1984).
10. J.E. Wertz and J.R. Bolton. Electron Spin Resonance., McGraw-Hill. New York. 1972. Chapt. 6.
11. A.M.de P. Nicholas and D.R. Arnold. Can.J.Chem., **64**, 270 (1986).
12. W.J. Leigh, D.R. Arnold, R.W.R. Humphreys, and P.C. Wong. Can.J.Chem., **58**, 2537 (1980).
13. R.W.R. Humphreys and D.R. Arnold. Can.J.Chem., **57**, 2652 (1979).
14. (a) X. Creary. J.Org.Chem., **45**, 280 (1980); (b) X. Creary and M.E.M.-Mohammadi. J.Org.Chem.,**51**, in press June,1986); (c) X. Creary and M.E.M.-Mohammadi. unpublished results.
15. (a) S. Dincturk, R.A. Jackson, H. Townson, H. Agirbas, N.C. Billingham, and G.J. March. J.Chem.Soc.,Perkin Trans.2, 1121 (1981); (b) S. Dincturk, R.A. Jackson. J.Chem.Soc.,Perkin Trans. 2, 1127 (1981).
16. D.M. Chipman. J.Chem.Phys., **78**, 3112 (1983).
17. (a) H.G. Korth, H. Trill, and R. Sustman. J.Am.Chem.Soc., **103**, 4483 (1981); (b) I.G. Green and J.C. Walton. J.Chem.Soc.,Perkin Trans. 2, 1253 (1984); (c) I. MacInnes and J.C. Walton. J.Chem.Soc.,Perkin Trans.2, 1073 (1985).
18. M.S. Conradi, H. Zeldes, and R. Livingston. J.Phys.Chem., **83**, 2160 (1979).

19. D.R. Arnold, A.M.de P. Nicholas, and M.S. Snow. Can.J.Chem., **63**, 1150 (1985).
20. N.B. Chapman and J. Shorter. Correlation Analysis in Chemistry, Plenum Press, New York (1978).
21. W.J. Muizebelt and R.J.F. Nivard. J.Chem.Soc.(B), 913 (1968).
22. S.W. Benson, K.W. Egger, and D.M. Golden. J.Amer.Chem.Soc., **87**, 468 (1965).
23. D.D.M. Wayner and D.R. Arnold. Can.J.Chem., **63**, 2378 (1985).
24. A. Pross and L. Radom. Prog.Phys.Org.Chem., **13**, 1 (1981).

DETERMINATION OF THERMODYNAMIC STABILISATION IN AMINO-SUBSTITUTED ALKYL RADICALS

Iain MacInnes and John C. Walton
Department of Chemistry, University of St. Andrews,
St. Andrews, Fife KY16 9ST, U.K.
Derek C. Nonhebel
Department of Chemistry, University of Strathclyde,
Glasgow, G1 1XL, U.K.

ABSTRACT. The rotational barriers about the C-N bonds in aminoalkyl radicals $H_2N\dot{C}HR$ were determined by line shape analysis of the exchange broadened EPR spectra. The stabilisation energies of the radicals were deduced from these rotation barriers and found to be significantly greater when R is an electron withdrawing group than when R is an alkyl group.

INTRODUCTION

Aminoalkyl radicals **1** containing electron withdrawing substituents (R) should be subject to capto-dative (C-D) stabilisation. In the transition state for rotation about the C-N bond, **2**, the unpaired electron cannot be delocalised onto nitrogen and any C-D stabilisation is lost. It follows that the C-D stabilisation can be estimated from the increase

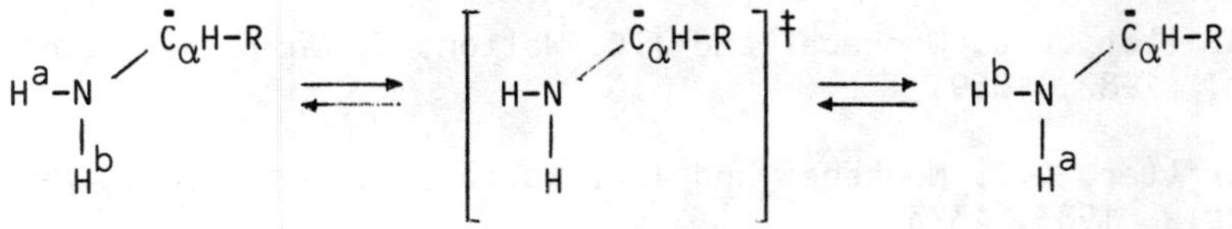

1 **2** **3**

in this rotational barrier in comparison with the barriers of radicals without C-D stabilisation. We have developed an EPR exchange broadening technique for determining rotational barriers in amino-substituted radicals[1-3] and have applied this to several substituted aminoalkyl radicals.

RESULTS AND DISCUSSION

The exchange broadened EPR spectra of the amino-substituted radicals were simulated at each temperature[4] and the rotational barriers V_0 were determined from an Arrhenius plot of the resulting rate constants:

H. G. Viehe et al. (eds.), Substituent Effects in Radical Chemistry, 189–190.
© 1986 by D. Reidel Publishing Co.

TABLE C-N Bond Rotation Barriers V_O in Amino-substituted Radicals. $H_2N\dot{C}HR$

R	V_O/kcal mol^{-1}	log(A/s^{-1})
Me	7.6 ± 0.4	13.2 ± 0.4
Me_2CH	7.3 ± 0.4	13.1 ± 0.4
Me_3C	7.5 ± 0.2	13.4 ± 0.4
CF_3	10.9 ± 0.6	13.4 ± 0.4
Bu^tOCO	14.9 ± 1.2	13.2 ± 0.5
HC≡C	10.5 ± 1.2	13.8 ± 0.7
N≡C	11 ± 2	—

The V_O values of the three alkyl substituted radicals are all identical to within the experimental error. The rotation barriers in the C-D radicals are all significantly greater than those of the amino-alkyl radicals. The increase, which is a measure of the thermodynamic stabilisation due to the C-D effect, varies from 3 kcal mol^{-1} for ***1***, R = C≡CH (the weakest acceptor) to *ca.* 7 kcal mol^{-1} for ***1***, R = Bu^tOCO. In the latter case intramolecular hydrogen bonding may contribute to the large barrier. Interestingly, the radical with R = CF_3, *i.e.* an inductively electron-withdrawing group, also shows an additional stabilisation of similar magnitude.

Stabilisation of this magnitude is in good agreement with that found from the rotation barriers of C-D substituted allyl radicals.[5]

REFERENCES

1. D. Griller, D.C. Nonhebel and J.C. Walton, *J. Chem. Soc. Chem. Comm.*, 1982, 1059.

2. D. Griller, D.C. Nonhebel and J.C. Walton, *J. Chem. Soc., Perkin Trans.2*, 1983, 1373.

3. I. MacInnes, J.C. Walton and D.C. Nonhebel, *J. Chem. Soc. Chem. Comm.*, 1985, 712.

4. J. Heinzer, *Quantum Chemistry Program Exchange*, No.209, University of Indiana, Indiana, 1972.

5. H-G. Korth, P. Lommes and R. Sustmann, *J. Am. Chem. Soc.*, 1984, **106**, 663.

PHOSPHINYLHYDRAZYLS $R_2N\dot{N}P(O)L_2$. INFLUENCE OF THE PHOSPHINYL GROUP ON THE 3-ELECTRON NN π BOND.

P.Tordo
SREP Université de Provence CNRS UA 126
13397 Marseille Cedex 4
France

ABSTRACT. 1,1-Dimethyl-2-phosphinylhydrazyls and 1-t-Butyl-2-phosphinyl hydrazyls have been generated from the corresponding hydrazines. Their radical structure, have been assigned on the basis of their ESR spectral parameters. In contrast with the behavior of trialkylhydrazyls or 1,2-dialkylhydrazyls it has been shown that in the phosphinylhydrazyls the larger nitrogen splitting is due to the tervalent nitrogen. Phosphinylhydrazyls have been shown to exist in equilibrium with their dimer form. The radical $Me_2N\dot{N}P(O)(OEt)_2$ is particularly persistent and the free activation of energy for the rotation about the NN bond has been measured ($\Delta G^{\neq}$ = 9.8 ± 0.5 kcal/mole).

INTRODUCTION

The electronic distribution in the 3-electron bond of hydrazyls appears to be very sensitive to the electronic effect of the groups attached to the nitrogen (1). Thus, for 1,1-dialkylhydrazyls, Y=R, X=H, it has been shown by ^{15}N labelling that the smaller nitrogen splitting comes from the divalent N(1) nitrogen.

Scheme 1

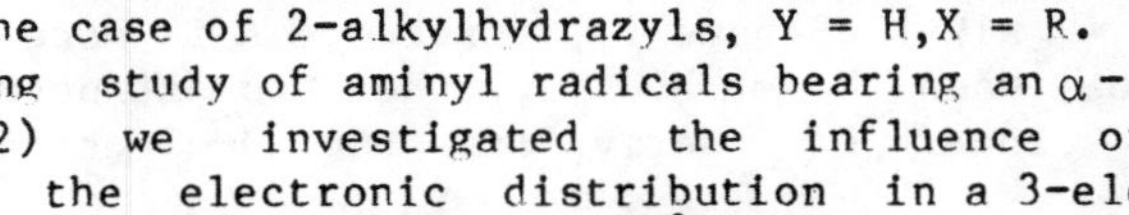

A B

This observation was rationalized as being due to the inductive effect of the alkyl groups which stabilize resonance structure **B** relative to **A** . The same argument accounts for the reversal of the nnitrogen splittings in the case of 2-alkylhydrazyls, Y = H, X = R.

In a continuing study of aminyl radicals bearing an α- or β-phosphorus substituent (2) we investigated the influence of a phosphinyl substituent on the electronic distribution in a 3-electron bond by studying some phosphinylhydrazyls, $\mathbf{R_2N\dot{N}P(O)L_2}$ **1** , (R = H or alkyl). It was found that the radicals could be successfully generated by hydrogen atom abstraction from the corresponding phosphonylhydrazines **1H** with tert- butoxyl radicals.

H. G. Viehe et al. (eds.), Substituent Effects in Radical Chemistry, 191–194.
© 1986 by D. Reidel Publishing Co.

RESULTS AND DISCUSSION

- **ESR parameters.** The ESR parameters for radicals **1•** and related radicals are listed in the Table. For hydrazyls **1c•**, **1d•** and **1e•**, the distinctions between the phosphorus and the hydrogen (N-H) hyperfine splittings (h.f.s.) were established by studying the corresponding N-deuterated hydrazyls.

^{15}N-labelling experiments (3) have shown that for trialkylhydrazyls like **2•** (Table), A(1) > A(2). Comparison of the h.f.s. for radical **2•** (A(2)/A(1) = 0.89) and **1a•** (A(2)/A(1) = 1.88) clearly shows that on changing the alkyl substituent of the divalent nitrogen to a phosphinyl group, A(2) becomes much larger than A(1) while A_H(**Me**N(2) is also strongly increased ($\langle A_H \rangle$ = 4.42 G).
These dramatic changes in the electron density distribution of the three electron N-N bond are, we suggest a consequence of the electron-withdrawing effect of the phosphinyl group. As a result, resonance structure **B** will be stabilized relative to **A** and the unpaired elctron density on N(2) will be increased, and on N(1) decreased, relative to **2•**.

Two different kinds of hydrazyl could be generated by hydrogen abstraction from **1cH**, **1dH** and **1eH**, moreover, restriction of rotation about the N-N bond could give rise to conformationnal isomers. Nevertheless, in each case, only one spectrum was observed.

Scheme 2

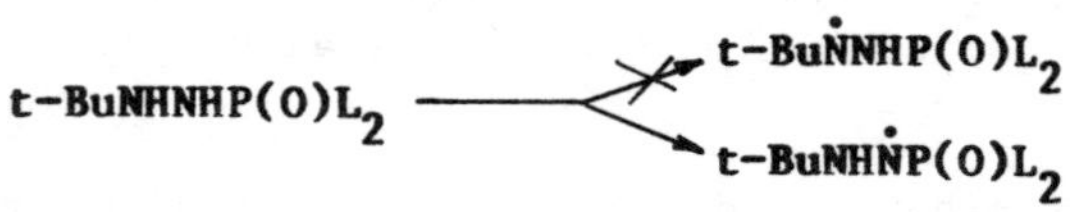

The comparison between the h.f.s. of hydrazyls **1c•** , **1d•** , **1e•** and those of hydrazyl **3•** (Table) clearly indicated that hydrogen abstraction on **1cH**, **1dH** and **1eH** has occured at N(1).
As for **1a•** and **1b•** , the electron-withdrawing phosphinyl group should stabilize resonance structure **B** relative to **A** (Scheme 1).
This implies that N(2) will now produce the larger splitting, and, as a consequence of its increased spin density, the splitting by the hydrogen attached to N(2) will also increase (from 3.0 G for **3•** to 9-10 G for **1c•** , **1d•** or **1e•** , Table).

According to the valence-bond theory, the 3-electron bond in hydrazyls is expected to become stronger when structures **A** and **B** become closer in energy.
When Y and X = alkyl (Scheme 1), structure **A** is more stable than structure **B** , but when Y = $P(O)L_2$, the two structures are expected to be closer in energy and the 3-electron bond should be stronger.
These electronic effects are a consequence of the capto-dative (3) (push-pull (4)) structure of phosphonylhydrazyls, which can also be considered aminyl radicals bearing both a good elctron-donnor (R_2N-) and a good electron-acceptor (-$P(O)L_2$) group.

Table. ESR parameters (G) for some hydrazyls $R_2N(2)$ $\dot{N}(1)P(O)L_2$ in solution

R	L		T,°C	A(1)	A(2)	A(P)	A(R)
Me	OEt	**1a•**	21	7.17	13.5	10.2	11.0(3H);11.95(3H).
"	"		126	7.75	13.44	8.85	10.98 (6H).
"	Et	**1b•**	0	7.75	13.15	11.25	9.90(3H);11.25(3H).
"	"		70	7.90	13.05	10.15	10.70 (6H).
t-Bu;H	OEt	**1c•**	25	8.22	12.30	12.90	9.06 (1H).
idem in MeOD			25	8.0	12.25	12.25	1.50 (1D).
t-Bu;H	OPh	**1d•**	25	7.75	12.52	11.52	10.28 (1H).
"	Et	**1e•**	-55	8.60	11.91	16.97	9.1(1H);1.6(2H).
$Me_2\dot{N}NMe$		**2•**(1a)	rt	11.7	10.5		5.9 ; 8.2
t-BuN(H)Ṅt-Bu		**3•**	rt	12.8	9.45		3.0
$H_2N\dot{N}C(O)OEt$		**4•**(1c)	0	7.8	10.2		9.6(2H);1.3(2H)

- **Temperature effects.** Hydrazyls **1a•** and **1b•** are particularly persistent and their ESR signals were easily detected over an unusually large temperature range (− 30 to + 126°C for **1a•** and − 30 to + 100°C for **1b•**). The highest concentration of **1a•** was observed in the temperature range 20 to 40°C. Radical **1a•** is destroyed irreversibly by greater warming, but its concentration can be reversibly decreased and increased by lowering and raising the temperature between 20 and -20°C.

These observations are in agreement with the existence of an equilibrium such as:

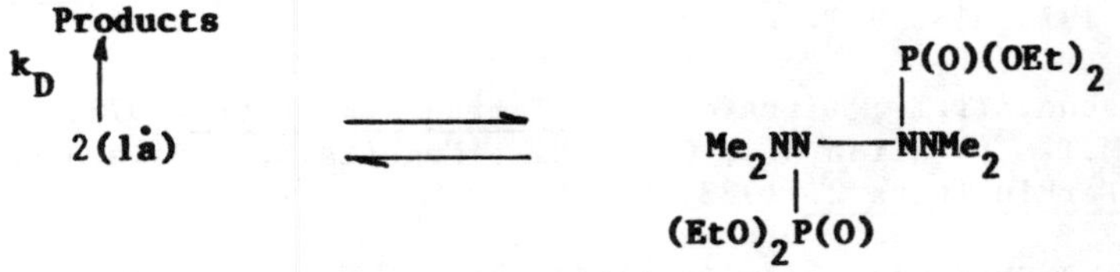

and also to a relatively low value for k_D, the rate of the irreversible decay of **1a•** to give the products P.

As mentioned above, the enhanced stability of **1a•** and **1b•** compared to that of a trialkylhydrazyl is certainely attributable to their capto-dative structure. Furthermore, the absence of β- hydrogen atoms relative to the divalent nitrogen prevents a disproportionation process. Hydrazyls **1c**• , **1d**• and **1e**• are less perssistent than **1a**• and **1b**• , and their ESR spectra become undetectable above **ca** 70°C. The highest concentration of **1e**• was detected at **ca** -50°C and it also can be reversibly decreased and increased by lowering and raising the temperature between -70 and -50°C. The lesser persistence of radical **1e**• relative to that of **1a**• may be reasonably attributable to the occurence of a disproportionation involving the amino hydrogen.

Another interesting observation is that the ESR spectra of **1a**• and **1b**• undergo marked and reversible changes on varying the temperature. For radical **1a**• at 100°C and above the two methyl groups are equivalent. Below 40°C, the spectra were satisfactorily simulated assuming that the two methyl groups are non-equivalent.
In the temperature range, + 40 to + 100°C, owing to a distinct broadening of different lines, the spectra cannot be satisfactorily simulated using "static" parameters
The line broadening effect can be explained in terms of the rotation around the N-N bond, which exchanges the coupling constants, A(Me1) and A(Me2) of the CH3 groups. Analysis of the excess width of appropriate lines, leads to $\Delta G^{\#}$ = 9.8 ± 0.5 kcal/mole for **1a**• and 8± 0.5 kcal/mole for **1b**• .
Hindered rotation around the NN bond has been observed for a large number of hydrazyls however the $\Delta G^{\#}$ was not measurable. Only the barrier of rotation for 1-benzoyl-2,2-bis-(2,5-di-t-butylphenyl)hydrazyl, has been determined,and was reported to have a $\Delta G^{\#}$ of **ca** 7.4 kcal/mole, while for **H2NN•H** ab initio calculation was reported to give a value of 10 kcal/mole (5).

REFERENCES.

(1) **(a)** Malatesta,V. ; Ingold,K.U. J.Am.Chem.Soc. 1973, 95, 6110. **(b)** Lunazzi,V. ; Ingold,K.U. ibid. 1974, 96, 5558. **(c)** Malatesta,V. ; Lindsay,D. ; Horsville,E.C. ; Ingold,K.U. Can. J. Chem. 1974, 52, 864.

(2) **(a)** Tordo,P. ; Boyer,M. ; Friedmann,A. ; Santero,O. ; Pujol,L. J.Chem.Phys. 1978, 82, 1742. **(b)** Berchadsky,Y. ; Negareche,M. ; Tordo,P. ibid. 1982, 86, 1742.

(3) Viehe,H.G. ; Merenyi,R. ; Stella,L. ; Janousek,Z. Angew.Chem., Int.Ed.Engl. 1979, 18, 917.

(4) **(a)** Balaban,A.T. ; Itratoiu,R. Tetrahedron Lett. 1973, 1879. **(b)** Caproiu,M.T. ; Elian,M. ; Grecu,N. ; Negoita,N. ; Balaban,A.T. J.Chem.Soc., Perkin Trans 2. 1983, 591.

(5) Baird,N.C. J.Chem.Ed. 1977, 54, 1977.

SUBSTITUENT EFFECTS ON THE DIMERIZATION OF VINYL NITROXIDES AND ON THE DISSOCIATION OF THEIR TRICYCLIC DIMERS

Hans Günter Aurich, Otto Bubenheim and Klaus-Dieter Mogendorf
Fachbereich Chemie of the University Marburg
Hans Meerwein Straße
D-3550 Marburg
FR Germany

ABSTRACT. Vinyl nitroxides 3 are formed by oxidation of hydroxylamines 1. They dimerize to yield bicyclic compounds 5 or/and tricyclic compounds 7. Whereas dimers 7 dissociate at slightly elevated temperatures to give radicals 3, dimers 5 do not. Compounds 7 rearrange in refluxing methanol to afford the tetracyclic compounds 8. In refluxing benzene or toluene compounds 5 are additionally formed by conversion of 7 in some cases. The substituent effects on the dimerization of 3 and on the dissociation and conversion of 7 are discussed.

I. THE DIMERIZATION OF VINYL NITROXIDES

Oxidation of the hydroxylamines 1 by lead dioxide afforded vinyl nitroxides 3, which were detected by ESR spectroscopy.

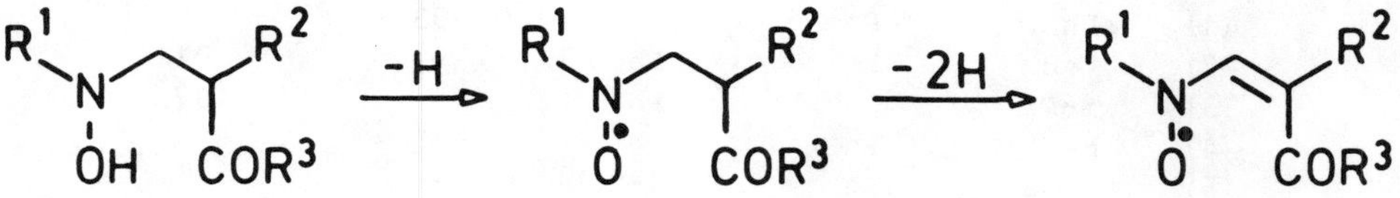

1 2 3

From the reaction mixture the bicyclic dimers 5[1)] and/or the tricyclic dimers 7[2)] could be isolated.

The dimerization of vinyl nitroxides is governed by several factors:

a) The steric strain in intermediate 6 is relatively high (hexasubstituted ethane!). However, in one of the staggered conformations of the d.l-form of 6 one of the nitrone groups and the two carbonyl groups are forced together so tightly that the activation barrier for the intramolecular (3+2+2)-cycloaddition forming the tricyclic dimer 7 should be low.
On the other hand, one can assume that a higher activation barrier must be overcome to place the nitrone group and the olefinic group of 4 into positions suitable for the intramolecular (3+2)-cycloaddition

H. G. Viehe et al. (eds.), Substituent Effects in Radical Chemistry, 195–198.
© 1986 by D. Reidel Publishing Co.

The two routes of dimerization:

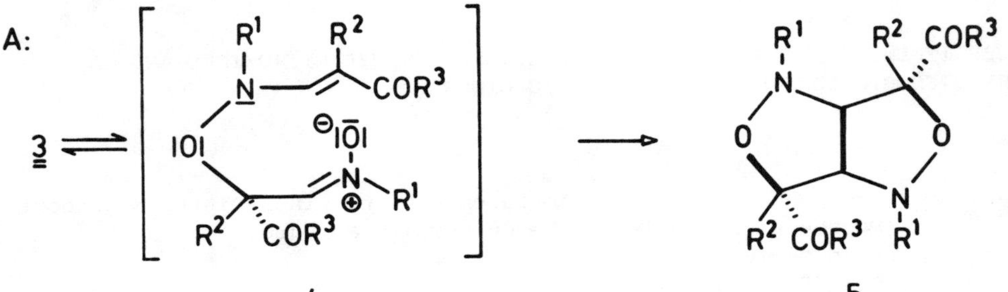

4 5

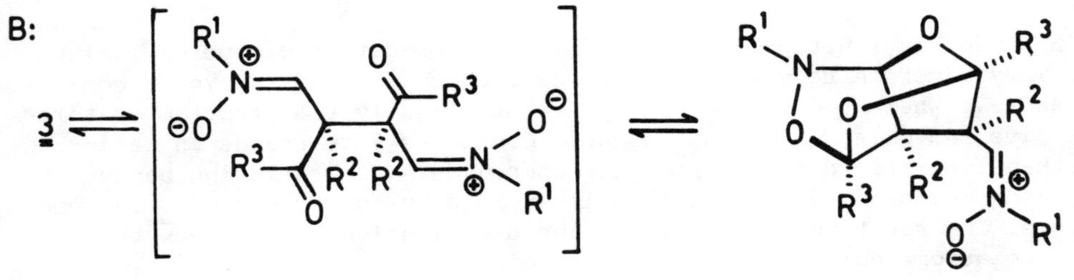

6 7

Examples:

	R^1	R^2	R^3	5 (% yield)	7 (% yield)
(1)	Ph	Ph	Ph	93	-
(2)	t-Bu	Ph	Ph	-	91
(3)	i-Pr	Ph	Ph	-	87
(4)	$Me_2C(CN)$	Ph	Ph	55	12
(5)	t-Bu	Ph	$4\text{-}O_2N\text{-}Ph$	-	61
(6)	t-Bu	$4\text{-}O_2N\text{-}Ph$	Ph	-	90
(7)	$c\text{-}C_6H_{11}$	$4\text{-}O_2N\text{-}Ph$	Ph	-	72
(8)	Me	$4\text{-}O_2N\text{-}Ph$	Ph	42	-
(9)	$Ph\text{-}CH_2$	$4\text{-}O_2N\text{-}Ph$	Ph	90	-

and to perform this reaction.

b) The relative stabilization of the intermediates 4 and 6 is markedly affected by the substituents R^1:
Electron donating substituents R^1 (normal alkyl groups) stabilize the two nitrone moieties in 6, thus favoring formation of dimer 7 (examples (2), (3) and (5).
Substituents R^1 exerting less electron-donating properties (phenyl, substituted phenyl, $Me_2C(CN)$) stabilize the oxyenamine moiety in 4.

In addition, the nitrone group due to its reduced stabilization is more reactive, facilitating the intramolecular (3+2)-cycloaddition of 4. Thus formation of dimer 5 is favored by such substituents R^1 (examples (1), and (4)).

c) A more electron-attracting substituent R^2 (p-nitrophenyl) can compensate the inherent instability of intermediate 4 R^1 = alkyl to some extent. In this case intermediates 4 and 6 are of comparable energy. Thus the reaction is directed to dimer 5 by the more reactive N-methyl and N-benzyl nitrone moieties ((8) and (9)), whereas formation of dimer 7 remains the more favored reaction pathway with the sterically more hindered N-tert butyl and N-cyclohexyl nitrone groups ((6), and (7)).

II. DISSOCIATION OF DIMERS 7

Most of the dimers 7 dissociate in solution at slightly elevated temperatures to afford radicals 3. The degree of dissociation is strongly solvent-dependent. Donor solvents favor the dissociation, since they destabilize the dimers 7 which show a considerable accumulation of free electron pairs and stabilize the radicals 3. In acetone radicals 3 should be stabilized in particular by dipolar interactions. The effect of substituents on the dissociation of dimers 7 is shown in table I.

TABLE I. Equilibrium constants and degree of dissociation α in acetone (10^{-3} m solution)

	R^1 = t-Bu		40°C		50°C	
	R^2	R^3	K	α(%)	K	α(%)
(1)	Ph	Ph	$8.53\cdot10^{-6}$	4.5	$1.79\cdot10^{-4}$	19.0
(2)	Ph	4-MeOPh	$2.73\cdot10^{-5}$	7.9	$1.72\cdot10^{-3}$	47.6
(3)	Ph	4-O_2NPh	-	<0.1	$1.41\cdot10^{-4}$	17.1
(4)	4-MeOPh	Ph	$1.58\cdot10^{-5}$	6.1	$1.19\cdot10^{-4}$	15.8
(5)	4-MeOPh	4-MeOPh	$2.08\cdot10^{-5}$	6.95	$2.32\cdot10^{-4}$	21.3
(6)	4-O_2NPh	4-MeOPh	-	<0.1	$3.85\cdot10^{-4}$	26.6

For dimer 7 R^1 = $Me_2C(CN)$, R^2 = R^3 = Ph dissociation is only observed at about 100°C. The following dimers 7 do not dissociate at all up to 100°C:

R^1 = t-Bu, R^2 = 4-O_2NPh, R^3 = Ph - R^1 = t-Bu, R^2 = Me, R^3 = Ph -

R^1 = t-Bu, R^2 = Ph, R^3 = Me.

Donor-substituted phenyl groups as for instance 4-MeOPh at position R^3 facilitate dissociation by stabilization of the radical due to their enhanced stabilization of the carbonyl group in 3 (compare (1) and (2)).

III. ISOMERIZATION OF THE DIMERS 7

The dimers rearrange to tetracyclic compounds 8 at room temperature in the presence of acids. (Execptions are the dimers with $R^2 = 4\text{-}O_2NPh$, which rearrange only at 65°C).

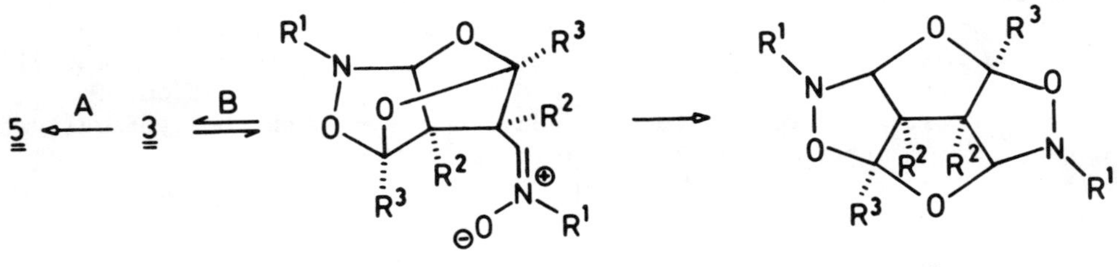

7 8

In the absence of acids most of the dimers 7 rearrange to compounds 8 in refluxing methanol, whereas in aprotic solvents no rearrangement occurs at about 65°C (e.g. in refluxing acetone or tetrahydrofuran). At higher temperatures in aprotic solvents (refluxing benzene or toluene) compounds 8 are formed along with compounds 5 in some cases. The latter must be formed via radicals 3 which under these conditions may partly overcome the higher activation barriers of pathway A.

TABLE II. Conversion of dimers 7 to 8 and 5 (yields in %)

R^1 = t-Bu R^2	R^3	in benzene(80°C) 8	 5	in toluene(110°C) 8	 5	reaction time(h)
Ph	Ph	60	30	70	16	12
Ph	4-MeOPh	45	30	-	45	3
4-MeOPh	Ph	a)		50 b)	-	15
4-MeOPh	4-MeOPh	50	23	58	30	3
$4\text{-}O_2NPh$	Ph	a)		a)		48
$4\text{-}O_2NPh$	4-MeOPh	a)		-	76	3
Ph	Me	>80	-	c)	c)	8
Me	Ph	>80	-	c)	c)	8

a) no isomerization occured b) 50% unchanged 7 was isolated c) not performed

References

1) H.G. Aurich, K.-D. Mogendorf, and M. Schmidt, J. Org.Chem. 49, 2654 (1984).
2) H.G. Aurich, G. Baum, W. Massa, K.-D. Mogendorf, and M. Schmidt, Chem. Ber. 117, 2615 (1984).

SUBSTITUENT EFFECTS ON THE CC-BOND STRENGTH.

H. Birkhofer, H.-D. Beckhaus and C. Rüchardt*
Institut für Organische Chemie und Biochemie
Universität Freiburg, D-7800 Freiburg i.Brsg., FRG.

ABSTRACT. Stabilization energies for a series of carbon radicals substituted in the α-position by -CN, $-OCH_3$ and $-SCH_3$ were determined by a combination of kinetic and thermochemical methods and EFF calculations. The data are discussed in the context of substituent effects in free radical chemistry, in general.

INTRODUCTION.

Substituent effects and structure reactivity relationships have been discussed in free radical chemistry for a long time on the basis of the presumed influence of a substituent or of a structure variation in general on the "stability" of the radical species taking part in a reaction or in a reaction series.[1] The distinction between kinetic and thermodynamic stability was often ignored and a separation of steric and electronic effects was seldom achieved. It is not unexpected, therefore, that conclusions frequently were drawn by circular argumentation. A typical example is shown on the first slide.1.

Clearly, no interpretation of reactivity data is achieved in this way. Strangely, this type of reasoning is persisting in many textbooks and, occasionally, even in the most recent literature, despite the fact that convincing experimental results and arguments disproving the relationships between reactivity and radical stability can be found even in the older literature. A few examples for this are shown on the next slides.

The selectivity of chlorine atoms in the example shown is opposite to the expectation from a thermochemical basis.[2]

The alternating sequence in certain copolymers, which is due to differences in the selectivity of the two types of radical end groups, cannot be explained on the basis of the different "radical stabilities" of these two types of radicals.[3]

H. G. Viehe et al. (eds.), Substituent Effects in Radical Chemistry, 199–218.
© 1986 by D. Reidel Publishing Co.

Circular — Argumentation

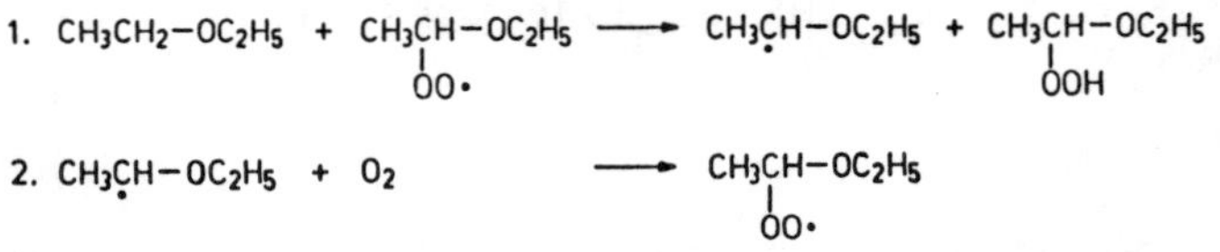

<u>A says:</u> "Autoxidation of ethers in the α-position is a fast and efficient process because α-ethoxyethyl radicals are stabilized by resonance".

<u>B says:</u> "α-Ethoxyethyl radicals are stabilized by resonance. This is obvious from their efficient and preferential formation in the autoxidation chain".

Relative Reactivities of C—H Bonds in Chlorinations

$(C_6H_5)_2CH{-}H$	$C_6H_5CH(CH_3){-}H$	$(CH_3)_2CH{-}H$	
2.6	3.3	4.3	$k_{rel}(Cl\cdot)$

Cl• attacks at aliphatic positions faster than at benzylic hydrogens.

G. Rüssell (1963)

Copolymerisation

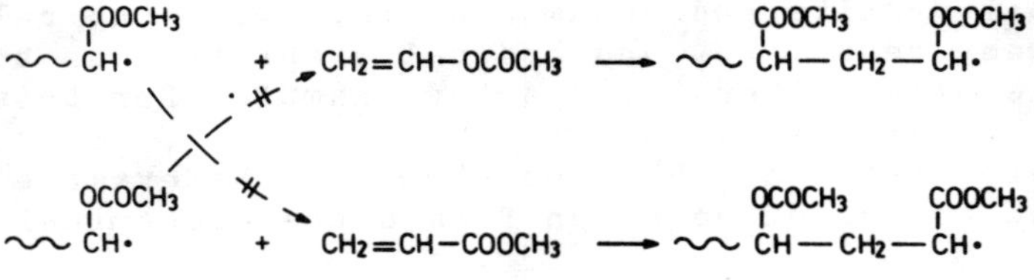

The alternating sequence cannot be due to differences in "stability" of the attacking radicals or the radicals being generated.

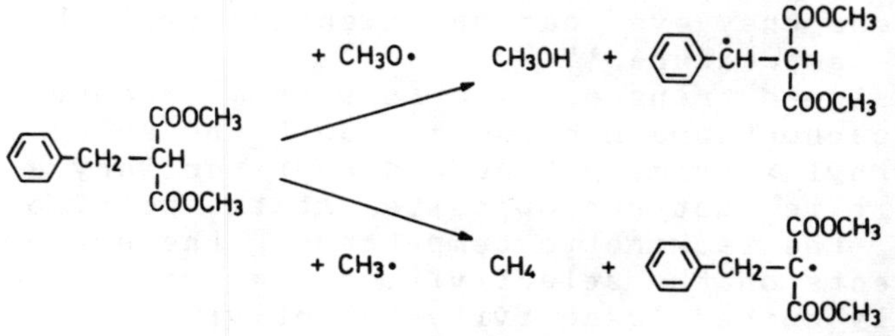

The point of attack is specific for the attacking radical and not at the weakest C—H bond.

H.C. McBay et al. (1954)

Benzyl malonic ester is attacked at different sites by different radicals.[4] "Nucleophilic" methyl radicals attack at the point of lowest electron density. "Electrophilic" methoxy radicals, in contrast, prefer the point of higher electron density in the benzylic position. The rule that "the most stable radical is generated with preference" is definitely not in accord with these results.

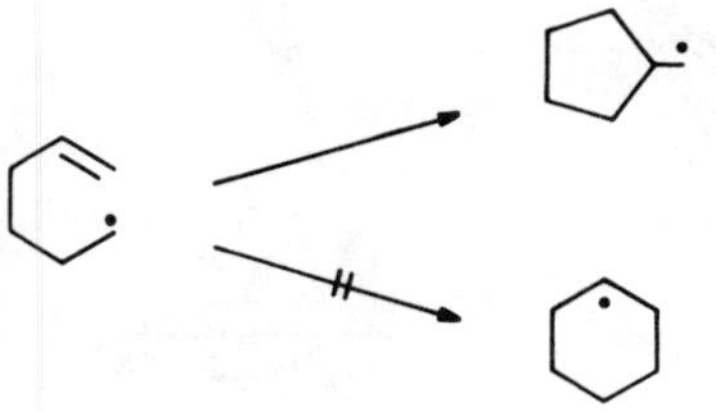

Under kinetic control primary and not secondary radicals are formed by cyclisation.

In cyclizations of 5-hexenyl radicals primary cyclopentylmethyl radicals are formed in preference to the presumably more stable cyclohexyl isomers. In recent years this result has been explained convincingly by stereoelectronic control of the kinetic competition.[5]

Finally, it has been pointed out, that relative selectivities of a series of radicals are dependent on temperature and they even can be inverted simply by changing the temperature.[6]

In the halogen transfer reactions from bromotrichloromethane and carbontetrachloride at 130°C the selectivity series is methyl > primary > secondary > tertiary alkyl while at 0° it is just the opposite. At the point of intersection, the isokinetic temperature, the entropy and enthalpy effects on the selectivities are just cancelling. The frequently quoted "reactivity-selectivity" relationship accordingly is also not on safe grounds in radical chemistry. This is a final warning against inferring radical stability from selectivity data as has frequently been done.

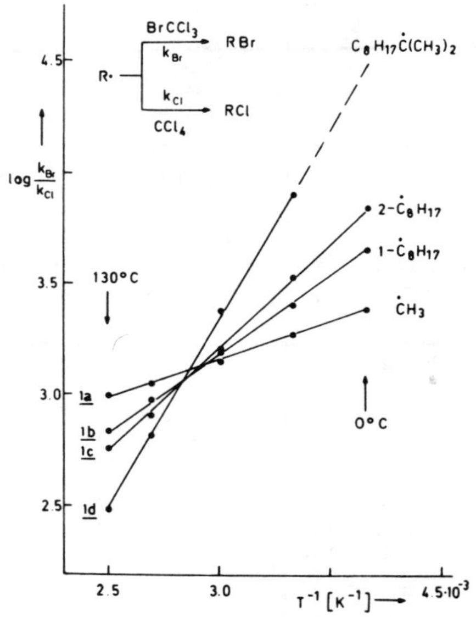

Figure Temperature dependence of the selectivity of alkyl radicals 1 in the competition system $BrCCl_3/CCl_4$. (Ref.6)

SUBSTITUENT EFFECTS AND THE REACTION COORDINATE.

We know today that substituent effects in radical chemistry are highly dependent upon the type of reaction coordinate of the reaction steps in which the overall reactivity or selectivity are determined.[1] In endothermic dissociation processes the thermodynamic stabilities of the radical products a. and b. are, of course, decisive for the rate of the reaction. In exothermic steps, on the other hand, the thermodynamic stability of the radical product x-a-b. has little if any influence on the rate.

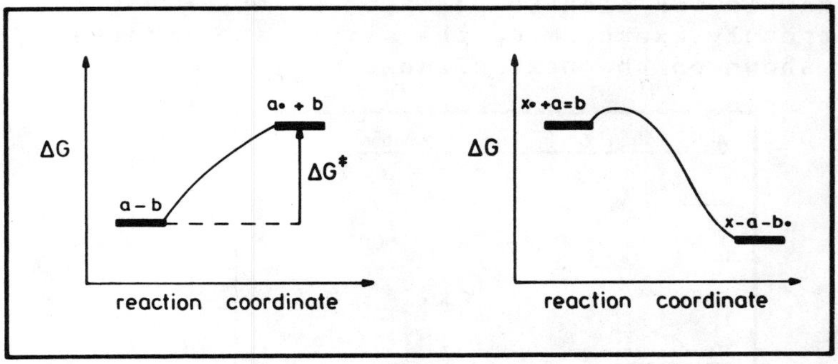

Most free radical reactions are chain reactions. It is a basic principle of chains that their chain steps must be fast and usually exothermic processes in order to compete successfully with the termination reactions.[7] This is in particular true for radical addition reactions in which a weak π-bond is replaced by a σ-bond. In radical addition-reactions reactivity and stereoselectivity, according to the recent literature,[8,9] are controlled mainly by steric effects and polar effects, i.e. by frontier orbital interactions.[10] FMO interaction seems to be the favored control of reactivity in reactions with early transition states on the reaction coordinate.

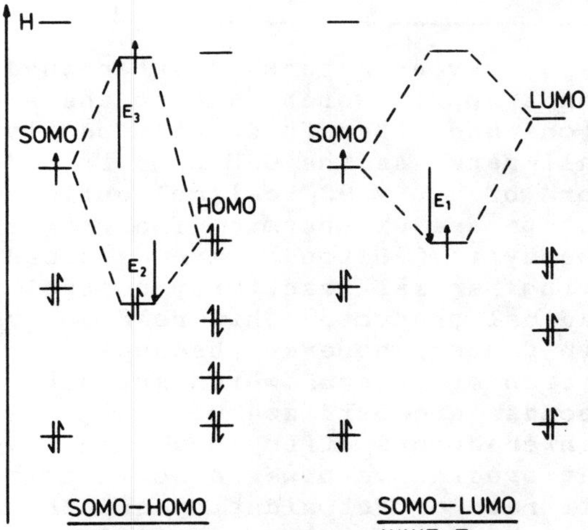

As an example for a chain reaction in which not all steps are strongly exothermic, the simple autoxidation chain[11] is shown on the next slide.

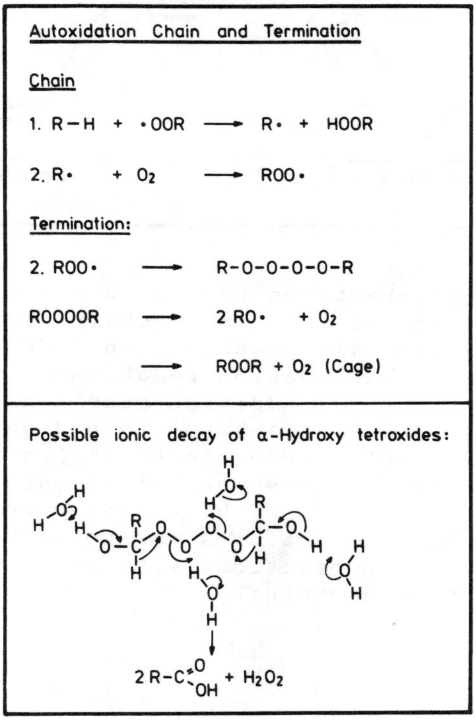

The selectivity of oxygen attack is determined in the hydrogen abstraction step, 1, which is also the slower step under usual conditions and, therefore, also partly responsible for the overall rate. As the O-H bond in hydroperoxides is a weak bond of $D = 90$ kcal/mol only few hydrogen transfer steps are exothermic. The weakest bonds, i.e. tertiary and benzylic C-H bonds, are oxidized with preference, indicating partial reactivity control by the stability of the radical products. This need not be necessarily the main factor, however, because also C-H bonds in the α-position of ethers, which are not particularly weak bonds, are oxidized with high rates. Favored SOMO-HOMO interactions offer an alternative explanation of great predictive power also in this example.

The high overall rate of autoxidation of ethers, which is frequently quoted as additional evidence for the stability of α-alkoxyalkyl radicals, is in fact due to the lack of an efficient chain termination step. The fastest dimerization process in this system leads to tetroxides, which mostly decompose again into radicals. α-C-H bonds of alcohols, which probably have quite similar bond strengths, are much more resistant to autoxidation . Alcoholic beverages, starch and cellulose, keep well, at least in principle, even if exposed to air. It has been shown that in aqueous solution a more efficient chain termination is available in this system.[12] The termination mechanism may be a fast ionic decomposition of the tetroxide which is favored by hydrogen bonding and by the possibility of fast proton transfer to and from the α-hydroxy groups. A possible reaction path is proposed on the slide.

Overall rate data and selectivity data of chain reactions, therefore, are not generally suited to obtain information about thermodynamic radical stabilities and, as a rule do not often depend on them.

Also kinetic stabilities, for which the term persistence was coined very appropriately[13] depend largely on the surroundings, i.e. the presence or absence of reaction partners. ESR spectra of reactive radicals, which are known to dimerize or disproportionate with diffusion controlled rates can be observed only under proper conditions of concentration and surroundings. The rates of recombination or disproportionation of conjugated radicals, like benzyl, allyl or capto dative substituted radicals, have also been shown to be diffusion controlled as long as the radicals are not sterically hindered.[14] Selectivities in radical-radical reactions generally are controlled by differences in activation entropy and not in activation enthalpy.[14b,c] Examples are the ratio of disproportionation to combination or the regio-selectivity or stereoselectivity of radical recombinations.[14c,15]

THE CC-CLEAVAGE REACTION.[16]

For obtaining information about the thermodynamic stability of radicals from reactivity, we have to choose a reaction in which the radicals are generated in an endothermic process. The thermal cleavage of CC-bonds is the prototype as far as carbon radicals are concerned.

In the birth hour of free radical chemistry M. Gomberg already has used this process for the generation of trityl radicals from their dimers and later on others studied extensively substituent effects on the equilibrium of the presumed hexaaryl ethanes with triarylmethyl radicals. For a long time they were interpreted by the stabilizing effects of the substituents in the triaryl methyl radicals.[17]

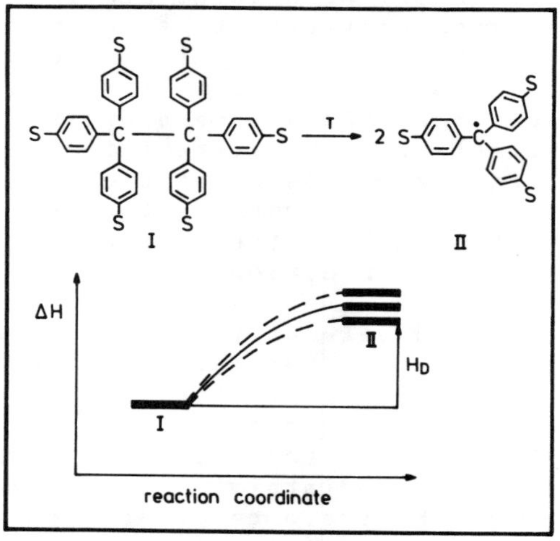

The data had to be reinterpreted, however, when it was recognized later that the dimers are indeed α-p-dimers and not α-α-dimers.[16a)]

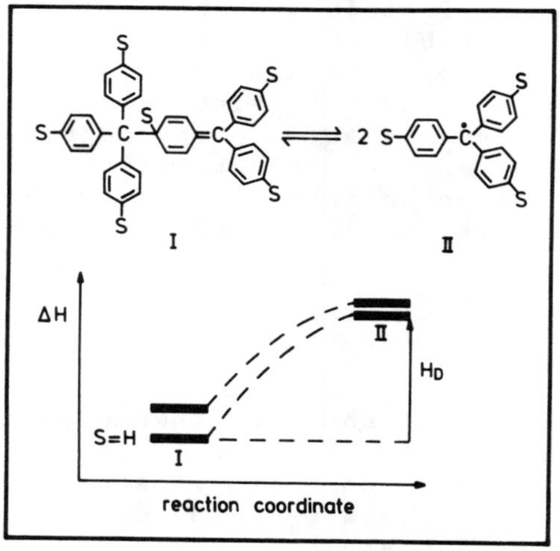

The fact that the largest disssociation constants were found when p-substituents were present has now become much better understood. Large substituents S destabilize the α-p-dimer by a steric effect, i.e. by repulsive interactions with the second trityl group in the dimer. The substituent effects on the radical stability turned out to be much smaller than originally thought. Steric and electronic effects were hardly separated from each other. Ziegler has confirmed the combined action of steric ground state effects of substituents and their radical stabilizing potential in a series of important papers on the kinetics of the CC-bond cleavage reactions in the fourties.[16a)]

It took 30 more years till the experimental and theoretical tools became available to achieve a quantitative separation of the steric and the electronic part of the substituent effects in the CC-cleavage reaction. This opened the way for determining substituent effects on the radical stability of carbon radicals. As a recent review is available,[16b)] I will only briefly outline how this was achieved.

For a series of strained hydrocarbons with a weak central CC-bond between two quaternary carbons a linear relationship between the free enthalpy of activation and their strain enthalpy was observed.

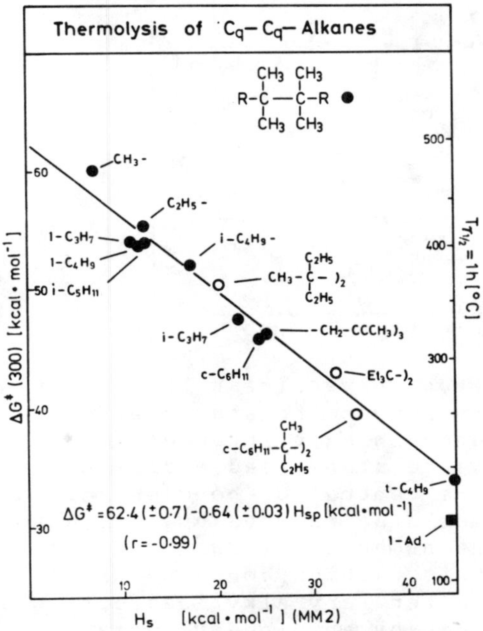

The former was determined from the kinetics and the latter from thermochemical work and force field calculations. The slope of the correlation was increased when a correction for residual strain in the radical was introduced. As we know from independent work that even triethyl methyl radicals dimerize with diffusion controlled rates, the activation enthalpies of these reactions become identical with bond dissociation enthalpies. The influence of strain on CC-cleavage of this type can be predicted, therefore, quantitatively from this relationship.

When stabilizing phenyl groups were introduced into the system the plot shown on the next slide, was obtained by a similar investigation.

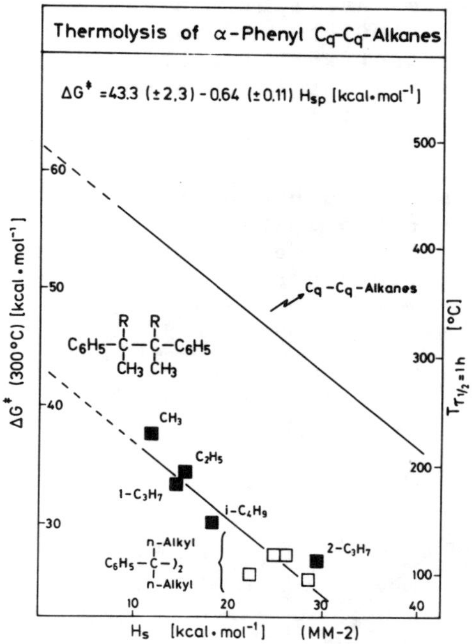

The two correlations have almost the same slope and, therefore, apparently respond to steric effects in the same fashion. The difference in axis intercept must be due to the fact, that resonance stabilized radicals are formed in the phenyl series. The central CC-bond energy and accordingly ΔH^{*}, should be reduced by twice the benzyl resonance energy. The numerical value of the resonance energy obtained in this fashion corresponds to the increase in stabilization of a tertiary alkyl radical, when an α-alkyl group is replaced by an α-phenyl group. It amounts to 8.4 kcal·mol^{-1} in our system and agrees perfectly with the difference in tertiary C-H bond energies of cumene and 2-methyl-propane (7.1-8.8 ± 1.5 kcal·mol^{-1}). A similar analysis with similar results (7.8 kcal·mol^{-1} stabilization) was performed for a series of hydrocarbons with a central CC-bond between two tertiary carbons.

We believe that this method of determining thermodynamic resonance energies of carbon radicals has several advantages.

1. Resonance energies are obtained from a reaction series and not from a single experiment.
2. The CC-cleavage reaction corresponds as closely as possible to the definition reaction for determining bond dissociation energies.
3. Steric effects are not ignored but are eliminated by experiment.

In the same way as shown for phenyl, resonance energies were also determined for other substituents.

Stabilisation Energies of Radicals

	H_r [kcal • mol^{-1}]	n
$CH_3-\dot{C}R_2$	≡0	13
$C_6H_5-\dot{C}R_2$	8.4	17
$R-C(=O)-\dot{C}R_2$	6.5	3
$N{\equiv}C-\dot{C}R_2$	5.5	7
$CH_3O-C(=O)-\dot{C}R_2$	3.5	6
$CH_3O-\dot{C}R_2$	1.3	7

The resonance stabilization decreases in the series α-keto-, α-cyano- and α-carbethoxy substitution and becomes smallest for α-methoxy groups. The latter are frequently hardly distinguished from methly groups in plots of the type shown before. For the following discussion I will show the correlations used for the analysis of the α-cyano and α-methoxy substituent effects.

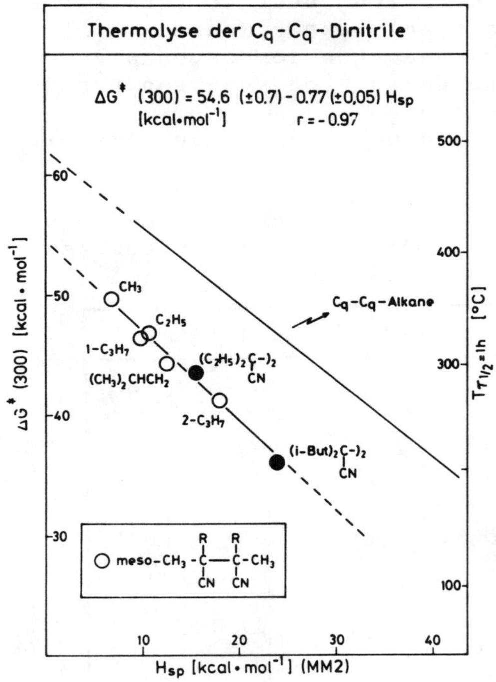

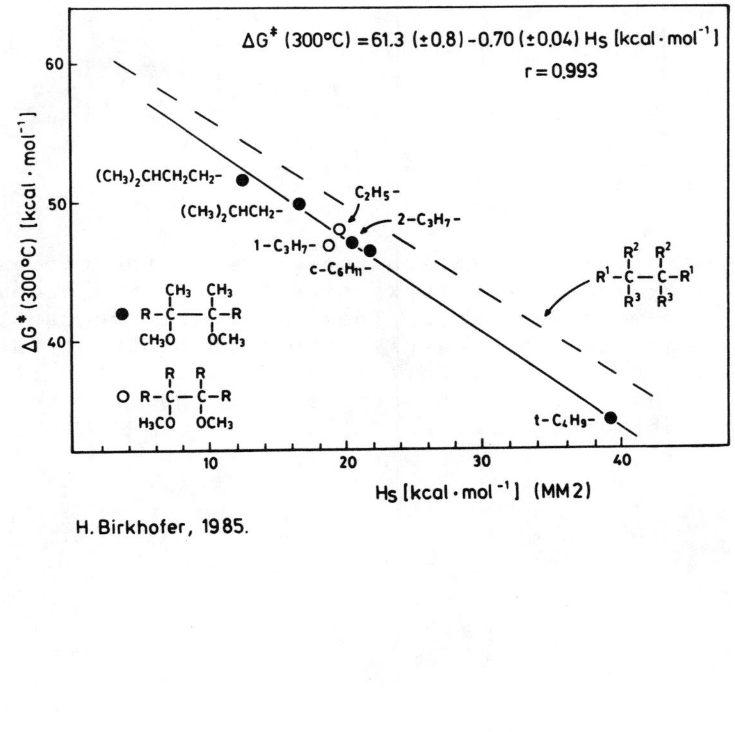

H. Birkhofer, 1985.

THE QUESTION OF ADDITIVITY OF SUBSTITUENT EFFECTS.[16b)]

It is a long standing question, whether the combined interaction of more than one substituent on a radical center results in an additive stabilization effect, or less, or possibly even more. In order to get some information from experiment we used the CC-cleavage reaction again as probe. Results are shown on the next slide.

Stabilisation Energies of Radicals [kcal • mol⁻¹]			
Radical	H_r(exp)	H_r(calc)	n
$(CH_3)_3C\cdot$	≡0	≡0	
$(C_6H_5)_2\dot{C}-R$	12	16.8	3
$(C_6H_5)_3C\cdot$	19	25.2	3
$R-\dot{C}(CN)_2$	9±5*	11	5
$C_6H_5-\dot{C}(CH_3)(CN)$	15	14	1
$C_6H_5-\dot{C}(CH_3)(OCH_3)$	9.4	9.7	5
$C_6H_5-\dot{C}(OCH_3)(CN)$	14	15.2	1 (2)
$(\triangleright)_3C\cdot$	3.2	—	1
$(\triangleright)_2\dot{C}-CN$	8.5	7.7	1

*corrected for change in dipol interaction

For the benzhydryl radical the CC-cleavage in about 7 compounds of different structure has been investigated now[18)] and the mean value for the stabilization energie is 11-12 ± 1 kcal/mol. In the trityl case, of course no symmetrical dimers could be analyzed. Instead, assymmetrical cleavage reactions were used, e. g. of 1.1.1-triphenyl-neopentane. By proper analysis of the steric contribution and by elimination of entropy influences, stabilization energies could be obtained. Our best value, which was obtained from 6 different model compounds, is presently 16 ± 2 kcal·mol⁻¹.[18)] The non additivity of the effect of more than one α-phenyl group must be mainly due to the non planarity of benzhydryl and trityl radicals, which also has been established by spectroscopy. In the benzhydryl and the trityl case we find 5.5-6.0 kcal/mol per phenyl, which, indeed, may signalize additivity for the non planar phenyl groups.

In the other radicals of the table carrying two hetero-substituents steric hindrance of resonance is not expected and is also not observed. The resonance energy for the malodinitrile radicals has a much greater error limit than the other data. The strain enthalpy H_s used for estimating the steric effect contains as a large component the dipolar interaction energy, which may be different for the starting material (which was not the dimer of the radical) and the radical products. In addition $\Delta S^{\neq}$ was not nearly constant in the series. A clear answer to wether or not additive stabilization is found for this radical cannot be given presently. The other entries in the table clearly support the addivity of radical stabilization by substituents.

In the meantime, we have investigated several further compounds which, on thermolysis, generate radicals bearing a donor and an acceptor substituent.[16b,19]. The aim was to get more reliable information about their thermodynamic stabilization energies, and in this way to make a contribution to the mechanistic interpretation of the reactions in which these types of radicals take part.

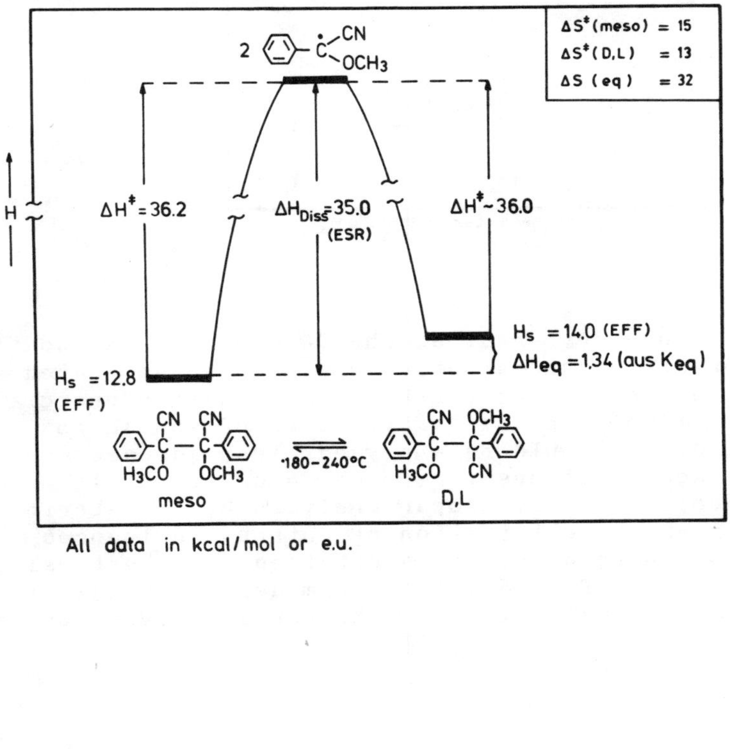

All data in kcal/mol or e.u.

These systems offer the great advantage that it is possible to determine not only the activation parameters of the decomposition reaction by experiment (rate of meso/D,L equilibrations, NMR), but also the dissociation enthalpies (by ESR) and the difference in heat of formation (by equilibration) for the two diastereomers. The system shown on the slide and most of our other model systems generate radicals, which cannot disproportionate. Therefore, the dissociation process and the accompanying meso/D,L-equilibration can be studied quite undisturbed. In this way, a deep insight into the reaction coordinate was possible, which was supported by EFF calculations. The two diastereomers, shown on the slide, differ very little in strain enthalpy. The near identity of the activation enthalpies and the dissociation enthalpy (which applies to the equilibrium mixture) support the assumption that even the dimerization of highly stabilized radicals is a process without activation enthalpy, as long as the strain enthalpy built up during the dimerization process is not high. Therefore, capto dative radicals, in general, have no particular kinetic stability. This has been supported by direct kinetic rate data from Sustmann's and Viehe's laboratory.[14a)]

It is interesting to note, however, that the entropies of activation are much smaller than the entropies of dissociation. At the transition state of a CC-dissociation process the bond enthalpy is reduced to zero, but the radicals have not yet achieved complete freedom of motion.

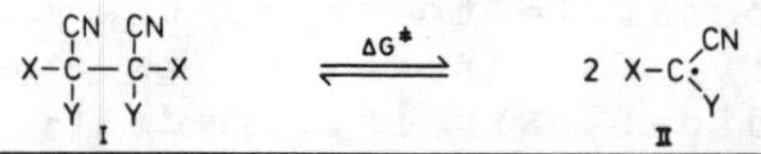

ΔH [kcal / mol]
ΔS [e.u.]

X	Y	meso/D,L	Kinetics (ΔT~ 40–50°C)[a)] ΔG (300°C)‡	ΔH‡	ΔS‡	ESR (ΔT~ 50°C)[b)] ΔH(diss)	ΔS(diss)	EFF (MM2) Hs (I)	Hs (II)[c)]
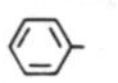	$-OCH_3$	meso	27.9	36.6	15.1	35.0±2	31.8±4	12.8 (12.8)[d)]	3.0
		D,L	28.4	36.0	13.2			14.0	
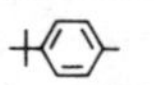	$-OCH_3$	meso	27.3	38.4	19.3	32.7±1	26.4±2	17.2	5.7
		D,L	27.8	37.9	17.5			18.6	
$(CH_3)_3C-$	$-OCH_3$	meso	28.6 29.4[e)]	34.8 34.5[e)]	10.9 10[e)]	34.3±1	41.2±1	33.7 ( )[d)] 36.3	3.0
		D,L	27.6	34.8	12.6				
$(CH_3)_2CH-$	$-OCH_3$	meso	36.6[f)]	44.2[f)]	13.2[f)]	—	—	23	0.3
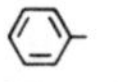	$-SCH_3$	meso	19.4	30.7	19.7	28.6±·1	32.8±·2	—	—
		D,L	19.1	31.7	22.0				

a) NMR b) refering to meso/D,L equilibrium mixture at T

c) Hs of XYCH–CN d) experimental value e) DSC f) GC

This points to the existence of a more or less loose but structured primary radical pair. More information concerning this point in other systems has been obtained by studying the selectivities or radical radical reactions, which are, apparently, generally entropy controlled.[15]

Very similar results for other systems are shown on the next slide.

Only in the p-t-butylphenyl substituted system a significant difference between the activation enthalpies and the enthalpy of dissociation is found. This could point to a small enthalpy of activation for the radical dimerization in this case, but no independent support for this is available. The larger entropies of activation than in the unsubstituted case (and the smaller entropy of dissociation) point to a looser structure of the radical pair intermediate, as expected for bulkier radicals.

Due to it's large ground state strain, H_s, the aliphatic t-butyl substituted system (entry 3) is of similar stability as the aryl substituted ones are. ΔH^* and $\Delta H(diss)$ agree even in this heavily strained system. The higher entropy of dissociation must be due to heavy steric hindrance in the ground state, which, because ΔS^* is small, is probably released only to a small part at transition state. Apparently also in aliphatic capto dative systems, tight radical pairs are the first intermediates. Freedom of rotation around bonds is only partially available at this stage. Very similar is the situation for the isopropyl derivative in entry 4. The smaller strain enthalpy of this dimer leads to a rather high enthalpy of activation and, therefore, it was not possible to study the dissociation equilibrium successfully. ΔS^*, however, is again small, pointing to a tight radical pair intermediate.

Finally, we investigated the first example of a system with participation of a thiomethyl group. ΔH^* and $\Delta H(diss)$ are about 6 $kcal \cdot mol^{-1}$ smaller than for the corresponding methoxy compound (entry 1), while the entropy situation is closely comparable. This could be due to about 3 $kcal \cdot mol^{-1}$ better stabilization of the radical fragments by the thiomethyl than the methoxy functional group. As long as the strain enthalpy of the thiomethyl system is not known, however, no safe conclusion can be drawn.

For estimating the stabilizing effect of the capto dative substitution pattern in relation to unsubstituted alkyl radicals, the strain enthalpies of all compounds are required. The data on the table are results of EFF calculations. One of them has been studied and another one is presently being checked by thermochemical work. In order to test if there is much residual strain in the radical products, the strain enthalpies of the corresponding hydrocarbons were also calculated. Due to the difference in hybridization, they must even overestimate the residual strain. This is supported by additional EFF calculations for the radicals themselves with the Beckhaus version of the MM2 force field for radicals.[16b)]

An experimental test for the reliability of these EFF-calculations are the meso/D,L equilibrium constants and the enthalpies of these equilibria determined from them. The latter agree perfectly with the differences in calculated strain enthalpies of the pairs of diastereomers. The stabilization energies obtained from these results are summarized on the next slide. Their error limit is probably ∿ ± 1 kcal·mol^{-1}.

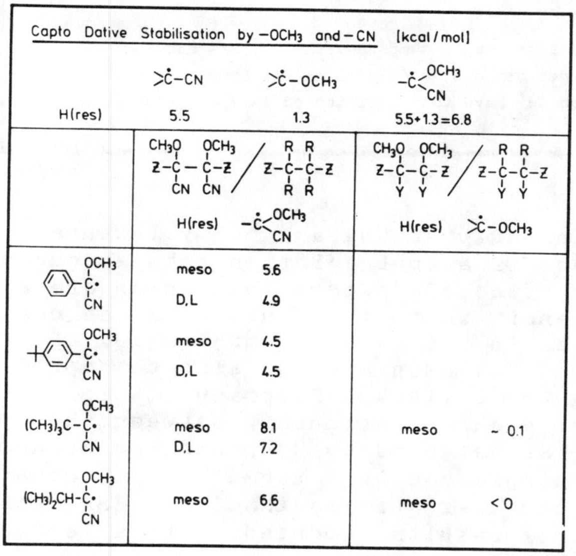

The results which I have presented today suggest that the radical center is stabilized by attached substituents in an additive manner. The validity of this statement depends, of course, on the precision of our methods for the determination of the stabilization energies.

On the slide the three main factors influencing this precision are listed once more: The correct estimation of changes in strain enthalpies and polar interaction energies in the dissociation process and the correctness of eliminating the temperature effect on ΔH^{*} via the compensation effect. I, myself, am quite convinced of the significance of the conclusion that additivity of substituent effects is the rule, despite the fact, that the error in the determination of the stabilization energies is probably ± 1 kcal/mol in most cases.

Problems of the determination of resonance energies of substituted alkyl radicals

1. H(res.) depends on the correct estimation of the strain enthalpy, H_s, of the starting material and the radical products.

2. Dipolar interactions are incorporated in H_s and must be estimated.

3. For different series of compounds thermolysis studies have to be performed at widely different temperatures. T-effects on $\Delta H^{\neq}$ have to be eliminated (compensation effect).

SUMMARY.

Let me address briefly at the end of my lecture, the question if there is a contradiction between our conclusions and other types of information, which are available.

1. A very sensitive tool for detecting delocalization of spin densities in radicals is, of course, ESR-spectroscopy. This method has been applied also to capto dative radicals mainly by Stella and Sustmann.[20] To the best of my knowledge, no clear relationship between the spin delocalization, as detected by ESR, and thermochemical stabilization energies has been formulated. I have the suspicion that the energetic basis of the ESR method and of the thermochemical results reported in this lecture is grossly different.

2. Quantum chemical calculations from several groups[21-23] indicate that stabilization of a radical center by capto dative substitution is larger than additive. These results, as far as I understand them, have been formulated in a qualitative way and I do not know of a quantitative prediction of stabilization enthalpies in $kcal \cdot mol^{-1}$.

3. Finally, let me comment on whether our results are in opposition to reactivity data, e.g. to the concept of capto dative reactivity in radical chemistry, which has been propagated and developed strongly by Viehe and his group in recent years,[24] or e.g. to the concept of stereoelectronic effects for radical cyclizations or to the reactivity in the α-position of hetero elements.

I have discussed in the early part of this lecture the arguments which contradict the proposition that "radical stability" is the main factor determining reactivity in radical chemistry in general, and in chain reactions in particular. Besides steric effects in many cases, frontier orbital interactions seem to be the decisive factor determining reactivity in radical chemistry and in particular for exothermic reaction steps which pass an early transition state.[10] Again, I am not aware of a direct relationship between the quality of FMO interactions and thermochemical radical stabilization energies.

I would consider it a great advantage if the group of experts present at this symposium could find some kind of agreement in answering these questions.

ACKNOWLEDGEMENT. Support of this work by the Deutsche Forschungsgemeinschaft and the Fonds der Chemischen Industrie is gratefully acknowledged.

REFEENCES

1) For a general discussion see e.g. a) C. Rüchardt, Angew. Chem. 82, 845, (1970); Angew. Chem. Int. Ed. Engl. 9, 830 (1970). b) C. Rüchardt et al., XXIII Int. Congress of Pure and Applied Chemistry, Vol. 4, special lectures, Butterworths, London, 1971 p. 223.

2) G. Russell in Free Radicals (J.K. Kochi, editor), Vol. I, p. 275, J. Wiley - Interscience, New York 1973.

3 A.D. Jenkins in Advances in Free Radical Chemistry (G.H. Williams, editor) Vol. II, p.139, Logos Press - Academic Press, London, 1967.

4) H.C. McBay, O. Tucker, A. Milligan, J. Org. Chem. 19, 1003, (1954).

5) A.L.J. Beckwith, K.U. Ingold in Rearrangements in Ground and Transistion States, Vol. I, (P. de Mayo, editor), p. 162, Academic Press, New York, 1980.

6) B. Giese, Acc. Chem. Research 17, 438 (1984) and previous publications.

7) see e.g. E.S. Huyser, Free Radical Chain Reactions, Wiley - Interscience, New York, 1970.

8) J.M. Tedder, Angew. Chem. 94, 433 (1982); Angew. Chem. Int. Ed. Engl. 21, 401 (1982).

9) B. Giese, Angew. Chem. 95, 771 (1983); Angew. Chem. Int. Ed. Engl. 22, 753 (1983).

10) see e.g. I. Fleming, Frontier Orbitals and Organic Chemical Reactions, 1. edition, J. Wiley, London, 1976, p. 182 ff.

11) F.R. Mayo (editor), Oxidation of Organic Compounds, Vol. I, American Chemical Society, Advances in Chemistry Series 75, 1968.

12) see e.g. C. von Sonntag, Int. J. Radiat. Biol. 46, 507 (1984) and references given.

13) D. Griller, K.U. Ingold, Acc. Chem. Research 9, 13 (1976).

14) for recent literature see H.G. Korth, P. Lommes, W. Sicking, R. Sustmann, Int. J. Chem. Kinetics 15, 267 (1983).
H.G. Korth, R. Sustmann, R. Merényi, H.G. Viehe, J. Chem. Soc. Perkin Trans. II, 1983, 67.

b) K. Münger, H. Fischer, Int. J. Chem. Kin. 16, 1213 (1984) and previous papers.

c) K.H. Eichin, H.-D. Beckhaus, S. Hellmann, H. Fritz, E.-M. Peters, K. Peters, H.-G. v. Schnering and C. Rüchardt, Chem. Ber. 116, 1787 (1983).

15) A. Peymann, Diplomarbeit Univ. Freiburg, 1983.
H.-J. Hähnle, Diplomarbeit Univ. Freiburg, 1984.

16a) C. Rüchardt and H.-D. Beckhaus, Angew. Chem. 92, 417 (1980); Angew. Chem. Int. Ed. Engl., 19, 429 (1980) and references given there.

b) C. Rüchardt and H.-D. Beckhaus, Topics in Current Chem. 130, 1 (1985).

17) see e.g. G.W. Whieland, Advanced Organic Chemistry, 2nd edition, J. Wiley, New York, 1949, p. 647 ff.

18) J. Schaetzer, Dissertation Univ. Freiburg, 1986 in preparation.

19) For earlier work see A. Zamkanei, J.H. Kaiser, H. Birkhofer, H.-D. Beckhaus and C. Rüchardt, Chem. Ber. 116, 3216 (1984).

20) L. Sylvander, L. Stella, H.-G. Korth, R. Sustmann, Tetrahedron Lett. 1985, 749.

21) D. Crans, T. Clark, P. von R. Schleyer, Tetrahedron Lett. 1980, 3681

22) M. Klessinger, Angew. Chem. 92, 937 (1980); Angew. Chem. Int. Ed. Engl. 19, 908 (1980).

23) G. Leroy, J. Mol. Struct. 120, 91 (1985) and previous publications.

24a) L. Stella, Z. Janusek, R. Merényi, H.G. Viehe, Angew. Chem. Int. Ed. Engl. 17, 691 (1979).

24b) H.G. Viehe, Z. Janusek, R. Merényi, L. Stella, Acc. Chem. Res. 18, 148 (1985).

THE DIFFERENT RECOMBINATIONS OF DIPHENYL METHYL RADICALS $Ph_2C^{\bullet}$-R, R = tBu, CN, COOR', COR'.

W. P. Neumann and R. Stapel
Universität Dortmund
Lehrstuhl für Organische Chemie I
Postfach 500500
D-4600 Dortmund 50
Federal Republic of Germany

ABSTRACT. The dimerization behaviour of the radicals $Ph_2C^{\bullet}$-R has been investigated. It depends on the kind of R. With R = CN they equilibrate with ethane type dimers, but with quinonoid ones, when R = tBu or COOR'. In the latter case, substituted para positions (with tBu) give rise to an ethane. New and unexpected dimers of an enol ether type have been found with R = COR', R' = H, Me, Ph. Tetraphenyl succinic acid (esters) reported in the literature in fact have quinonoid or, after 1,5-H migration, benzenoid structures. ESR data and dissociation enthalpies are given.

1. INTRODUCTION

Mesomeric free radicals $A^{\bullet} \rightleftharpoons A'^{\bullet}$ can form, in principle, three types of dimers, or even more if additional mesomeric forms have to be considered:

$$A\text{-}A \rightleftharpoons 2\,(A^{\bullet} \leftrightarrow A'^{\bullet}) \rightleftharpoons A\text{-}A'$$
$$\downarrow\uparrow$$
$$A'\text{-}A'$$

The classical example for this is Gombergs triphenyl methyl forming an A-A' type dimer (quinonoid) having been assumed to be an A-A type one (ethane like) for nearly seventy years. In fact, an A-A type dimer is formed if the meta and para positions are substituted by larger groups[1)], and also if the steric strain in A-A is diminished by using a 9-fluorenyl instead of a Ph_2C group[2)].

An early example of this behaviour is the isobutyro nitrile radical giving a metastable A-A' dimer, the ketenimine, and finally a stable A-A dimer, the tetramethyl succinodinitrile[3)].

2. RESULTS AND DISCUSSION

We report here on the corresponding dimerization behaviour of diphenylmethyl radicals, $Ph_2C^{\bullet}$-R, R = tBu, $SiMe_3$, CN, COOR', COR', depending on

H. G. Viehe et al. (eds.), Substituent Effects in Radical Chemistry, 219–222.
© 1986 by D. Reidel Publishing Co.

the kind of R, and describe a number of new radicals of this type.

R = tBu[4] and $SiMe_3$[5] cause, as we have found recently, a quinonoid dimer A-A' (I) because of the steric strain of R, even with the less bulky $SiMe_3$ residue, and the strain release, if R is situated at a sp^2-C atom.

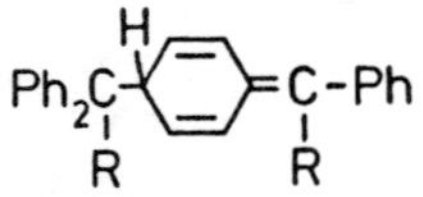
I

But, with R = CN an ethane type of dimer A-A (II)[6] has been reconfirmed now, in spite of the higher mesomeric stabilization of this radical (e.g., a^N = 2.0), and no (intermediate) ketenimine or other A-A' type dimer could be detected. The reversible dissociation of this dimer begins at 50°C.

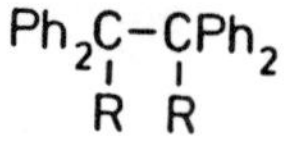
II

It was surprising, therefore, to find quinonoid dimers A-A' of type I with $Ph_2C^{\cdot}$-COOR', R' ranging from Me to tBu, see table 1.

This is contradictory to the numerous reports on the preparation of tetraphenyl succinic acid or its esters in the literature[7]. We repeated all of them to the best of our knowledge, but in all cases we got either the quinonoid I, R = COOR', or the product of the 1,5-H shift III generated from I very easily by traces of acids:

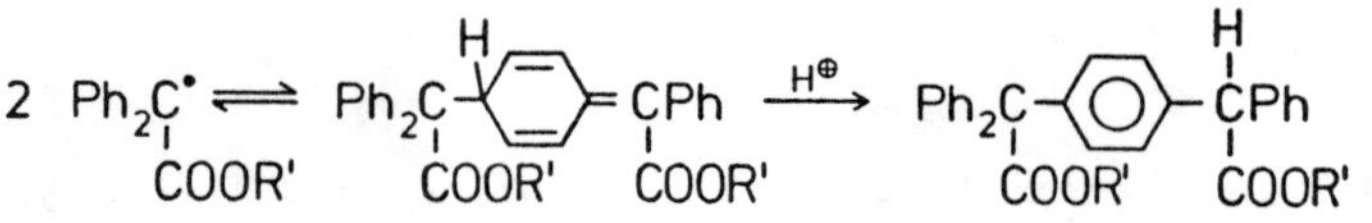
III

A real tetraphenyl succinic acid derivative - the first one to our knowledge - could be prepared when we blocked the four p-positions necessary for formation of the quinonoid dimer by tBu groups: An A-A dimer of the ethane type IV could be isolated. Its dissociation enthalpy is surprisingly close to that of the quinonoid I, see table 1.

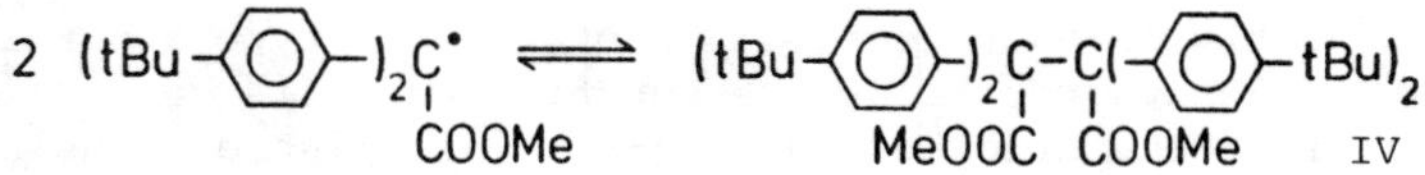
IV

Quite an unexpected dimer structure has been found when we investigated the radical $Ph_2C^{\cdot}$-COPh. It was neither the ethane II (R = COPh) as described earlier[8], nor the quinonoid I (R = COPh), but an A-A" typed enol ether V (R" = Ph) dissociating reversibly already below room temperature:

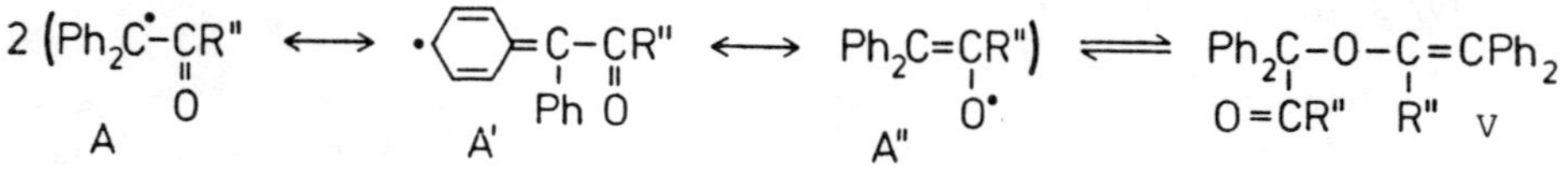
V

Table 1. ESR Data of Radicals $Ar_2C^{\cdot}-R$ and Dissociation Enthalpies of Their Dimers

Ar	R	a_o^H(G)	a_m^H(G)	a_p^H(G)	a^R(G)	T (°C)	Solv.	Dimer	H_{Diss} (kcal/mol)	Solv.
Ph	tBu	2.70	1.10	2.80	$a^H_{CH_3}$ = 0.24	20	PhH	I	18.5±0.2	PhH
Ph	$SiMe_3$	2.86	1.20	3.02	$a^H_{CH_3}$ = 0.17	21	PhH	I	19.6±2.0	Xylene
Ph	CO_2Me	2.80	1.20	3.02	$a^H_{CH_3}$ = 0.80	22	CCl_4	I	21.8±0.1	CCl_4 a)
Ph	CO_2Et	2.80	1.20	3.05	$a^H_{CH_2}$ = 0.75	22	PhH	I		
Ph	CO_2iPr	2.80	1.20	3.05	a^H_{CH} = 0.45	22	PhH	I		
Ph	CO_2tBu	2.83	1.22	3.10	$a^H_{CH_3}$ = 0.035	50	CCl_4	I	21.6±0.1	CCl_4
Ph	CO_2SiMe_3	2.72	1.20	2.95		22	PhH	I		
Ph	CO_2CH_2Ph	2.80	1.20	3.03	$a^H_{CH_2}$ = 0.75	40	PhH	I		
Ph	CHO	2.50	1.10	2.75	a^H_{CHO} = 0.98	100	Ph_2O	V	18.6±0.4	Ph-tBu
Ph	$COCH_3$	2.50	1.10	2.69	$a^H_{CH_3}$ = 0.70	100	Ph-tBu	V	20.0±0.2	Ph-tBu
Ph	COPh	2.82	1.20	3.09		22	PhH	V	17.7±0.2	Ph-tBu
Ph	CO-p-C_6H_4-tBu	2.86	1.20	3.17		22	CCl_4	V		
Ph	CN	3.00	1.20	3.20	a^N_{CN} = 2.00	100	Ph-tBu	II	29.1±1.5	PhCl b)
p-Anisyl	CN	3.00	1.00	a^H_{OMe}=0.50	a^N_{CN} = 2.00	60	Ph-Cl	II	24.7±1.2	PhCl
p-tBu-phenyl	CO_2Me	2.80	1.20	a^H_{tBu}=0.105	$a^H_{CH_3}$ = 0.80	22	CCl_4	IV	21.5±0.1	CCl_4

a) 21.3±0.1 in Ph-tBu; 25.4±0.1 in PhCl; b) 23.6±1.1 in Ph-tBu.

Thus, we have established a late verification of G. Wittigs old assumption[9] that this radical has some oxyl type character (A"). This is not a single exception, because with $Ph_2C^{\bullet}$-COMe and $Ph_2C^{\bullet}$-CHO we also found enol type dimers V, R" = Me, H. The reason for this different behaviour of the esters on the one, and the aldehyde, methyl and phenyl ketone on the other side is the easier enolization of the latter three types of radicals[10].

The diphenyl methyl radicals mentioned mostly are new, have been prepared by bromine abstraction from the corresponding methyl bromide by Cu-powder in good yields, and have been - also mostly for the first time - investigated by ESR, see table 1. The dimers have been identified by IR, ^1H, ^{13}C NMR at corresponding low temperatures. Their dissociation enthalpies are given in table 1.

We are grateful to the Deutsche Forschungsgemeinschaft for supporting this work.

3. REFERENCES

1) M. Stein, W. Winter, A. Rieker, Angew. Chem. **90**, 737 (1978)
2) H.A. Staab, K.S. Rao, H. Brunner, Chem. Ber. **104**, 2634 (1971)
3) G.S. Hammond, O.D. Trapp, R.T. Keep, D.L. Neff, J. Am. Chem. Soc. **81**, 4878 (1959)
4) A.K. Zarkadis, W.P. Neumann, R. Marx, W. Uzick, Chem. Ber. **118**, 450 (1985)
5) H. Hillgärtner, W.P. Neumann, W. Schulten, A.K. Zarkadis, J. Organometal. Chem. **201**, 197 (1980)
6) G. Wittig, U. Pockels, Ber. Dtsch. Chem. Ges. **69**, 790 (1936)
7) H. Bickel, Ber. Dtsch. Chem. Ges. **22**, 1537 (1889); H. Gilman, R.V. Young, J. Org. Chem. **1**, 315 (1936); B. Witten, F.Y. Wiselogle, J. Org. Chem. **6**, 584 (1941); W.G. Kofron, C.R. Hauser, J. Org. Chem. **35**, 2085 (1970); O.A. Kruglaya, L.I. Belousova, I.D. Kalikhman, N.S. Vyazankin, Zh. Obshch. Khim. **51**, 716 (1981)
8) A. Löwenbein, L. Schuster, Liebigs Ann. Chem. **481**, 106 (1930)
9) G. Wittig, V. Kairies, W. Hopf, Ber. Dtsch. Chem. Ges. **65**, 767 (1932)
10) The difference of energy between the keto and enol form is about 10 kcal/mol for aldehydes, 16 kcal/mol for ketones but 24 kcal/mol for esters: F. Klages, Lehrbuch der Organischen Chemie, 3. Aufl., Bd II, W. DeGruyter, Berlin 1962
On the other hand the enol form can be favoured by the steric and resonance effects of aryl groups: H. Gero, J. Org. Chem. **19**, 1960 (1962); H. Hart, Chem. Rev. **79**, 515 (1979).

SOME CHEMISTRY OF RADICAL IONS

J. M. Tedder
University of St. Andrews,
Department of Chemistry
St. Andrews
Fife KY16 9ST, Scotland

ABSTRACT. The paper is concerned with reactions involving radical cations in the gas phase. Three separate systems have been studied. These are: alkane radical cations with alkanes (*e.g.* $CH_4^{+\cdot} + C_4H_{10} \rightarrow$); alkene radical cations with alkenes (*e.g.* $C_2H_4^{+\cdot} + C_2H_4$) and the formation of aromatic radical cations (*e.g.* $C_6H_6^{+\cdot}$) in the reactions of alkyl cations with aromatic nuclei.

(a) THE REACTIONS OF $CH_4^{+\cdot}$ WITH LINEAR, BRANCHED AND CYCLIC ALKANES

SUMMARY. The reactions of $CH_4^{+\cdot}$ with linear, branched and cyclic alkanes have been studied in the gas phase using a triple quadrupole mass spectrometer. The initial reaction with $CH_4^{+\cdot}$ is charge exchange but the resulting cation-radical can fragment in two different ways, one involving the formation of a carbocation and a radical, and the other to the formation of a shorter chain alkane and a new cation--radical.

Discussion

The initial reaction of $CH_4^{+\cdot}$ with neutral linear alkanes C_nH_{2n+2} is electron transfer yielding methane and an excited alkane radical cation which fragments to yield a wide variety of products (Scheme 1).

$$C_nH_{2n+2} + CH_4^{+\cdot} \longrightarrow (\ddagger)^{+\cdot} \longrightarrow (C_nH_{2n+2})^{*+\cdot} + CH_4$$

$$\downarrow$$

fragmentation products

Scheme 1

The fragmentation patterns we have obtained when $CH_4^{+\cdot}$ reacts with normal alkanes are similar, but not identical to the 'cracking patterns'

H. G. Viehe et al. (eds.), Substituent Effects in Radical Chemistry, 223–244.
© 1986 by D. Reidel Publishing Co.

reported in electron impact studies.[1] The unpaired electron of the initial secondary ion $(C_nH_{2n+2})^{*+\cdot}$ is not likely to be centred at one of the terminal methyl groups. Typical secondary ion fragmentation can therefore be depicted as in Scheme 2. The carbocations $RCH_2CH_2^+$ or

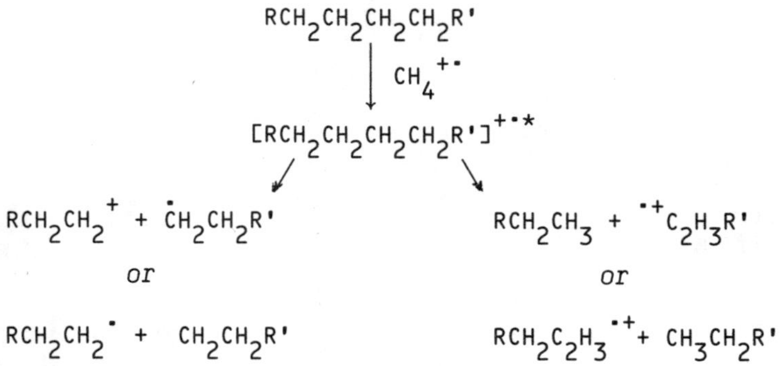

Scheme 2

$R'CH_2CH_2^+$ formed can undergo further fragmentation. The total spectrum (Figure 1) shows evidence of step-wise degradation of the alkyl cations yielding ethylene and a new carbocation of 28 mass units less (Scheme 3).

$$R{-}CH_2{-}CH_2^+ \longrightarrow R^+ + CH_2{=}CH_2$$

Scheme 3

This fragmentation will terminate either in $C_{2l}H_{2l+1}^+$ (*e.g.* $C_2H_5^+$ or $C_3H_7^+$; CH_3^+ and H^+ are less likely). There is a second fragmentation pathway, also initiated by charge exchange and this involves the formation of short chain alkanes C_lH_{2l+2} and the simultaneous formation of radical ions of the general formula $C_mH_{2m}^{+\cdot}$ (where $l + m = n$).

Tables I and II depict the fragmentation patterns formed when $CH_4^{+\cdot}$ and CH_3^+ react with the branch chain alkanes isobutane (2-methylpropane) and isopentane (2-methylbutane). Tables I and II show that the fragment ions for the reactions of $CH_4^{+\cdot}$ and CH_3^+ with n-butane and with isobutane (or with n-pentane and with isopentane) are the same. However, the relative proportions of the secondary ions derived from the linear alkanes are different from those derived from branched chain alkanes when $CH_4^{+\cdot}$ is the primary ion but are almost identical when CH_3^+ is the primary ion. The initial secondary cations in the reaction of straight chain or branched chain alkanes with CH_3^+ have the composition $(C_nH_{2n+1})^+$. Such carbonium ions can rearrange to give a more stable structure. In other words the carbonium ion formed by hydride abstraction from a straight chain ion will rearrange to a branched chain ion, *i.e.* similar to the ions formed by hydride abstraction from the corresponding branched chain alkane. In contrast the radical cation $[C_nH_{2n+2}]^{+\cdot}$, formed by electron transfer, rearranges as it fragments so that the cracking patterns from the straight chain alkane radical ions are different from the cracking patterns from the branched chain radical

TABLE I. Ions formed in the reaction of $CH_4^{+\cdot}$ and CH_3^+ with n-butane and 2-methylpropane (%)*

	m/e	27	38	29	41	42	43	55	56	57	58
$CH_4^{+\cdot}$	n-C_4		3	5	1	8	53		2	6	21
	iso-C4				1	32	51		2	6	8
CH_3^+	n-C_4	5		21	27		13	1		39	0
	iso-C4	1		26	25	1	18	1		29	1

*Figures rounded off to nearest whole number

TABLE II. Ions formed in the reaction of $CH_4^{+\cdot}$ and CH_3^+ with n-pentane and 2-methylbutane (%)*

	m/e	27	29	41	42	43	55	56	57	70	71	72
$CH_4^{+\cdot}$	n-C_5		1	1	32	37		2	7	5	3	12
	iso-C_5		1	1	31	18		13	16	4	6	10
CH_3^+	n-C_5	1	6	3	1	70	1		2		14	1
	iso-C_5	1	4	2	1	67	2	1	6		15	1

*Figures rounded off to nearest whole number

ions. These conclusions are consistent both with the rearrangements reported by Gaumann[2] and by Wolkoff[3] and his co-workers.

Figure 2 depicts the fragmentation patterns formed when $CH_4^{+\cdot}$ and CH_3^+ react with simple cycloalkanes. Neither series of mass spectra shown in Figure 2 consist of stepwise fragmentation characteristic of the corresponding fragmentations of the linear alkanes described above. The parent ions (from the $CH_4^{+\cdot}$ series) or the parent less one mass unit (from the CH_3^+ series) are relatively much more important than in the spectra of the linear alkanes. In the spectra derived from the CH_3^+ primary ions with methylcyclopentane and methylcyclohexane, the secondary ion $C_4H_7^+$ (*m/e* 55) is the major charged fragment. The ion $C_3H_5^+$ (*m/e* 41), possibly the allyl cation, is also an important ion in all the CH_3^+-initiated reactions.

The fragmentation patterns observed for the linear alkanes are consistent with a step-wise loss of C_2H_4 from the initially formed radical cation ($C_nH_{2n+2}^{+\cdot}$ from $CH_4^{+\cdot}$) or carbocation ($C_nH_{2n+1}^+$ from CH_3^+). However there are other possible fragmentation pathways and for each sequence the observed secondary ions have been formed in the ion source, isolated in the first quadrupole and there subjected to collision-induced decomposition (c.i.d.) using nitrogen as the inert molecule. A degradation scheme of this kind for $C_8H_{18}^{+\cdot}$ is shown in Scheme 4. (Similar schemes have been worked out for all the other linear radical cations.) The particular fragmentation pathway will be partly determined by the neutral species produced. Thus the formation of $C_3H_5^+$ in the fragmentation of $C_8H_{18}^{+\cdot}$ (Scheme 4) is associated with the

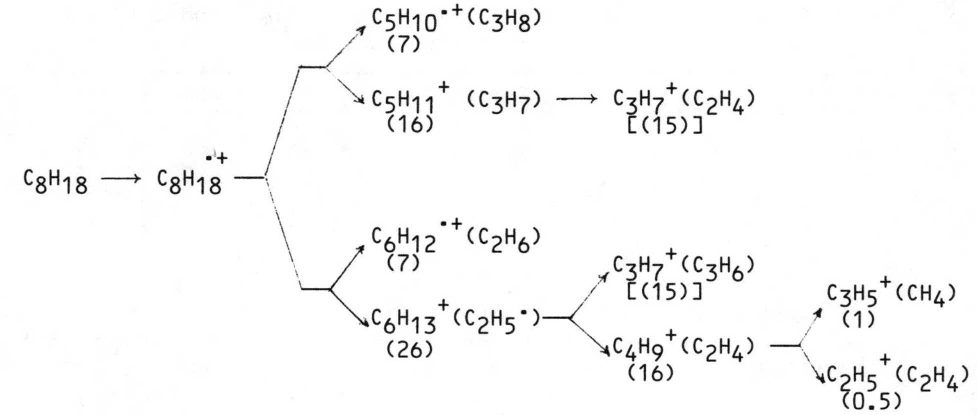

Scheme 4 The successive c.i.d. spectra (nitrogen the collision gas) of the ionic fragments occurring in the reaction between $CH_4^{+\cdot}$ and C_8H_{18} (relative abundance).

production of CH_4 and the production of $C_2H_5^+$ with the coproduction of C_2H_4. The principal ions produced by the c.i.d. of $C_6H_{14}^{+\cdot}$ and $C_6H_{13}^+$ are compared in Scheme 5.

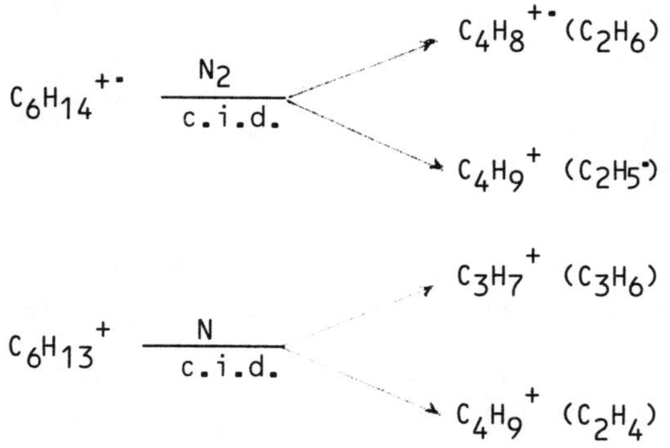

Scheme 5 The principal fragment ions formed in collision with nitrogen

The relative proportion of daughter ions derived from a particular fragment ion formed from different neutral molecules can be compared. Thus $C_4H_9^+$, on collision with nitrogen, fragments to $C_3H_6^+$, $C_3H_5^+$, and $C_2H_5^+$ and their relative proportions from different precursors are shown in Table III. Similar schemes can be drawn for the fragmentation of $C_6H_{13}^+$ and $C_5H_{11}^+$.

The fragmentation pattern observed when $CD_4^{+\cdot}$ replaces $CH_4^{+\cdot}$ as primary ion differed little, but there were additional peaks in the fragmentation pattern observed when CD_3^+ replaced CH_3^+. New ions observed (in addition to those in the ions induced by CH_3^+) in the reaction of CD_3^+ with n-octane included *m/e* 30, 31, 32, 34, 46, 48, 56, 73, 74, 84, 86, 88, 98, and 100. The majority of these additional ions

TABLE III. % Fragmentation of *m/e* 57 formed from different alkanes

$RH + CH_4^+ \longrightarrow$

C_4H_9 (*m/e* 57) $\xrightarrow[\text{c.i.d.}]{N_2}$	$C_3H_6^+$(42)	$C_3H_6^+$(41)	$C_2H_5^+$(29)
R = C_6H_{13}		79	21
C_7H_{15}		83	17
C_8H_{17}	1	24	75
C_9H_{19}		87	13

were present in low concentrations and while they are evidence for hydrogen atom scrambling, it was not a major process.

The reaction of $CH_4^{+\bullet}$ with n-heptane was also investigated using the tandem mass spectrometer described in earlier papers. The primary ions $(CH_4)^{+\bullet}$ were injected into the collision chamber with *ca.* 8 V. The predominant secondary ion was the parent (*m/e* 100). The increase in the proportion of the parent ion (*m/e* 100) in the tandem mass spectrometer illustrates the different discrimination of the two instruments (Table IV). The tandem mass spectrometer preferentially selects ions formed by 'long range' transfer since ions formed by actual collision will tend to move out of the extraction zone.

TABLE IV Ions formed by the reaction of $CH_4^{+\bullet}$ with n-heptane (a) in the triple quadrupole and (b) in the tandem mass spectrometer

m/e	43($C_3H_7^+$)	56($C_4H_8^+$)	57($C_4H_9^+$)	70($C_5H_{10}^+$)	71($C_5H_{11}^+$)	100(P^+)
Triple quadrupole	16	11	20	13	24	9
Tandem	14	7	24	12	15	28

The present results with the triple quadrupole were all taken when the difference in potential between the ion source and the first quadrupole was 0.3 V. However, very little difference in the relative proportions of the various fragmentation ions was observed when the potential difference between the source and the first quadrupole was varied from 0 to 2 V.

Conclusions

Although the reactions of the alkyl radical ions $(C_nH_{2n+2}^{+\bullet})$ with alkanes has received considerable study, there has been remarkably little reported about the chemistry of isolated $CH_4^{+\bullet}$and still less about the reactions of CH_3^+ with alkanes. The present work is an attempt to rectify this omission. The results obtained with the triple quadrupole mass spectrometer are completely consistent with the earlier chemical ionisation studies. The principal fragmentation pathways of $C_nH_{2n+2}^{+\bullet}$ and $C_nH_{2n+1}^+$ are very different and even more striking are the differences of cyclo-$C_nH_{2n}^{+\bullet}$ and cyclo-$C_nH_{2n-1}^+$. The important feature of the present study is that the fragmentation pathways of alkyl radical cations and carbocations have been clearly identified. This information is essential if the ion-molecule chemistry involving ions and molecules containing heteroatoms attached to alkyl chains is to be understood.

References

1. J.H. Beynon, R.A. Saunders, and A.E. Williams, 'The Mass Spectra of Organic Molecules', Elsevier, London, 1968, p.93.
2. R. Liardon and T. Gäumann, *Helv. Chim. Acta*, 1969, **52**, 1042.
3. P. Wolkoff and J.L. Holmes, *J. Am. Chem. Soc.*, 1978, **100**, 7346; J.L. Holmes, P.C. Burgers, M. Yousuf, A. Mollah, and P. Wolkoff, *ibid.*, 1982, **104**, 2879; P. Wolkoff, S. Hammerum, and J.L. Holmes, *Int. J. Mass Spectrom Ion Phys.*, 1983, **47**, 343; J.F. Wendelboe, R.D. Bowen, and D.H. Williams, *J. Chem. Soc., Perkin Trans.2*, 1981, 958.

(b) THE REACTIONS OF ALKENYL RADICAL CATIONS $C_nH_{2n}^{+}$ WITH ALKENES

SUMMARY. The gas phase reactions of ethenyl ($C_2H_4^{+\cdot}$), propenyl ($C_3H_6^{+\cdot}$) butenyl ($C_4H_8^{+\cdot}$) with each of ethene, propene, but-1-ene, *cis*-but-2-ene and 2-methylpropene have been reinvestigated using a triple quadrupole mass spectrometer.

Discussion

<u>The Interaction between the Ethenyl Radical Cation $C_2H_4^{+\cdot}$ with Ethene</u>

The principal reaction between the ethenyl radical ions $C_2H_4^{+\cdot}$ and ethene is disproportionation to yield $C_3H_5^+$ (*m/e* 41) and a neutral methyl radical:

$$C_2H_4^{+\cdot} + C_2H_4 \longrightarrow [C_4H_8]^{*+\cdot} \longrightarrow C_3H_5^{+} + CH_3^{\cdot}$$

The only other secondary ion formed in appreciable yield is $C_4H_7^+$ (*m/e* 55) with hydrogen atoms being ejected:

$$C_2H_4^{+\cdot} + C_2H_4 \longrightarrow [C_4H_8]^{*+\cdot} \longrightarrow C_4H_7^{+} + H^{\cdot}$$

The principal tertiary ion is $C_5H_9^+$ (*m/e* 69). In separate experiments the ion $C_3H_5^+$ (prepared from a variety of sources) was reacted with ethene and gave $C_5H_9^+$ in high yield. Hence in the present study this ion is chiefly formed by the addition of $C_3H_5^+$ ions with ethene:

$$C_3H_5^{+} + C_2H_4 \longrightarrow [C_5H_9]^{*+} \xrightarrow{M} C_5H_9^{+}$$

The corresponding tertiary ion $C_6H_{11}^+$ (*m/e* 83) derived from $C_4H_7^+$ (*m/e* 55 is also observed but in very low concentration. As the pressure of ethene increased so the proportion of $C_6H_{11}^+$ increased while the proportion of $C_4H_7^+$ decreased. In separate experiments the reaction of $C_4H_7^+$ with ethene gave $C_6H_{11}^+$ in high yield.

$$C_4H_7^+ + C_2H_4 \longrightarrow [C_6H_{11}]^{*+} \xrightarrow{M} C_6H_{11}^+$$

The alternative fate of the two secondary ions is fragmentation.

The c.i.d. spectrum of $C_3H_5^+$ consists of a single fragment $C_3H_3^+$ but the c.i.d. spectrum of $C_4H_7^+$ consists of $C_2H_5^+$, $C_3H_3^+$ and $C_4H_5^+$, in relative proportions which vary slightly depending on the nature of the collision gas (0.4 : 0.3 : 0.3).

The remaining product ions occurring in appreciable concentration were quaternary ions and their origin was established by a series of independent experiments:

(i) $C_5H_7^+$ whose concentration increased as the concentration of $C_5H_9^+$ increased, was formed in the c.i.d. of $C_5H_9^+$

$$C_5H_9^+ + M \longrightarrow C_5H_7^+ (H_2)\ 25\%, \quad C_3H_5^+ (C_2H_4)\ 75\%$$

[The $C_5H_9^+$ ions were prepared by electron impact on cyclopentane and cyclohexane.]

(ii) $C_7H_{13}^+$ was formed by the addition of $C_5H_9^+$ to C_2H_4. Other ions observed included $C_3H_3^+$ 62%, $C_5H_2^+$ 9%.

$$C_5H_9^+ + C_2H_4 \longrightarrow C_7H_{13}^+ \quad 6\%$$

(iii) C_8H_{15} was formed similarly

$$C_6H_{11}^+ + C_2H_4 \longrightarrow C_8H_{15}^+$$

We can now draw a reaction scheme for the principal forward interactions of $C_2H_4^{+\cdot}$ with C_2H_4 (many of the steps are reversible but these are not shown).

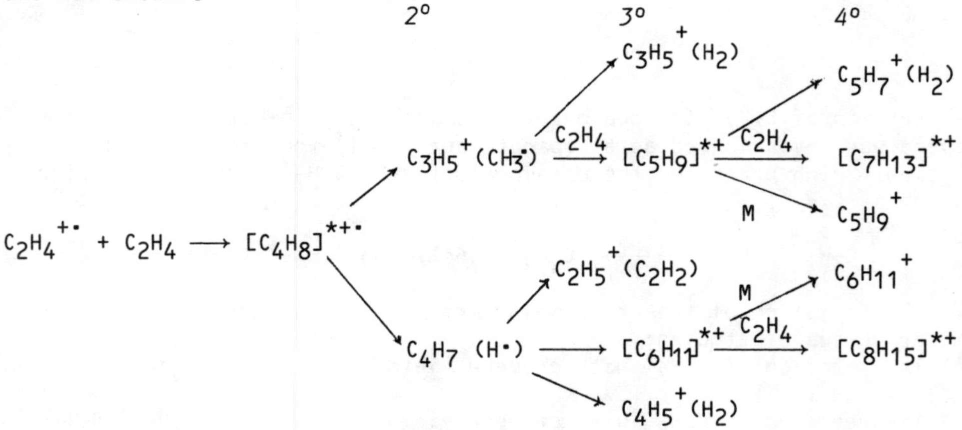

In addition there were trace amounts (*i.e.* less than 10^{-2} of the principal secondary ions) of odd electron ions *m/e* $C_4H_6^{+\cdot}$, $C_3H_4^{+\cdot}$, and $C_2H_2^{+\cdot}$. These three ions are probably secondary ions derived directly from $[C_4H_8]^{*+\cdot}$:

$$C_4H_8^{*+\bullet} \longrightarrow \begin{cases} C_2H_2^{+\bullet} \ (C_2H_6) \\ C_3H_4^{+\bullet} \ (CH_4) \\ C_4H_6^{+\bullet} \ (CH_2) \end{cases}$$

Another very minor ion is $C_2H_3^+$ which may also be derived from the initial adduct. Unlike the odd electron ions whose concentration

$$C_4H_8^{*+\bullet} \longrightarrow C_2H_3^+ + C_2H_5\bullet$$

decreases rapidly with increasing ethylene pressure, the concentration of $C_2H_3^+$ only decreases slowly.

The Interaction between Propenyl Radical Cations $C_3H_6^{+\bullet}$ and Propene

The very striking feature of the series of reactions initiated by the interaction of $C_3H_6^{+\bullet}$ with C_3H_6 is the occurrence of two reaction sequences, one involving even electron ions and the other involving odd electron ions. This is in sharp contrast with the ethenyl/ethene system where all the major ions were even electron species. We will consider the even electron reactions first.

There are three secondary even electron ions $C_3H_7^+$ (*m/e* 43), $C_4H_7^+$ (*m/e* 55) and $C_5H_9^+$ (*m/e* 69). The latter ion, $C_5H_9^+$, is derived from two separate routes, one of which is directly from the initial complex.

$$C_3H_6^{+\bullet} + C_3H_6 \longrightarrow [C_6H_{12}]^{*+\bullet} \longrightarrow \begin{cases} C_3H_7^+ \\ C_4H_7^+ \\ C_5H_9^+ \leftarrow \text{also from another route} \end{cases}$$

The reaction of $C_3H_7^+$ (formed by electron impact on n-butane) with propene was investigated as a separate series of experiments and at relatively high pressures the main product from $C_3H_7^+$ with the starting propene is $C_4H_9^+$.

$$C_3H_7^+ + C_3H_6 \longrightarrow C_4H_9^+ \ (66\%) \text{ and } C_6H_{13}^+ \ (16\%)$$

The formation of the other main tertiary ions was elucidated in a similar series of studies:

(i) The reaction of $C_4H_7^+$ with propene yielded $C_7H_{13}^+$ and $C_5H_9^+$ (+ C_2H_4).
(ii) The reaction of $C_5H_9^+$ with propene yielded $C_8H_{15}^+$ and $C_7H_{13}^+(C_2H_2)$.
(iii) $C_7H_{13}^+$ ions underwent c.i.d. to provide another source of $C_5H_9^+(+C_2H_4)$.

We can now draw a reaction scheme of the principal forward reactions arising from the reaction of $C_3H_6^+$ with C_3H_6.

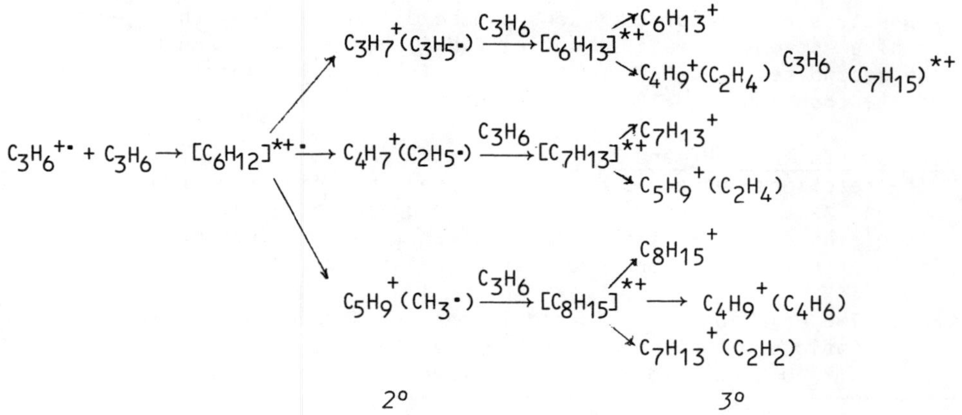

In the ethene reactions all the predominant ions were even electron ions, but odd electron reactions represent important processes in the propene system. Not only is $C_4H_8^{+\cdot}$ an important secondary ion, but $C_5H_{10}^{+\cdot}$ is also a major ion. The concentration dependence of this latter ion shows it to be a tertiary ion.

$$C_3H_6^{+\cdot} + C_3H_6 \rightarrow [C_6H_{12}]^{*+\cdot} \rightarrow C_4H_8^{+\cdot}(C_2H_4) \xrightarrow{C_3H_6} C_5H_{10}^{+\cdot}(C_2H_4)$$

There seems no doubt that $C_3H_7^{+}$ and $C_4H_8^{+\cdot}$ are derived from a common precursor, and both are associated with the formulation of stable species, the allyl radical and the ethene molecule respectively.

$$[C_6H_{12}]^{*+\cdot} \longrightarrow C_4H_8^{+\cdot}\ (C_2H_4)$$
$$[C_6H_{12}]^{*+\cdot} \longrightarrow C_3H_7^{+}\ (C_3H_5\cdot)$$

The Interaction between Butenyl Radical Cations with But-1-ene, *cis*-But-2-ene, and 2-Methylprop-1-ene

The principal product ions from the interaction of the butenyl radical cations with the corresponding butenes are listed below in Table V. The first feature of this table is that all the product ions are fragment ions (2° and 3°); and no ions with more than seven carbon atoms were detected in appreciable concentration.

TABLE V Principal Product Ions from the Interaction between Butenyl Radical Ions and the Corresponding Butene: % Secondary Ion Signal at *ca.* 5 x 10^{-3} mbar (3.75 x 10^{-3} torr)

Olefin	$C_3H_5^+$	$C_4H_7^+$	$C_4H_9^+$	$C_5H_9^+$	$C_5H_{10}^{+\cdot}$	$C_5H_{11}^+$	$C_6H_{11}^+$	$C_6H_{12}^{+\cdot}$
$C_2H_5CH{=}CH_2$	7	8	16	8	12	17	9	10
$CH_3CH\overset{C}{=}CHCH_3$	4	6	8	6	8	5	33	26
$(CH_3)_2C{=}CH_2$	2	1	87	1	2	5	0.5	0.4

Two separate series of product ions can be distinguished with an even number of electrons (the first series $C_3H_5^+$, $C_4H_7^+$, $C_5H_9^+$ and $C_6H_{11}^+$ and the second series $C_4H_9^+$ and $C_5H_{11}^+$). There is in addition a series of odd electron ions $C_4H_8^{+\cdot}$, $C_5H_{10}^{+\cdot}$ and $C_6H_{12}^{+\cdot}$.

Cross Products (a) C_2H_4 and C_3H_6

The reactions between a neutral alkene and a different radical cation can easily be studied in a multiple quadrupole mass spectrum. The results of the cross reaction between ethenyl cations into propene and propenyl cations with ethene are listed in Table VI.

TABLE VI The Principal Ions formed in the Cross Reactions of Ethenyl Cations with Propene and Propenyl Cations with Ethene at 5×10^{-3} mbar (3.75 torr) : % Secondary Ion Signal

	$C_2H_4^+$	$C_3H_5^+$	$C_3H_6^{+\cdot}$	$C_3H_7^+$	$C_4H_7^+$	$C_4H_8^{+\cdot}$	$C_4H_9^+$	$C_5H_7^+$	$C_5H_9^+$	$C_5H_{10}^{+\cdot}$	$C_6H_{11}^+$	$C_8H_5^+$
$C_2H_4^{+\cdot} + C_2H_4$	P	21	-	-	5	-	-	11	52	-	2	-
$C_3H_6^{+\cdot} + C_2H_4$	-	8	P	-	49	-	-	2	6	12	16	-
$C_2H_4^{+\cdot} + C_3H_6$	P	2	5	9	6	9	29	1	14	8	1	1
$C_3H_6^{+\cdot} + C_3H_6$	-	2	P	8	6	8	28	1	14	10	1	3

The principal reaction pathways for the interaction of ethene and propene with the corresponding cations can be summarised in the following sequences:

$$C_2H_4^{+\cdot} + C_2H_4 \longrightarrow [C_4H_8]^{*+\cdot} \longrightarrow C_3H_5^+(CH_3\cdot) \xrightarrow{C_2H_4} C_5H_9^+$$

$$C_3H_6^{+\cdot} + C_2H_4 \longrightarrow [C_5H_{10}]^{*+\cdot} \longrightarrow C_4H_7^+(CH_3\cdot) \xrightarrow{C_2H_4} C_6H_{11}^+$$

$$\left.\begin{array}{l} C_3H_6^{+\cdot} + C_3H_6 \\ C_2H_4^{+\cdot} + C_3H_6 \longrightarrow C_3H_6^{+\cdot} + C_3H_6 \end{array}\right\rangle \longrightarrow [C_6H_{12}]^{*+\cdot} \longrightarrow$$

$$C_5H_9(CH_3\cdot) \xrightarrow{C_3H_6} [C_8H_{15}]^{*+} \longrightarrow C_4H_9^+(C_4H_6)$$

The chief secondary ion in the reaction of an ethene radical cation with a neutral ethene molecule is $C_3H_5^+$ (*m/e* 41), but as the pressure of the ethene is raised the yield of the tertiary ion $C_5H_9^+$ (*m/e* 69) increases until at high pressures it becomes the predominant ion. The interaction between the propene radical cation and ethene is endothermic and the extent of reaction is much less, but the predominant secondary ion is $C_4H_7^+$ (*m/e* 55) which at the higher ethene pressure yields $C_6H_{11}^+$ (*m/e* 83) as a tertiary ion. The product ions from the reaction of the propene radical cation with propene are the same as those derived from the reaction of the ethene radical cations with propene, and the relative yields are the same, because the principal reaction of the ethene radical cation with propene is charge exchange.

$$C_2H_4^{+\cdot} + C_3H_6 \longrightarrow C_2H_4 + C_3H_6^{+\cdot}$$

Cross Products (b) C_2H_4 and C_4H_8

Table VII shows the principal reaction of the ethene radical cation with but-2-ene and but-2-ene is simple charge exchange. The

TABLE VII The Cross Reactions of Ethenyl Cations and Butenyl Cations with Ethene and two Butenes at 5 x 10^{-3} mbar (3.75 x 10^{-3} torr): % Secondary Ion Signal

	$C_3H_5^+$	$C_3H_7^+$	$C_4H_7^+$	$C_4H_8^{+\cdot}$	$C_4H_9^+$	$C_5H_7^+$	$C_5H_9^+$	$C_5H_{10}^{+\cdot}$	$C_5H_{11}^+$	$C_6H_{10}^{+\cdot}$	$C_6H_{11}^+$	$C_6H_{12}^{+\cdot}$
$C_2H_4^{+\cdot}$ + $C_2H_5CH{=}CH_2$	5	2	9	41	11	1	6	6	7	2	6	4
$C_2H_4^{+\cdot}$ + $CH_3CH{=}CHCH_3$	1	1	2	38	10	1	5	3	3	1	18	16
$C_2H_4^{+\cdot}$ + $CH_2{=}C(CH_3)_2$	2	1	2	8	76	-	2	2	5	1	1	1
$(C_2H_5CH{=}CH_2)^{+\cdot}$ + C_2H_4	18	5	21	P	-	6	29	5	2	-	10	5
$(CH_3CH{=}CHCH_3)^{+\cdot}$ + C_2H_4	15	1	11	P	-	11	37	1	1	-	3	19
$(CH_2{=}C(CH_3)_2)^{+\cdot}$ + C_2H_4	39	4	18	P	-	9	19	2	-	-	-	9

yields of secondary ions from the reactions of the two butene radical cations with ethene are low as would be expected for endothermic processes. The principal ions are all fragments derived from the simple adduct:

$$C_4H_8^{+\cdot} + C_2H_4 \longrightarrow [C_6H_{12}]^{*+\cdot} \longrightarrow \begin{cases} C_3H_5^+ \ (C_3H_7\cdot) \\ C_4H_7^+ \ (C_2H_5\cdot) \\ C_5H_9^+ \ (CH_3\cdot) \end{cases}$$

The reactions of the but-1-enyl and but-2-enyl cations, formed by charge exchange with ethenyl ions, are those already observed in the reactions of butenyl cations with the neutral butenes (see Table V). In particular the high yields of $C_6H_{11}^+$ and $C_6H_{12}^{+\cdot}$ in the reaction of ethenyl cations with but-2-ene should be noted.

Three studies of the reaction of the ethenyl radical cation with ethene are summarised in Table VIII. The most striking feature of the Table is the rapid decrease in the yield of the $C_3H_5^+$ (*m/e* 41) ion as the pressure of ethene increases. This decrease in the yield of $C_3H_5^+$ is matched by a dramatic increase in the yield of $C_5H_9^+$ (*m/e* 69). The importance of being able to study the reactions over a wide range of

TABLE VIII The Reaction of Ethenyl Radical Cations with Ethene Molecules expressed as % of Total Secondary Ion Current

m/e	Tiernan & Futrell 5×10^{-6} torr	Tiernan & Futrell 6×10^{-5} torr	Field 2×10^{-4} torr	Present Work 4.3×10^{-4} torr	Present Work 2.5×10^{-3} torr	Present Work 7.8×10^{-3} torr	Wexler & Marshall 2×10^{-2} torr
26	-	-	0.9	4.5	1.1	-	0.1
27	-	-	7.7	1.8	0.6	0.5	0.2
29	2.3	1.9	19.3	-	1.7	0.6	0.7
39	0.6	0.7	4.2	2.8	2.3	3.6	5.2
40	0.9	1.1	1.2	2.2	0.7	0.1	0.1
41	84	80	52	77	49	11	5
53	-	0.2	3	0.5	0.8	1.4	3.1
54	0.4	0.4	0.3	0.3	0.5	0.4	0.2
55	10	9.4	4.2	7.0	7.4	3.7	2.0
56	0.2	0.5	1.5	0.2	0.2	0.7	6.0
57	-	-	0.2	-	0.1	0.4	5.5
67	-	0.5	0.6	1.5	8.1	12	2.8
69	0.1	3.1	2.1	1.8	27	62	60
79	-	-	0.2	-	-	0.3	1.2
81	-	-	0.01	-	-	-	1.3
83	-	-	0.05	-	0.6	2.7	3.4
97	-	-	-	-	-	0.3	1.0

pressures illustrates one of the advantages of a multiple quadrupole. The proportion of secondary ions will in general decrease as the pressure rises (*e.g.* $C_2H_5^+$ *m/e* 29; $C_3H_5^+$ *m/e* 41; $C_4H_7^+$ *m/e* 55; $C_3H_3^+$ *m/e* 39) while tertiary and higher order ions will increase (*e.g.* $C_4H_5^+$ *m/e* 53; $C_5H_7^+$ *m/e* 67; $C_5H_7^+$ *m/e* 69; C_5H_9, $C_6H_{11}^+$ *m/e* 83). The present results bridge the data obtained previously. They show that the large cross section for *m/e* 29 (see Table VIII) reported by Field must have been an artefact and the presence of *m/e* 42 reported by Tiernan and Futrell (not included in the Table) was also probably an instrumental error. On the whole, however, now that the present results are available to cover the important intermediate pressure range, a very concordant picture is obtained.

The four principal secondary ions from the reaction of the propenyl radical cation with propene molecules are listed in Table IX. The agreement between the different techniques is satisfactory especially as they represent studies at very different pressures.

Steric factors suggest that the intermediate adduct ion was the result of a linear transition state. We can depict the formation of the even electron ions as follows

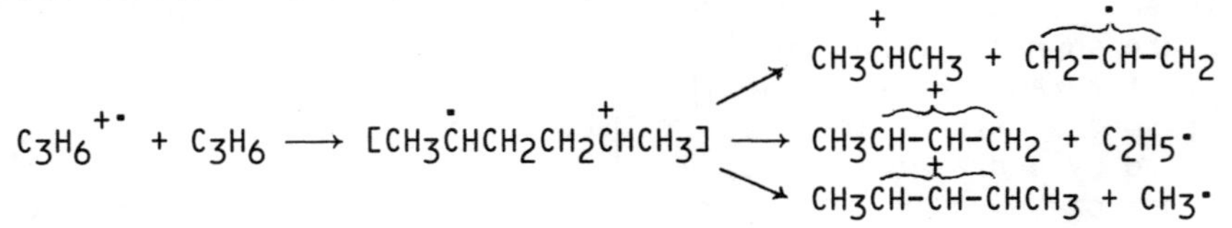

TABLE IX The Reaction of Propenyl Radical Cations with Propene Molecules: % of Principal Secondary Ions

m/e	Abramson & Futrell	Koyana *et al*	Harrison	MacKenzie-Peers	Herod *et al*	Present Work 7×10^{-5}	Present Work 4×10^{-3}	Bowers *et al*
$C_3H_7^+$ 43	27	30	27	26	21	23	23	24
$C_4H_7^+$ 55	18	25	21	19	11	20	16	13
$C_4H_8^{+\cdot}$ 56	34	22	29	33	42	33	23	43
$C_5H_9^+$ 69	21	15	23	22	26	23	39	20

The $C_5H_9^+$ may have the alternative linear structure $C_2H_5\overset{+}{\overbrace{CH\text{-}CH\text{-}CH_2}}$. There is also the odd electron butene radical cation $CH_3\overset{+}{C}H\text{-}\overset{\cdot}{C}HCH_3$. The major tertiary ion is $C_4H_9^+$ derived from two sources ($C_6H_{13}^{+*} \rightarrow C_4H_9^+ + C_2H_4$ and $C_8H_{15}^{+*} \rightarrow C_4H_9^+ + C_4H_6$). Unlike the secondary ions it is probably branched and its relative lack of reactivity, as instanced by its accumulation at higher pressures suggest that it is either the 2-methylpropyl cation $(CH_3)CHCH_2^+$ or more likely the *tert*-butyl cation $(CH_3)_3C$.

The principal reaction of all three butenes was hydrogen transfer

$$C_4H_8^{+\cdot} + C_4H_8 \longrightarrow C_4H_9^+ + C_4H_7\cdot$$

but the carbon skeleton of these products probably varied from butene to butene. With but-1-ene and but-2-ene the butyl cation is unlikely to be the 2-butyl ion $CH_3\overset{+}{C}HCH_2CH_3$ but with *iso*-butene (2-methylpropene) the product cation is almost certainly the t-butyl cation $(CH_3)_3C^+$. With but-1-ene at high pressures the principal product ions are likely to be linear $C_4H_9^+$ and $C_5H_{11}^+$.

$$C_2H_5CH{=}CH_2^{+\cdot} + C_2H_5CH{=}CH_2 \longrightarrow [C_2H_5\overset{\cdot}{C}HCH_2CH_2\overset{+}{C}HCH_3]^* \begin{cases} \rightarrow C_4H_9^+ + CH_3\overset{\cdot}{\overbrace{CH\text{-}CH\text{-}CH_2}} \\ \rightarrow C_5H_{11}^+ + \overset{\cdot}{\overbrace{CH_2\text{-}CH\text{-}CH_2}} \end{cases}$$

The principal ions derived from *cis*-but-2-ene at high pressure are $C_6H_{11}^+$ and $C_6H_{12}^{+\cdot}$.

$$CH_3CH{=}CHCH_3^{+\cdot} + CH_3CH{=}CHCH_3 \longrightarrow [\overset{C}{C_8H_{16}}]^{*+\cdot} \begin{cases} \rightarrow C_6H_{11}^+ + C_2H_5^\cdot \\ \rightarrow C_6H_{12}^{+\cdot} + C_2H_4 \end{cases}$$

The difference between the high pressure reactions of but-1-ene and but-2-ene can be attributed to steric effects which prevents a linear transition state with the latter and hence leads to a cyclic transition state.

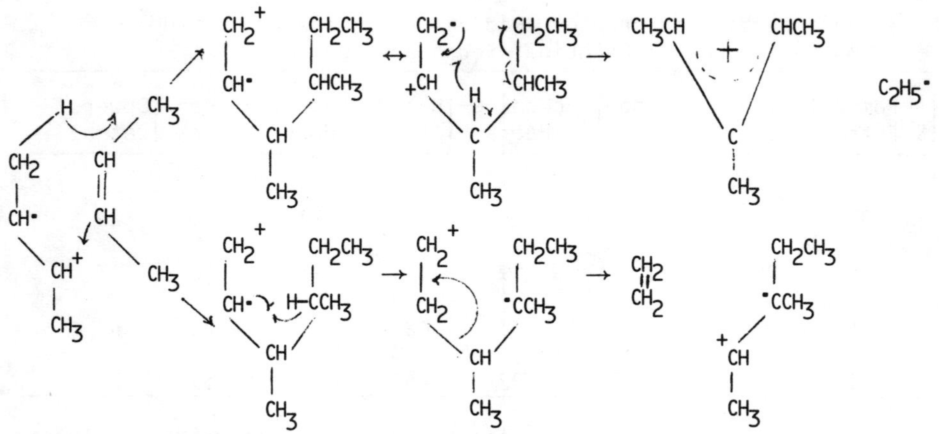

The large difference between the low and high pressure reactions of but-2-ene can be attributed to a change in mechanism. At low pressure hydrogen atom transfer through space is the major pathway whereas at high pressure the principal reactions involve actual collision.

$$CH_3CH{=}CHCH_3^{+\cdot} + CH_3CH{=}CHCH_3 \begin{cases} \nearrow \text{low pressure, } C_4H_7^{\cdot} + C_4H_9^{+} \quad \textit{H atom transfer} \\ \searrow \text{high pressure - activated complex formed } [C_8H_{16}]^{*+\cdot} \end{cases}$$

$$[C_8H_{16}]^{*+\cdot} \longrightarrow C_6H_{11}^{+} + C_2H_5^{\cdot} \qquad [C_8H_{16}]^{*+\cdot} \longrightarrow C_6H_{12}^{+\cdot} + C_2H_4$$

A similar change in mechanism is not observed with but-1-ene because there is no appreciable steric hindrance in the formation of the linear activated complex.

2-Methylpropene behaves very differently, yielding almost only one product $C_4H_9^{+}$ which is undoubtedly the t-butyl cation.

$$(CH_3)_2C{=}CH^{+\cdot} + (CH_3)_2C{=}CH_2 \longrightarrow (CH_3)_3C^{+} + \overbrace{CH_2{-}\dot{C}{-}CH_2}(CH_3)$$

See Table X overleaf.

Conclusions

The interaction of an alkenyl radical cation with an alkene of the same or lower ionisation energy results in the formation of an excited adduct which fragments to yield new cations, radicals and alkenes.

$$C_lH_{2l}^{+\cdot} + C_mH_{2m} \longrightarrow [C_nH_{2n}]^{*+\cdot} \longrightarrow \begin{cases} (C_{n-x}H_{2n-2x-1})^{+} + (C_xH_{2x+1})^{\cdot} \\ (C_{n-x}H_{2n-2x})^{+\cdot} + (C_xH_{2x}) \\ (C_{n-x}H_{2n-2x+1})^{+} + (C_xH_{2x-1})^{\cdot} \end{cases}$$

where $l = 2,3\ldots$; $m = 2,3\ldots$;
$n = (l+m)$ $x = 0,1,2\ldots$

TABLE X The Reactions of But-1-enyl, But-2-enyl and 2-Methylpropenyl Cations with their Parent Molecules: % of Principal Product Ions

	$C_4H_9^+$	$C_5H_9^+$	$C_5H_{10}^{+\cdot}$	$C_5H_{11}^+$	$C_6H_{11}^+$	$C_6H_{12}^{+\cdot}$	P (Torr)
$CH_3CH_2CH{=}CH_2^{+\cdot}$ + $CH_3CH_2CH{=}CH_2$							
	25	16	31	6	16	7	5×10^{-6}
	14	4	38	3	22	15	5×10^{-5}
	22	11	16	24	13	14	4×10^{-3}
$CH_3CH_2{=}CHCH_3^{+\cdot}$ + $CH_3CH\ CHCH_3$							
	55	45	-	-	-	-	5×10^{-6}
	70	4	10	-	10	6	5×10^{-5}
	9	14	9	6	35	28	4×10^{-3}
$(CH_3)_2C{=}CH_2^{+\cdot}$ + $(CH_3)_2C{=}CH_2$							
	86	14	-	-	-	-	5×10^{-6}
	93	3	2	-	-	-	5×10^{-3}
	90	1	2	6	-	-	4×10^{-3}

Reactions of this kind were extensively studied in the late 1960s using tandem magnetic mass spectrometers and latterly ion cyclotron resonance was developed. Both these techniques involved studies of ion-molecule reactions at 'low pressure' (*i.e.* 10^{-6} - 10^{-5} torr) when the mean free path was much greater than the reaction chamber. Studies were also made using a 'high pressure' (*i.e.* 10^{-2}) source. The triple quadrupole can be used in the intermediate range (10^{-5} - 10^{-2} torr). This is important because many of the reaction pathways change with changing pressure. Thus we have already cited the low pressure reaction of $C_2H_4^{+\cdot}$ with the neutral C_2H_4 to give $C_3H_5^+$ (*m/e* 41), but at high pressure this secondary ion reacts as fast as it is formed to yield $C_5H_9^+$ (*m/e* 69) [Table VIII]. Other examples of the effect of changing pressure are illustrated in the reactions of the $C_4H_8^{+\cdot}$ radical ions with the corresponding alkenes (Table X).

References to earlier work

A.L. Mitchell and J.M. Tedder, *J. Chem. Soc.*, *Perkin Trans.2*, 1984, 667.
C.E. Melton and P.S. Rudoph, *J. Chem. Phys.*, 1960, **32**, 1128.
S. Wexler and R. Marshall, *J. Amer. Chem. Soc.*, 1964, **86**, 781.
J.J. Myher and A.G. Harrison, *Canad. J. Chem.*, 1968, **46**, 101.
T.D. Tiernan and J.H. Futrell, *J. Phys. Chem.*, 1968, **72**, 3080.
M.T. Bowers, D.D. Elleman and J.L. Beauchamp, *J. Phys. Chem.*, 1968, **72**, 3599.
F.P. Abramson and J.H. Futrell, *J. Phys. Chem.*, 1968, **72**, 1826.
F.P. Abramson and J.H. Futrell, *J. Phys. Chem.*, 1968, **72**, 1994.

A.A. Herod and A.G. Harrison, *J. Phys. Chem.*, 1969, **73**, 3189.
A. MacKenzie-Peers, *J. Phys. Chem.*, 1969, **73**, 4141.
A.A. Herod, A.G. Harrison, R.M. O'Malley, A.J. Ferrier-Correia and K.R. Jennings, *J. Phys. Chem.*, 1970, **74**, 2720.
I. Koyana, I. Omura, and J. Tanaka, *J. Phys. Chem.*, 1966, **44**, 3850.
I. Koyana, *J. Phys. Chem.*, 1966, **45**, 706.
A.G. Harrison, *Canad. J. Chem.*, 1962, **41**, 236.
M.T. Bowers, D.H. Aue and D.D. Elleman, *J. Amer. Chem. Soc.*, 1972, **94**, 4255.

(c) THE FORMATION OF RADICAL CATIONS IN THE REACTION OF XCH_2^+ WITH BENZENE AND HETEROCYCLIC COMPOUNDS

SUMMARY. The reactions of XCH_2^+ ions with benzene has been studied in the gas phase where "X" has been: H, Cl, F, CN and OH. The principal product ion was $C_6H_6^{+\cdot}$ but others included $C_5H_5^+$, $C_6H_5^+$, $C_6H_7^+$ and $C_7H_7^+$. The study was extended to the reactions of the same XCH_2^+ ions with the five membered heterocyclic molecules pyrrole, furan and thiophene.

Discussion

The Reactions of CH_2X^+ with Benzene

Table XI lists the principal product ions formed in the reaction of CH_2X^+ with benzene (where X = H, F, Cl, C≡N. PH).

TABLE XI The Reaction of XCH_2^+ with Benzene (% of total secondary ion current)

CH_2X^+	+ C_6H_6 →		$C_5H_5^+$	$C_6H_5^+$	$C_6H_6^{+\cdot}$	$C_6H_7^+$	$C_7H_7^+$
		m/e	65	77	78	79	91
	Ip x 10^6	Ip x 10^5 torr					
CH_3^+	2.3	2.3	-	34	44	0	14
	23	4.5	9	25	40	0.5	3
CH_2F^+	3	2.3	-	-	63	1.8	20
	30	4.5	2	2	57	2.2	27
CH_2Cl^+	2.3	3	-	-	64	-	37
	30	3	2	1	36	2.6	51
CH_2CN^+	3	1.5	5	1	71	9	10
	15	4.5	3	1	72	3	10
CH_2OH^+	3	1.5	-	1	-	92	1
	23	4.5	-	1	1	91	0.6

NOTE: All values have been corrected for ^{13}C and ^1H isotopes.

The principal reactions of methyl cations with benzene are either abstraction of a hydride ion (H^-) or the transfer of an electron.

$$CH_3^+ + C_6H_6 \longrightarrow C_6H_5^+ + CH_4 \quad \sim 30\%$$
$$CH_3^+ + C_6H_6 \longrightarrow C_6H_6^{+\cdot} + CH_3^{\cdot} \quad \sim 43\%$$

Hydride transfer which represents approximately 30% of the secondary ion current with CH_3^+, is unimportant when the primary ion XCH_2^+ contains an electronegative substituent. The fluoromethyl cation (FCH_2^+) also accepts an electron from benzene as its main process. The other major reaction of the fluoromethyl cation is the formation of $C_7H_7^+$ (probably tropylium). The important feature of this reaction is the formation of HF.

$$CH_2F^+ + C_6H_6 \longrightarrow C_7H_7^+ + HF \quad \approx 22\%$$
$$CH_2F^+ + C_6H_6 \longrightarrow C_6H_6^{+\cdot} + CH_2F^{\cdot} \quad \sim 66\%$$

At first sight it seems surprising that the tropylium route is the more important when the primary ion is the chloromethyl cation (CH_2Cl^+), since H-F is a much stronger bond than H-Cl. However, it is probable that the transition state is very early and the transfer of an electron

$$CH_2Cl^+ + C_6H_6 \longrightarrow C_7H_7^+ + HCl \quad \sim 52\%$$
$$CH_2Cl^+ + C_6H_6 \longrightarrow C_6H_6^{+\cdot} + CH_2Cl^{\cdot} \quad \sim 37\%$$

to fluoromethyl cation occurs earlier along the reaction co-ordinate than to the chloromethyl cation.

The other very striking feature of Table XI is the almost exclusive formation of $C_6H_7^+$ in the reaction of CH_3O^+ with benzene.

$$H_2C{=}OH^+ + M \longrightarrow H_2C{=}O + HM^+ \quad \approx 92\%$$

Undoubtedly the reason for this is the very acidic character of the protonated formaldehyde ion.

The Reactions of $X\text{-}CH_2^+$ with Pyrrole

The principal secondary ion formed by the reaction of the $X\text{-}CH_2^+$ (X = H, F, Cl and CN) with pyrrole is that resulting from charge exchange.

$$XCH_2^+ + C_4H_5N \longrightarrow C_4H_5N^{+\cdot} + XCH_2^{\cdot}$$

The reaction of (N_3CO^+) ions is exceptional and the principal product ion is the result of transfer of a proton.

$$CH_2=\overset{+}{O}H + C_4H_5N \longrightarrow C_4H_6\overset{+}{N} + CH_2O$$

An ion *m/e* 53 is observed in all the reactions listed in Table XII as a minor product, and it is probably $C_4H_5^+$ except when CH_2CN^+ is the primary ion. In the latter example the ion *m/e* 53 represents nearly

TABLE XII The Reaction of XCH_2^+ with pyrrole (% of total secondary ion current)

CH_2X^+ + C_4H_5N →		*m/e*	$C_4H_5^+$ 53	$C_3H_6N^+$ 56	$C_4H_5N^{+\cdot}$ 67	$C_4H_6N^+$ 68	$C_4H_7N^+$ 69	$C_5H_6N^+$ 80
	$Ip\times10^6$	$Is\times10^5$ torr						
CH_3^+	3.8	2.3	1	-	68	3	-	1
	11.3	5	1	-	73	5	-	1.4
CH_2F^+	3.8	2.3	1	-	68	3	-	1
	11.3	6.8	1	-	85	4	-	2
CH_2Cl^+	4.5	2.3	1	-	87	1	-	4
	11.3	6	1	-	88	2	-	4
CH_2CN^+	4.5	4.5	8	-	77	7	-	2
	7.5	6.8	8	-	78	7	-	1
CH_2OH^+	3	2.3	1	6	8	49	9	-
	7.5	11.3	1	3	13	64	2	1

10% of the secondary ion current and nearly an order of magnitude greater than that for the other XCH_2^+ listed in Table XII. It therefore seems possible that it is $C_3H_3N^+$ as well as $C_4H_5^+$.

The ion *m/e* 80 is present in all the mass spectra, but it is never very intense.

A possible structure of this ion is protonated pyridene:

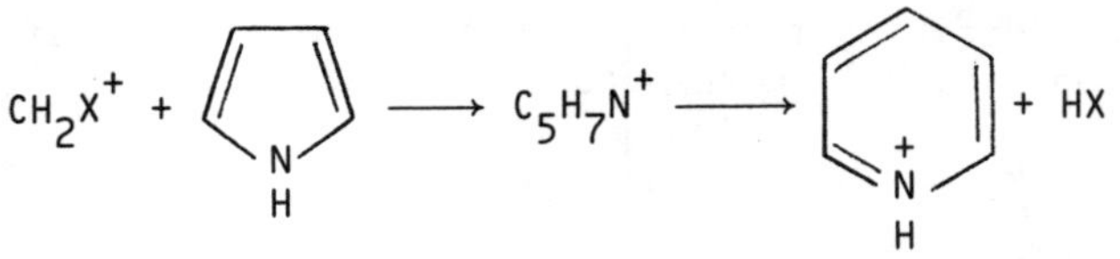

X = F 2%

X = Cl 4%

The Reactions of $X{\cdot}CH_2^+$ with Furan

The principal secondary ion, with all the primary ions except $CH_2=\overset{+}{O}H$ was $C_4H_4O^+$, *i.e.* simple charge exchange. The $CH_2=\overset{+}{O}H$ transferred a proton:

$$CH_2X^+ + C_4H_4O \longrightarrow CH_2X^{\cdot} + C_4H_4O^{+\cdot} \quad (X = H, F, Cl \text{ and } CN)$$

$$CH_2=\overset{+}{O}H + C_4H_4O \longrightarrow CH_2 + C_4H_5O^+$$ (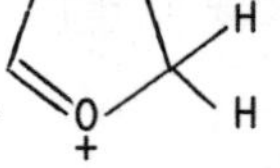 probably)

The *m/e* 81 ion corresponds to the insertion of "CH" into the

TABLE XIII The Reaction of XCH_2^+ with Furan (% of total secondary ion current)

CH_2X^+ + C_4H_4O →	Ip x 10^6	Is x 10^5 torr	$C_3H_5^+$ *m/e* 41	$C_4H_5^+$ 53	$C_4H_4O^{+\cdot}$ 68	$C_4H_5O^+$ 69	$C_5H_5O^+$ 81
CH_3^+	3.8	2.3	2	4	46	5	-
	11.3	5	2	3	56	8	-
CH_2F^+	3.8	3.8	1	13	63	3	0.5
	15	23	1	7	71	9	0.9
CH_2Cl^+	15	23	0.3	17	55	7	7
	30	38	-	11	49	12	12
CH_2CN^+	2.3	2.3	-	15	77	6	-
	6	7.5	-	11	82	4	0.3
CH_2OH^+	5	4.5	17	1	2	76	0.8
	3.0	3.8	4	0.5	2	89	2

furan ring and is observed for all the XCH_2^+ ions except when X = H.

$$CH_2Cl^+ + C_4H_4O \longrightarrow HCl + C_5H_5O^+ \text{ (probably }$$ 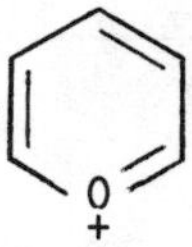 $$)$$

X = F 0.9%; X = Cl 12%

The other ion which deserves mention is *m/e* 41, *i.e.* $C_3H_5^+$ which is a major product ion in the reaction of $\overset{+}{C}H_2OH$ ion with furan.

$$CH_2OH^+ + C_4H_4O \longrightarrow C_3H_5^+ + CO + CH_2O$$

<u>The Reaction of $X\text{-}CH_2^+$ with Thiophene</u>

The spectra of ions from the reaction of XCH_2^+ with thiophene shows some substantial differences to the spectra from pyrrole and furan. The biggest difference is the presence of CHS^+ as a major ion in all of the spectra. *M/e* 85 ($C_4H_5S^+$) is almost the only ion from CH_2OH^+ and *m/e* 97 is only observed when X is Cl, CN or F (very small).

$$CH_2Cl^+ + C_4H_4S \longrightarrow C_5H_5S^+ + HCl$$

X = F 1.3%
X = Cl 11%

The ring expansion reaction is significantly more important when X is Cl than when X is F with furan and thiophene and the extent of reaction is greater.

See Table XIV overleaf.

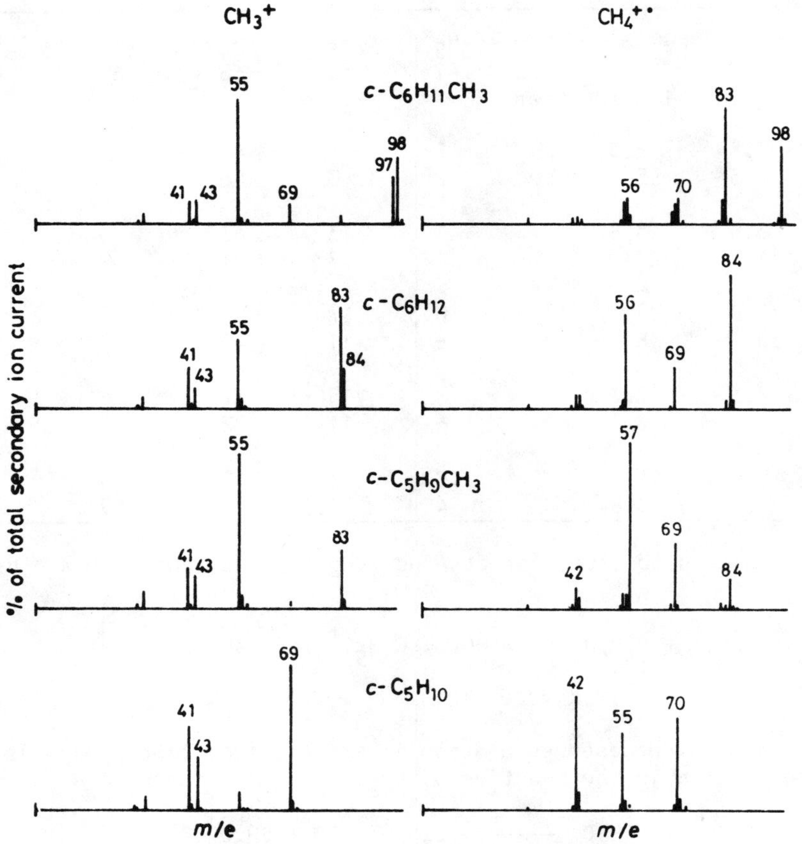

Figure 2. Fragmentation patterns from the reaction of CH_3^+ and $CH_4^{+\bullet}$ ions with cycloalkanes

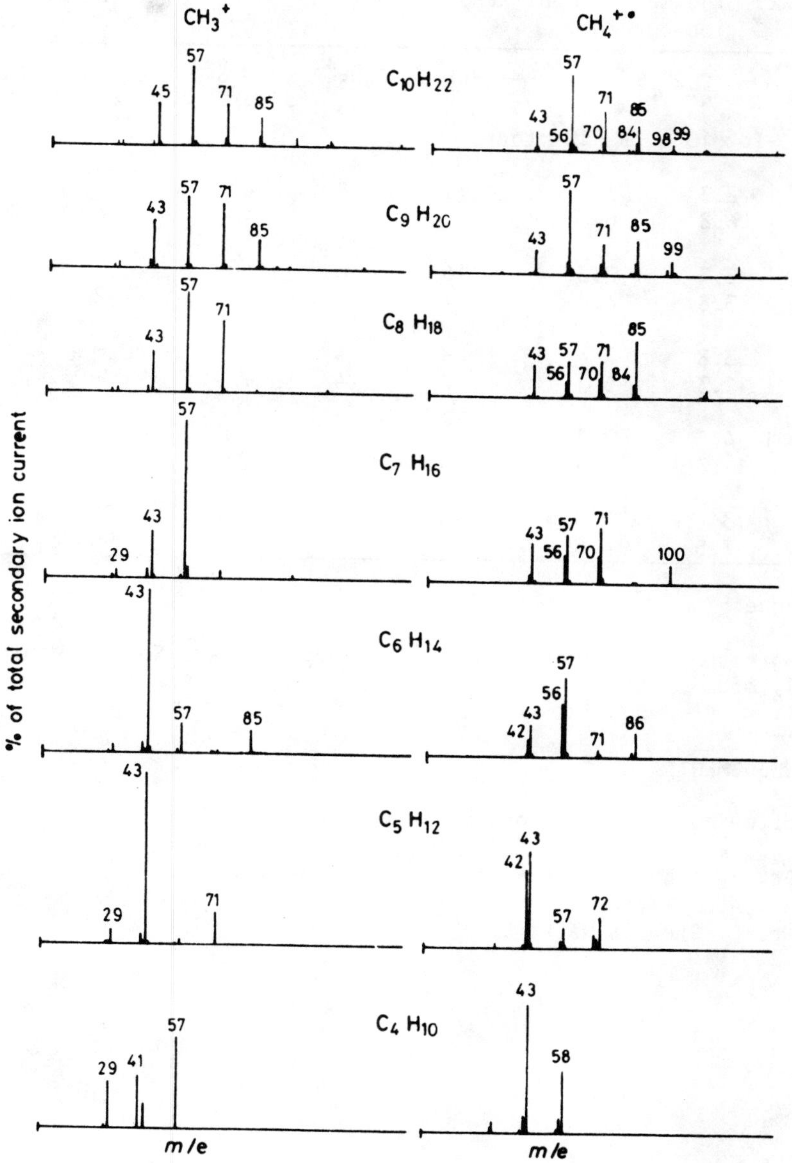

Figure 1. Fragmentation patterns from the reaction of CH_3^+ and $CH_4^{+\cdot}$ ions with normal alkanes

TABLE XIV The Reaction of XCH_2^+ with Thiophene (% of total secondary ion current)

CH_2X^+ + C_4H_4S →	$Ip \times 10^6$	$Is \times 10^5$ torr	HCS^+ 45	$C_4H_5^+$ 53	$C_4H_4S^{+\cdot}$ 84	$C_4H_5S^+$ 85	$C_5H_5S^+$ 97
CH_3^+	2.3	3	19	5	40	-	-
	6.8	6	16	4	48	-	-
CH_2F^+	2.3	2.3	16	6	60	1	-
	3.8	6.8	11	5	64	2	1.3
CH_2Cl^+	3.8	3	9	4	67	-	12
	11.3	5	7	4	74	-	11
CH_2CN^+	4.5	2.3	7	7	80	-	3
	11.3	4.5	6	9	78	-	4
CH_2OH^+	3	2.3	10	-	4	78	-
	7.5	4.5	7	-	3	90	-

Reference

J.D. Morrison, K. Stanney and J.M. Tedder, *J. Chem. Soc., Perkin 2*, 1981, 838.

ACKNOWLEDGMENTS

The following have collaborated in this work:

Dr. Jonathan H. Batey of V.G. Gas Analysis Ltd.
Mr. Alan Mitchell
Mr. G. Stewart Walker.

Substituent Effects on the Methylenecyclopropane Rearrangement. A Probe for Free Radical Effects.

Xavier Creary
Department of Chemistry
University of Notre Dame
Notre Dame, IN 46556
U.S.A.

ABSTRACT. The thermal rearrangement of 3-aryl-2,2-dimethylmethylenecyclopropanes, **7**, is described. This reaction, which proceeds by way of a singlet trimethylenemethane biradical , is used as a probe for free radical substituent effects. The effect of third row elements is described. The methylenecyclopropane rearrangement of related systems is subject to captodative rate enhancements. The captodative effect appears to be a useful criterion for establishing the donor or acceptor properties of various groups which can act as free radical stabilizing groups.

The area of substituent effects on free radicals has continued to hold the interest of mechanistic chemists over the years. This is partly because the effect of substituents is not as well understood as in carbocationic reactions. One problem is the separation of true radical stabilizing effects from polar effects which are known to operate in many radical reactions. Because of the magnitude of polar effects, rates of many radical reactions correlate with Hammett σ or Brown σ^+ values. Classic examples of this are free radical brominations of substituted toluenes using bromine,[1] N-bromosuccinamide,[1] or bromotrichloromethane,[2] all of which give correlations with σ^+. Hydrogen atom abstractions by alkoxy radicals,[3] chlorine atoms,[4] and peroxy radicals[4] also give correlations with σ^+. Much of the literature involving hydrogen atom abstractions from toluenes has been summarized.[5] Much of this data has been interpreted in terms of stabilization of a transition state with polar character as in **1b**. The free radical bromination of substituted toluenes therefore does not give an accurate picture of true radical stabilizing effects despite the intermediacy of a benzyl radical.

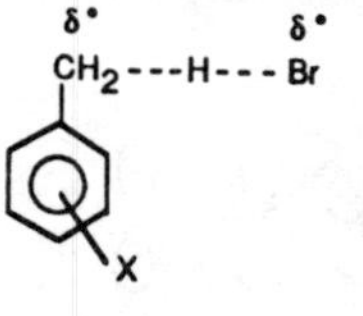

1a

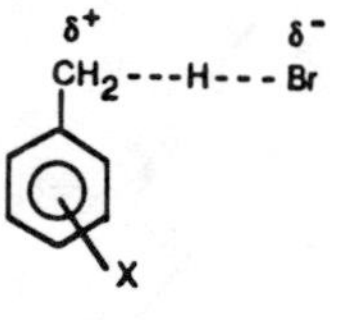

1b

In order to gain truer insights into radical stabilizing effects, such effects in a radical reaction which was deviod of significant extraneous polar effects would have to be evaluated. Timberlake[6] has developed a probe based on pyrolysis of azo compounds **2** in which polar effects are apparently not large. Others have tried to set up various free radical substituent constants in the form of $\sigma^{\cdot}$ values with the goal of separation of polar factors from true free radical stabilizing effects. One such attempt by Fisher and Meierhoefer[7] uses the reaction of 4-substituted-3-cyanotoluenes, **3**, with N-bromosuccinamide in an attempt to minimize polar contributions to this free radical bromination. Another attempt by Jackson et al.[8] looks at substituent effects on thermal decomposition rates of dibenzyl mercury compounds, **4**. These probes all give an improved evaluation of free radical substituent effects.

H. G. Viehe et al. (eds.), Substituent Effects in Radical Chemistry, 245–262.
© *1986 by D. Reidel Publishing Co.*

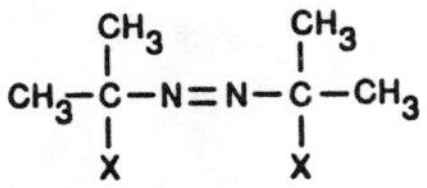

2

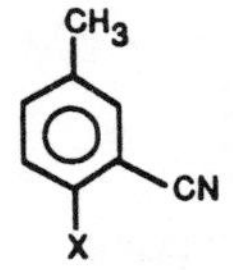

3

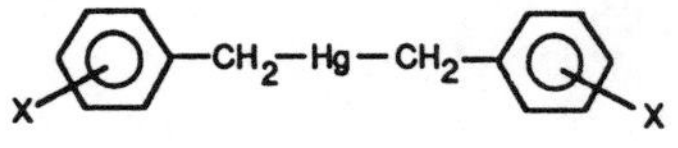

4

A number of years ago,[9] we also sought to evaluate the effect of a wide range of substituents on a free radical reaction. The reaction chosen was the thermal rearrangement of a methylenecyclopropane which involves fragmentation of a nonpolar carbon-carbon bond.[10] The transition state for this rearrangement has free radical character,[10d-k] and hence offered a chance to evaluate substituent effects on a radical-like transition state in which extraneous polar effects do not overwhelm true free radical effects. We have found that a wide variety of substituted methylenecyclopropanes **7** can be prepared by arylcarbene

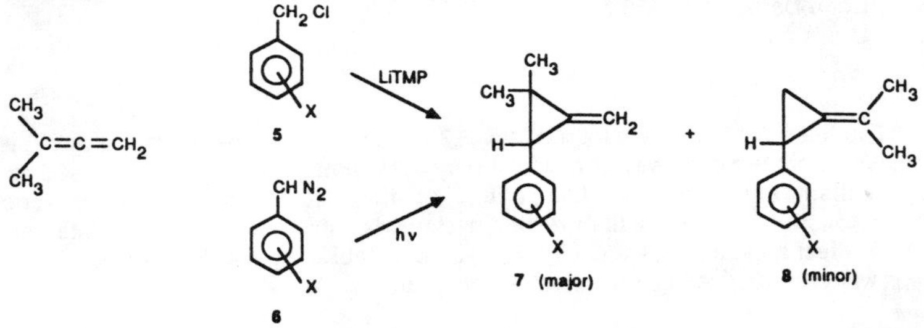

additions to 1,1-dimethylallene. Benzyl halides, **5**, or aryldiazomethanes, **6**, served as carbene precursors. We have also found that **7** rearranges quantitatively to **8** at a convenient rate at 80-100 °C in a reaction easily monitored by gas chromatography or NMR. The mechanism of this and other related methylenecyclopropane rearrangements is suggested to involve a biradical such as **9**. In the transition state for rearrangement of **7**, radical character develops at the benzylic carbon as the cyclopropane bond fragments. It was felt that charge separation in this transition state should be minimal and therefore this reaction should allow an evaluation of true radical stabilizing effects.

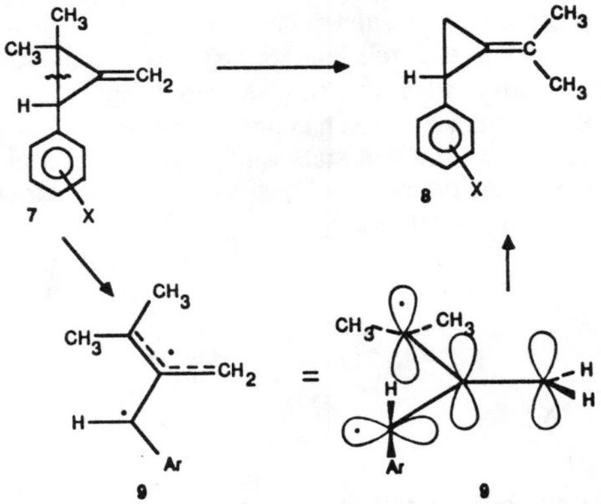

An extensive study of substituent effects on the rearrangement rate of **7** has been carried out. Data in isooctane are given in Table 1. While rate constants do not show the wide range that many cationic reactions do, nonetheless there are real and discernable differences. The first thing apparent from the data (Figure 1) is that there is no single polar substituent constant that correlates all or even most of the data. All substituents do not enhance the rate of this rearrangement. Attempts to correlate the data in Table 1 with the free radical substituent constants of Meierhoefer and Jackson gave low correlation coefficients. Neither were attempts at correlation with E_R[11] or E_D[12] values more fruitful. This was due in part to a lack of a significant number of similar substituents, especially in the meta position.

Table 1. Rearrangement rates of **7** in isooctane at 100 °C.

Substituent	10^4 k, s^{-1}	Substituent	10^4 k, s^{-1}
p-CN	10.28	m-SiMe$_3$	3.87
p-SCH$_3$	9.53	m-CH$_3$	3.82
p-CO$_2$CH$_3$	8.09	p-H	3.58
p-OCH$_3$	6.16	m-OCH$_3$	3.40
p-SiMe$_3$	5.24	m-Cl	3.30
p-Br	4.88	m-F	3.17
p-t-Bu	4.78	m-CF$_3$	3.08
p-Cl	4.75	p-F	2.98
p-CH$_3$	4.65	3,5-Cl$_2$	2.87
p-CF$_3$	4.25	m-CN	2.69
3,5-Me$_2$	4.23		

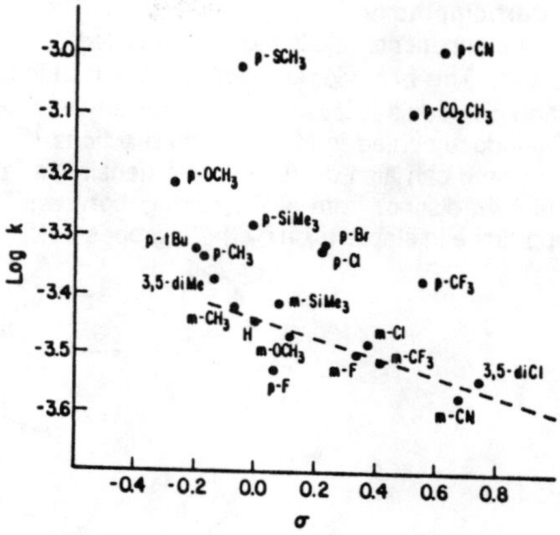

Figure 1. log k for **7** vs. σ

Arnold and Dust[13] have recently developed a free radical substituent constant, termed $\sigma_{\alpha}^{\cdot}$, which is based on benzyl radical hyperfine coupling constants. These $\sigma_{\alpha}^{\cdot}$ values are a measure of spin density at the benzylic carbon and therefore indicate the ability of various substituents to delocalize spin. Arnold and Dust have plotted our rate data for **7** versus their $\sigma_{\alpha}^{\cdot}$ values. Figure 2 shows this plot. The relatively good correlation lends support to our use of this rearrangement as a probe for free radical substituent effects, in spite of the relatively small rate effects.

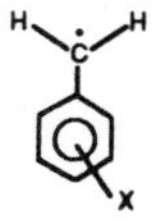

$$\sigma_{\alpha}^{\cdot} = 1 - \frac{\text{hfc } \alpha\text{–H}_x}{\text{hfc } \alpha\text{–H}_0}$$

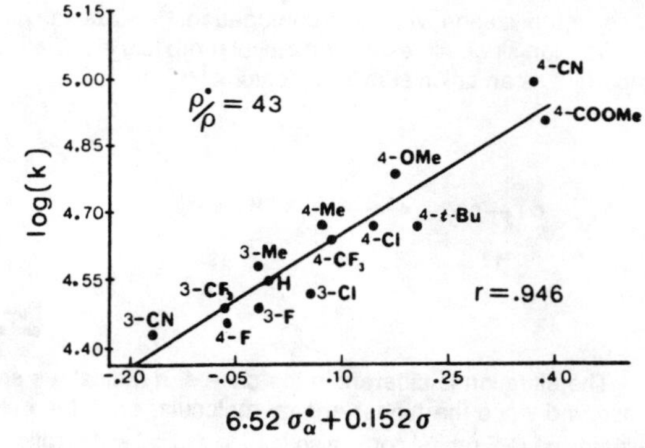

Figure 2. log k for **7** vs. f($\sigma_{\alpha}^{\cdot}$).

Electronegative Meta Substituents. As can be seen from Figure 1, meta substituents with positive σ values all lead to slower rearrangement rates than unsubstituted **7**. The least squares line (correlation coefficient 0.97) using eight meta substituents is shown in Figure1. Although the effect is small (ρ = -0.16), it is nonetheless real. This suggests that **9** behaves as an electron deficient species, being slightly destabilized by nonconjugating electron withdrawing groups. A similar type of substituent effect was seen in the decomposition of dibenzyl mercury compounds **4** and was interpreted in terms of ground state stabilization of **4** by electronegative substituents due to the polar nature of the carbon-mercury bond.[8] Our results show that there is some rate retardation due to electronegative substituents even when the bond fragmenting is nonpolar. The implication is that radicals are slightly destabilized by nonconjugating electron withdrawing groups, i.e. all groups do not stabilize free radicals.

Carbomethoxy and Cyano Groups. These substituents in the para position lead to relatively large rate enhancenents relative to unsubstituted **7**. In fact, the largest rate difference is between **7**-m-CN and **7**-p-CN. This behavior is readily rationalized in terms of π bond conjugation effects of the carbonyl and cyano groups as illustrated in the allylic type structures **10b**, **10c** and **11b**,**11c**. This conjugation effect is well documented in other radical reactions.[14] In addition to being conjugating groups, carbomethoxy and cyano can also delocalize spin density as "acceptor" groups. This type of stabilization mechanism (which is distinct from allylic conjugation) would imply that forms such as **10d and 11d** have some importance in stabilization by these groups.

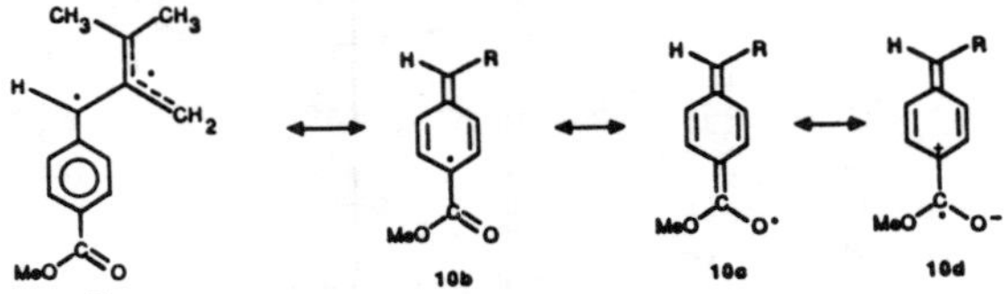

10a 10b 10c 10d

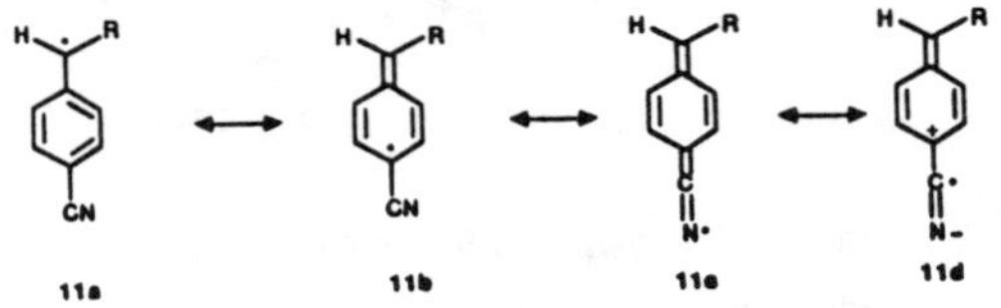

11a 11b 11c 11d

Trifluoromethyl Group. The p-trifluoromethyl substituent in **7**-p-CF_3 results in a rate enhancement relative to **7**-p-H and **7**-m-CF_3. This is consistent with transition state stabilization by the p-CF_3 group. As a rationale, the possibility of C-F hyperconjugation comes to mind. The question of C-F hyperconjugation has been controvertial since Roberts first suggested such a possibility.[15] The importance of forms such as **12a** in stabilization of the anion **12** has been disputed. From a molecular orbital viewpoint, it has been pointed out that C-F hyperconjugation as a mode of anion stabilization is fundamentally different from cation stabilization via hyperconjugation.[16] Qualitatively, this is shown in Figure 3. There is no net stabilization since all resultant molecular orbitals are filled. Hence C-F hyperconjugation is not considered important as an anion stabilizing feature.[17]

12 12a

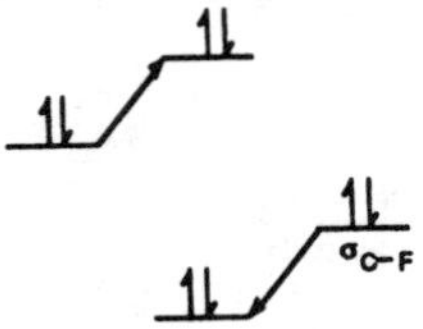

Figure 3. Interaction of a doubly occupied orbital and a vicinal C–F σ orbital.

The situation is different in the case of a radical, as shown in Figure 4. A net stabilization should be observed since the higher energy molecular orbital is only singly occupied. Additionally, an interaction with the σ* C-F orbital could also lead to radical stabilization, as shown in Figure 5. While the magnitudes of these interactions are uncertain, they represent potential mechanisms by which the electronegative CF_3 group could impart stabilization to the transition state leading to **9**.

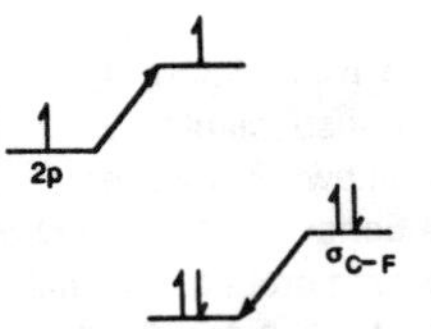

Figure 4. Interaction of a singly occupied 2p orbital and a vicinal C–F σ orbital.

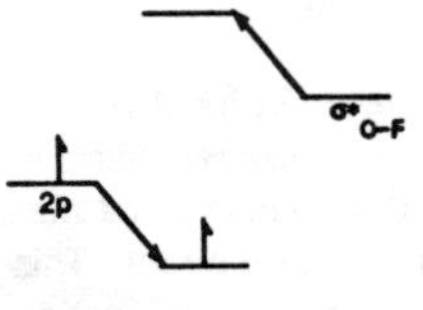

Figure 5. Interaction of a singly occupied 2p orbital and a vicinal C–F σ* orbital.

Trimethylsilyl Group. The p-trimethylsilyl group also shows radical stabilizing properties. This effect has also been previously demonstrated by Wilt,[18] and has been suggested to have its origin in (d-p) π overlap. A valence bond rationale is given in **13b**. and **13c**. This type of stabilization implies that the vacant d-orbitals of silicon are of fundamental importance in radical stabilization by silicon. A stabilizing interaction of the radical center with an adjacent C-Si σ bond (as represented by **13d**) cannot be ruled out.

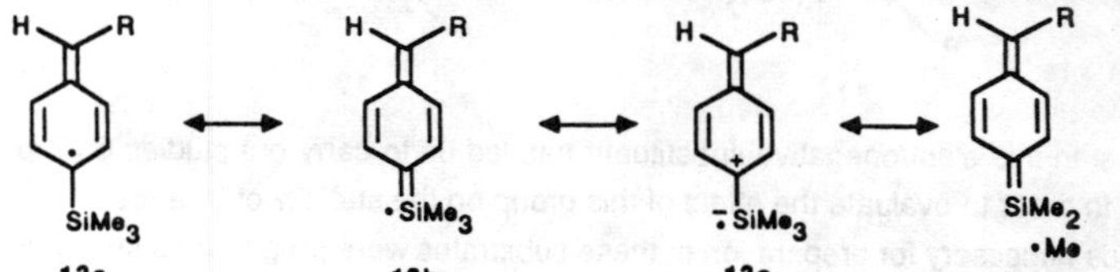

13a 13b 13c 13d

Halogen Substituents. Our results on rearrangement of **7** suggest that p-chloro and p-bromo substituents stabilize **9**, but a p-fluoro substituent is destabilizing. A valence bond description, as illustrated in **14a**, has been used to account for this halogen stabilizing effect.[7] This type of delocalization results in an expanded octet and is not possible with fluorine. With the p-fluoro substituent, inductive destabilization of the radical center now becomes important, accounting for the decreased rate of **7**-p-F. The nonbonding electrons associated with chlorine and bromine may also be of importance in stabilization of **9**, i.e. **14b**, where the halogen acts as a "donor" group, may impart substantial stabilization.

14a 14b

Methoxy Group. Our results indicate that the p-methoxy group is stabilizing to the intermediate **9**. Kirmse has seen the same effect in a methylenecyclopropane rearrangement.[10k] From a valence bond viewpoint, stabilization is a result of interaction of the radical center with the nonbonding electron pair as in **15b**. The analogous form **16b** would be of less importance because of the greater electronegativity of fluorine. This valence bond description would account for the stabilizing ability of p-methoxy vs. the destabilizing p-fluoro influence.

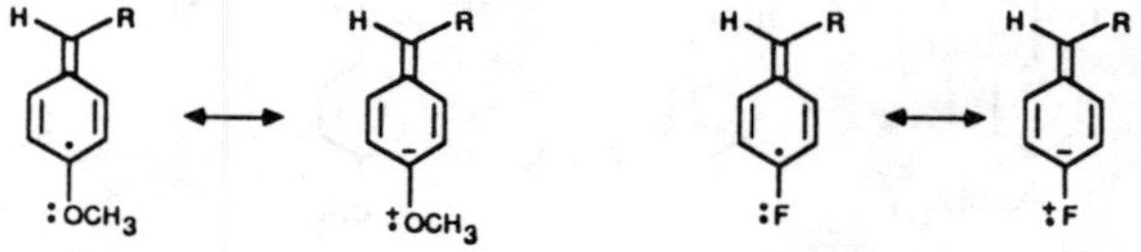

15a 15b 16a 16b

From a PMO viewpoint, stabilization by methoxy could be derived from an interaction of the radical center with a filled oxygen nonbonding orbital as in Figure 6. The suggestion[19] that less stabilization results as the nonbonding orbital energy is lowered seems applicable here. Less stabilization is seen for the p-flouro substituent due to the lower energy of the filled 2p fluorine orbitals relative to the oxygen 2p orbitals.

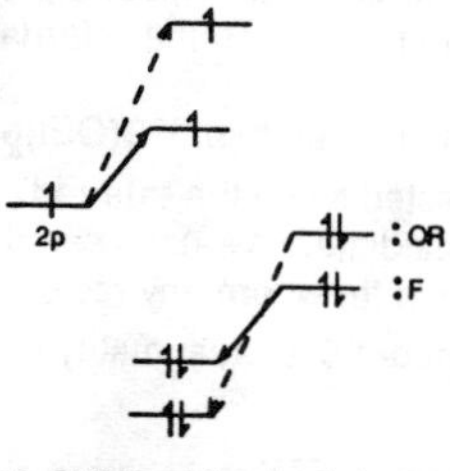

Figure 6. PMO description of radical stabilization by oxygen and fluorine nonbonding electrons.

The Diethylphosphonate and Diethylthiophosphonate Groups. The diethylphosphonate group, $PO(OEt)_2$, is stabilizing to the carbanion **17**.[20] Our recent studies[21] suggest that, contrary to naive expectations, this group is only weakly destabilizing when placed directly on a cationic center as in

17 **18**

18. Our interest in this electronegative substituent has led us to carry out studies on 7-p-$PO(OEt)_2$ and 7-m-$PO(OEt)_2$ in order to evaluate the effect of this group on the stability of free radicals.[22] The requisite diazocompounds necessary for preparation of these substrates were prepared as shown below.

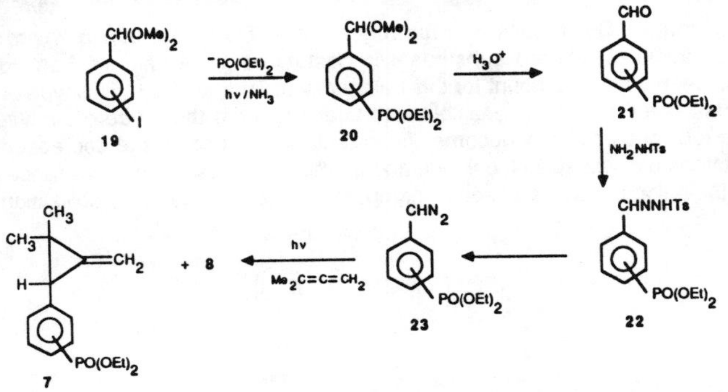

Rate data for rearrangement of 7-p-$PO(OEt)_2$ and 7-m-$PO(OEt)_2$ in C_6D_6 are summarized below. A p-diethylphosphonate group increases the rearrangement rate by a factor of 1.51 while this substituent in

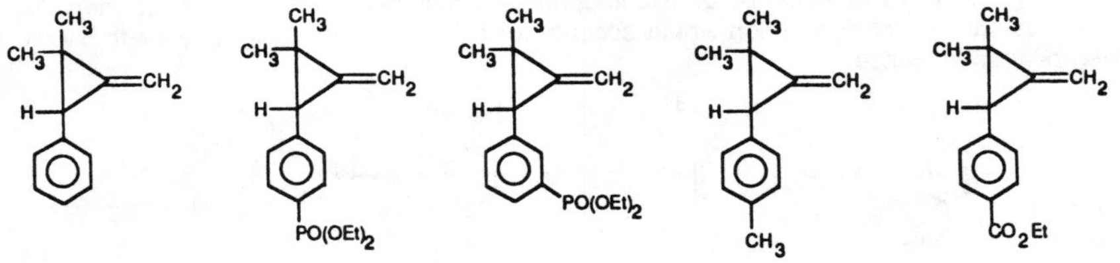

relative rearrangement rates in C_6D_6 at 80 °C

1.00 1.51 0.79 1.30 2.45

the meta position slightly retards the rate. As before, this is attributed to the fact that the intermediate radical is still an electron deficient species which can be slightly destabilized by a purely electronegative substituent. The rate enhancing effect of the diethylphosphonate substituent is therefore a conjugative effect which increases the rate of formation of **9** despite the retardation expected for an electronegative substituent.

While the data suggest a radical stabilizing property of $PO(OEt)_2$, as before, there is only a small rate enhancement observed. This is not unexpected since the true radical stabilizing effect of a substituent is attenuated by the aromatic ring in the intermediate. We have sought confirming data on the influence of $PO(OEt)_2$ attached directly to a free radical center where any radical stabilizing effect would not be greatly attenuated. Rearrangement rates of **24** (where **25** is presumably the intermediate) to the corresponding

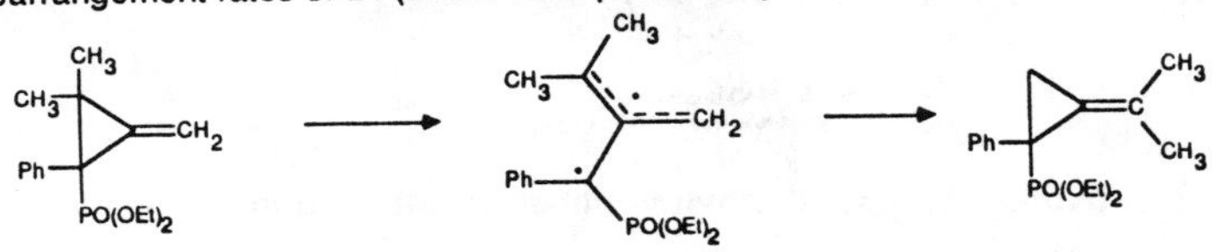

24 **25** **26**

isopropylidenecyclopropane **26** were therefore determined as well as rates of the analogs **27** and **28**. The rate trends parallel those seen in the p-substituted systems but with a much larger rate spread. The diethylphosphonate substituted system **24** rearranges faster than the unsubstituted analog **7**-p-H but slower than the carboethoxy substituted analog **27**. However this data is much more difficult to interpret since steric factors are undoubtedly present and not a constant factor throughout the series. In any event, this data is consistent with a radical stabilizing effect of $PO(OEt)_2$ if one assumes that steric factors are not the only major factors influencing rearrangement rates in this series.

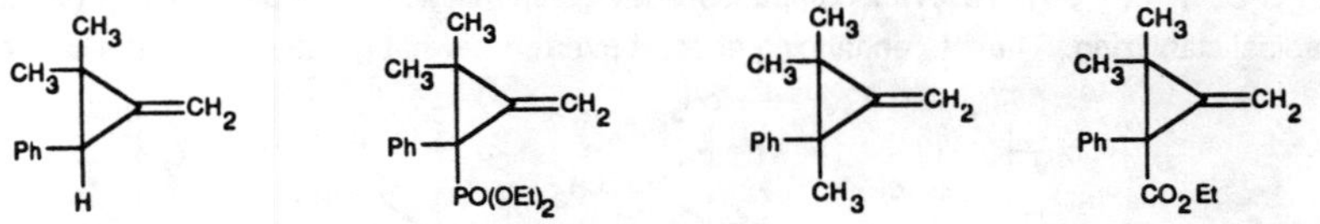

7 **24** **27** **28**

relative rearrangement rates in C_6D_6 at 60 °C

1.0 73 7.5 491

In an attempt to reduce steric factors in an unattenuated system, a third series of substrates (**29**-**31**) was subjected to thermal methylenecyclopropane rearrangement. Rates again implicate the diethylphosphonate group as free radical stabilizing, but somewhat less effective than carboethoxy. In systems **29**-**31**, steric effects are undoubtedly smaller than in the phenyl analogs **7**-p-H, **24**, and **27**. One should therefore have more confidence in using this data as evidence for the radical stabilizing nature of diethylphosphonate. Finally, a recent thermolysis study by Timberlake and Luedtke[6d] on substituted azo compounds supports the conclusion that $PO(OEt)_2$, is radical stabilizing.

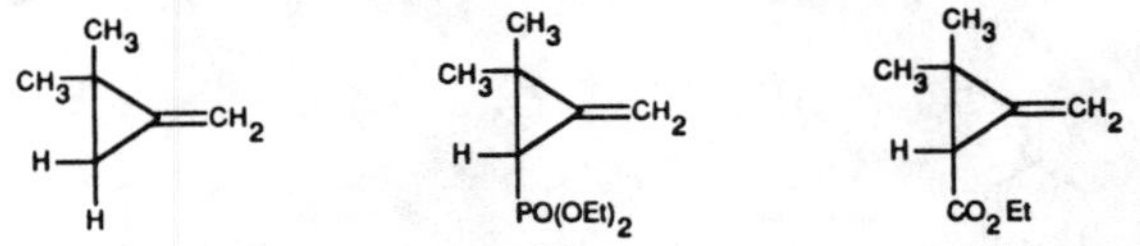

29 **30** **31**

relative rearrangement rates in toluene-d_8 at 170 °C

1.0 32 560

By what mechanism does the diethylphosphonate group, which contains an oxidized third row element, stabilize a free radical? The effect is presumably conjugative in that it is observed in the para but not in the meta position of an aromatic ring. A valence bond description of spin delocalization utilizing the vacant phosphorus d-orbitals is shown below. Forms such as **32b**, **32c**, and **32d** predict spin delocalization onto phosphorus and to oxygen.

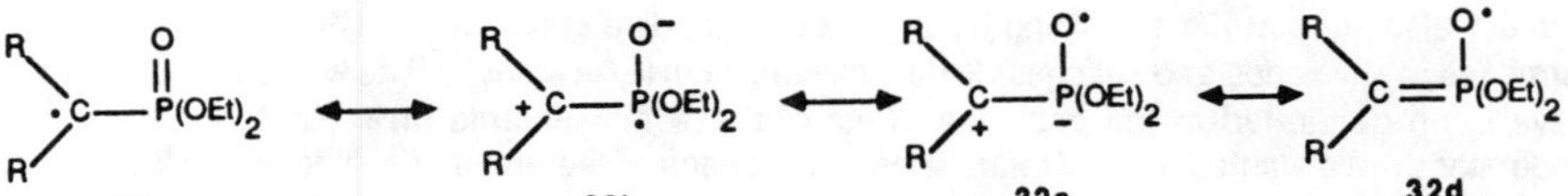

32a **32b** **32c** **32d**

A PMO description would involve stabilization by interaction of the singly occupied carbon orbital of the radical with any vacant or any filled orbital of $PO(OEt)_2$ of appropriate symmetry and energy. One such potential interaction, using a vacant phosphorus d-orbital is shown. These valence bond and molecular orbital descriptions suggest that $PO(OEt)_2$ stabilizes radicals by an acceptor type of mechanism.

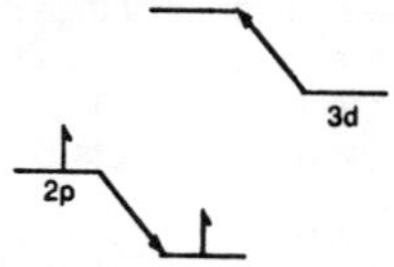

PMO description of radical stabilization by an adjacent vacant d- orbital

The diethylphosphonate group is therefore remarkably similar to the carbonyl and the cyano groups with respect to carbanions, carbocations and free radicals. All three groups are carbanion stabilizing. All three appear to interact with carbocationic centers in a fashion which minimizes the destabilizing inductive effect.[21,23,24] Finally, all three groups are free radical stabilizing as judged by their effect on the methylenecyclopropane rearrangement.

Attention was next turned to systems 7-p-PS(OEt)$_2$ and 7-m-PS(OEt)$_2$ for comparison with our data on the O-phosphoryl analogs. The thermal rearrangement of 7-p-PS(OEt)$_2$ proceeds at a faster rate than that of 7-m-PS(OEt)$_2$ or 7-p-H. Relevant comparisons are given below. The thiophosphoryl group is therefore quite radical stabilizing. The rate enhancing effect is even somewhat greater than in the O-phosphoryl

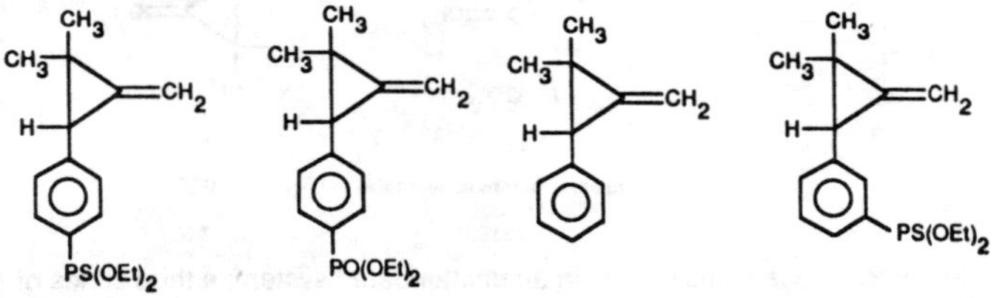

relative rates in C_6D_6 at 80 °C	1.95	1.51	1.00	0.88

analog 7-p-PO(OEt)$_2$. These data support the importance of spin delocalization involving sulfur as in **33c**. Such delocalization would necessitate interaction of the radical center of **33** with a vacant phosphorus d-orbital. The faster rate of 7-p-PS(OEt)$_2$ relative to the O-phosphoryl analog 7-p-PO(OEt)$_2$ could be a result of the increased spin density on sulfur (as in **33c**) relative to the oxygen analog.

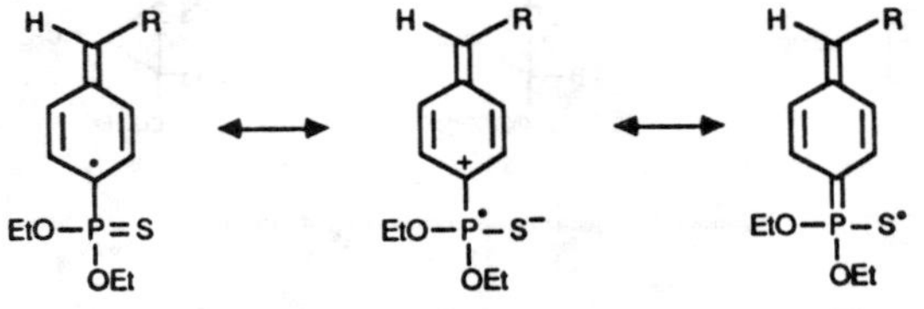

33a **33b** **33c**

Sulfide, Sulfinyl and Sulfonyl Groups. The effect of sulfur containing groups on free radicals is an area of current interest. Neighboring sulfide substituents are thought to stabilize free radicals. Evidence for this comes from a variety of studies including vinyl sulfide co-polymerization data,[25] azocumene pyrolyses,[26] hydrogen atom abstraction data,[27a] and ESR data.[28] Our data on 7-p-SCH$_3$ in Table 1 also support the stabilizing influence of the suldide substituent. The effect of sulfur in oxidized forms (as in sulfoxides and sulfones) is not as clear. While α-sulfinyl[27a] as well as α-sulfonyl radicals[27b] have been generated in the past, the effect of these substituents on radical stability is not as well understood. We wanted a more quantitative comparison of the effect of sulfide, sulfoxide, and sulfone on free radicals.

We have now prepared the additional substituted methylenecyclopropanes 7-m-SCH$_3$, 7-p-SOCH$_3$, 7-m-SOCH$_3$, 7-p-SO$_2$CH$_3$, and 7-m-SO$_2$CH$_3$ by the addition of the appropriate carbene to 1,1-dimethylallene. These substrates allow one to compare the effect of sulfur in various oxidation states on the methylenecyclopropane rearrangement. Pertinent rate data in C_6D_6 for rearrangement of these substrates to the corresponding isopropylidenecyclopropanes **8** is given below. The first thing that is apparent from the data is that the para isomers for all of these sulfur containing substrates rearrange faster than the meta isomers, as well as the parent unsubstituted system 7-p-H. The thiomethyl substituent in the meta position slightly retards the rate relative to 7-p-H, as expected from our earlier studies (in isooctane) where electronegative substituents in the meta position slightly decreased rates.

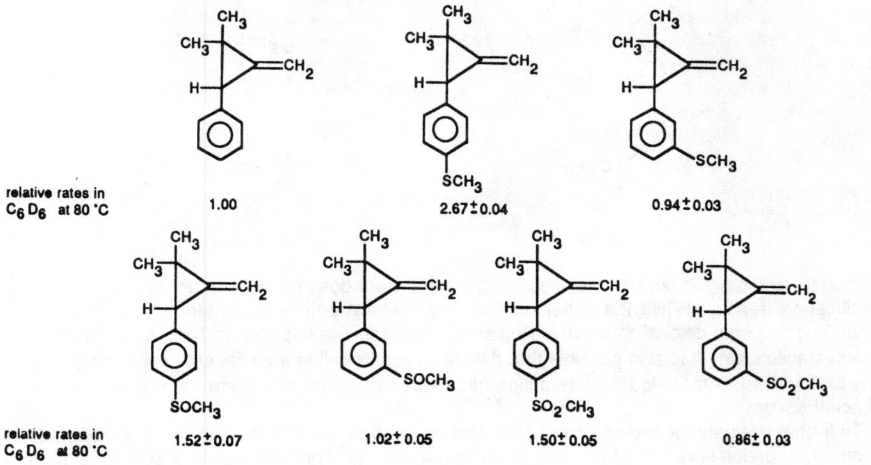

The 7-p-SCH_3 / 7-m-SCH_3 rate ratio of 2.83 verifies that the radical stabilizing effect of the thiomethyl group is a conjugative effect. Arnold's $\sigma_{\alpha}^{\cdot}$ value[13b] also supports this conjugative stabilization. This stabilization has been discussed in terms of interaction of the radical center with the doubly occupied nonbonding molecular orbital of the sulfur.[19,13d] A net stabilizing interaction occurs. An equivalent valence bond description of spin delocalization by sulfur is shown in **34a-34b**. Our study, as well as

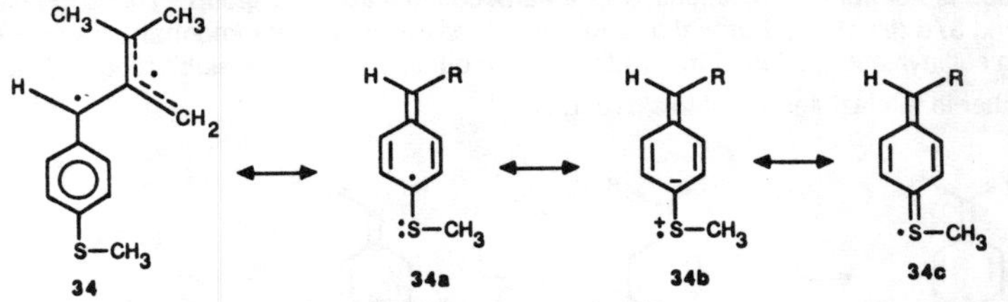

Arnold's $\sigma_{\alpha}^{\cdot}$ value,[13b] and earlier studies,[26] show that thiomethoxy is much more effective than methoxy as a radical stabilizing group. The greater stabilization of sulfur relative to oxygen has been rationalized in terms of a more favorable interaction of the sulfur nonbonding electron pair with the radical center.[19]

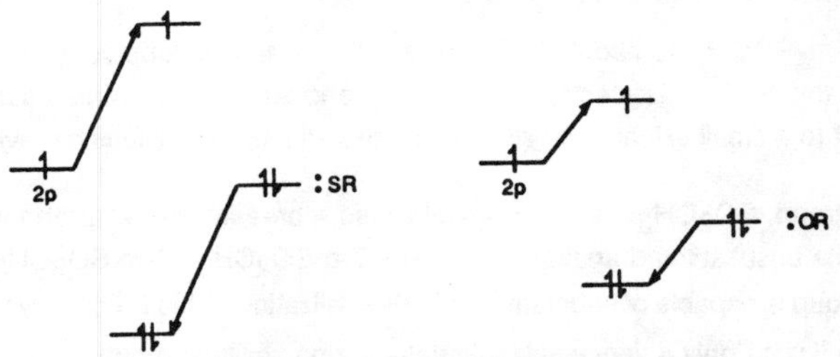

PMO rationale of greater radical stabilization by sulfur relative to oxygen.

However an unanswered question concerns the importance of sulfur d-orbitals in radical stabilization. Theoretical calculations have questioned the importance of d-orbitals in spin delocalization.[29] If interaction of the radical center with the heteroatom nonbonding electron pair were the only stabilizing factor, then thiomethoxy should also be much more effective than methoxy in stabilization of cationic centers. This is not the case. Experimental data suggests that, in solution, methoxy substituted cations **36** are more stable than their thiomethoxy analogs **35**.[30,31] The ability of a carbon 2p orbital to interact

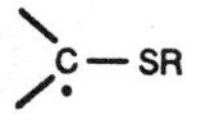

more stable

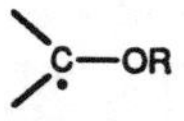

less stable

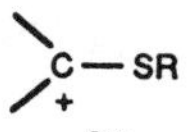

35

less stable

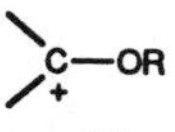

36

more stable

with sulfur and oxygen nonbonding electron pairs therefore does not explain both radical and carbocation stabilization trends. While the sulfur nonbonding electron pair is undoubtedly important, some other factor, such as spin delocalization involving sulfur d-orbitals (as depicted in **34c**), may also be involved in radical stabilization. It is also possible that delocalization involving an adjacent antibonding C-S σ^* orbital may be involved. An analogous interaction has been suggested to account for stabilization of anions by adjacent sulfur.[32]

To further evaluate the importance of other factors (such as d-orbital delocalization) in radical stabilization by sulfur, attention was turned to more oxidized substrates. The sulfoxide **7**-p-$SOCH_3$ is more reactive than either **7**-p-$SOCH_3$ or the unsubstituted analog **7**-p-H. It is, however, less reactive than the sulfide **7**-p-SCH_3. This is again consistent with Arnold's $\sigma_\alpha^{\bullet}$ value for $SOCH_3$ which indicates that this group is radical stabilizing. The mechanism of spin delocalization by $SOCH_3$ is uncertain, although it has been suggested[13b] that this group behaves as a spin doner (i.e. **37** derives stabilization by interaction with the sulfur nonbonding pair as in **37b**). However our unpublished data on α-sulfinyl carbocations indicate that the sulfinyl group is not particularly effective as a carbocation stabilizing group. This suggests that forms such as **37c** and **37d** (involving d-orbital delocalization) may well be more important than the nonbonding electron pair in sulfinyl stabilization of free radicals. The question of radical stabilization by $SOCH_3$ will be addressed further in the last section of this paper.

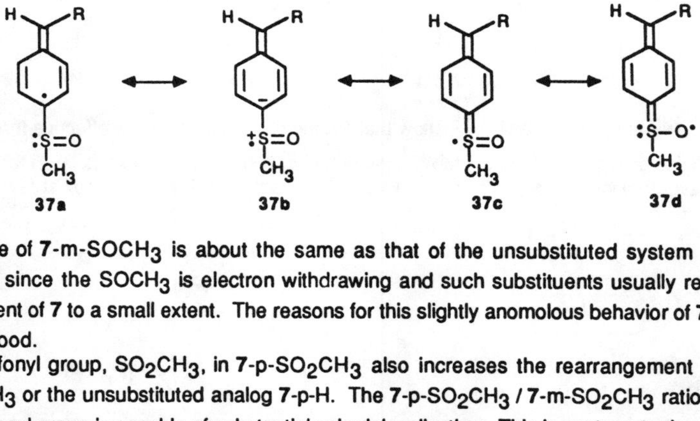

37a **37b** **37c** **37d**

The rate of **7**-m-$SOCH_3$ is about the same as that of the unsubstituted system **7**-p-H. This is unexpected since the $SOCH_3$ is electron withdrawing and such substituents usually retard the rate of rearrangement of **7** to a small extent. The reasons for this slightly anomolous behavior of **7**-m-$SOCH_3$ are not understood.

The sulfonyl group, SO_2CH_3, in **7**-p-SO_2CH_3 also increases the rearrangement rate relative to **7**-m-SO_2CH_3 or the unsubstituted analog **7**-p-H. The **7**-p-SO_2CH_3 / **7**-m-SO_2CH_3 ratio of 1.74 implies that the sulfonyl group is capable of substantial spin delocalization. This is contrary to the $\sigma_\alpha^{\bullet}$ value for this group[13b] which indicates only a very weak spin delocalizing ability in a benzyl radical. Our results imply that delocalization involving sulfur d-orbitals (or some other form of transition state stabilization) is

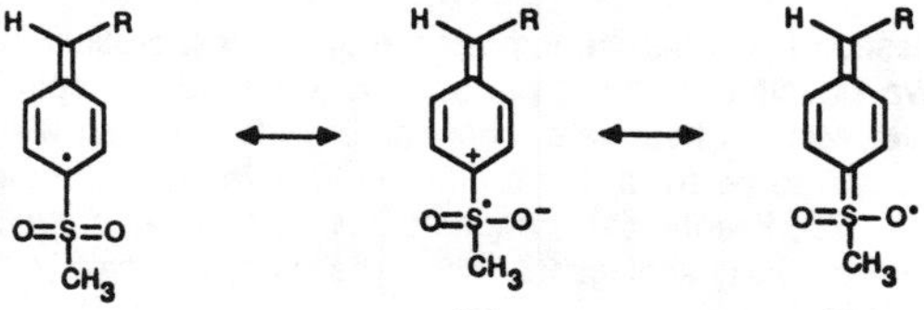

38a **38b** **38c**

important. Forms such as **38b** and **38c** would account for the stabilizing effect of the sulfonyl group in which there are no nonbonding electron pairs associated with sulfur. This implies that sulfur d-orbital effects (**34c**) may also be important in stabilization of the unoxidized sulfide system **34**. We do not understand the reason for the conflicting conclusions regarding the ability of SO_2CH_3 to stabilize benzylic radicals based on our present data and the $\sigma_{\alpha}^{\bullet}$ value. Our data is qualitatively in agreement with Timberlake's pyrolysis study[6b] on **39** where SO_2CH_3 appears to be capable of stabilizing the transition state leading to the radical **40**.

39 **40**

Captodative Effects in the Methylenecyclopropane Rearrangement. The captodative effect is a free radical stabilizing feature recently reviewed by Viehe,[33,34] in which the combination of electron donor and electron acceptor substituents attached to a radical center exert a synergistic stabilizing effect. Donor substituents (such as NR_2, OR) stabilize a free radical by interaction of the nonbonding electron pair of the donor group with the radical center (as in **41b**), while electron acceptor

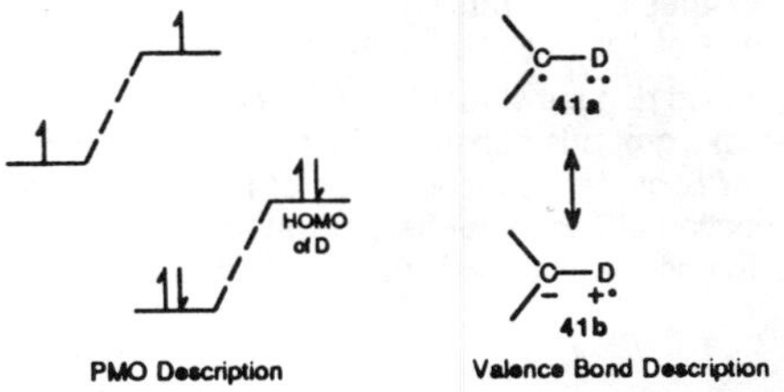

Free radical stabilization by a donor group D

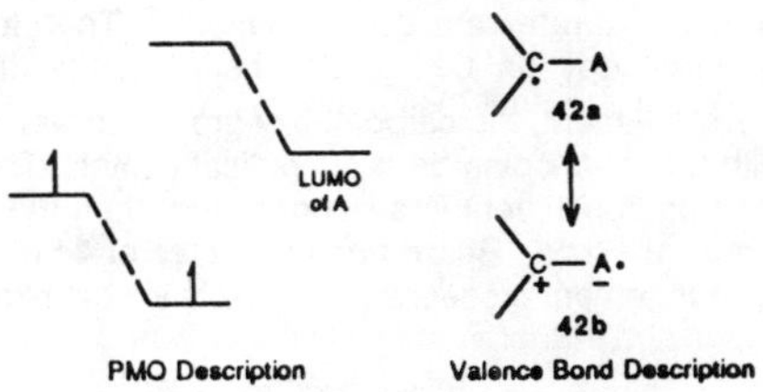

Free radical stabilization by an acceptor group A

substituents can also stabilize by interaction of a vacant acceptor orbital with the radical center (as in **42b**). As Viehe has shown,[33] the actual effect of having both donor and acceptor groups attached to the same radical center is greater than the sum of the individual effects. This captodative effect has been rationalized in terms of a PMO approach in which the donor ability of the group D is enhanced because of the presence of the acceptor group A, and vice versa. A valence bond rationalization leads to the same conclusion. A captodative radical receives enhanced stabilization as reflected in resonance forms **43d** and **43e**. The donor ability of D, as reflected by form **43c**, is enhanced due to the presence of the electron acceptor group, as reflected in **43d**. The acceptor ability of the group A, as reflected by form **43b**, is enhanced by the presence of the donor group, as reflected in **43e**.

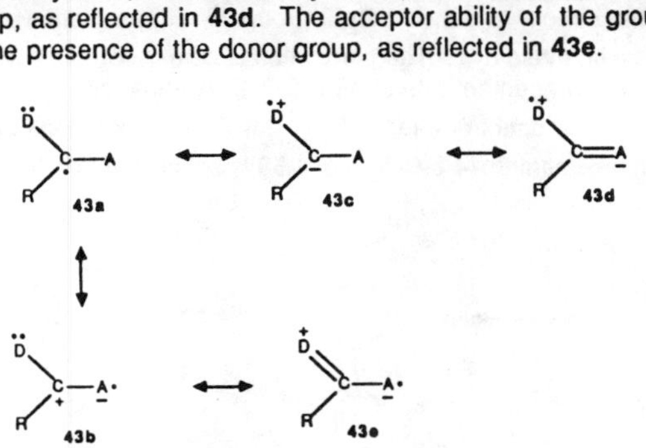

We have now prepared a series of methylenecyclopropanes **44** and monitored their rearrangement to the isopropylidenecyclopropanes **45**. These substrates **44** were prepared by addition of the appropriate carbenes to 1,1-dimethylallene, with the diazocompounds shown serving as the carbene precursors.

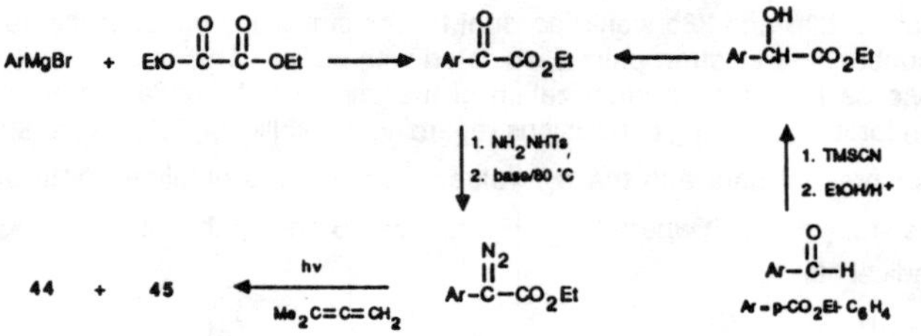

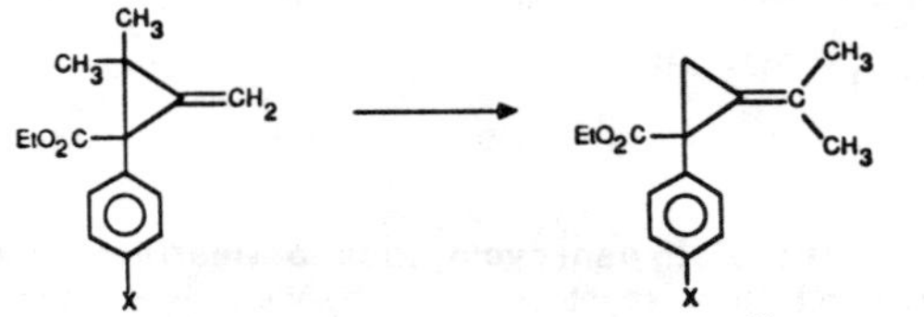

44 45

Rates of thermal rearrangement of **44** to **45** in isooctane were monitored spectrophotometrically. In the cases of **44**-p-SCH_3, **44**-p-$SOCH_3$ and **44**-p-SO_2CH_3, the absorbance change was not large enough to permit accurate rate determinations. Therefore rates of these substrates were monitored by NMR spectroscopy (in C_6D_6) by observing the disappearance of the olefinic methylene signal. In this rearrangement, the carboethoxy group, as well as the aromatic ring substituent can, in principle, interact with the developing benzylic radical center. The carboethoxy group is considered an acceptor group. If the ring substituent is a donor group, then the potential for captodative stabilization of the intermediate biradical exists. Rearrangement rates of **44** to **45** as a function of substituent are given in Table 2. This rearrangement proceeds presumably via the biradical **46**. In principle, the demand for stabilization by the

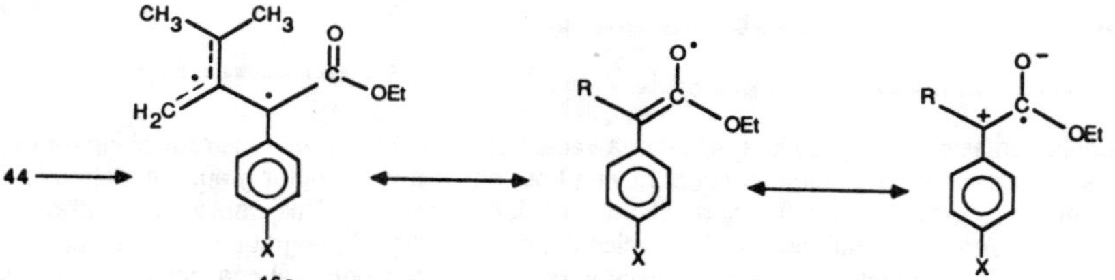

46a 46b 46c

substituent X in **46** should be <u>less than</u> in the "parent" system **6** since the benzylic radical center in **46** is additionally stabilized by the directly attached CO_2Et group.[35] Figure 7 shows an attempt to correlate rearrangement rates of **44** with those of **7**. The correlation is quite poor. Consider first the effect of the p-CO_2Et substituent. The rearrangement rate of **44**-p-CO_2Et is enhanced by a factor of 1.58 relative to **44**-p-H. This enhancement is no doubt due to further stabilizing interactions of the p-CO_2Et group as shown in **47**. The rate enhancement in **44**-p-CO_2Et of 1.58 is somewhat less than the factor of 2.45 seen

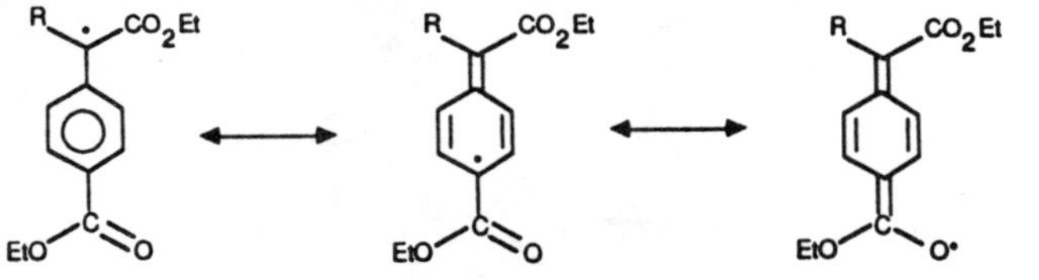

47a 47b 47c

Table 2. Rearrangement rates of **44** in isooctane at 50 °C.

Substituent	k, s^{-1}	k_{rel} for 44	k_{rel} for 7
p-H	5.61×10^{-4} 3.61×10^{-5a}	1.00	1.00
p-CO_2Et	9.81×10^{-4}	1.63	2.45^b
p-CH_3	1.06×10^{-3}	1.89	1.30^c
p-OCH_3	2.44×10^{-3}	4.35	1.72^c
p-SCH_3	1.38×10^{-4a}	3.83^a	2.67^b
p-F	7.96×10^{-4}	1.42	0.83^c
p-$SOCH_3$	4.41×10^{-5a}	1.22^a	1.52^b
p-SO_2CH_3	4.21×10^{-5a}	1.17^a	1.50^b

a rate in C_6D_6 at 25 °C. b relative rate in C_6D_6 at 80 °C. c relative rate in isooctane at 100 °C.

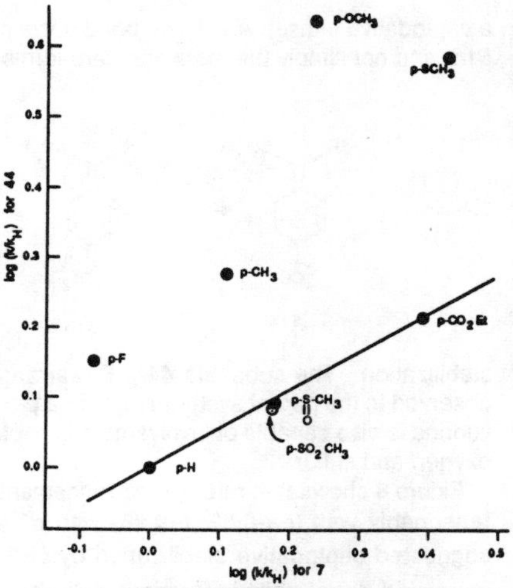

Figure 7. A plot of log (k/k_H) for rearrangement of **44** vs. log (k/k_H) for rearrangement of **7**.

in the parent system **7**-p-CO_2Et (Table 2). This behavior is not unexpected since the demand for p-CO_2Et stabilization in **9**-p-CO_2Et should be larger than in **46**. The rate behavior of **44**-p-CO_2Et is therefore considered quite "normal". A line has therefore been arbitrarily drawn between the p-H and p-CO_2Et substituents in Figure 7.

Rates for **7**-p-OCH_3 and **7**-p-SCH_3 show relatively large positive deviations from "normal" behavior as defined by the line in Figure 7. The rate enhancing effect of p-OCH_3 and p-SCH_3 substitution in **44** is greater than in **7**, despite the expected decreased demand for radical stabilization in **44**. This is suggestive of enhanced transition state stabilization and is presumably due to the captodative effect. Analogous rate enhancing captodative effects were observed earlier by Viehe [36] in the thermal cis-trans isomerization of cyclopropanes **48** and by Arnold[37] in the thermal cis-trans isomerization of the tetraphenylethylene **49**. This effect is apparently also manifested in the methylenecyclopropane rearrangement, as depicted in **50c** and **50d**, despite the fact that the donor group is insulated from the radical center by the aromatic ring.

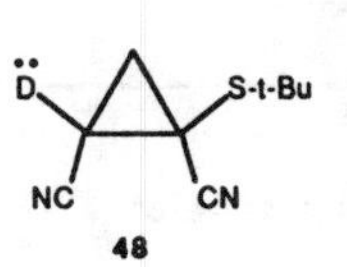

48

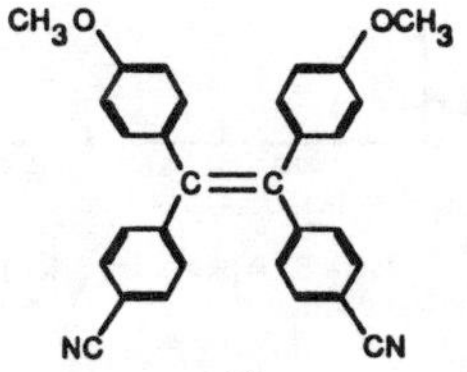

49

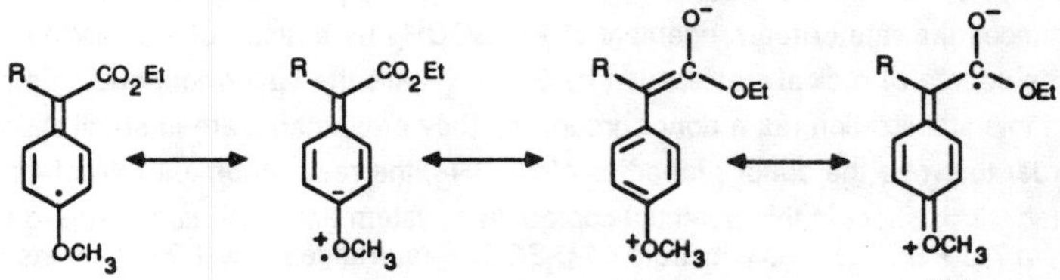

50a **50b** **50c** **50d**

The rate of **44**-p-CH_3 also deviates from the arbitrary line in Figure 7, i.e. p-CH_3 substitution is more effective in stabilization of **46** than in stabilization of **9** despite the decreased demand for stabilization in **46**. This suggests that the methyl group, in conjunction with carboethoxy, is capable of captodative stabilization. To account for this observation, the methyl group must be behaving as a true donor group in a captodative sense. A valence bond description would require the involvement of forms such as **51c** and **51d**, and not simply the more standard forms such as **51a** and **51b**, to account for this increased

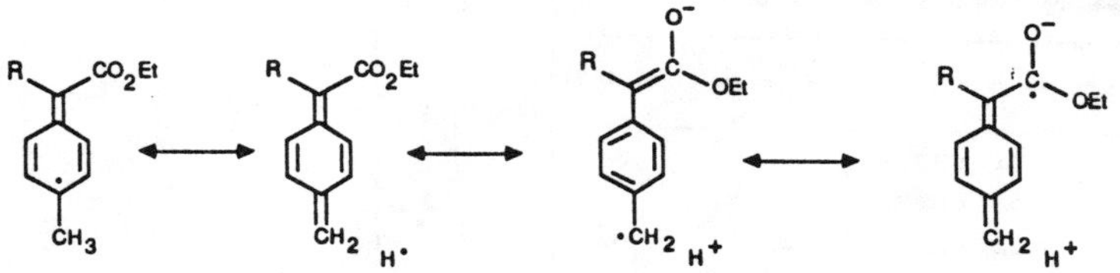

51a **51b** **51c** **51d**

stabilization. The substrate **44**-p-F rearranges faster than **44**-p-H. This trend is opposite that which is observed in the parent system **7**, where a p-F substituent is slightly rate retarding. These data indicate that fluorine is also capable of involvement in captodative stabilization, as are the less electronegative elements oxygen and sulfur.

Figure 8 shows that rates of rearrangement of **44**, where captodative stabilization is postulated, correlate reasonably well (ρ=-0.82; r=0.98) with σ^+ values. This can be viewed as evidence in support of the suggested captodative stabilization by OCH_3, SCH_3, CH_3 and F. σ^+ Values are one measure of the increased donor ability of various substituents under conditions of electron demand. Since the donor ability of various substituents is enhanced in captodative systems, it is not surprising that some correlation is observed with the σ^+ substituent constant.

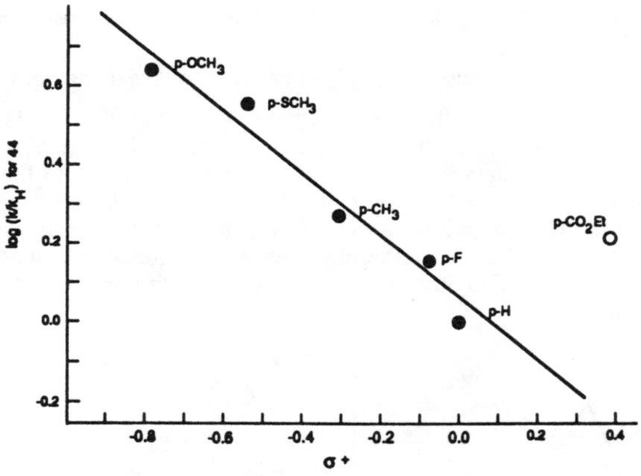

Figure 8. A plot of log (k/k_H) for rearrangement of **44** vs. σ^+

Attention was next turned to the substrate **44**-p-$SOCH_3$. As previously discussed, the sulfinyl group, $SOCH_3$, enhances the rate of rearrangement of **7**-p-$SOCH_3$ by a factor of 1.52. We now return to the question of the source of radical stabilization by $SOCH_3$. Are the sulfur nonbonding electrons of great importance in this stabilization (as a donor group), as they presumably are in stabilization by the SCH_3 group? In order to probe the donor properties of $SOCH_3$, the rearrangement rate of **44**-p-$SOCH_3$ has been measured. The effect in this potential captodative system has been compared to the effect in the "parent" system **7**-p-$SOCH_3$. This substrate **44**-p-$SOCH_3$ rearranges only 1.22 times faster than **44**-p-H. These data suggest that there is no unusual rate enhancement in the potentially captodative system **44**-p-$SOCH_3$. The $SOCH_3$ group therefore does not exhibit increased donor capabilities despite the sulfur nonbonding electron pair, i.e. forms such as **52d**, which utilize the nonbonding electron pair, do not appear to have much importance in stabilization of α-sulfinyl radicals. We have reached analogous conclusions in studies where cationic intermediates are generated. Cation **53**, which contains the sulfinyl group forms 7.4 x 10^{-6} times less readily than the α-H analog **54**. In contrast, the cation **55** forms 10^{11}

times more readily than **54**. The sulfur nonbonbing electron pair in **53** therefore does not appear to lend

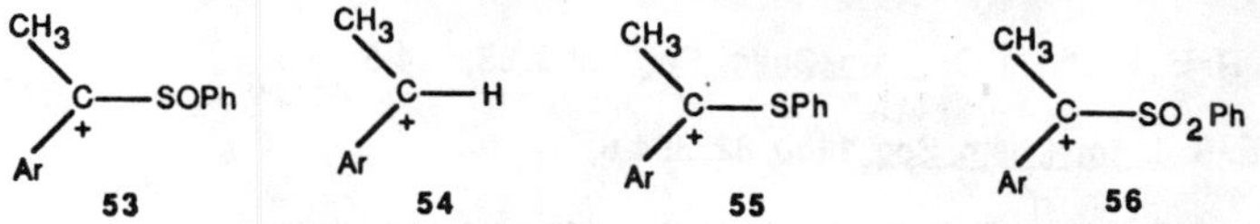

53 **54** **55** **56**

7.4×10^{-6} 1.0 10^{11} 3.8×10^{-9}

Relative rates of carbocation formation from bromides in trifluoroethanol. Ar = p-$CH_3C_6H_4$

much stabilization to cation **53**. We conclude that the sulfinyl group therefore stabilizes free radicals by an acceptor rather than a donor type of mechanism. Utilization of vacant sulfur d-orbitals, as in **52b** and **52c**, would account for the stabilizing properties of this acceptor group.

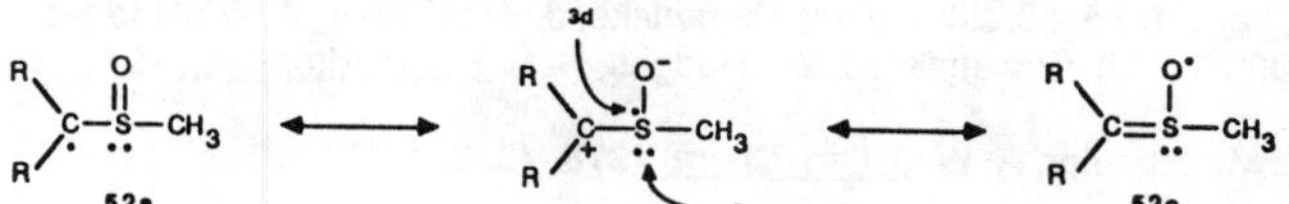

52a **52b** **52c**

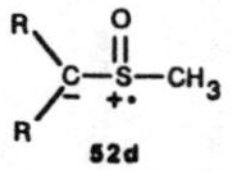

52d

Finally, attention was focused on the sulfonyl group, SO_2CH_3, which increases the rearrangement rate of **7**-p-SO_2CH_3 by a factor of 1.50. Stabilization of the transition state leading to **9** by the sulfonyl group is presumably via an acceptor type of mechanism. The rate behavior of **44**-p-SO_2CH_3 is quite similar to that of **44**-p-$SOCH_3$. Both substrates react at essentially the rates predicted by the line in Figure 7. This is consistent with the fact that there can be no captodative stabilization in the rearrangement of **44**-p-SO_2CH_3. It also supports our contention that no captodative stabilization operates for $SOCH_3$ and that $SOCH_3$ therefore does not stabilize radicals by a donor mechanism.

Summary and Conclusions. The thermal rearrangement of the substituted methylenecyclopropanes **7** offers a useful probe for free radical substituent effects. The major advantages of this probe are the ability to easily prepare and examine a wide variety of substituents in the absence of complicating steric and extraneous polar effects. The major disadvantage of this probe is the relatively small substituent effects. Despite this potential limitation, this probe has proven quite useful in analyzing free radical substituent effects. The biradical intermediate formed in this rearrangement is slightly destabilized by electronegative groups in the meta position but is stabilized by conjugating or third row elements in the para position. The captodative effect also operates in the thermal rearrangement of the related substrate **44**. No captodative effect operates in the rearrangement of **44**-p-$SOCH_3$, which suggests that the sulfinyl group is not a donor group. This "captodative test" appears to be a useful criterion for establishing the donor or acceptor properties of various radical stabilizing groups.

References and Notes

1. Pearson, R. E.; Martin, J. C. J. Am. Chem. Soc. **1963**, 85, 354-5.

2. Huyser, E. S. J. Am. Chem. Soc. **1960**, 82, 394-6.

3. (a) Walling, C.; Jacjnow, B. B. J. Am. Chem. Soc. **1960**, 82, 6113-5. (b) Gilliom, R. D.; Ward, B. F., Jr. Ibid. **1965**, 87, 3944-8. (c) Kennady, B. R.; Ingold, K. U. Can. J. Chem. **1966**, 44, 2381-5.

4. Russell, G. A.; Williamson, R. C., Jr. J. Am. Chem. Soc. **1964**, 86, 2357-64.

5. Pryor, W. A.; Lin, T. H.; Stanley, J. P.; Henderson, R. W. J. Am. Chem. Soc. **1973**, 95, 6993-8.

6. (a) Timberlake, J. W.; Dube, M. F. Tetrahedron, **1980**, 36, 1753-6. (b) Leudtke, A. E.; Timberlake, J. W. J. Org. Chem. **1985**, 50, 268-70. (c) Timberlake, J. W.; Garner, A. W.; Hodges, M. L. Tetrahedron Lett. **1973**, 309-12. (d) Timberlake, J. W.; Hodges, N. L. Tetrahedron Lett. **1970**, 4147-4150.

7. Fisher, T. H.; Meierhoefer, A. W. J. Org. Chem. **1978**, 43, 224-228.

8. (a) Dincturk, S.; Jackson, R. A.; Townson, M. Chem. Commun. **1979**, 172-174. (b) Dincturk, S.; Jackson, R. A.; Townson, M.; Agirbas, H.; Billingham, N. C.; March, G. J. Chem. Soc., Perkin Trans. 2, **1981**, 1121-1126. (c) Dincturk, S.; Jackson, R. A. Ibid. **1981**, 1127-1131. (d) Agirbas, H.; Jackson, R. A. Ibid. **1983**, 739-742.

9. Creary, X. J. Org. Chem. **1980**, 45, 280-4.

10. For a discussion of this general class of rearrangement, see: (a) Kon, G. A. R.; Nanji, H. R. J. Chem. Soc. **1932**, 2557-68. (b) Ettlinger, M. G. J. Am. Chem. Soc. **1952**, 74, 5805-6. (c) Ullman, C. F. Ibid. **1960**, 82, 505-6. (d) Chesick, J. P. Ibid. **1963**, 85, 2720-3. (e) Shields, T. C.; Shoulders, B. A.; Krause, J. F.; Osborn, C. L.; Gardner, P. D. Ibid. **1965**, 87, 3026-57. (f) Gajewski, J. J. Ibid. **1968**, 90, 7178-9. (g) Noyari, R.; Takaya, H.; Nakanisi, Y.; Nozaki, H. Can. J. Chem. **1969**, 47, 1242-5. (h) Gilbert, J. C.; Butler, J. R. J. Am. Chem. Soc. **1970**, 92, 2168-9. (i) Doering, W. E.; Roth, H. D. Tetrahedron **1970**, 26, 2825-35. (j) Gajewski, J. J. J. Am. Chem. Soc. **1971**, 93, 4450-8. (k) Kirmse, W.; Marawski, H.-R. J. Chem. Soc.. Chem. Commun. **1977**, 122-3. (l) Hehre, W. J.; Salem, L.; Willcott, M. R. J. Am. Chem. Soc. **1974**, 96, 4328-30.

11. Yamamoto, T.; Otsu, T. Chem. Ind. (London) **1967**, 787-9.

12. Sakurai, H.; Hayaski, S.; Hosomi, A. Bull. Chem. Soc. Jpn. **1971**, 44, 1945-9.

13. (a) Dust, J. M.; Arnold, D. R. J. Am. Chem. Soc. **1983**, 105, 1211-1221, 6531. (b) Wayner, D. D. M.; Arnold, D. R. Can. J. Chem. **1984**, 62, 1164-8. (c) Arnold, D. R.; Nicholas, A. M. De P.; Snow, M. S. Ibid. **1985**, 63, 1150-5. (d) Wayner, D. D. M.; Arnold, D. R. Ibid. **1985**, 63, 2378-83.

14. Walling, C. "Free Radicals in Solution"; Wiley: New York, 1957.

15. Roberts, J. D.; Webb, R. L.; McElhill, E. A. J. Am. Chem. Soc. **1950**, 72, 408-11.

16. Holty, D. "Progress in Physical Organic Chemistry"; Streitweiser, A., Jr., Taft, R. W., Eds; Wiley: New York, 1971; Vol 8, pp 1-74.

17. The importance of C-F hyperconjugation employing the σ^* C-F orbital has been discussed. See: Bingham, R. C. J. Am. Chem. Soc. **1975**, 97, 6743-6.

18. (a) Wilt, J. W.; Aznavoorian, P. M. J. Org. Chem. **1978**, 43, 1285-6. (b) Wilt, J. W.; Kolewe, O. J. Am. Chem. Soc. **1965**, 87, 2071-2. (c) Wilt, J. W.; Kolewe, O.; Kramer, J. F. Ibid. **1969**, 91, 2624-31. (d) Kochi, J. K.; Krusic, P. J. Ibid. **1969**, 91, 6161-4.

19. Bernardi, F.; Epiotis, N. D.; Cherry, W.; Schlegel, H. B.; Whango, M.-H.; Wolfe, S. J. Am. Chem. Soc.

1976, 98, 469-78.

20. Wadsworth, W. S., Jr.; Emmons, W. D. J. Am. Chem. Soc. **1961**, 33, 1733-38. The σ value for this group is 0.52. See Tsvetkov, E. N.; Lobanov, D. I.; Isosenkova, L. A.; Kabachnik, M. I. J. Gen. Chem. USSR (engl. Transl.) **1969**, 39, 2126-32.

21. (a) Creary, X.; Geiger, C. C.; Hilton, K. J. Am. Chem. Soc. **1983**, 105, 2851-8. (b) Creary, X.; Underiner, T. J. Org. Chem. **1985**, 50,2165-70.

22. Previous polymerization studies on diethyl vinylphosphonate imply that the diethylphosphonate group has some free radical stabilizing properties. See (a) Arcus, C. L.; Matthews, R. J. S. J. Chem. Soc.**1965**, 4607-12. (b) Pike, R. M.; Cohen, R. A. J. Polymer Sci.**1960**, 44, 531-8. CA 54:23410g. See also (c) Zenftman, H.; Calder, D. British Patent 812983. CA53:P15647f.

23. Creary, X. Acc. Chem. Res. **1985**, 18, 3-8.

24. Gassman, P. G.; Tidwell. T. T. Acc. Chem. Res. **1983**, 16, 279-85.

25. Price, C. C.; Zomfefer, J. J. Am. Chem. Soc. **1950**, 72, 14-17.

26. (a) Timberlake, J. W.; Garner, A. W.; Hodges, M. L. Tetrahedron Lett. **1973**, 309-12. (b) Ohno, A.; Kito, N.; Ohnishi, Y. Bull Chem. Soc. Japan 1971, 44, 463-467.

27. For a summary, see (a) Block, E. "Reactions of Organosulfur Compounds" Academic Press, New York, 1978 p. 183. (b) Ibib. p.194-5. (c) Ibid. p. 196-8.

28. Biddles, I.; Hudson, A.; Wiffen, J. T. Tetrahedron **1972**, 28, 867-867.

29. (a) Bernardi, F.; Csizmadia, I. G.; Schlegel, H. B.; Tiecco, M.; Whangbo, M. H.; Wolfe, S. Gazz. Chim. Ital. **1974**, 104, 1101-8. See also (b) Bernardi, F.; Csizmadia, I. G.; Mangini, A.; Schlegel H. B.; Whangbo, M. H.; Wolfe, S. J. Am. Chem. Soc. 1975, 97, 2209-18.

30. σ^+ values support this contention. See (a) Brown, H. C.; Rao, C. G.; Ravindranathan, M. J. Am. Chem. Soc. **1977**, 99, 7663-7. (b) Brown, H. C.; Okamoto, Y.; Inukai, T. Ibid. **1958**, 80, 4964-8. See also (c) McClelland, R. A.; Leung, M. J. Org. Chem. **1980**, 45, 187-189. (d) McClelland, R. A. Can. J. Chem. **1977**, 55, 548-51. (e) Jensen, J. L.; Jencks, W. P. J. Am. Chem. Soc. **1979**, 101, 1476-88. (f) Bohme, H.; Fisher, H.; Frank, R. Justus Liebigs Ann. Chem. **1949**, 563, 54-72. (g) Jones, T. C.; Thornton, F. R. J. Am. Chem. Soc. **1967**, 89, 4863-7. (h) Modena, G.; Scorrano, G.; Venturello, P. J. Chem. Soc., Perkin Trans. 2 **1979**, 1-6.

31. Theoretical calculations suggests that $+CH_2$-SH is more stable than $+CH_2$-OH (in contrast to most experimental data in solution). See (a) Bernardi, F.; Csizmadia, I. G.; Schlegel, H. B.; M. H.; Wolfe, S. Can. J. Chem. **1975**, 53, 1144-53. (b) Bernardi, F.; Csizmadia, I. G.; Epiotis, N. D. Tetrahedron **1975**, 31, 3085-8.

32. (a) Epiotis, N. D.; Yates, D. L.; Bernardi, F.; Wolfe, S. J. Am. Chem. Soc. **1976**, 98, 5435-9. (b) Barbarella, G.; Dembech, P.; Garbesi, A.; Bernardi, F.; Bottoni, A.; Fava, A. Ibid. **1978**, 100, 200-3.

33. Viehe, H. G.; Janousek, Z.; Merenyi, R. Acc. Chem. Res. **1985**, 18, 148-54.

34. The captodative effect is also referred to as merostabilization or "push-pull" stabilization. For further leading references, see Leigh, W. J.; Arnold, D. R.; Humphreys, R. W. R.; Wong, P. C. Can. J. Chem. **1980**, 58, 2537-49.

35. Form **12b** shows carbonyl stabilization by a standard conjugative mechanism while form **12c** emphasizes the acceptor properties of the carbonyl group. From a molecular orbital viewpoint, the interaction of the radical nonbonding orbital with the carbonyl π^* orbital is more effective than with the π orbital. Hence the carbonyl group can also stabilize a free radical by an acceptor type of mechanism.

36. Merenyi, R.; DeMesmaeker, A.; Viehe, H. G. Tetrahedron Lett. **1983**, 2765-2768.

37. Leigh, W. J.; Arnold, D. R. Can. J. Chem. **1981**, 59, 609-620.

2-OXOMORPHOLIN-3-YL RADICALS

Tad H. Koch
Department of Chemistry and Biochemistry
University of Colorado
Boulder, Colorado 80309-0215

ABSTRACT. 2-Oxomorpholin-3-yl radicals are formed when their dimers bi(2-oxomorpholin-3-yl) are dissolved. The dimer radical equilibrium constant for 3,5,5-trimethyl-2-oxomorpholin-3-yl varies between 1×10^{-9} and 6×10^{-16} as a function of medium, consistent with a dipolar radical structure. Variously substituted oxomorpholinyls and diiyls have been synthesized and studied. They are mild one electron reducing agents and are useful for in vitro and in vivo reduction of anthracycline anti-tumor drugs.

INTRODUCTION

Dissolution of meso and dl-bi(3,5,5-trimethyl-2-oxomorpholin-3-yl (1 and 2) at ambient temperature establishes an equilibrium with the persistent free radical 3,5,5-trimethyl-2-oxomorpholiny-3-yl (3).[1] Radical 3, assigned the acronym TM-3, is an important example of a class of radicals now described as capto-dative,[2] merostabilized,[3] or push-pull stabilized.[4] Recent calculations suggest that TM-3 has an unusually high stabilization energy within this class of radicals.[5] The persistence of TM-3 was immediately intriguing because TM-3 is a derivative of the alanyl radical. The effect of structure and medium on dimer radical equilibrium, radical reactivity, and applications of this reactivity have been investigated. An important goal has been the discovery of biological uses for these radicals and biological implication of their persistence.

RESULTS AND DISCUSSION

The meso and dl-dimers of TM-3 are prepared by irradiation of 5,6-dihydro-3,5,5-trimethyl-1,4-oxazin-2-one (4) in 2-propanol solvent at -30°C with a medium pressure mercury lamp through Pyrex.[6] They are isolated in 67% yield by suction filtration; separation of the meso and dl-isomers is achieved by low temperature alumina flash

H. G. Viehe et al. (eds.), Substituent Effects in Radical Chemistry, 263–270.
© 1986 by D. Reidel Publishing Co.

chromatography.[7] In solution the separated stereoisomers equilibrate to the thermodynamic mixture via bond homolysis and radical combination. The oxazinone is prepared in 50% isolated yield by condensation of ethyl pyruvate with 2-amino-2-methylpropanol in refluxing toluene.[6]

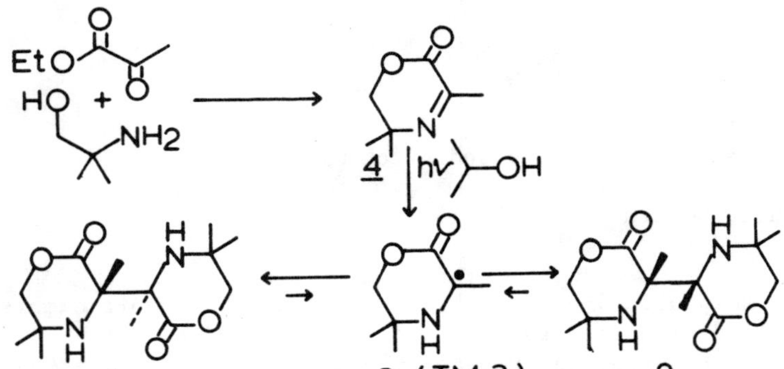

1 3 (TM-3) 2

A number of derivatives of TM-3 dimer have been prepared with appropriate modifications of this procedure.[8] Derivatives were designed and synthesized to establish the effect of substitution on radical formation and reactivity and dimer solubility and toxicity. These derivatives include bi[5,5-bis(acetoxymethyl)-3-methyl-2-oxomorpholin-3-yl] (5), bi[5,5-bis(acetoxymethyl)-3-ethyl-2-oxomorpholin-3-yl (6), bi[3-(acetoxymethyl)-5,5-dimethyl-2-oxomorpholin-3-yl (7), and bi[3,5,5-trimethyl-2-oxopiperazin-3-yl] (8).

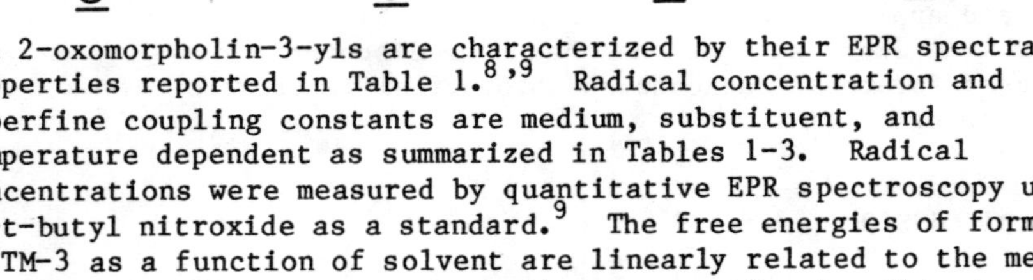

5 6 7 8

The 2-oxomorpholin-3-yls are characterized by their EPR spectral properties reported in Table 1.[8,9] Radical concentration and hyperfine coupling constants are medium, substituent, and temperature dependent as summarized in Tables 1-3. Radical concentrations were measured by quantitative EPR spectroscopy using di-t-butyl nitroxide as a standard.[9] The free energies of formation of TM-3 as a function of solvent are linearly related to the methyl hyperfine coupling constant according to equation 1. This correlation suggests that radical stability is inversely related to the spin density at the 3-position. Equation 1 gives the free energy change at the temperature at which the coupling constant is measured.

Equation 1 $\Delta G = a_{CH_3} \times 2.16 - 8.09$

The free energy change also correlates with Koswer Z-values moderately well. A plot of ΔG° versus Z-value follows equation 2 with a correlation coefficient of 0.992.[9]

Equation 2 $\Delta G° = 34 - (0.25) \times \text{Z-value}$

Table 1. EPR Hyperfine Coupling Constants for the Oxomorpholinyl and Oxopiperazinyl Radicals in Chloroform Solvent

Radical Dimer	T (°C)	a_{CH_3} (G)	a_{CH_2} (G)[a]	a_N (G)	a_{NH} (G)	a_{CH_2} (G)[b]
7	60		5.2	7.5	5.2	0.4
1+2	60	10.9		6.4	4.2	0.3
5	60	11.9		5.7	3.7	0.4
6	60		8.3	5.5	3.6	0.3
8	90	11.7		5.1	3.7	

[a]CH_2 substituent at the 3-position. [b]CH_2 at the 6-position of the ring.

Table 2. EPR Hyperfine Coupling Constants for the TM-3 Radical as a Function of Medium and Temperature[a]

Medium	Temperature (°C)	a_{CH_3} (G)	a_N (G)	a_{NH} (G)
[b,c]EtOH/Mg^{++}	30	8.8	6.5	6.5
[b]EtOH/Na^+	40	9.8	6.4	5.2
MeOH	60	9.9	6.5	5.3
EtOH	60	10.1	6.5	5.2
Me_2SO	60	10.5	6.4	4.4
$CHCl_3$	60	10.9	6.4	4.2
$CHCl_3$	90	11.1	6.4	3.8
glyme	60	11.4	6.2	3.8
benzene	80	12.1	6.2	2.8

[a]g-Values are in the range 2.00355 ± 0.00015. [b]Ionic strength = 0.3μ. [c]Hyperfine coupling constants for TM-3 radical complexed with Mg^{++}.

Table 3. Thermodynamic Parameters for Bond Homolysis of TM-3 Dimers (1 and 2) as a Function of Medium[a]

Medium	K (mol/ℓ)	ΔG° (kcal/mol)	ΔH° (kcal/mol)	ΔS° (cal/mol·°K)
[b]EtOH/Mg^{++}	1.3×10^{-9}	12	23	36
[b]EtOH/Na^{+}	7.0×10^{-11}	14	22	28
EtOH	3.2×10^{-11}	14	22	24
glyme	6.8×10^{-14}	18	30	40
benzene	5.8×10^{-16}	21	36	51

[a]Values of K and ΔG° are reported at 25°C. [b]Ionic strength = 0.3μ.

TM-3 and the other 2-oxomorpholin-3-yls react as mild one electron reducing agents.[10] Reduction of tetracyanoethylene to its radical anion, paraquat to its radical cation, and silver cation to silver metal have been achieved. Electron transfer coupled with proton transfer occurs in the reduction of benzil to benzoin, oxygen to hydrogen peroxide, diphenylpicrylhydrazyl to diphenylpicrylhydrazine, di-t-butylnitroxide to di-t-butylhydroxylamine, and isatin to isatide. In the reductions, the oxomorpholinyls are transformed back to the dihydrooxazinones from which they were synthesized. The latter four reduction reactions occur faster than the rate of combination of the oxomorpholinyls and can thus be used to measure the rate of bond homolysis of radical dimers.[7,8] Kinetic parameters for the bond homolysis as a function of substituent and solvent are summarized in Table 4.

Table 4. Kinetic Parameters for Bond Homolysis as a Function of Dimer Structure and Solvent

Dimer	Solvent	ΔH≠ (kcal/mol)	ΔS≠ (cal/mol·°K)
7	$CHCl_3$	21	-6
6	$CHCl_3$	25	8
2	EtOH	20	-3
2	glyme	24	-2
2	$CHCl_3$	27	7
2	benzene	28	6
5	$CHCl_3$	30	11
8	MeOH	28	6

In general polar, hydrogen bonding solvents and Mg^{++} facilitate bond homolysis. The solvent and substituent effects are consistent with the proposed polar nature of the radicals resulting from intramolecular charge transfer resonance, illustrated with resonance structure 1a. Complexation of TM-3 radical with Mg^{++} is evident from the EPR spectrum which shows complexed and uncomplexed signals. Substituents which stabilize the enolate part of 1a facilitate bond homolysis, and substituents which destabilize the aminium ion part of 1a inhibit bond homolysis. The importance of structure 1a is also consistent with equations 1 and 2.

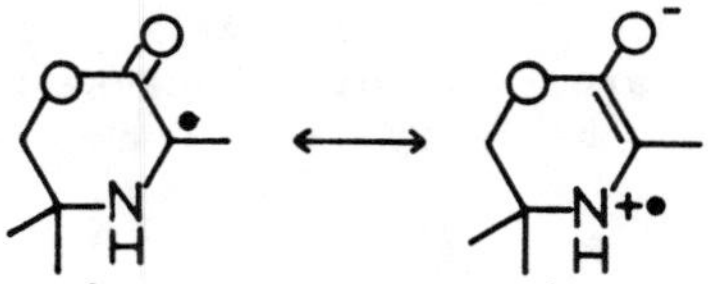

1 1a

Bond homolysis is also influenced by steric effects. Steric crowding of the six methyl groups of the dl-dimer of TM-3 is evident from the single crystal X-ray analysis.[11] Furthermore, bi(3-methyl-2-oxomorpholin-3-yl), the dimer of the oxomorpholinyl radical without the gem dimethyl, undergoes bond homolysis with reluctance.[8]

The kinetic and thermodynamic data for bond homolysis of TM-3 dimer give kinetic parameters for combination of TM-3 radicals reported in Table 5.[9] A striking result is the apparent negative enthalpy of activation for radical combination in benzene and glyme solvents. The enthalpy of activation is coupled with a large negative entropy of activation. Houk has recently proposed an explanation for the negative enthalpy of activation for the cycloaddition of halocarbenes to alkenes which may be applicable here.[12] With the enthalpy and entropy changes observed, the maximum in the free energy change, which defines the transition state, can occur at a point at which the enthalpy is below that of the starting materials. Consequently, the rate of radical combination is entropy controlled.

Table 5. Kinetic Parameters for Combination of TM-3 Radicals[a]

Solvent	k (ℓ/mol·s)	$\Delta G^{\neq}$ (kcal/mol)	$\Delta H^{\neq}$ (kcal/mol)	$\Delta S^{\neq}$ (cal/mol·°K)
EtOH	4×10^7	7	-1	-28
glyme	1×10^8	6	-6	-42
benzene	6×10^8	6	-8	-45

[a]Values of k and $\Delta G°$ are reported at 25°C; all values are calculated from the data in Tables 3 and 4.

Diradical oligomers based upon the 2-oxomorpholin-3-yl unit have also been prepared.[13] Photoreduction of dl-bi(5,6-dihydro-3,5,5-trimethyl-1,4-oxazin-2-on-6-yl) (10) at -40°C gives oligomers of dl-bi(3,5,5-trimethyl-2-oxomorpholin-6-yl)-3,3'-diyl. End group analysis and osmometric molecular weight measurements indicate that the oligomers are macrocycles. The average molecular size of kinetically formed material is five monomeric units and thermodynamic material formed by equilibration of the kinetic material, three monomeric units. CPK models suggest that steric interactions in the trimer are minimized when the methyl groups are located on the outside of the macrocycle and the lactone functional groups on the inside. Such a structure places these oligomers in the family of coronands. These macrocycles are intriguing because they can potentially change size and stereochemistry in response to a template.

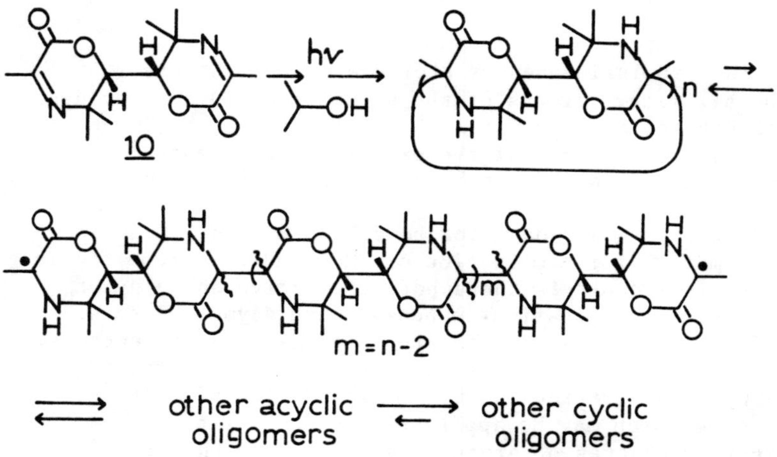

TM-3 dimers readily reduce quinone anti-tumor drugs such as the clinically important anthracyclines, adriamycin and daunomycin (11).[14] Reduction ultimately results in glycosidic cleavage to give the deoxy aglycones. Again reduction is fast relative to the rate of TM-3 radical combination. Consequently, TM-3 dimer releases the reducing agent, TM-3 radical, at the known first order rate. TM-3 has proven to be useful for the generation and characterization of many of the redox states of the anthracyclines, especially the previously unknown quinone methide state 12, as illustrated in Scheme 1.[15] The overall reduction process in general transforms the anthracycline from an active and toxic form to an inactive and non-toxic form. Because some of the oxomorpholinyl dimers are non-toxic they have potential as pharmaceutical materials for the *in vivo* manipulation of the anthracyclines to therapeutic advantage and as antidotes. A particularly important example is the water soluble radical dimer bi[3,5-dimethyl-5-(hydroxymethyl)-2-oxomorpholin-3-yl] (13, DHM-3). High adriamycin dose DHM-3 rescue therapy of mice with L-1210 leukemia results in up to a factor of five improvement in test/control response and long term survivors.[16] DHM-3 is also potentially an effective antidote for anthracycline extravasation necrosis incident to therapy.[17]

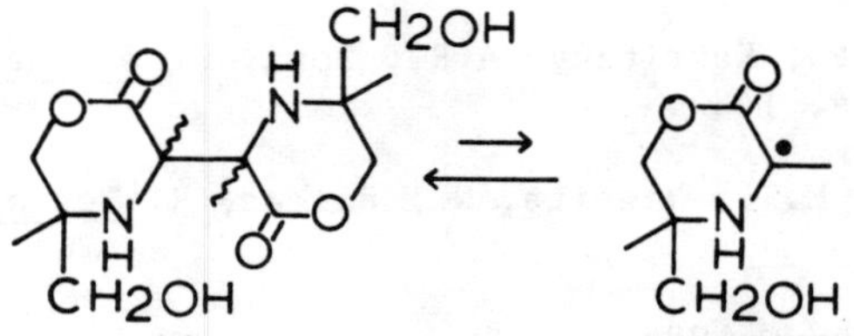

13 DHM-3

Scheme 1

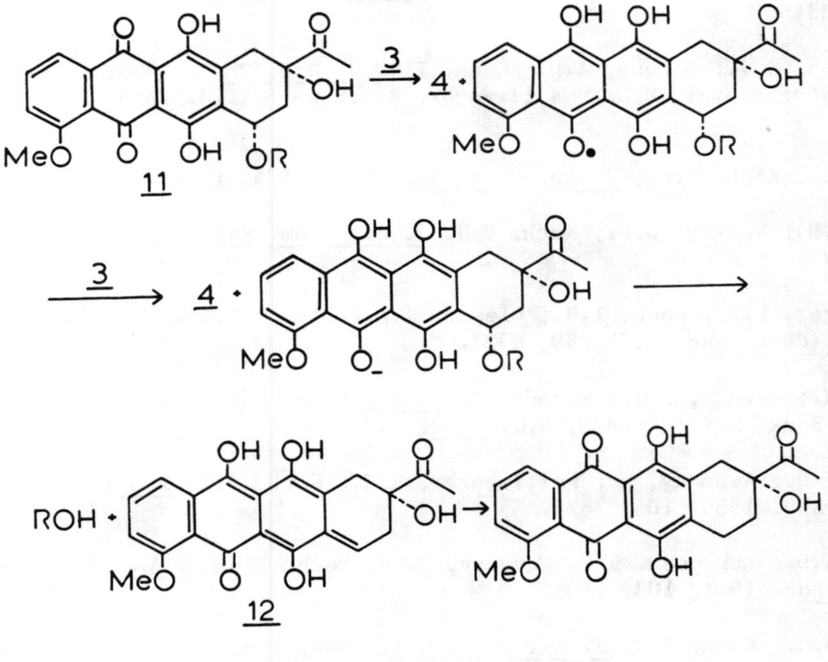

ACKNOWLEDGMENT. Research support was received from the National Institutes of Health (CA-24665), the National Science Foundation (CHE-8419718), the Petroleum Research Fund, and the 3M Co.

REFERENCES

[1]Koch, T.H.; Olesen, J.A.; DeNiro, J. J. Am. Chem. Soc. 1975, **97**, 7285.

[2]Viehe, H.G.; Merenyi, R.; Janousek, Z. Angew. Chem. Int. Ed. Engl. 1979, **18**, 917.

[3]Baldock, R.W.; Hudson, P.; Katritzky, A.R.; Soti, F. J. Chem. Soc., Perkin Trans. 1 1974, 1422.

[4]Balaban, A.T.; Caprois, M.T.; Negoita, N.; Baican, R. Tetrahedron 1977, **33**, 2249.

[5]Leroy, G. Adv. Quant. Chem. 1985, **17**, 1.

[6]Koch, T.H.; Olesen, J.O.; DeNiro, J. J. Org. Chem. 1975, **40**, 14; Barone, A.D.; Ph.D. Thesis, University of Colorado 1982.

[7]Bennett, R.W.; Wharry, D.L.; Koch, T.H. J. Am. Chem. Soc. 1980, **102**, 2345.

[8]Himmelsbach, R.J.; Barone, A.D.; Koch, T.H. J. Org. Chem. 1983, **48**, 2989; Kleyer, D.L.; Haltiwanger, R.C.; Koch, T.H. ibid, 147.

[9]Olson, J.B.; Koch, T.H.; J. Am. Chem. Soc. 1986, **108**, in press.

[10]Burns, J.M.; Wharry, D.L.; Koch. T.H. J. Am. Chem. Soc. 1981, **103**, 948.

[11]Haltiwanger, R.C.; Koch, T.H.; Olesen, J.A.; Kim, C.S.; Kim, N.K. J. Am. Chem. Soc. 1977, **99**, 6327.

[12]Houk, K.N.; Rondan, N.G.; Mareda, J. J. Am. Chem. Soc. 1984, **106**, 4291; Houk, K.N.; Rondan, N.G. ibid., 4291.

[13]Gaudiano, G.; Sweeney, K.; Haltiwanger, R.C.; Koch, T.H. J. Am. Chem. Soc. 1984, **106**, 7628.

[14]Barone, A.D.; Atkinson, R.F.; Wharry, D.L.; Koch, T.H. J. Am. Chem. Soc. 1981, **103**, 1606.

[15]Kleyer, D.L.; Koch, T.H. J. Am. Chem. Soc. 1984, **106**, 2380.

[16]Averbuch, S.D.; Gaudiano, G.; Koch, T.H.; Bachur, N.R. Cancer Res., 1985, **45**, 6200.

[17]Averbuch, S.D.; Gaudiano, G.; Koch, T.H.; Bachur, N.R. J. Clin. Oncol. 1986, **4**, 88.

AZOALKANE DECOMPOSITIONS AS MEASURES OF RADICAL SUBSTITUENT EFFECTS

Jack W. Timberlake
Department of Chemistry
University of New Orleans
New Orleans, LA 70122

ABSTRACT. The thermal decomposition of substituted azopropanes is shown to be a good model for determining radical substituent effects. In this series steric and inductive contributions are minor with the most significant portion of the 10^9 rate acceleration (16 kcal/mol) differences being the result of electronic (resonance) stabilization. These results have been correlated with other systems of generating radicals and have been related to ESR studies and molecular orbital calculations.

Introduction

Knowledge of stabilization effects in carbocations and carbanions are well characterized and from a linear free energy (LFE) standpoint, they give good agreement from one system to another.[1] By contrast, substituent effects in radicals are not as completely understood and show conflicting results when attempting to compare one model to another.[2] In some cases this is due to the intervention of polar transition (and/or ground) state effects that mask true radical character thus rendering it difficult to relate kinetic results to true thermodynamic stability.[3] Intuitively, if one considers cations and anions as a dichotomy, then radicals might be expected to be at a midpoint between these extremes (a one electron reduction of cations and a one electron oxidation of anions), and they might behave more like one or the other depending on their mode of generation.

Discussion

For a number of years we have used azopropanes as a method of kinetically generating substituted radicals which we have argued

H. G. Viehe et al. (eds.), Substituent Effects in Radical Chemistry, 271–281.
© 1986 by D. Reidel Publishing Co.

should not involve polar transition states and should therefore mimic true radical character.[4-11] Table I lists some of the substituted azopropanes whose rates we have determined or have been taken from the literature.[6] Before discussing the specifics it is important to note that the relative rates span a range of 10^9 or a free energy difference of greater than 16 kcal/mol. If they indeed monitor radical stabilities, it is clear that this model is a sensitive one and is capable of detecting small differences. We will examine five different aspects of this system: 1) the mechanism of decomposition, 2) ground state steric effects, 3) inductive contributions, 4) electronic stabilizing effects, and 5) their correlation or lack of correlation with other systems.

For decades a controversy has raged over whether azoalkanes decompose thermally via a concerted two bond cleavage or by a two

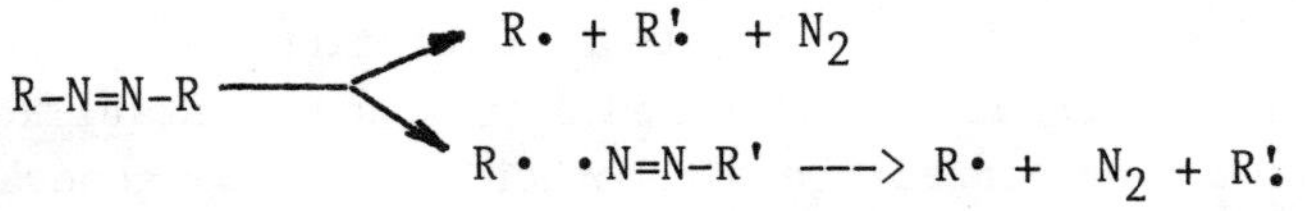

step, one bond at a time, mechanism. Many carefully conceived methods have been employed to answer this question and numerous reviews have appeared summarizing these results.[14] Until recently it was believed that for symmetrical systems the concerted mechanism prevailed but that for R and R' groups with greatly different radical stabilities stepwise C-N bond cleavage occurs with a continuum between the two resulting when the R groups' stabilities approach each other. However, by no means is this a universally held platform. Recent MNDO calculations by Dannenberg[15], pressure dependence studies by Neumann[16], and product and thermochemical results by Engel[17] have called into question the concerted path even for symmetrical cases. It is nonetheless difficult to dismiss all previous experimental studies[14], and in general, we adhere to the concerted mechanism for symmetrical analogs, although the interpretation of our results is not affected provided the mechanism, concerted or stepwise, is the same in all cases.

Another aspect of the mechanism question has been raised by Firestone.[18] Using the Linnett theory of bonding, he has predicted that azo transition states should contain an element of polar

Table I. Rate Data for Azopropanes

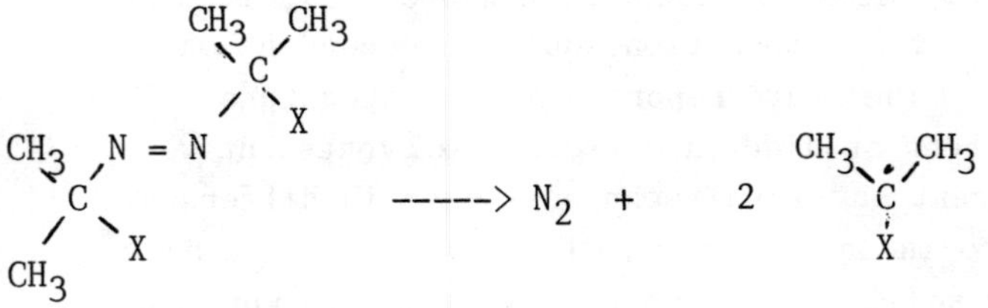

1

	X	Rel rate 100°	$\Delta G^{\ddagger}$, kcal/mol 100°	ref
a	H	1.0	40.8	lit.,6
b	CH_3	$5.0x10^2$	36.2	6,12
c	$CH_3CO_2CH_2$	$5.1x10^2$	36.1	lit.,6
d	CH_3COCH_2	$1.6x10^3$	35.3	13
e	CH_3SO_2	$4.6x10^3$	34.5	11
f	CH_3O	$5.2x10^3$	34.4	4,6
g	Cl	$5.6x10^3$	34.4	6,13
h	$(CH_3)_3C$	$6.3x10^3$	34.3	6
i	Cyclo-C_3H_5	$2.6x10^4$	33.2	12
j	O=C=N	$1.7x10^5$	31.8	10
k	CH_3 COS	$3.2x10^5$	31.4	6
l	$(CH_3)_3C$ CH_2	$6.5X10^5$	30.8	6,7,8
m	CH_3S	$9.3x10^6$	28.9	11,13
n	NH_2 CO	$1.6x10^7$	28.5	lit.6
o	$C_2H_5O_2C$	$1.3x10^8$	26.9	lit.6
p	N≡C-	$1.4x10^8$	26.8	lit.6
q	C_6H_5CO-	$5.4x10^8$	25.9	9
r	CH_2=C=CH-	$1.0x10^9$	25.4	11
s	C_6H_5-	$2.0x10^9$	24.9	lit.6
t	CH_2=CH-	$4.4x10^9$	24.3	14
u	HC≡C	$4.9x10^9$	24.2	14

character described by $^{\delta\delta+}R$ ---$^{\delta\delta+}N{=}N.....R^{\delta-}$. While it is virtually impossible to completely discount such a proposal, it might be expected that the rates of decomposition would be dependent on solvent polarity changes. There are reports in the literature[17] where azo compounds have been studied in different solvents but we were concerned that different sampling techniques used in different laboratories might lead to inconsistencies. For this reason we have systematically examined a select series where polar effects would be expected to be most pronounced. Compounds 1 (X=Ph, PhCO, CN, MeO_2C) thermalized in cumene, cyclohexanone, DMF, and cyclohexanol were shown to give no more than a three-fold difference and even then not in a predictable way with any known solvent polarity parameter.[19] We can only conclude that if there is polar character it can not be very substantial.

How sensitive are azopropanes to steric effects on their way from tetrahedral ground states to transition states that are presumably more trigonal in nature? The difference in radical stability between **1b** (X = CH_3) and **1h** (X =$(CH_3)_3C$) should not be large except for a loss of some small hyperconjugation contribution.[6] The rate factor of 13 reflects a modest acceleration (compared to an overall $\Delta k_{rel} = 10^9$) with respect to such a drastic change in steric bulk. Therefore, if groups are limited in size, relative to tert-butyl, this effect will be minimal. This is not to say that these effects can always be discounted as **1l** (X = [$(CH_3)_3C\ CH_2$-] relative to **1b** (X =CH_3) shows a rate acceleration of 1,300. In addition we have found one branched tert-alkyl group where, relative to methyl, Δ_{rate} = 57,000.[8] These are extreme cases, and with the aid of molecular models, these steric effects can be qualitatively predicted employing Newmans rule of six.[6,20] Therefore, with the judicious choice of substituents, we feel justified in concluding that the steric contributions of the entries in table I are relatively minor and are not the major cause of the pronounced rate accelerations.

A second reason for discounting steric effects comes from evaluation of the data in table II. In both series the increase in rate, $H<CH_3<CH_3O<Cl<C_6H_5$, follows the same order as that for azopropanes. Steric contributions in phenylazomethanes (**2**)[6] and arylpyrazolines[21] (3, table II) should be minimal for they are para substituted. It would be somewhat fortuitous if all three were to

parallel each other so closely if steric effects were the determining factor. It is, however, interesting to note the pronounced diminution of substituent effects in the two phenyl substituted series. We have attributed this to the lesser electronic demand of the two stabilized benzyl series which is consistent with what is expected according to the Hammond postulate for a transition state for 1 that is further developed along the reaction coordinate.[6]

Inductive contributions are somewhat more difficult to evaluate in this series. With the substituent directly attached to the incipient radical center, it is hard to separate pure electron withdrawing or donating effects from resonance contributions. Nevertheless, a comparison of **1**, **c**, **d**, **f** and **g** to **1b** (a choice made

Table II. Rate Data for Phenylazomethanes and Arypyrazolines

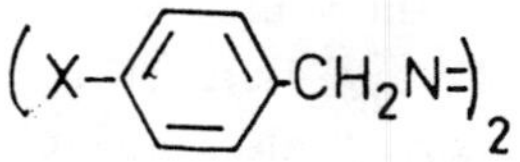

2

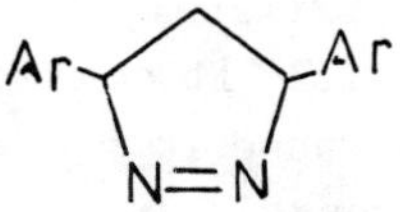

3

X	2 rel rate 150°	3 rel rate 80°
H	1.0	1.0
CH_3	1.1	1.4
CH_3O	1.3	1.5
Cl	1.5	1.9
C_6H_5	2.0	---

so all are essentially tertiary) is informative. With respect to **1b**, compounds **1f** (X=CH_3O) and **1g** (X=Cl) show only a modest increase, a factor of 10, corresponding to a stabilization energy of approximately 2 kcal/mol. If this is the result of pure induction, then it is indeed minor. Furthermore, the intervention of CH_2 units (1c,d) diminishes the effect even further.

Clearly the most pronounced stabilization results from pi type substituents, eg 1 **r**,**s**,**t** and **u** (X=CH_2=C=CH, C_6H_5, CH_2=CH, and HC≡C). This is not surprising since all reactive intermediates show similar effects. However, the acceleration by carbonyl groups (1 **n**, **o**,and **q**) and cyano (1**p**) is a little unusual in that these groups, while stabilizing towards anions,[22] are destabilizing towards cations.[23-25] The latter has been reported to be the result of the unfavorable inductive withdrawing nature of these groups which overides the resonance stabilization. To a first approximation then groups that are good carbanion stabilizers can qualitatively be predicted to be the same for radicals.

The effect of heteroatoms adjacent to the radical also makes an interesting comparison. The stabilization of oxygen (**1b to 1f**), mentioned earlier to be approximately 2 kcal/mol, increases with sulfur to a range of 4.8 to 7.3 (**1b** compared to **1k** and **1m**). The limits are hard to define because **1m** gives an unexplainable negative $\Delta S^{\ddagger}$ [6,11] which is not reasonable for fragmentation reactions like these. For compound **1k** $\Delta S^{\ddagger}$ = + 12 e.u. In any event, the additional 3 to 5 kcal stabilization of sulfur relative to oxygen is what is expected for better carbon p_z orbital interaction with sulfur than oxygen as theory predicts.[26] In an oxidized state (1e, X=CH_3SO_2-) this effect is essentially washed out as X=CH_3O and CH_3SO_2 are almost identical.[11]

We have also examined substituents with two pi systems orthogonal to each other **1j** (X=N=C=O)[10] and **1r** (X=CH_2=C=CH)[11] and find little difference between the latter and **1t** (X=CH_2=CH)[11] but a substantial difference of 6.4 kcal between X=NCO and CH=C=CH_2.

To what extent does this system correlate with other free radical models? The answer is simply that for traditional radical abstraction reactions on substituted toluenes they do not.[3] This is undoubtedly the result of polarization effects. However, more recent work by Creary[27] and Arnold[28] provides a more realistic test. The Creary system follows the thermal rearrangement of arylmethylenecyclopropanes and Arnold's looks at the methylene hyperfine coupling constants of substituted benzyl radicals. Arnold defines a dual substituent parameter Hammett relationship using his $\sigma_\alpha^\bullet$ parameters and conventional σ parameters to determine the amount of polar contribution. Using $\log k_{rel} = \sigma_\alpha^\bullet \rho^\bullet + \sigma\rho$ the Creary vs Arnold correlation gives $\rho^\bullet = 8.39$ and $\rho = 0.039$ with an r = 0.963 for fourteen points. Arnold believes the ratio of $\rho^\bullet/\rho$ for a reaction assesses the importance of spin delocalization versus polar effects. The Arnold vs Creary ratio of 215[29] is thought to reflect the pureness of both systems with little polar character in either. A comparison of our azopropane series with the Arnold $\sigma_\alpha^\bullet$ parameters is unfortunately limited to eight substituents common to both. However, the $\rho^\bullet/\rho$ ratio of 346 (r=0.943) is satisfying to the extent that the magnitude truly measures pure radical character.[19]

There can be little question that the Arnold parameters provide a major breakthrough in the study of radical substituent effects. However, $\sigma_\alpha^\bullet$ values by themselves do not allow stabilization energies to be determined and by necessity requires cross comparison with some other system that measures the energetics of radical formation.

We have also attempted to correlate our rate values with methyl hyperfine coupling constants for the radical $(CH_3)_2\dot{C}-X$. According to the McConnell equation[30] this hyperfine coupling should be proportional to the spin density of the carbon $2P_z$ orbital. As X increases in its ability to delocalize the odd electron, spin density at methyl will decrease resulting in a lower a Me_{hfs} value. The relationship here is less satisfactory.[31] Figure 1 seems to reflect either no correlation or a temptation to force a two tier relationship. In this fashion one could argue that the line of lesser slope (X=$CONH_2$, CO_2R, CN, COC_6H_5, C≡CH, and C_6H_5 with r = 0.939) contains the polar groups and that the line of greater slope (r=0.979) contains the non-polarizing better-behaved substituents. However, this would require explaining why NCO is not a polar group, why C≡CH and C_6H_5 are polar and why $CH=CH_2$ has such a low aMe_{hfs} compared to C_6H_5 and C≡CH. We have no definitive explanation for the lack of correlation in the two systems but simply acknowledge that they may measure different effects.

One final attempt at cross-comparison resulted in a molecular orbital calculation in collaboration with Professor Ken Houk. We had earlier found that a plot of $\log k_{rel}$ for azo decompositions versus % spin density at atom 3 for the series $(CH_3)_2$ C-X-Y) where X-Y = CH_3,
1 2 3

NCO, CN, CH≡C, CH_2=C=CH and CH_2=CH was surprisingly good (modified STO-3G, r=0.987).[10,13] Using *ab initio* UHF-SCF calculations (geometry optimization with STO-3G and single point determination at the 4-31G level) the results, while encouraging for a limited set, table III, X = H, CH_3, CN, HC≡C, CH_2=CH, r=0.964 for the series $(CH_3)_2\dot{C}-X$), breaks down for other conjugating substituents; particularly carbonyl groups. We believe this is a consequence of spin contamination which occurs in conjugated systems as is known to be the case for UHF calculations.[32] Calculations correcting for spin

Table III Kinetic Data vs Calculated BDE

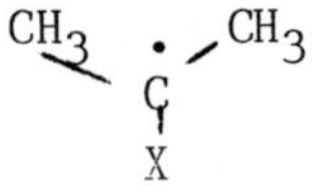

X	$\Delta G^{\ddagger}$, kcal	BE cal 4-31, kcal	
H	0	0	
CH_3	-4.6	-2.2	
CN	-13.9	-11.4	r = 0.964
HC≡C	-16.5	-15.6	
H_2C=CH	-16.5	-19.7	

contamination (ROHF) will have to be done for the results to be more meaningful.

Conclusion

Within certain limits we believe that the rates of decomposition of substituted azopropanes provide good models for determining radical substituent effects. These rates correlate fairly well with stabilization energies calculated by molecular orbital methods and with $\sigma_{\alpha}^{\bullet}$ values determined by Arnold. It provides a method of obtaining values for stabilization energies of various substituents in the absence of polarization effects and has shown that the most significant radical stabilizing groups are those which are also good anion stabilizers.

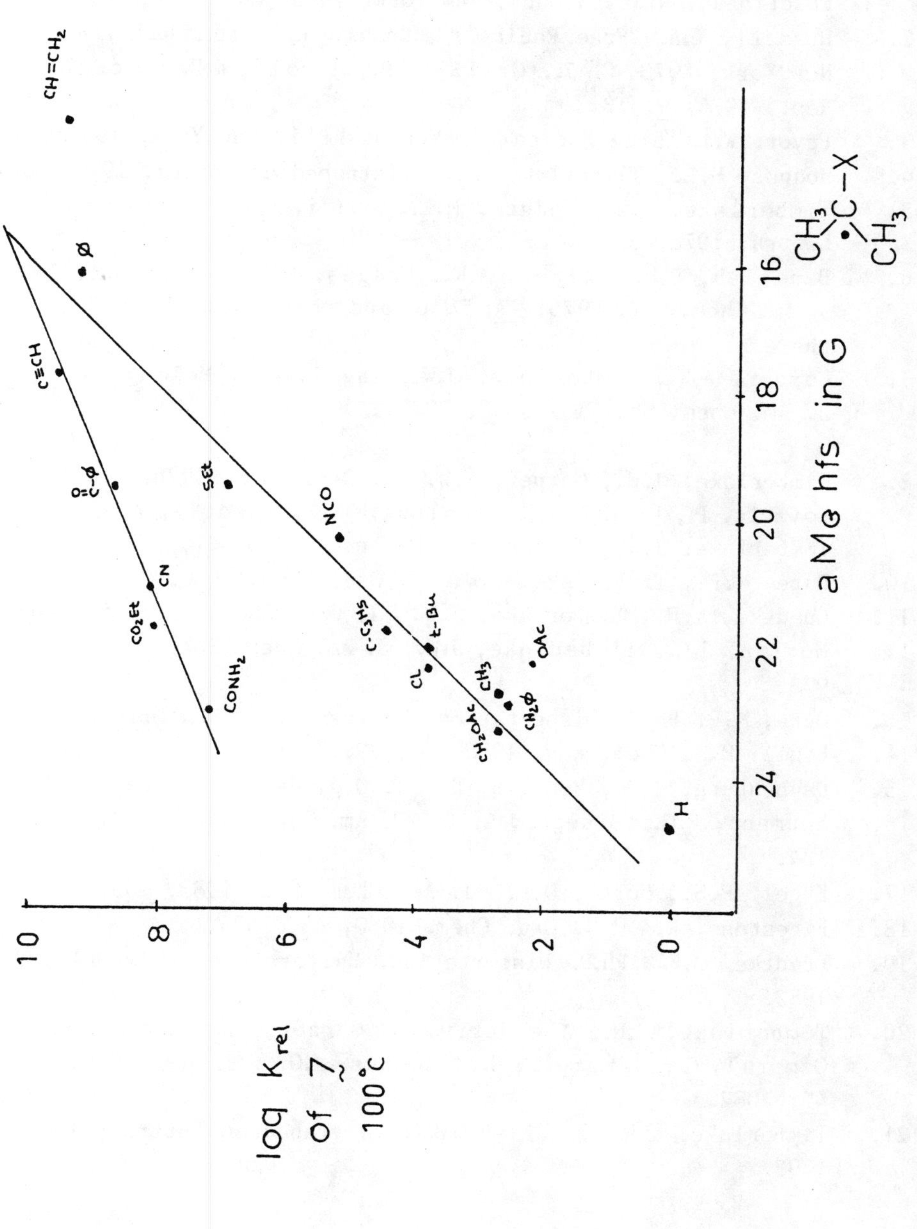
log k_rel
of 7
100°C
10
8
6
4
2
0
24
22
20
18
16
a Mg hfs in G
CH=CH2
Ø
C≡CH
SEt
NCO
CN
CO2Et
CONH2
c-C3H5
t-Bu
CL
CH3
OAc
CH2OAc
CH2Ø
H

References and Notes

1. Leffler, J.E., Grunwald, E. "Rates and Equilibrium of Organic Reactions", Wiley & Sons, New York, 1963 Ch 7.
2. Russell, G.A. "Free Radicals", Kochi, J.K. Ed., Wiley & Sons, New York, 1973, Ch 7., Griller, D., Ingold, K.U. Acc. Chem. Res., 1976, 9, 13.
3. Pryor, W.A. "Free Radicals", McGraw-Hill, New York, 1966.
4. Hodges, M.L., Timberlake, J.W. Tetrahedron Lettrs. 1970, 4147.
5. Timberlake, J.W., Hodges, M.L., Garner, A.W. Tetrahedron Lettrs. 1973, 309.
6. Bandlish, B.K., Garner, A.W., Hodges, M.L., Timberlake, J.W. J. Am. Chem. Soc. 1975, 97, 5856, and references listed therein.
7. Garner, A.W., Timberlake, J.W., Engel, P.S., Melaugh, R.A. J. Am. Chem. Soc. 1975, 97, 7377.
8. Timberlake, J.W., Garner, A.W. J. Org. Chem. 1976, 41, 1666.
9. Kovacic, P., Lisiak, M., Zawalski, R.C., Luedtke, A.E., Timberlake, J.W. Tetrahedron Lettrs. 1980, 21,, 425.
10. Dube, M.F., Timberlake, J.W. J. Org. Chem. 1980, 36, 1753.
11. Luedtke, A.E., Timberlake, J.W. J. Org. Chem. 1985, 50, 268.
12. Martin, J.C., Timberlake, J.W. J. Am. Chem. Soc. 1970, 92, 978.
13. Dube, M.F. Ph.D. Dissertation, University of New Orleans, 1978.
14. Engel, P.S. Chem. Rev. 1980, 80, 99.
15. Dannenberg, J.J., Rocklin, D. J. Org. Chem. 1982, 47, 4529.
16. Neumann, R.C., Binegar, G.A. J. Am. Chem. Soc. 1983, 105, 134.
17. Engel, P.S., Gerth, D.B. J. Am. Chem. Soc. 1983, 105, 6849.
18. Firestone, R.A. J. Org. Chem. 1980, 45, 3604.
19. Luedtke, A.E. Ph.D. Dissertation, University of New Orleans, 1983.
20. Timberlake, J.W., Alender, J.A., Garner, A.W., Hodges, M.L. Ozmeral, C., Szilagyi, S., Jacobus, J.O. J. Org. Chem. 1981, 45, 2082.
21. Timberlake, J.W., Bandlish, B.K. Tetrahedron Lettrs. 1971, 1393.

22. Stowell, J.C. "Carbanions in Organic Synthesis", John Wiley and Sons, Interscience, N.Y. 1979.
23. Gassman, P.G., Doherty, M.M. J. Am. Chem. Soc. 1982, 104, 3742 and references cited
24. Creary, X., Geiger, C.C. J. Am. Chem. Soc. 1982, 104, 4151 and references cited.
25. Paddon-Row, M.N., Santiago, C., Houk, K.N. J. Am. Chem. Soc. 1980, 102, 6561.
26. Bernardi, F., Epiotis, N.D., Cherry, W., Schlegel, H.B., Wangbo, M.H., Wolfe, S. J. Am. Chem. Soc. 1976, 98, 469.
27. Creary, X. J. Org. Chem. 1980, 45, 280.
28. Durst, J.M., Arnold, D.R. J. Am. Chem. Soc. 1983, 105, 1221.
29. Revised $\sigma_{\alpha}^{\bullet}$ values from D.R. Arnold, private communication.
30. McConnell, H.M. J. Chem. Phys. 1956, 24, 764.
31. Values reported in Gauss from Berndt, A., Fischer, H., Paul, H. "Numerical Data and Functional Relationships in Science and Technology", Landolt-Berstein, Springer-Verlag, New York, 1971, Part B, Vol 9.
32. Baird, N.C., Taylor, K.F. Can J. Chem. 1980, 58, 733 and references listed therein.

ORBITAL EFFECTS IN CARBOHYDRATE RADICALS

B. Giese*, J. Dupuis, K. Gröninger, T. Haßkerl,
M. Nix and T. Witzel
Institut für Organische Chemie
Technische Hochschule, Petersenstraße 22
D-6100 Darmstadt
Germany

ABSTRACT. In cyclic six membered radicals equatorial substituents adjacent to the radical center induce equatorial attack and axial substituents induce axial attack of alkenes if the radical adopt a chair conformation. This is not the case with glucosyl radicals **19** and **24**, which change to a boat conformation, so that the ß-C-O bond is axial. This conformational change is favored by an interaction between the SOMO of the alkoxyalkyl radical and the LUMO of the adjacent C-O bond. Equatorial attack at the boat conformation leads to an axial substituted product. This stereoelectronic effect explains also the rel. rates of attack, the easy rearrangement of adjacent acyloxy groups and the rel. rate of chlorine abstraction from glycosyl chlorides. With very flexible systems the conformation of the intermediate, detected by ESR, may change on the way to the transition state.

INTRODUCTION

The stereochemistry of the C-C bond formation between cyclohexyl radicals and alkenes like acrylonitrile occurs with a very low stereoselectivity. Thus, radical **1** yields equatorial and axial substituted products **2** and **3** in a 55:45 ratio.

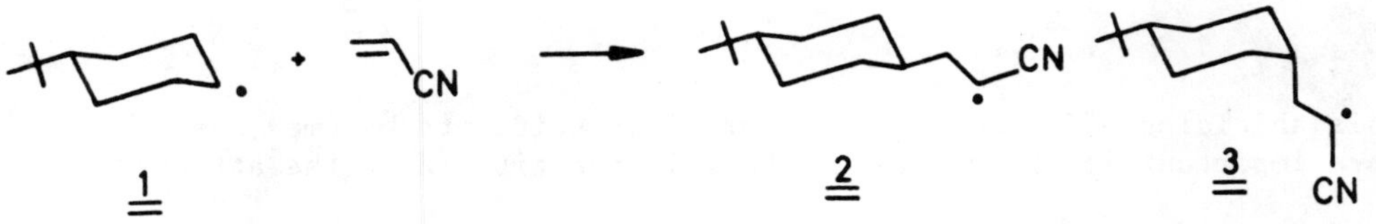

55 : 45

H. G. Viehe et al. (eds.), Substituent Effects in Radical Chemistry, 283–296.
© 1986 by D. Reidel Publishing Co.

But substituents adjacent to the radical center increase the stereoselectivity. It has turned out, that equatorial substituents favor equatorial attack [1]. The sugar radical **4**, for example, reacts with acrylonitrile predominately to the equatorial substituted product **5**.

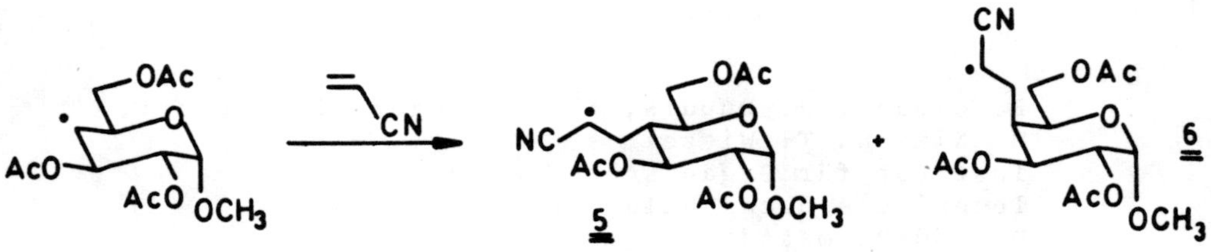

4 5 6

85 : 15

Similar results are obtained with hexopyranosyl radicals with the radical center at C-2 and C-3 and with substituted cyclohexyl radicals. The enhancement of this equatorial attack might be caused by a slight distortion of the cyclic radicals, so that the neighboring substituent X shields the cis-side of the semi-occupied p-orbital, as shown in radical **7**. The ß-coupling constants in ESR-spectra of **7** are about 35 G, compared to 40 G of radical **8**, in which the flattening is hindered by the axial substituent.

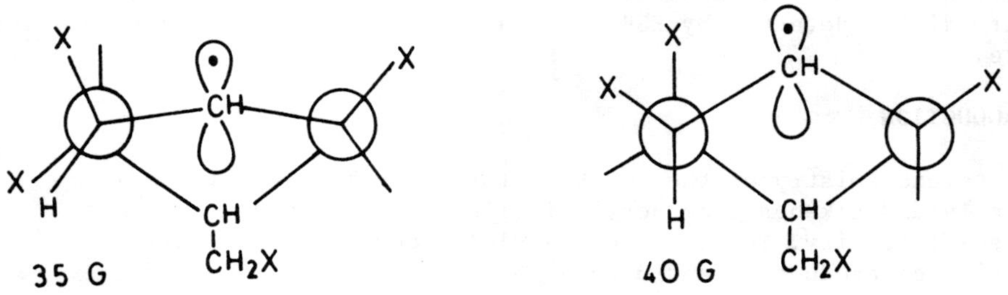

7 8

This shielding effect of the equatorial substituents becomes even more important if the reacting alkene is substituted at the attacked carbon atom. Thus fumarodinitrile gives exclusively trans-product **9** with radical **4**.

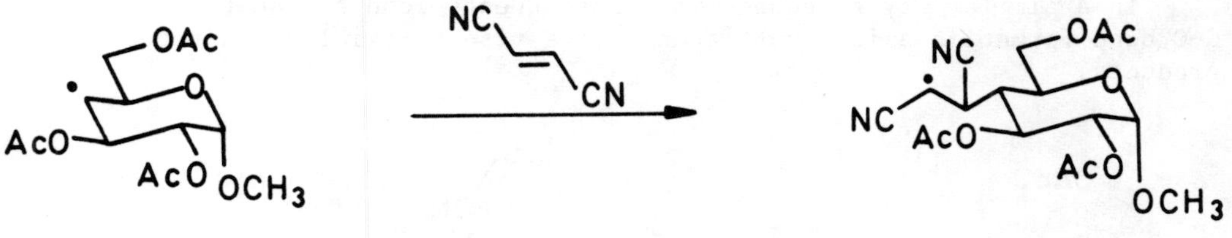

4 9

This increase in the stereoselectivity in going from a monosubstituted to a 1,2-disubstituted alkene can also be seen in rate measurements.

	rel. rates axial		
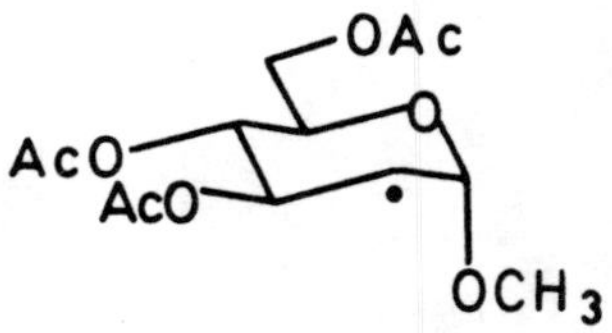 10	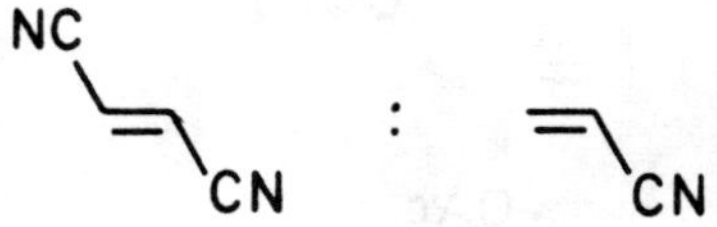	:	
	1	:	20
 1	10	:	1

Fumarodinitrile reacts 20 times slower than acrylonitrile with the axial position of radical **10**. This is a remarkable result, because the cyclohexyl radical **1** reacts about 10 times faster with fumarodinitrile than with acrylonitrile. The selectivity change by a factor of 200 for the axial attack is not caused by opposite polarities of radicals **1** and **10**. Both radicals behave like nucleophiles[2)] as the selectivities of 9 for radical **1** and 5 for radical **10** in reactions with the alkene pair α-chloroacrylonitrile/ acrylonitrile show. It is a steric effect that reduces the axial attack by fumarodinitrile.

In a similar way as equatorial substituents lead to equatorial C-C bond formation axial substituents give rise to axial substituted products.

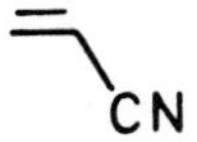

equatorial : axial

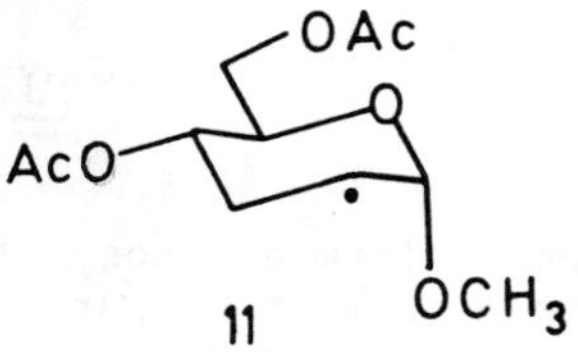

11

30 : 70

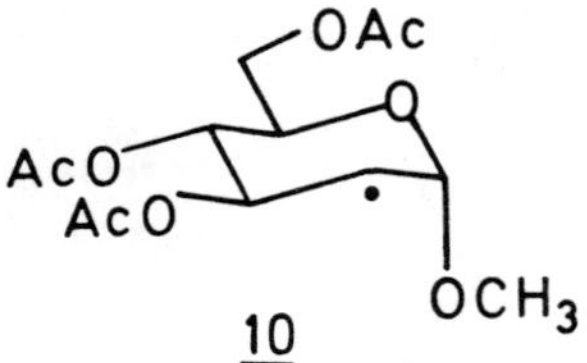

10

52 : 48

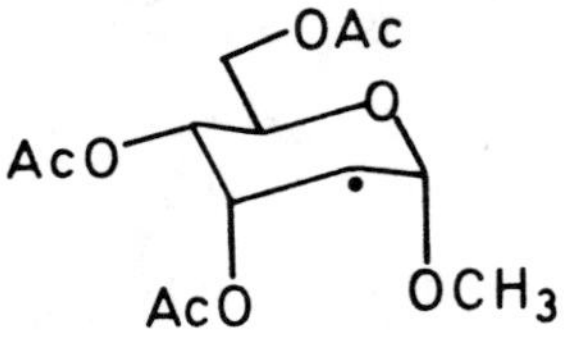

12

15 : 85

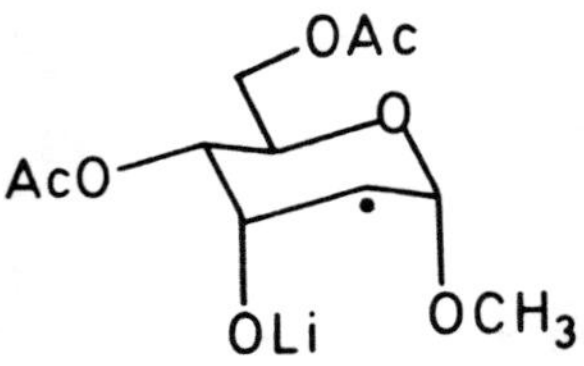

13

3 : 97

Thus radical **11** reacts with acrylonitrile predominately at the axial position. In **10** the axial and the equatorial substituents cancel their effects, but the two axial substituents in **12** and **13** lead predominately or nearly exclusively to axial attack of acrylonitrile.

Conclusion: Equatorial substituents lead to equatorial and axial substituents to axial C-C bond formation in six-membered cyclic radicals.

CONFORMATION OF GLYCOSYL RADICALS

According to the rule radicals generated by bromine abstraction from **14** should lead to equatorial C-C bond formation. But in contrast to expectation the axial substituted product **15a** is formed predominately and the formation of this stereomer is faster with fumarodinitrile than with acrylonitrile.

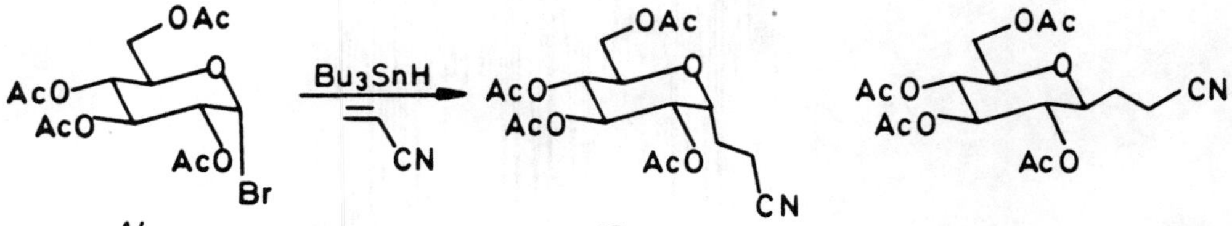

14 15a 15b

93 : 7

rel. rate : 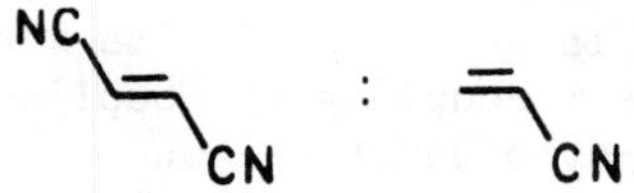

= 4 : 1

This apparent contradiction could be solved by ESR-measurements that had been carried out in the laboratories of H.Fischer and R.Sustmann [3]. Photolysis of ketone **16** or attack of trimethyltin radicals at **18** gives a radical **17** with a small ß-coupling (10.6 G) an two large γ-coupling constants.

Deuteration experiments show that the γ-coupling constant of 2.2 G belongs to the hydrogen at C-3. This ESR-spectrum is not in accord with a $^4C_1$-chair conformation in which the C-H bonds of the ring are axial.

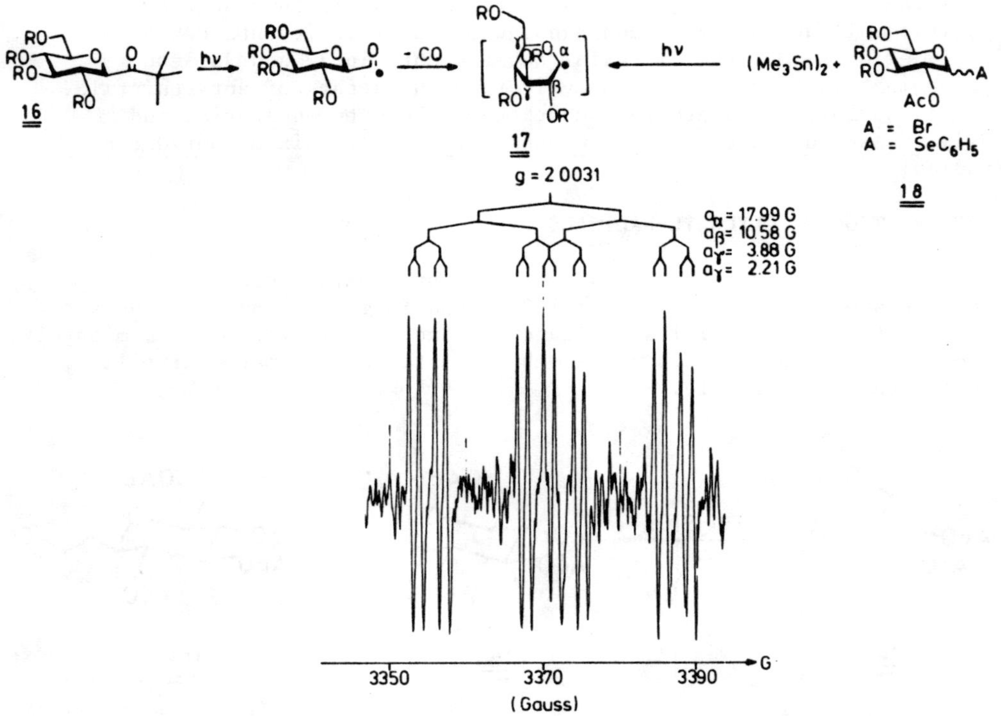

A $^{1}C_{4}$-chair with all C-H bonds being equatorial could be excluded because the bicyclic radical **24** with a trans-decalin linkage of the rings exhibits an ESR-spectrum that is very similar to that of the monocyclic glucosyl radical **17**. In good accord with the small ß-coupling (equatorial ß-C-H) and the large γ-coupling (W-coupling) is the boat conformation **19**, in which the C-H bonds at C-2 and C-3 are equatorial.

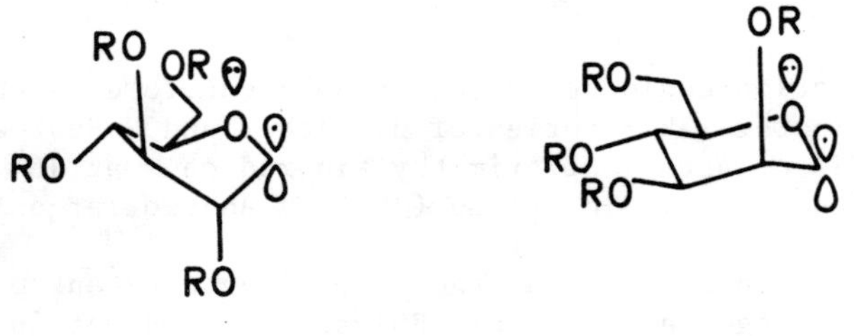

The formation of the glucosyl radical is therefore accompanied by a change from a chair to a boat conformation. In contrast to the glucosyl radical **19** the mannosyl radical **20** remains in the chair conformation of its precursor. In **19** and **20** the C-O bonds adjacent to the radical center are axial. This reminds of the anomeric effect in carbohydrate chemistry, in which the interaction between the non-bonding electrons of the oxygen and the LUMO of the C-O bond stabilizes the conformation.

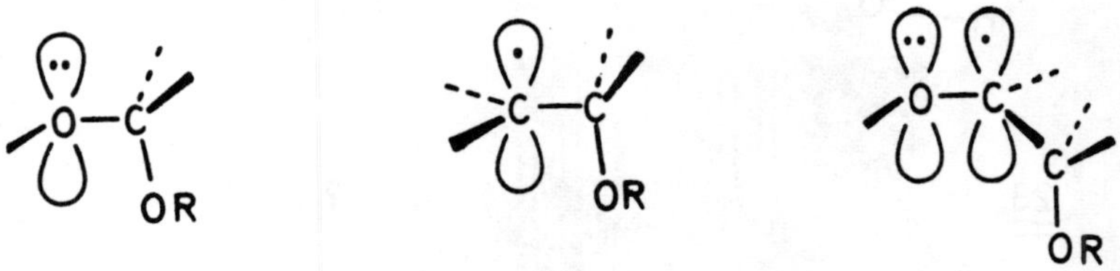

21 **22**

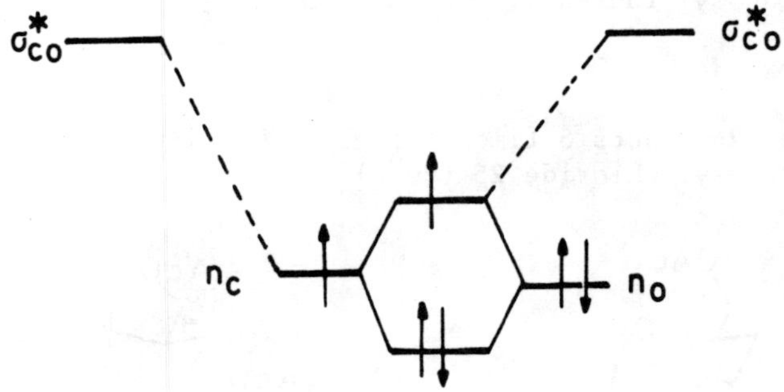

A similar stereoelectronic effect would favor the conformation **21** of an alkyl radical. But this interaction is not large enough to change the conformation of pyranosyl radicals with radical centers at C-2, C-3 and C-4, for example **4**. They remain in the chair conformation even if both adjacent C-O bonds are equatorial. In these radicals the SOMO-LUMO interaction is not strong enough to overcome the repulsive interaction of the conformational change. But in the glucosyl radical **19** the electronic interaction is larger than the steric repulsion in the boat conformation. The reason lies in the fact, that now an alkoxyalkyl radical with an energy-rich SOMO interacts with the LUMO. The energy gap is smaller and the interaction between the orbitals therefore larger. This orbital effect is just large enough to compensate the steric effects (~3 kcal/mol) because radical **23** remains in the chair conformation.

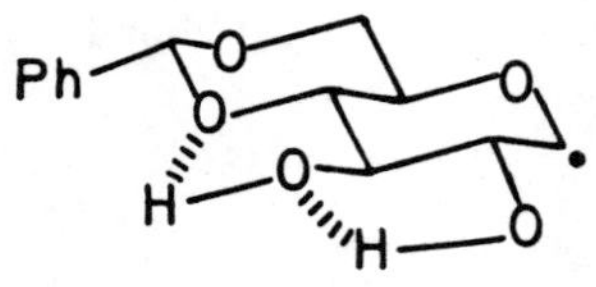

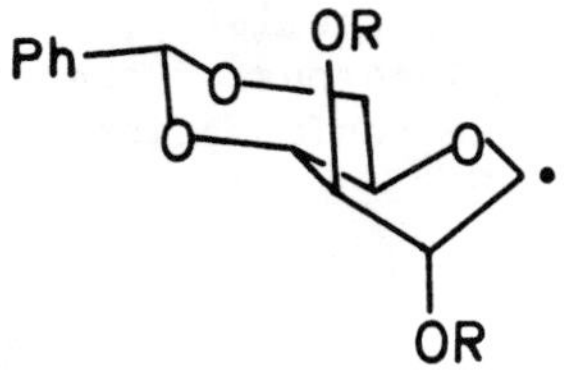

23 24

In going from **23** to a boat one has to break the intramolecular hydrogen bonds. The orbital interaction is not large enough to do this in addition to the steric effects arising from the conformational change. Without the possibility of intramolecular hydrogen bonds analogous bicyclic radicals adopt the chair conformation **24**.

This stereoelectronic effect explains the rate of formation, the easy acyloxy rearrangement and the stereochemistry of C-C bond formation of pyranosyl radicals.

RATE OF FORMATION

Mannosyl chloride **26** reacts 8 times faster with tri-n-butyltin radicals than glucosyl chloride **25** (80°C).

25

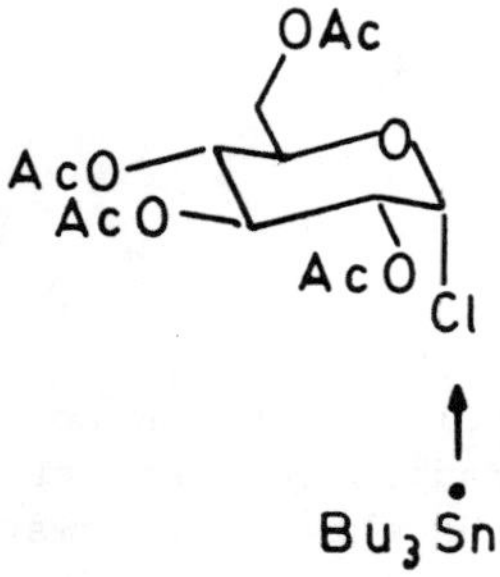

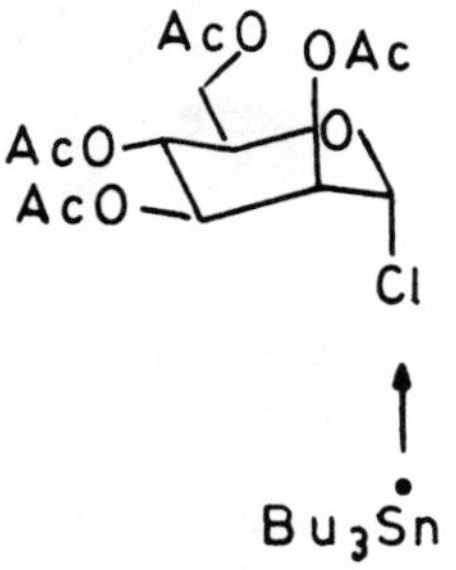

26

k_{rel} (80°C) 1.0 : 7.8

This can be explained by the stereoelectronic effect because during the halogen abstraction only in the mannosyl system the interaction between the radical center and the adjacent C-O bond becomes effective without changing the conformation of the sugar ring in the transition state.

ACYLOXY REARRANGEMENT

The reaction of acetobromoglucose **14** with allyltin in a radical chain reaction leads in dilute solution to the rearranged product **27**.

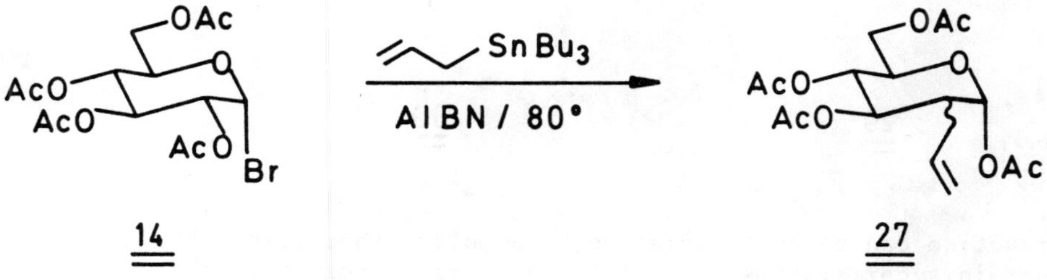

That this is a radical rearrangement can be concluded from ESR measurements which show under suitable conditions the appearance of the rearranged radical **28**.

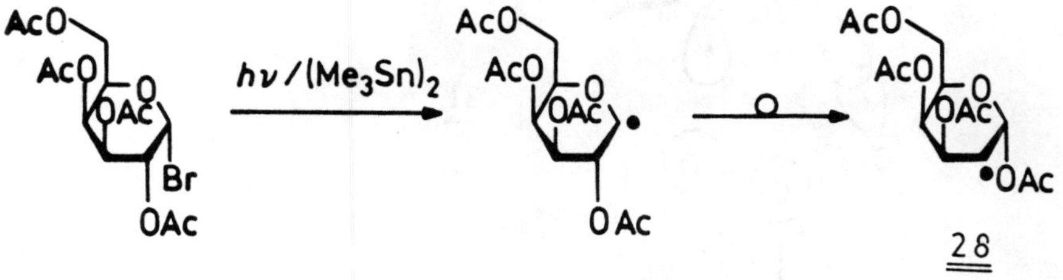

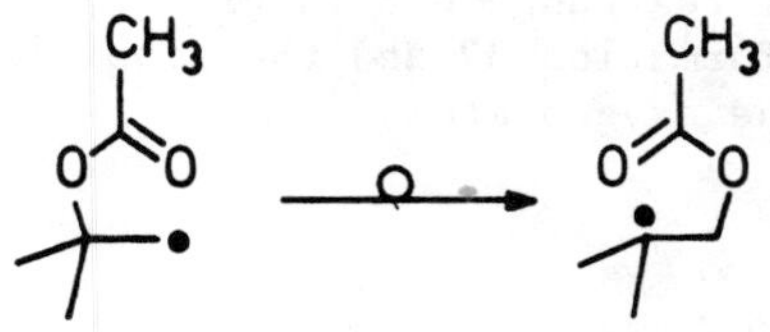

Studies by K. Ingold [4a)] and A.L.J. Beckwith [4b)] have shown that **29** is not an intermediate in this radical rearrangement.

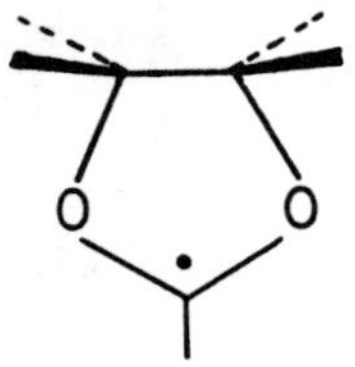

29

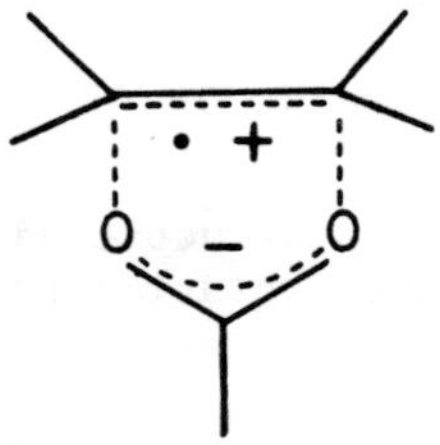

30

The reaction can be best explained by a polar transition state **30**. Whether in sugar systems radical **30** is a transition state or an intermediate shall be determined by trapping experiments with alcohols.

The simple orbital picture **31** explains why **29** is not an intermediate of the rearrangement. The route from **31** to **29** is forbidden by orbital symmetry.

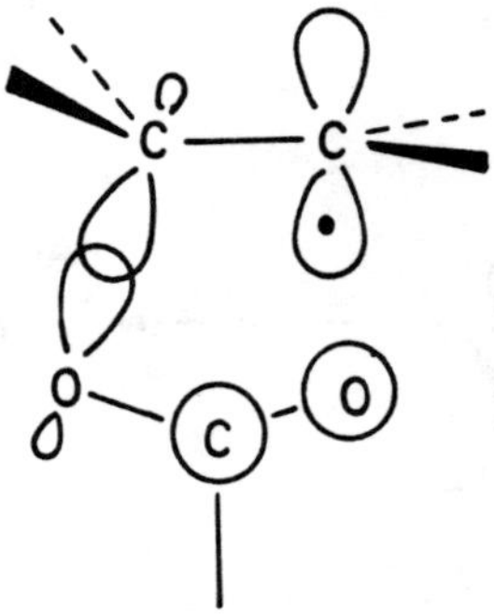

31

The relative high rate of rearrangement in glucosyl radicals is caused by their boat conformation **32** and the stabilizing effect of the positive charge by the oxygen atom.

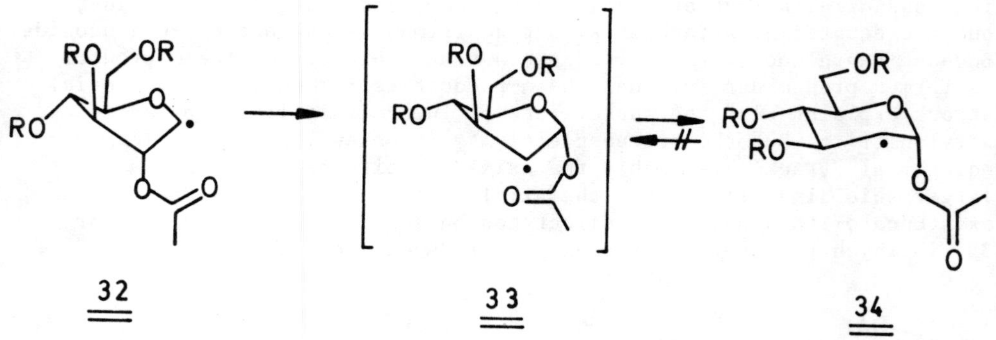

The rearrangement leads via **33** to **34**. A surprising result is that **32** goes in very diluted solution quantitatively to **34**, a reaction **34**→**32** could not be detected. The alkoxymethyl radical **32** seems to be less stable than the alkyl radical **34**. In **34** the stabilizing energy of the anomeric effect is gained but the stabilizing effect of alkoxy groups at the α- and ß-position at the radical center of **32** is lost.

STEREOCHEMISTRY OF C-C BOND FORMATION

The boat conformation of the glucosyl radical explains the stereochemistry of the C-C bond formation.

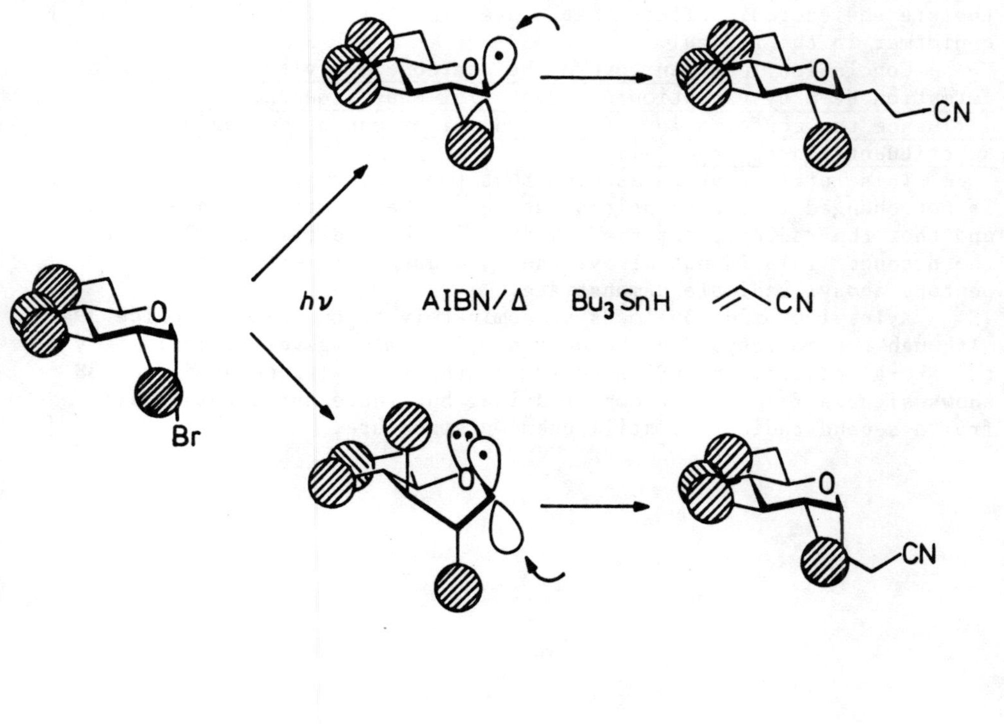

The equatorial attack at a chair conformer leads to a ß-C-glucoside, but the equatorial attack at a boat conformer leads to the α-C-glucoside. Because the glucosyl radical adopts a boat conformation the α-product is formed predominately. That the α-product is formed in an equatorial attack explains also the faster rate of fumarodinitrile compared to acrylonitrile. But the stereoselectivity is unusual high even for an equatorial attack. Presumably the axial substituent at C-3 has a larger shielding effect than the axial substituent at C-2. The existence of this shielding effect can be seen with the 3-deoxysugar **35**, in which the stereoselectivity is reduced from 13 to 2.3.

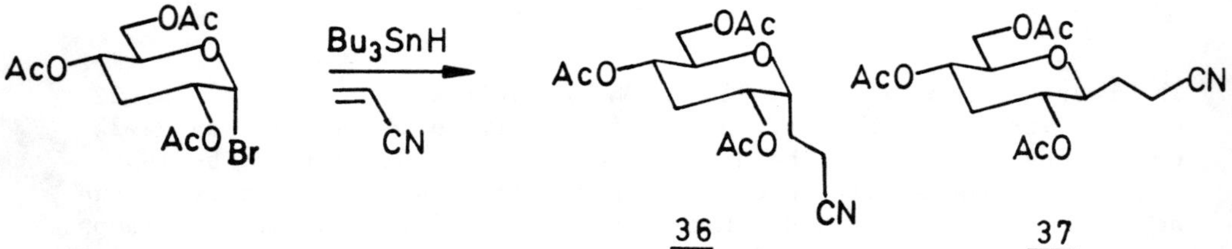

35 36 37

70 : 30

The predominant formation of the α-product can again be explained by the stereoelectronic effect that causes the formation of a boat conformer in the radical.

Conclusion: One can govern the stereoselectivity of the C-C bond formation even by variation of equatorial substituents, because they influence the conformation of the radical or can become axial substituents in the radical.

This interpretation assumes that the conformation of the radical is not changed in the transition state of the reaction with alkenes and that the radical, observed by ESR, is the radical that leads to the product. This is not always the case as experiments with pentopyranosyl radicals demonstrate.

Xylosyl bromide **38** leads predominately to the ß-C-glycoside **39a** although the corresponding hexopyranosyl bromide gives predominately the α-C-glycoside. The ESR spectrum of the radical generated from **38** shows signals from a boat conformation, but there are also signals from a second radical of still unknown structure.

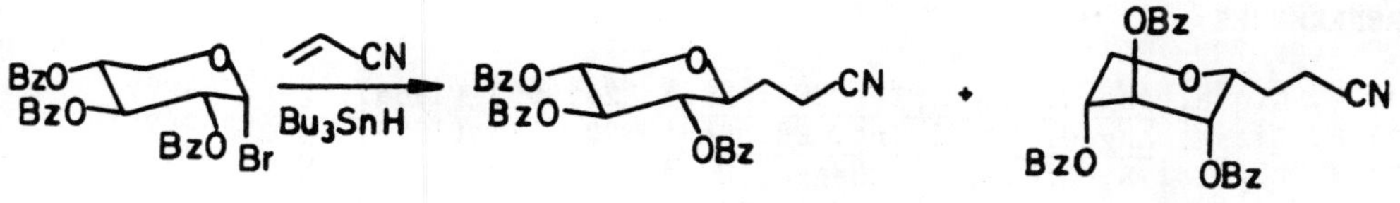

38 39a 39b

60 : 40

The boat conformation of a pentopyranosyl radical is much more flexible than that of a hexopyranosyl radical, because the "alkyl-anchor" at C-5 is absent. The radical could now be so flexible that a) several species of similar energy can exist and b) radicals can easily change their conformation during attack at alkenes, which prefers the formation of equatorial substituted products.

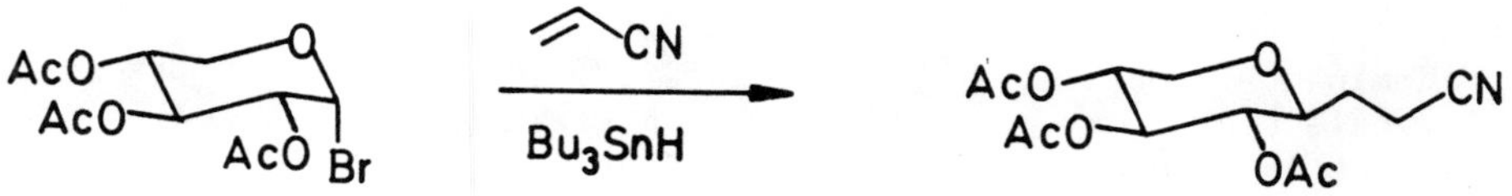

40 41

This effect should be larger the less bulky the substituents at the sugar ring are. As a consequence the tetraacetylxylosyl bromide **40** gives exclusively the ß-C-glycoside **41**.

ACKNOWLEDGEMENT. This work was supported by the Deutsche Forschungsgemeinschaft and the Fonds der Chemischen Industrie.

REFERENCES

1) B. Giese, Angew. Chem. Int. Ed. Engl. **24**, 553 (1985).
2) B. Giese, Angew. Chem. Int. Ed. Engl. **22**, 753 (1983).
3) J. Dupuis, B. Giese, D. Rüegge, H. Fischer, H.-G. Korth, R. Sustmann, Angew. Chem. Int. Ed. Engl. **23**, 896 (1984).
4) a) L.R.C. Barclay, J. Lusztyk, K.U. Ingold, J. Amer. Chem. Soc. **106**, 1793 (1984);
   b) S. Saebo, A.L.J. Beckwith, L. Radom, J. Amer. Soc. **106**, 5119 (1984).

ESR-SPECTROSCOPICAL INVESTIGATION OF CARBOHYDRATE RADICALS IN SOLUTION. RADICAL STABILIZATION BY β-C-O BONDS

H.-G. Korth and R. Sustmann
Institut für Organische Chemie
Universität - GHS Essen
D-4300 Essen, West Germany

J. Dupuis, K.S. Gröninger, T. Witzel, and B. Giese
Institut für Organische Chemie
der TH Darmstadt
D-6100 Darmstadt, West Germany

ABSTRACT. Carbohydrate free radicals were regiospecifically generated at C-1, C-2, C-3, and C-4 of alkylated and acylated deoxy-pyranoses and their configurations and conformations are deduced from the ESR hyperfine coupling constants. The preferred conformations of π-type C-1 1-deoxy-pyranosyl radicals are discussed in terms of stabilization by interaction with adjacent β-C-O bonds.

1. INTRODUCTION

In the free-radical CC-coupling reactions of carbohydrate derivatives with alkenes high diastereoselectivities for axial CC-bond formation have been observed /1/. In order to evaluate the connection between radical structure and stereoselectivity of CC-bond formation, we investigated by ESR spectroscopy the configurations and conformations of regiospecifically generated carbohydrate radicals /2/. Here we present the results for pyranosyl radicals derived from 1-deoxy-glucosyl- (**1**), 1-deoxy-xylosyl- (**2**), 1-deoxy-galactosyl- (**3**), 1-deoxy-mannosyl- (**4**), 1-deoxy-lyxosyl- (**5**), 2-deoxy-glucosyl- (**6**), 2-deoxy-mannosyl- (**7**), 2-deoxy-altrosyl- (**8**), 2-deoxy-galactosyl- (**9**), 3-deoxy-glucosyl- (**10**), 3-deoxy-allosyl- (**11**), and 4-deoxy-galactosyl (**12**) derivatives.

2. EXPERIMENTAL

Radicals **1 - 12** were generated in non-aqueous solution (benzene, toluene, THF; -30 to +95 °C) by halogen or phenylselenyl abstraction with photolytically generated tributylstannyl- or trimethylstannyl radicals, respectively, or by photolytical cleavage of the tert-butyl ketones in 2-propanol solution /3/. The conformations of the radicals have been de-

H. G. Viehe et al. (eds.), Substituent Effects in Radical Chemistry, 297–300.
© 1986 by D. Reidel Publishing Co.

duced from the $\cos^2\theta$-dependence of the β-H ESR hyperfine coupling (hfc) constants /4/. The configuration at the radical centre in C-1 pyranosyl radicals was deduced from the magnitude of the α-^{13}C-hfc constant in the α-^{13}C-enriched C-1 tetraacetyl-1-deoxy-glucosyl radical.

3. RESULTS

TABLE 1 Conformations and relevant hfc constants (mT) of C-1 1-deoxy-pyranosyl radicals (-30 to + 95 °C).

	substrate	radical	R	a(β-H)
1a	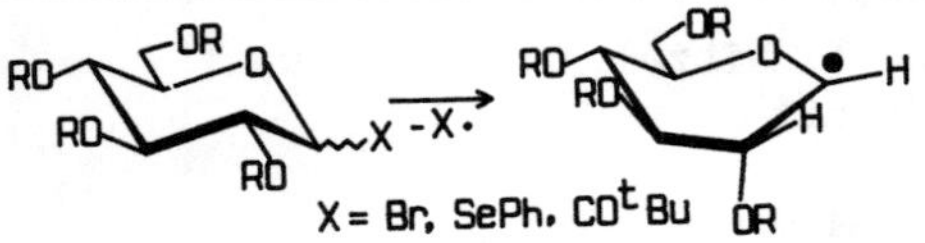		Ac Me	1.179 - 1.407 a(α-^{13}C) = 4.73 1.058 - 1.274
1b	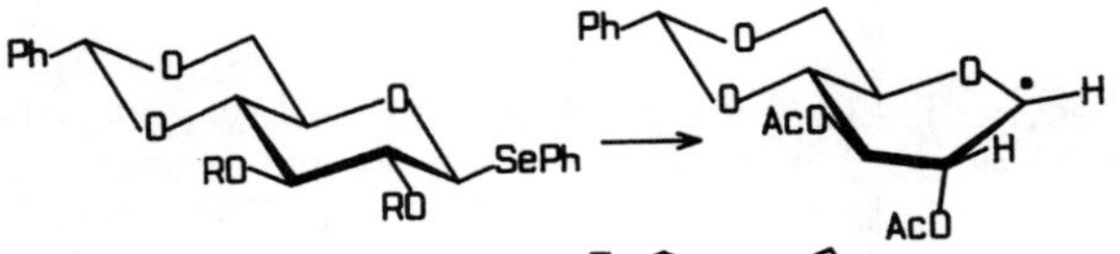		Ac	1.002 - 1.199
1c	"	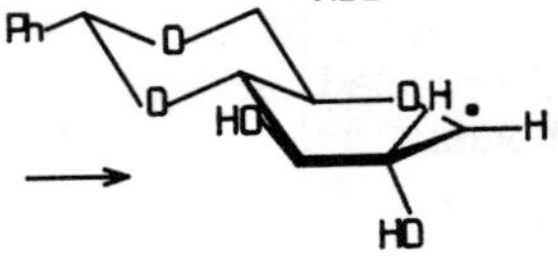	H	2.893
1d	"	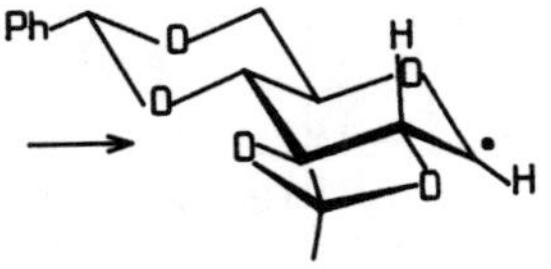	=C(CH_3)$_2$	3.445
2	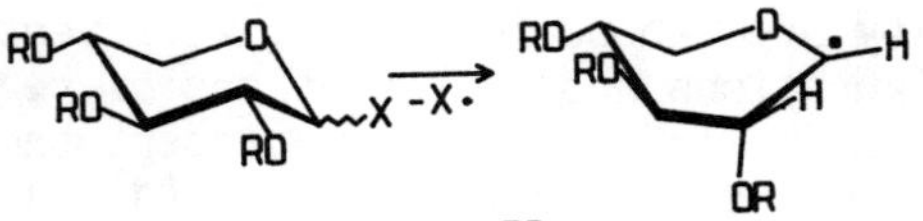		Ac Bz	1.189 - 1.221 1.250
3	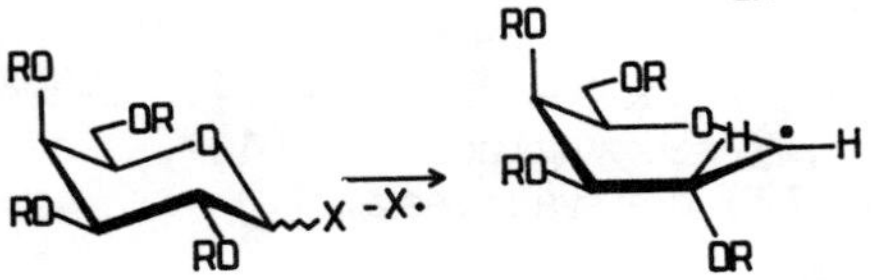		Ac Me	2.759 2.540 - 2.597
4	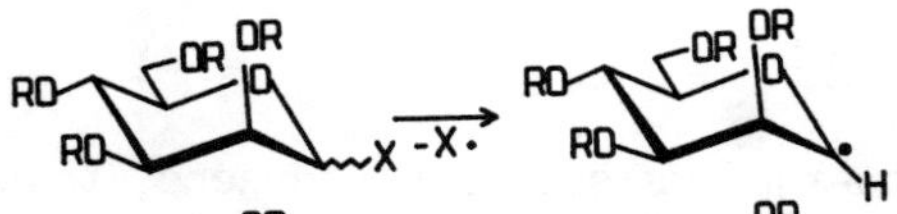		Ac	0.353 - 0.424
5	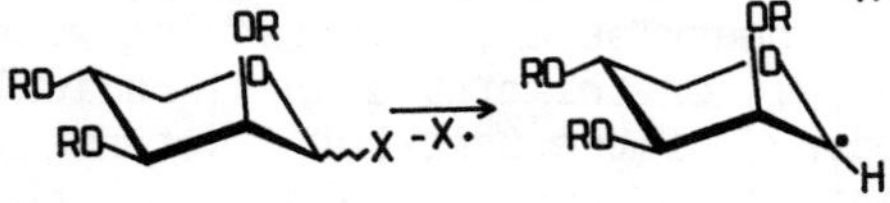		Ac	0.473 - 0.499

TABLE 2 Conformations and relevant hfc constants (mT) of C-2, C-3, and C-4 pyranosyl radicals at ambient temperature.

	substrate	radical		a (β_1-H)	a(β_2-H)
6	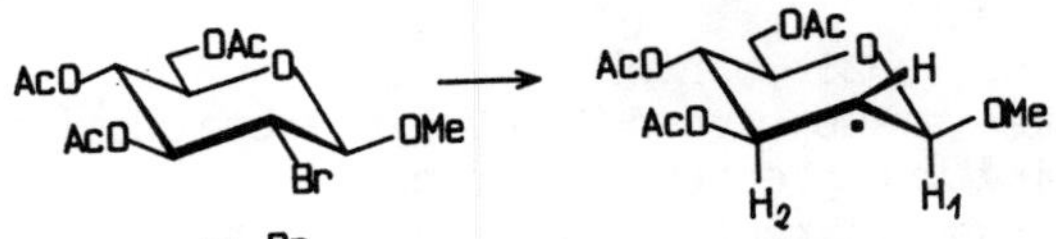			2.280	3.433
7	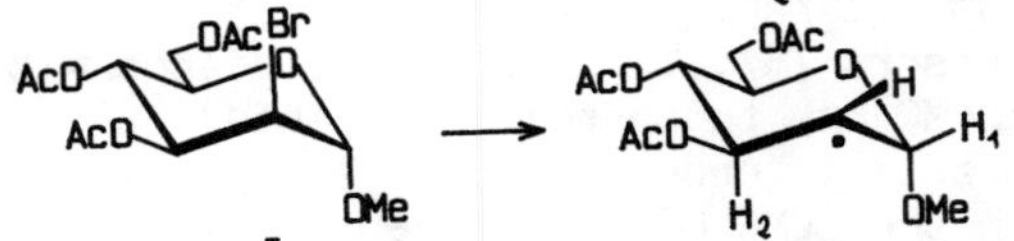			0.973	3.662
8	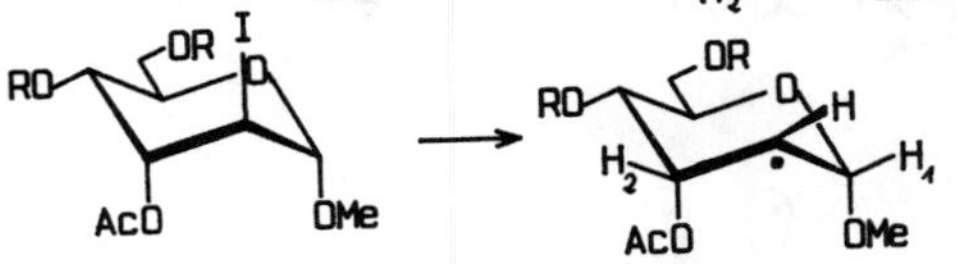		R = Ac	0.261	-
			R-R = PhCH	0.689	1.089
9	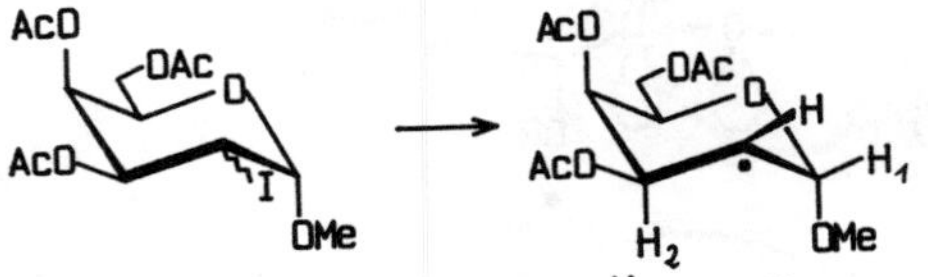			0.649	3.945
10 11	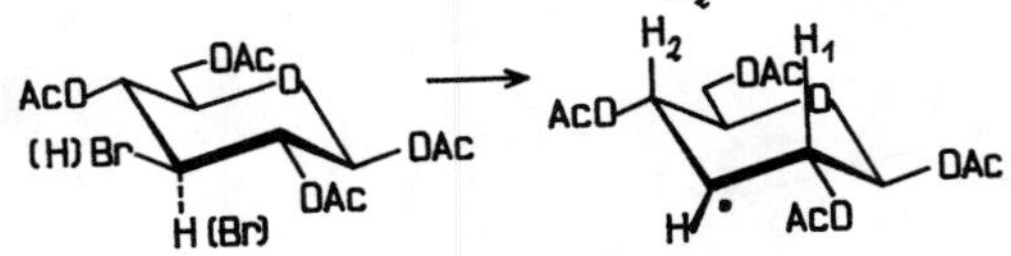			3.305	3.305
12	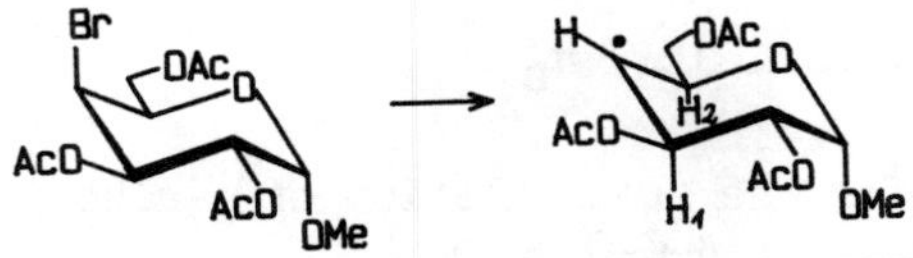			3.588	3.334

4. CONCLUSIONS

The α-^{13}C-hfc constant of the C-1 tetraacetyl-1-deoxy-glucopyranosyl radical reveals that C-1 pyranosyl radicals are π-type α-alkoxyalkyl radicals with almost planar radical centre. This is supported by the temperature dependencies of appropriate hfc constants.

O-alkylated and O-acylated C-1 1-deoxy-glucosyl- (**1a,b**) and 1-deoxy-xylosyl radicals (**2**) prefer to attain a (probably twisted) $B_{2,5}$ boat conformation leading to a coplanar arrangement of the singly-occupied p-orbital and the adjacent β-C-O bond. The transformation of the $^4C_1$ chair conformation of the radical precursors into the $B_{2,5}$ boat conformation is interpreted to occur by a stabilizing interaction of the radical SOMO and the σ*-LUMO of the β-C-O bond (Figure 1).

O-substituted C-1 1-deoxy-galactosyl radicals (**3**) prefer a ^{4}H half-chair equilibrium conformation, compromising steric repulsion of the substituents at C-3 and C-4 and stabilization by the β-C-O bond.

Since the singly-occupied p-orbital and the β-C-O bond are already in a coplanar arrangement in these systems, O-substituted C-1 1-deoxy-mannosyl- (**4**) and 1-deoxy-lyxosyl radicals (**5**) retain the $^{4}C_{1}$ chair conformation of their parent compounds.

The transformation of the $^{4}C_{1}$ chair conformation into the $B_{2,5}$ boat conformation in glucosyl systems is suppressed in cyclic acetalized derivates (**1d**) or compounds bearing unprotected OH groups (**1c**) /5/.

C-2, C-3, and C-4 deoxy-pyranosyl radicals generally retain the $^{4}C_{1}$ chair conformation of their precursors /5/. This is in accordance with the prediction that the radical SOMO has to be raised by interaction with a lone pair of the α-oxygen atom in order to achieve significant stabilization by interaction with the LUMO of an adjacent β-C-O bond.

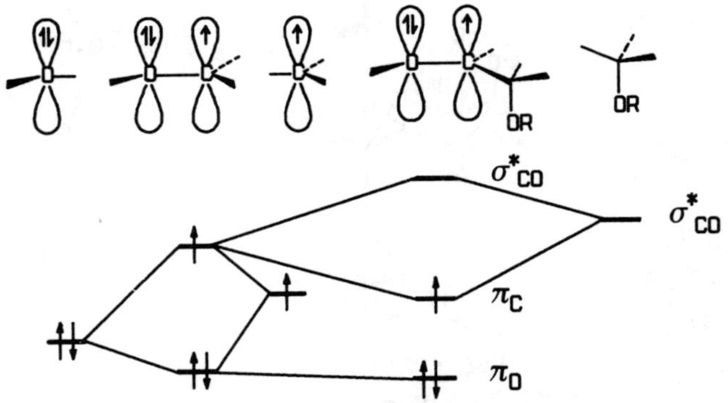

Figure 1. FMO scheme for the stabilization of α-alkoxy-alkyl radicals by β-C-O bonds.

4. REFERENCES

1) B. Giese, Angew. Chem. Int. Ed. Engl. **24**, 553 (1985).

2) J. Dupuis, B. Giese, D. Rüegge, H. Fischer, H.G. Korth, and R. Sustmann, Angew. Chem. Int. Ed. Engl. **23**, 896 (1984).
H.-G. Korth, R. Sustmann, J. Dupuis, K. S. Gröninger, T. Witzel, and B. Giese, J. Chem. Soc., Perkin Trans. 2, in the press.

3) Experiments by D. Rüegge and H. Fischer, University of Zürich.

4) J. K. Kochi, Adv. Free Radical Chem. **5**, 189 (1975).

5) B. C. Gilbert, D. M. King, and C. B. Thomas, J. Chem. Soc., Perkin Trans. 2 **1981**, 1186; ibid. **1982**, 169; ibid. **1983**, 675.

STUDIES ON THE CAPTODATIVE EFFECT.
ENTROPY/ENTHALPY COMPENSATION AS SOLVENT EFFECT IN RADICAL FORMING REACTIONS.
A RELATIVE RADICAL STABILISATION SCALE.

R. Merényi, Z. Janousek, H.G. Viehe,
Laboratoire de Chimie Organique, Université de Louvain,
Place L. Pasteur, 1, B 1348 Louvain-la-Neuve, Belgium.

Summary : In the first part of this paper, some synthetic, kinetic and thermodynamic results are reviewed showing evidence for the captodative (**cd**) radical stabilisation.

The second part gives several examples for manifestation of solvent effects as enthalpy-entropy compensation in radical forming reactions (homolytic dissociations, [2+2] cycloadditions). This phenomenon is particularly important for compounds with several polar substituents such as **cd**-radicals. Isokinetic temperatures are in/or near to the temperature domain of measurements leading to negligible solvent effects on rate. A hypothesis is developed proposing the explanation of this phenomenon by the change of polarisation when a radical is formed ("Radical Umpolung").

In the third part a Relative Radical Stabilisation Scale (RRS_x) is proposed, which gives an idea concerning the magnitude of the substituent effects and which can be used to estimate the degree of additivity if two or three substituents are present. It is based on average values of several correlating physicochemical measurements.

1. THE CAPTODATIVE EFFECT

During the last decade several research groups examined the hypothesis of captodative (**cd**) radical stabilisation (1). This concept, with different formulations, has been published before (2) by the papers of Dewar, Balaban and Katritzky, but apparently the majority of free radical chemists ignored this idea.

Whereas many questions remain open concerning the extent and the additivity of the radical stabilising effect of substituents (see p.313 and 320), several synthetic and physicochemical results support this **cd** concept today (3). We review here some results of the research group at Louvain-la-Neuve. In organic synthesis the selective dehydrodimerisation (1a, 4) and bridged dehydro dimerisation (1a, 4b, 5) and also the [2+2] cycloadditions (6) give so far the most important results in which the **cd** effect plays an important role.

H. G. Viehe et al. (eds.), Substituent Effects in Radical Chemistry, 301–324.
© 1986 by D. Reidel Publishing Co.

In the first case (4, Scheme 1) the **cd** substitution leads to selectivity, not previsible considering polar effects only. For example, in neat phase the methylmethoxyacetate (MMA) gives selectively by oxidation, even in competition with dicapto or didative analogs, the symmetrical captodative dimer.

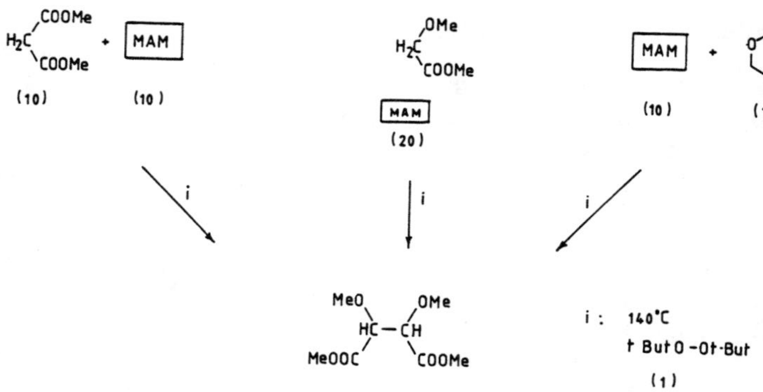

Scheme 1. Selectivity in dehydrodimerisation reactions. (In parenthesis : molar excess of substrate).

In the bridged dehydrodimerisation (5, Scheme 2), radicophilic olefins with 1,1-**cd**-substitution are used as efficient radical traps Again, nearly exclusive symmetrical dimerization is observed by capturing all kinds of radicals by these captodative olefins.

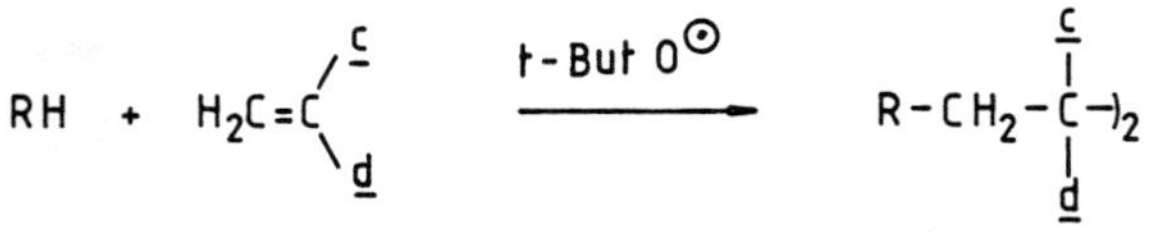

R =	alkanes	**c/d** =	SR/CN
	ethers		SR/COOR
	thioethers		NR_2/CN
	amines		OR/COOR
	ketones		etc.
	aldehydes		
	etc.		

Scheme 2. Bridged dehydrodimerisation.

The same radicophilic olefins are often good partners in [2+2] cycloadditions (1, 6), where a diradical like transition state is postulated (6e). This can be shown in the reaction with trifluorochloroethylene where, already at relatively low temperatures, good yields are obtained (6a, Table I).

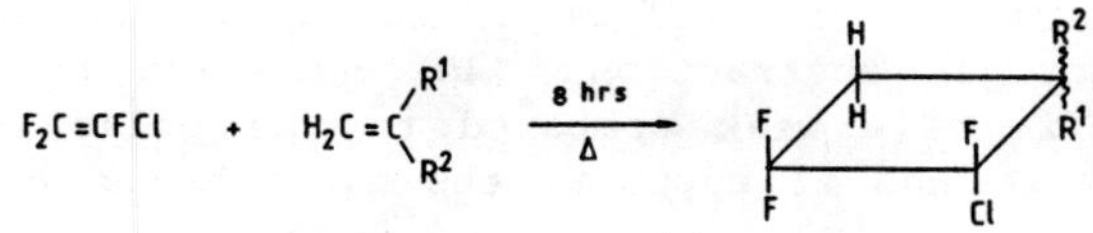

R^1	R^2	Temp.(°C)	Yield %
CN	H	160	65
S t-But	COOMe	120	68
Cl	CN	120	72
SePh	CN	120	70
SEt	CN	120	83
SPh	CN	120 (80)	85 (52)
St But	CN	120	90
OMe	COOMe	80	20

Table I. [2+2] Cycloadditions of chlorotrifluoroethylene with **cd** olefins (6a).

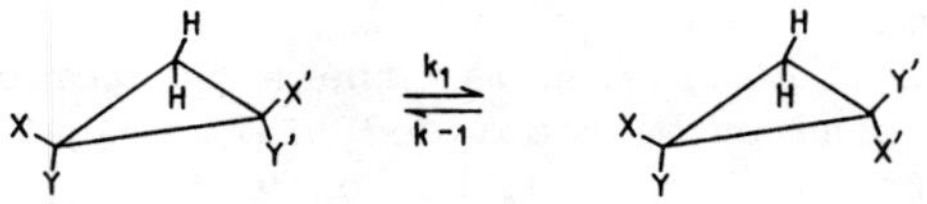

X	Y	X'	Y'	$\Delta G^{\neq}$ a) Kcal/mol	$t_{1/2}$ a)
D	H	D	H	59.1 b)	10^{14} years
CD_3	CH_3	CD_3	CH_3	50.8 c)	10^9 "
Ph	H	Ph	H	37.2 d)	18.7 "
N-morpholino	CN	Ph	H	34.0	93.4 days
S^tBu	CN	Ph	H	34.5	206 "
S^tBut	CN	COOMe	CN	31.0	40.6 hours
S^tBut	CN	Cl	CN	29.3	4.5 "
S^tBut	CN	OMe	CN	28.4	1.6 "
S^tBut	CN	SPh	CN	27.4	33 min
SMe	CN	SPh	CN	27.5	34 "
S^tBut	CN	Ph	CN	25.5	1.5 "
S^tBut	CN	Ph	COOMe	30.9	33 hours
S^tBut	CN	SPh	COOMe	29.3	4.5 "
OMe	CN	OMe	CN	28.3	55 min
Ph	CN	Ph	CN	25.5	1.5 "

a) Mean value of Z → E and E → Z at 373 K; b) ref. (9); c) ref. (10a); d) ref. (10b).

Table II. Homolytic Z to E isomerisation of cyclopropanes (8).

A captodative kinetic effect in radical additions to olefins (7) and in hydrogen abstractions (4c) has also been reported. Although the effect is weak compared to the polar effect, it is clearly observable and it supports the synthetic results mentioned above.

In order to get a quantitative evaluation of the captodative effect, several systems have been examined from which radicals are formed by bond dissociation. In one of these projects,to minimise the steric effect on bond homolysis, small ring compounds like tri- or tetrasubstituted cyclopropanes have been synthesised and the kinetics of their Z to E isomerisation have been measured (8, Table II, see also p. 307).

These measurements show for the series where X and Y are tert.butylthio and cyano a faster isomerisation when X' and Y' are **cd** substituents than for the analogous dicapto (**cc**) substituted compounds. Important decrease in the half-lifetime is observed for the bis-**cd** cyclopropanes (11) in contrast to 1,2-dicyano- 1,2-dicarbomethoxy-3,3-dimethylcyclopropane which do not isomerise till 200°C, where decomposition sets in.

In the two following systems, the same steric strain is maintained around the bond to be homolysed. This means that substituents are not changed around this bond, but only in vinylogous or phenylogous position. First, the meso to d,l isomerisation of the dimers of α-methyl-α-cyano substituted benzylic radicals with variation of the para-substitution was studied (12, Table III).

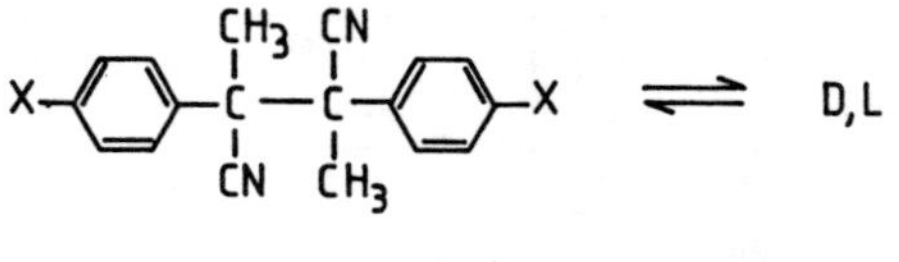

X	ΔG$^{\neq}$ (160°C) Kcal/mol	t 1/2 (160°) min
H	33.3	86
Cl	32.9	50
CN	32.8	45
COOMe	32.8	45
OMe	32.3	18
SMe	31.8	12
NH_2	30.7	4
NMe_2	30	2

Table III. Kinetics of the meso to dl isomerisation of p,p' substituted-2,3-diphenyl-2,3-dimethylbutanedinitriles (12) (o-dichlorobenzene, 120-160°C).

Kinetics show faster isomerisation for the **cd** couples of substituents (when the para-substitutent is a donor group), than for the dicapto substitution.

Instead of phenylogy, vinylogy was also tried for checking the **cd** effect. Thus equilibrium measurements have been made by ESR on the dissociation of dimers of tetrasubstituted allylic radicals (13, Table IV. See also p. 309). The dimers of the bis-**cd**-substituted allylic radicals show very low C-C bond dissociation energies.

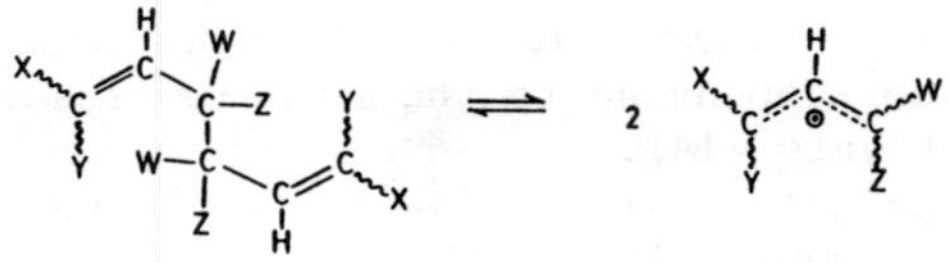

X	Y	W	Z	a^{β} GAUSS	ΔH_{DISS} (kcal/mol)	ΔS (cal/K.mol)	$\Delta G^{a)}$ at 300 K Kcal/mol
NMe_2	CN	NMe_2	CN		8.08 ± 0.25		(4.6)
S t-But	CN	S t-But	CN	3,67	13,50 ± 0,26	16.7 ± 4.0	8.49
SEt	CN	SEt	CN		13.40 ± 0.25		(8.4)
OMe	CN	OMe	CN	3.50	24.47 + 0.44	27.6 ± 2.9	16.19
OEt	CN	OEt	CN	3.55	24.70 ± 0.33		(16.4)
SMe	COOMe	SMe	COOMe		23.11 ± 0.46		(15.3)
SMe	COOMe	COOMe	COOMe	5.42	23.98 ± 0.42		(16.0)
OMe	COOMe	COOMe	COOMe	4.17	28.22 ± 0.15	30.7 ± 2.0	19.01
COOMe	COOMe	COOMe	COOMe	5.53	38.11 ± 0.11	40.5 ± 1.8	25.96
COOMe	COOMe	$S\text{-}CH_2\text{-}CH_2\text{-}S$		2,5	23.2 ± 0.2		

a) ΔG values in parentheses are estimated from the iso-equilibrium relationship (Fig. 3a).

Table IV. Thermodynamic data for the homolytic dissociation of octa-substituted 1,5-hexadienes (13).

It is difficult to estimate the contribution of the steric effect to the lowering of the C-C Bond Dissociation Energy but it is probably limited compared to the electronic effect induced by eight polar substituents. In three hexadienes of this series the substituents W,Z (COOMe) on the C-C bond to be homolyzed, do not change. The comparison of the free energy change, ΔG of these dissociation reactions leads to the same conclusion as obtained above from phenylogy : the couples of SR/Ester and OR/Ester stabilise the allylic radical better than the di-ester substitution on the same carbon. It is also to note that the combination with SR is more efficient than with OR.

Since in all these kinetic and thermodynamic studies, differences inbetween ground (diamagnetic) state and transition(radical)-state are measured, the substituent effects on these two states cannot be separated. Nevertheless the following trends are clearly revealed :

a) in all series the same relative order is observed in the decrease of ΔG or $\Delta G^{\neq}$ corresponding to the relative stabilisation of radicals by the substituents (14).

b) The free energy gap is always lower for the **cd** substitution than for **cc** substituted compounds. This is especially convincing in the example of the dissociation of the allylic radical dimers when only one vinylogous ester group is exchanged for a donor substituent (Table IV) : the free energy of dissociation is lowered for OMe with 7 Kcal/mol (and 10 Kcal/mol for SMe).

One can take account of two opposite effects (probably compensating each other) on the radical energy level : the higher stabilizing power of the ester than methoxy groups (see part 3 of this paper) and the slight non planarity of the tetraester substituted radical revealed by the ESR β Hfc (15). On the ground state level, the methoxy group is certainly not less stabilising for the double bond compared to the ester substituent (13, 16, 27). In conclusion, the lower ΔG corresponds in this case certainly to a synergism of the stabilizing effect of the substituents on the corresponding **cd** radical.

The same conclusion can be drawn from ESR data : a synergism has been found for the spin density delocalising effect on benzylic radicals for the **cd** substitution in contrast to the antagonism of two donor or two acceptor substituents (1b, 17).

2. SOLVENT EFFECTS IN RADICAL-FORMING REACTIONS.

A particularity of the kinetic and thermodynamic studies reported in the first part of this chapter is, that molecules with a relatively great number (2 to 8) of polar substituents are involved. In this second part, we discuss an important entropy/enthalpy compensation effect observed when solvents have been varied,which leads us to the conclusion, that rates ($\Delta G^{\neq}$) or equilibrium constants (ΔG) are the good parameters to compare substituent effects in radical forming reactions performed in solution.

At first we observed these surprising solvent effects in the kinetics of the Z $\rightarrow$ E isomerisation of 1,2-tetrasubstituted cyclopropanes (8, Table V). Although so far no extended study with more than 3 solvents has been made, even these limited cases show a perfect compensation (Figure 1). The isokinetic temperature is in the range of the measurements. This confirms the fact that generally

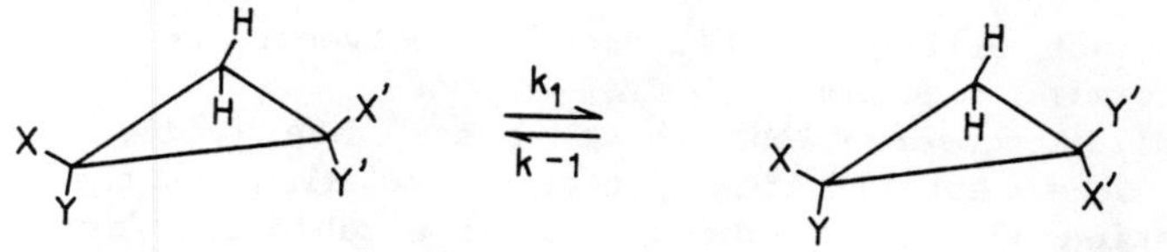

X	Y	X'	Y'	$\Delta G^{\neq}_{373K}$ a) Kcal/mol	$\Delta H^{\neq}$ Kcal/mol	$\Delta S^{\neq}$ cal/mol.K	Solvent
Ph	CN	Ph	CN	25.1	24.0	-3.0	$CDCl_3$
				25.9	21.9	-10.6	$PhNO_2$
St-But	CN	COOMe	CN	30.7	29.2	-4.1	$PhNO_2$
				30.6	31.1	+1.3	$CDCl_3$
St-But	CN	Ph	COOMe	30.8	29.6	-3.2	o-Xyl
				30.3	16.6	-36.6	$CDCl_3$
				29.2	30.4	+3.3	o-Xyl
St-But	CN	SPh	COOMe	29.2	27.2	-5.4	$PhNO_2$
				29.3	18.3	-29.4	$CDCl_3$

a) Only data deriving from Z → E are listed here (k_1).

Table V. Solvent effects observed on cyclopropane Z to E isomerisation (8c).

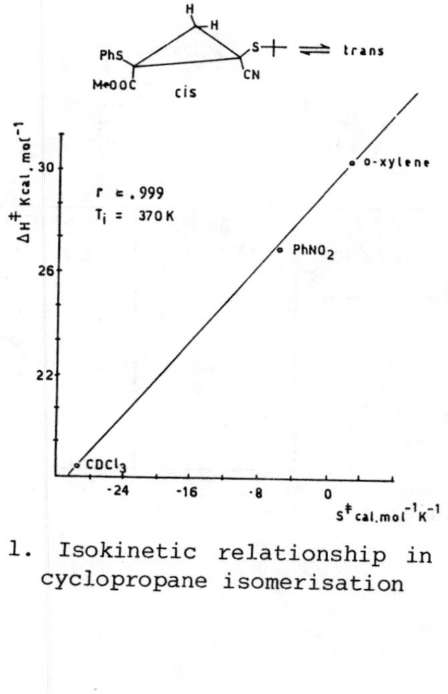

Figure 1. Isokinetic relationship in bis-**cd** substituted cyclopropane isomerisation

rates are not strongly influenced by solvents in carbon-centered radical forming reactions.

Similar observations have been made with cyclopropanes carrying other substitution patterns and also in the meso to dl isomerisation of benzylic dimers (12, see table III for $\Delta G^{\neq}$ values), but the effects involved there are less important.

The same phenomenon has been observed in [2+2] cycloaddition reactions where tetramethylene diradical type transition state have been postulated (6e). Kinetics of the reaction of α-phenylthioacrylo nitrile with 1,1-di-thioethyl-difluoroethylene was determined in

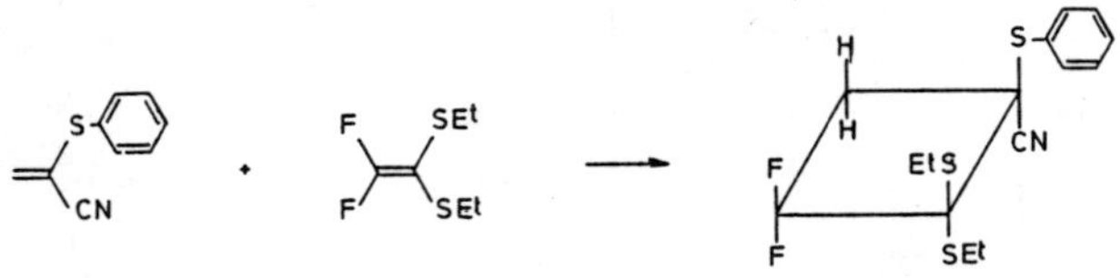

	DMF	Nitrobenzene
$\Delta H^{\neq}$(Kcal/mol)	12.5	16.8
$\Delta S^{\neq}$(cal/K.mol)	-39.7	-26.3
$\Delta G^{\neq}$(Kcal/mol) at 298 K	24.3	24.6

Table VI. Solvent effect on [2+2] cycloaddition reaction (18)

c	d	$\Delta H^{\neq}$ Kcal/mol	$\Delta S^{\neq}$ cal/K.mol	$\Delta G^{\neq}$ (300 K) Kcal/mol
-CN	-S t.But	24.9	-5.1	26.4
-CN	-N O	28.6	+7.9	26.2
-CN	$-OCH_3$	44.2	+36.1	33.4

Table VII. Kinetic data of methylenecyclobutane isomerisation (6d, 19).

several solvents (18) : Table VI shows some of these data which still have to be refined. It is to underline that the change in enthalpy is in general not related to solvent polarity parameters.

One observes compensation on the reverse reaction when methylene cyclobutane isomerises via diradical type transition state (6d, 19). Table VII shows, the influence of different substituents on the kinetics of this isomerisation measured in xylene solution.

The observed important variation of activation entropy for the same type of transition state is supposed to be caused by the variation of solvent/solute interaction here too. A recent study of the methoxy derivative is in agreement with this supposition by showing solvent effects comparable to those of Table VI (19).

The last example (13) analogously shows great variations in entropy change when the thermodynamic parameters of the homolytic dissotiation of the dimers of 1,1,3,3-tetrasubstituted allyl radicals are determined by ESR (see table IV and figures 2 and 3). Here also a large part of the enthalpy/entropy compensation shown in figure 3 should derive from solvent/solute interactions.

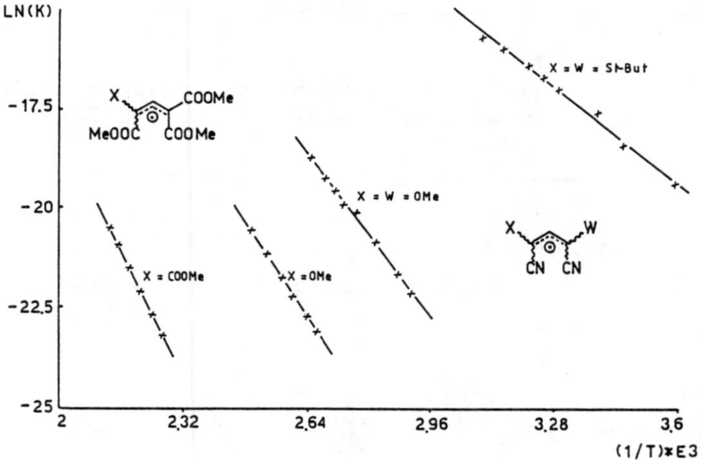

Figure 2. Dissociation of 1,5-hexadienes to allylic radicals : The dependence of equilibrium constants on temperature.

This kind of solvent effects with an isokinetic relationship (20) has been reported for ex. in the dissociation of ditert.butylperoxide (20c, d). As far as carbon centered radical formation is concerned one reaction which gives rise to a **cd** radical (21) and the decomposition of different azo-compounds (22) has been studied, in which case the extent of the variation is much sameller. There is an easy explanation for this difference : compounds with so many polar substituents have not been studied until now. Furthermore, there are scant reports in the literature on detailed solvent studies in this type of radical forming reactions. It might be of course suspected that all these results are artifacts as already recognized in some other systems (23). But it would be really surprising if such big change of ΔS values were due to the method and/or the errors of measurements.

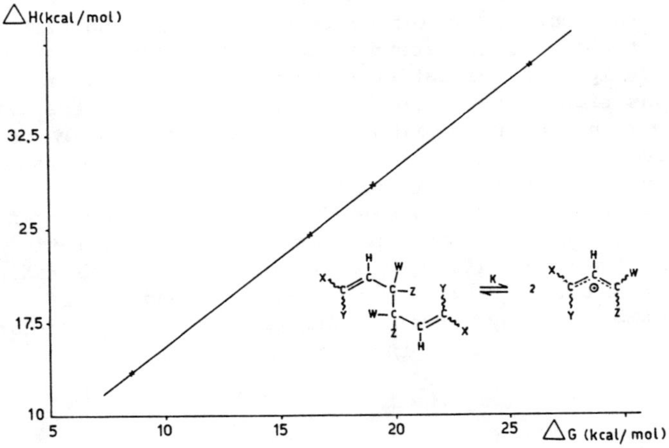

Figure 3. Dissociation of 1,5-hexadienes to allylic radicals : iso-equilibrium relationship.

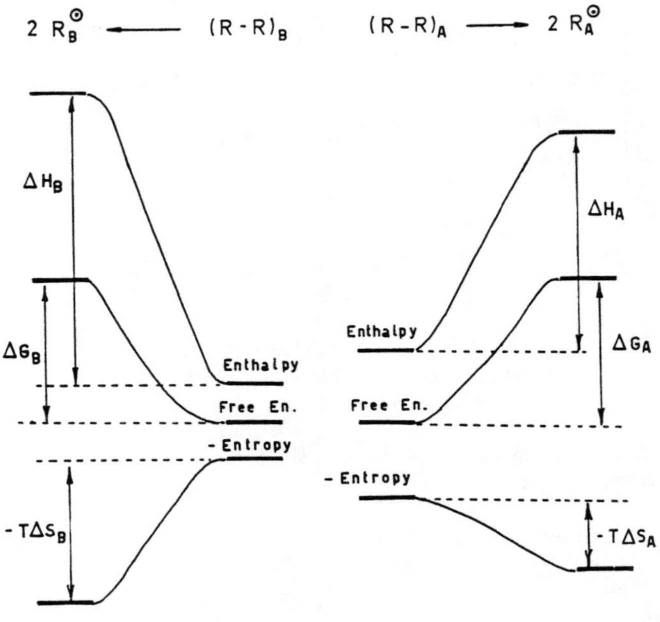

Scheme 3. Solvent effect on homolysis (in solvent A and B).

There are not yet sufficient results to present a final mechanistic picture of this solvent effect, but a hypothesis can be advanced. Examination of the energy-profile of a reaction in which radicals are formed from diamagnetic species, the dissociation of a dimer to radicals in two different solvents leads to an image depicted in Scheme 3 (cfr. ref. 20a).

Compared to solvent A, in solvent B a higher enthalpy change is observed. This might originate from a solvent effect on only one or on both of the states of this equilibrium. Let us suppose that both of them are affected : when the diamagnetic dimer is stabilized by the solvent B (relative to solvent A) the paramagnetic radical is destabilized. It is well known that entropy varies in the same sense as enthalpy on solvation. Stabilization of the dimer means less freedom, diminution of entropy, consequently -TS increases. In the radical state we have the opposite situation. The overall effect therefore leads by compensation to identical free energy changes (ΔG) in solvent A and B, and thus to identical rates or equilibrium constants.

A hypothesis to explain this phenomenon can be based on the fact that there is a change on polarisation, when a radical is formed (scheme 4)

For example, methylamine is polarised corresponding to the atomic electronegativities : the α-carbon is an electrophilic center, it becomes a nucleophilic radical.

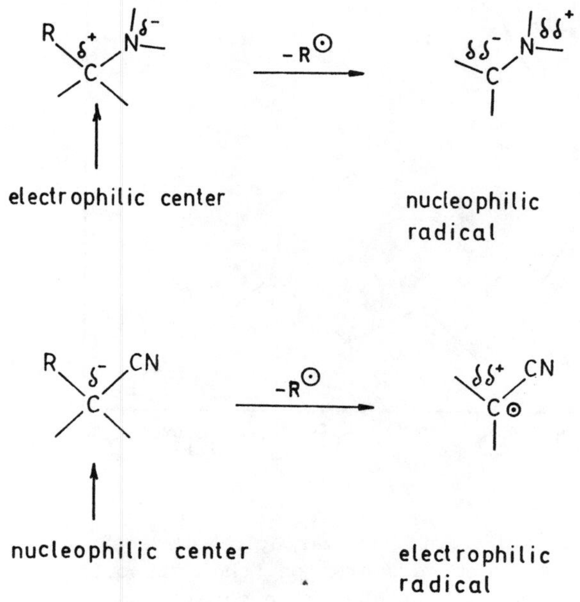

Scheme 4. "Radical Umpolung".

This does not mean necessarily that the original polarisation will be completely inversed by these $\delta\delta$ charges, but it will certainly be attenuated. In the same manner, the inverse situation can be illustrated for the cyanomethyl radical. This change in polarisation could be named "Radical Umpolung" (24). It can be expected that for captodative substitution this effect will even be enhanced because of **cd** conjugation via the radical.

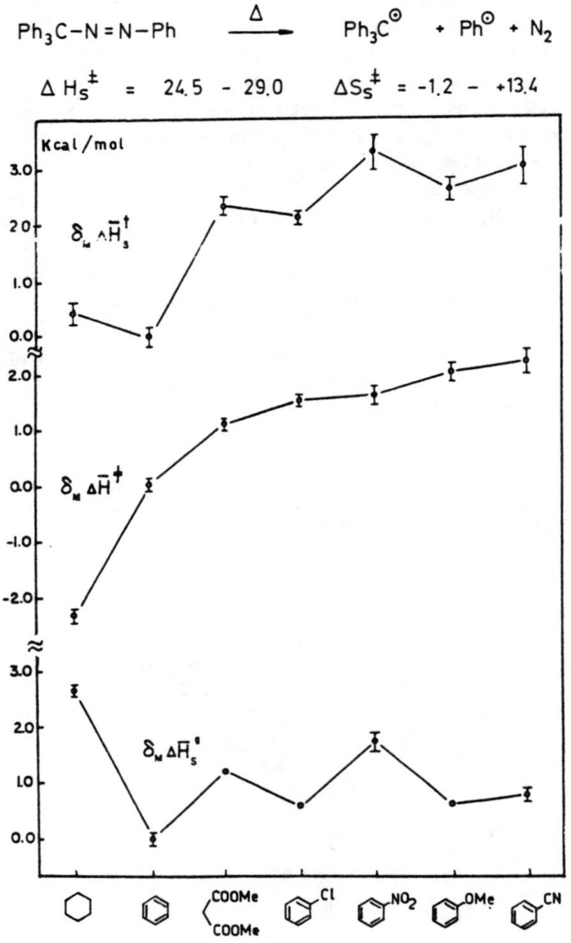

Figure 4. Dissection of the variation of activation enthalpy by solvent effects on the dissociation of azo-compounds. Reprinted with permission from Tetrahedron Letters ref. (22b), Copyright (1966), Pergamon Journals Limited.

This hypothesis on the energy profile and on "Radical Umpolung" could be experimentally controlled by measuring the relative heats of solvation of the diamagnetic compounds in different solvents and dissect then the enthalpy changes in ground state and transition state effects. This has already been done in at least one case when Bentrude and Mac Knight (22b) measured solvent effects in an azo-decomposition reaction studied by Alder and Leffler (22a) before (figure 4).

In these kinetics, the solvent effect is not as important as in the cases mentioned above, since the substituents have little polarity, but even there the same type of energy profile as in scheme 3 is observed. The modification of the activation enthalpy ($\delta\Delta H^{\neq}$) by solvents is dissected in ground state ($\delta\Delta H_s^g$) and transition state ($\delta\Delta H_s^{\neq}$) solvent effects : in benzene, compared to cyclohexane there is stabilization for both states but much more in the ground state than in the radicallike transition state. Complete inversion can be observed by comparison of several solvent pairs : stabilisation of the ground state and destabilisation of the transition state.

In conclusion, when the kinetics or thermodynamics are studied in solution to determine relative radical stabilisation by substituents one can have more confidence in ΔG values (or relative rates) than in enthalpy differences. These relative ΔG values are probably closer to gas phase ΔH values, than ΔH determined in solution.

3. RELATIVE RADICAL STABILISATION SCALE.

The organic and polymer chemists as well as the biochemists are highly interested in the extent of stabilization exerted by substituents on radicals. Many studies have been made (25-35), and several scales have been published but there is a lack of a simple comprehensive synthesis of these scales. In this paper, we try to make this synthesis for the monosubstitution on a carbon-radical center and to compose a Relative Radical Stabilisation (RRS_x) Scale for a large number of substituents. For two or three substituents on the radical center the problem of additivity of stabilising effects, synergism or antagonism will also be discussed.

The results of selected methods giving thermodynamic stabilisation energies or related data are reproduced in Tables VIII and IX. The selection of the methods might appear arbitrary, but we have selected only those which correlate well or which are supported by a number of reliable measurements or calculations.

The data derive from the following approaches : Stabilisation Energies calculated from thermochemical data by Leroy's group (27), decomposition of bis-dimethylated azo-compounds (31), Z to E isomerisation of 1,2-disubstituted cyclopropanes (32),"Resonance Enthalpies" calculated by Rüchardt's group (33), Rotation barriers (34), Arnold's sigma-α scale (29), Creary's methylene cyclopropane isomerisation (30), Jackson's secondary sigma-dot scale (28) and the method of Stella and Sylvander developped with Korth and Sustmann

Methode / X	SE• •CH_2X (27), a	AZO X–|–N=)$_2$ (31) a,b	X△X isomer. (32),a,b	H_r (33),a	Rotation barrier (34),a	$\sigma^{\bullet}_{\alpha}$ (29)	▷= isomer. (30),c	$\sigma^{\bullet}_{sec.}$ (28)	s^{exp} benzyl a^{β} (17)	RRS_X d	BDE XH_2C–H (35),a
I	-5.2 **-2.7**							.31 **3.1**		(0.2)	103.4 **0.7**
Br	-4.6 **-2.2**						4.69 **2.5**	.26 **2.4**		(0.9)	102.5 **1.3**
Cl	-1.8 **0.2**	34.4 **4.7**				.011 **2.5**	4.68 **2.3**	.25 **2.2**		2.4	99.2 **3.7**
F						-.011 **-2.5**	4.47 **-1.5**	.07 **-0.3**		-1.4	101.0 **2.4**
$SiMe_3$						.017 **3.8**	4.72 **3.1**			3.5	
CF_3						-0.09 **-2.0**	4.63 **1.4**			(- 0.3)	106.7 **-1.6**
NO_2								.76 **9.4**		(9.4)	
SO_2OMe						.013 **2.9**				(2.9)	
SOOMe						.016 **3.6**				(3.6)	
SO_2Ph		34.3 **4.8**				.018 **4.1**				4.5	
SO_2Me		34.5 **4.7**								(4.7)	
SOPh						.026 **5.9**				(5.9)	
SOMe						.018 **4.1**				(4.1)	
$CONH_2$		28.5 **9.1**								(9.1)	
COOMe		26.9 **10.3**		3.5 **5.8**	9 **7.5**	.043 **9.7**	4.91 **6.7**		.126 **7.3**	7.9	
COOH									.120 **7.0**	(7.0)	
COPh	3.4 **4.5**	25.9 **11.0**				.055 **12.4**				9.3	
COMe				6.5 **9.2**	9.4 **7.8**	.060 **13.6**				10.2	
CHO	4.7 **5.6**									(5.6)	94.7 **6.8**
CN	6.6 **7.2**	26.8 **10.4**	43.6 **7.7**	5.5 **8.1**		.040 **9.0**	5.01 **8.6**	.71 **8.7**	.150 **8.7**	8.6	92.9 **8.1**

a) Kcal/mol b) $\Delta G^{\neq}$ c) log k (100°) d) average value

Table VIII. Radical Stabilisation parameters (Normalised RRS values for $^{\bullet}CH_2X$ are given in boldface)

Methode / X	SE• •CH_2X (27),a	AZO X–┼–N=). (31)a,b	X△X isomer. (32),a,b	Hr (33),a	Rotation barrier (34),a	$\sigma_\alpha^\bullet$ (29)	▷= isomer. (30),c	$\sigma_{sec}^\bullet$ (28)	S exp. benzyl a^β (17)	RRS_x d	BDE XH_2C-H (35), a
H	-2.0 **0**	40.8 **0**	59.1 **0**	(-1.7) **0**	0 **0**	0 **0**	4.55 **0**	.009 **0**	0 **0**	**0**	104.4 **0**
Me	-0.5 **1.3**	36.2 **3.4**	55.3 **1.9**	0 **1.9**[e]		.015 **3.4**	4.67 **2.1**	.27 **2.5**	.034 **2.0**	**2,3**	100.3 **2.9**
Et	-0.3 **1.4**				0.4 **0.3**	.012 **2.7**				**1,5**	99.7 **3.3**
t-But		34.3 **4.8**				.008 **1.8**	4.68 **2.4**			**3,0**	93.9 **7.4**
Ph	11.6 **11.3**	24.9 **11.8**	37.2 **10.9**	8.4 **11.3**[e]	13.1 **10.9**					**11,2**	87.9 **11.6**
Vinyl	13.2 **12.7**	24.3 **12.2**	32.9 **13.1**		15.7 **13.1**					**12,8**	86.7 **12.4**
Ethynyl	10.6 **10.5**	24.2 **12.3**								**11,4**	89.0 **10.8**
OH	1.5 **2.9**				4.0 **3.3**					**3,1**	95.0 **6.6**
OMe	3.4 **4.5**	34.4 **4.7**	46.7 **6.2**	1.3 **3.4**	5.3 **4.4**	.018 **4.1**	4.78 **4.4**	.40 **4.4**	.074 **4.3**	**4,5**	93.3 **7.8**
OPh		33.2 **5.6**				.018 **4.1**				**4,9**	
OCOMe		37.0 **2.8**				-.005 **-1.1**				**0,9**	
OCOPh						0 **0**				**(0)**	
SMe		28.9 **8.8**				.063 **14.2**	4.98 **8.0**		.199 **11.6**	**10,7**	
SPh		29.6 **8.3**				.0.58 **13.1**				**10,7**	
SCOMe		31.4 **7.0**				.029 **6.5**				**6,7**	
SCN		30.1 **7.9**								**(7,9)**	
NH_2	7.5 **7.9**				7.6 **6.3**				.187 **10.9**	**8,4**	94 **7.3**
NHMe	11.1 **10.9**									**(10,9)**	87 **12.2**
NMe_2	10.6 **10.5**							.61 **7.3**		**(8,9)**	84 **14.3**
NCO		31.8 **6.7**								**(6,7)**	

a) Kcal/mol b) $\Delta G^{\neq}$ c) log k (100°) d) average value

Table IX. Radical Stabilisation parameters (Normalised RRS values for $^\bullet CH_2X$ are given in boldface)

based on benzylic β-couplings (17).

In Table VIII halogens and some special substituents together with electron acceptors, and in Table IX alkyl, multiple bond substituents and electron donors are listed. It can be seen, that the "white surface" of unexplored regions still dominates and leaves work and perspectives for research in this domain.

Figures 5 and 6 show two examples for correlations of different sets of data : $\Delta G^{\neq}$ of the Z to E isomerisation of 1,2-disubstituted cyclopropanes (32) correlated with Leroy's Stabilisation Energies (27) and with $\Delta G^{\neq}$ of Azo-decomposition (31). These data are already proportionally transformed into comparable RRS-values (to be defined below) and consequently the slope should be 1, as it is actually found in the first case.

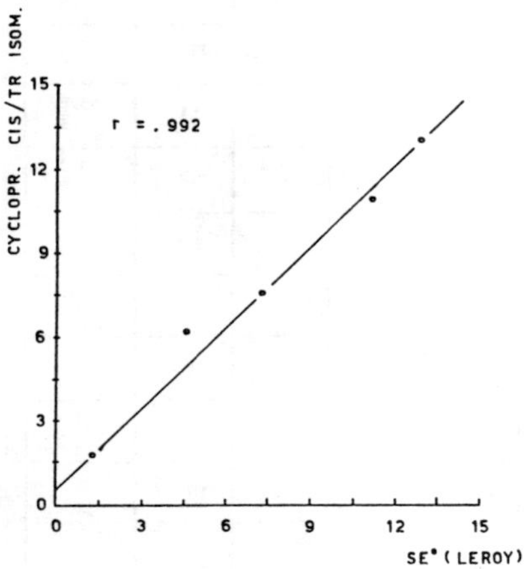

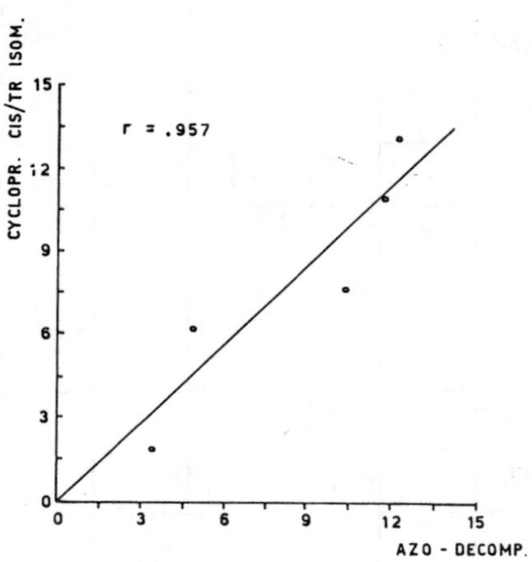

Fig. 5 and 6. Correlation of symmetrically 1,2-disubstituted cyclopropane Z to E isomerisation rates with Stabilisation Energies of $^{\bullet}CH_2X$ and with Azo-decomposition rates respectively (8c).

Why construct a scale, which would give average values deriving from all these measurements ? Beyond the arguments given above, namely the interest to give an evaluation on substituent effects, there is another reason. All these methods may be subjected to criticism and thus a scale deriving from a single method appears unreliable.

Generally speaking substituent effects on radical stabilisation are much less importants than on the corresponding ions. Consequently experimental or conceptional errors may even lead to inversion in a relative scale. An analysis of these errors, although not exhaustive can be outlined as follows : the Stabilisation Energy data (27) depend on the choice of the bond energies and of the resonance energies. Furthemore the precision of measurements of the heat of formation also influence these data : thus 6 Kcal difference between two values of heath of formation measurements can strongly modify the calculated stabilisation energy which is in the order of 12 Kcalories (36).

Bond Dissociation Energies are influenced by the ground state stabilisation energies of the diamagnetic compounds.

Concerning the kinetic measurements there also arises the question :when activation energy values are compared, what is the part in their difference ($\Delta\Delta G^{\neq}$) of ground state stabilisation or destabilisation and what is the part of transition state effects respectively ? Is the transition state really radical like and if so, is it so to the same degree, when substituents are varied ?

For ESR measurements, the question is if proportionality between variation of coupling constants and spin density delocalisation is always respected or not. What, in the value of coupling constants, is the part deriving from spin density induced by delocalisation and what is due to spin polarisation ? Is there really a linear correlation between spin delocalisation and stabilisation energies ?

In several systems which are used to determine the stabilizing effect of substituent X on the radical $^{\bullet}CH_2X$ the presence of a second or third stabilising substituent is necessary. Thus, when the effect of substituents in para-position on benzylic radicals is measured, the question arises whether the stabilizing effect of the phenyl itself can be considered as constant, while para substituents are varied ?

Therefore, in view of all the sources of error, there is a good chance that these errors, at least in part, will be statistically eliminated in an average RRS_x scale.

An examination of the data in Table VIII, indicates that only the cyano group is well documented. In Table IX there are sufficient data for phenyl (benzylic stabilisation), vinyl (allylic stabilisation) and also for the methoxy group. In order to compare these data a normalisation seems to be useful. As reference values those of allylic and benzylic radical stabilisation, as the most documented ones, have been chosen. Unfortunately by ESR and by two kinetic methods, no data are available yet for phenyl and vinyl substituents Consequently, in order to establish a Relative Radical Stabilisation Scale the normalisation has to be made in two steps : we define for $^{\bullet}CH_2X$ in all systems the stabilisation by hydrogen (X = H) as zero and the mean value of phenyl ($X = C_6H_5$) and vinyl ($X = CH{=}CH_2$) as 12. One of the reasons for this arbitrary choice of number 12 is the fact that stabilisation Energies in Kcalories are near to this value, but we insist, the numbers are dimensionless in the RRS scale. From the five methods which give values for phenyl and vinyl groups we can determine the corresponding RRS data of methoxy and cyano and we find their arithmetic mean as 6.53 (33c). This figure can then be used to define scales which have not included phenyl and vinyl, the hydrogen being always zero. These normalized values are now comparable and reveal similarities or contradictions of the different scales. In scheme 4, the definition of RRS is summarised.

By definition :

$$RRS_H = 0 \qquad \frac{RRS_{Ph} + RRS_{vinyl}}{2} = 12$$

By correlation :

$$\frac{RRS_{CN} + RRS_{OMe}}{2} = 6.53$$

Scheme 4. Definition of the RRS-Scale

The average values, of all the data concerning a given substituent, defined as RRS_x are also listed. If there is only one measurement or data are contradictory (I, Br) the mean value is unreliable. Taking only into consideration data, which derive from at least two measurements, we obtain the sequence of stabilisation of the $^{\bullet}CH_2X$ radical by X listed in Table X. Bond Dissociation Energies (BDE, ref. 35) are not included in the RRS_x average values but listed separately. A comparison gives some indication on the stabilisation of XCH_3 by X. This comparison shows that for substituents such as t.butyl or OMe the radical stabilising power would be overestimated if based on BDE-s.

X	RRS_X	Number of methods	X	RRS_X	Number of methods
$CH{=}CH_2$	12.8	4	OPh	4.9	2
C≡CH	11.4	2	OMe	4.5	9
Ph	11.2	5	SO_2Ph	4.5	2
SPh	10.7	2	$SiMe_3$	3.5	2
SMe	10.7	4	OH	3.1	2
COMe	10.2	3	t.Butyl	3.0	3
COPh	9.3	3	Cl	2.4	5
CN	8.6	8	Me	2.3	8
NH_2	8.4	3	Et	1.5	3
COOMe	7.9	6	H	0	-
SCOMe	6.7	2	F	-1.4	3

Table X. Average Relative Radical Stabilisation data : the RRS_x-scale for $^{\bullet}CH_2X$.

Based on the same normalisation procedure this scale can always be refined in the future as soon as new and more precise data become available. Even if certain data are questionable, there is good hope concerning the utility of this RRS_x scale.

Thanks to the RRS scale some interesting correlations became feasible. Thus it is interesting to show that the spin delocalisation parameters derived from Fischer's relation (37) (normalized here also to RRS scale) do not correlate with the RRS_x values (Figure 7); but if we eliminate acceptor substituents some correlation can be obtained (Figure 8) with a slope of 0.8 in place of 1. One may conclude from these correlations that the delocalisation is largely underestimated for acceptors by these increments, or that donors, on the contrary, are overestimated. The cause may lie in the different extent of spin polarisation effects on the β-coupling constants.

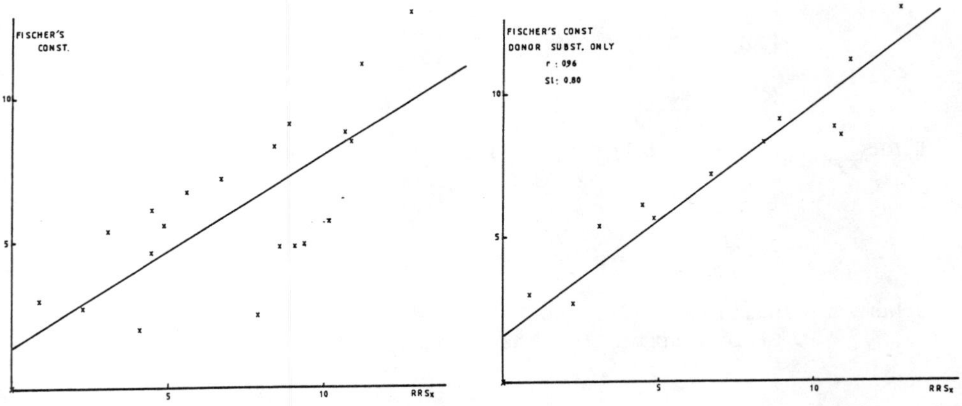

Figure 7 and 8. Correlation of RRS_x data with Fischer's constants .

Relative Radical Stabilisation (RRS_x) data providing an idea about the stabilizing effect of the substituent on monosubstituted radicals, they furthermore permit previsions on the extent of stabilization in case of di- or trisubstitution on radicals.

The problem for more than one substituent is : additivity or not. Is there a synergism or antagonism in the case of two substituents, especially if they are of the same or opposite polarity ? For the radical ˙CXYZ the question arises how to calculate the total stabilisation energy from individual substituent values.

Two kinds of calculation appear possible (scheme 5):

There is first what we would call arithmetic additivity, resulting from addition of stabilization energies, respectively the sum of the three RRS_x values.

The second possibility derives from the idea that spin delocalisation has some degree of proportionality to thermodynamic stabilisation. This calculation can be illustrated with the following consideration. One given substituent delocalises for example 20% of spin density present on a substituted methyl radical and another one 10%. When both these substituents are now attached to the same radical center ($^{\bullet}$CHXY radical), the first one delocalises 20% but the second one can only delocalise the 10% of the remaining 80% that means 8 %. This procedure leads us to a spin density like calculation of the same type as "Fischer's relation". This approach has been also applied by Stella, Sustmann and Korth to check for synergism or antagonism of two substituents. Scheme 5 shows how to calculte additivity. Scheme 6 and Table XI give a practical example.

ARITHMETICAL :

$$RRS_x + RRS_y + RRS_z = RRS_{xyz}$$

SPIN DENSITY LIKE :

$$\mathbf{f}.RRS_{xyz} = 1 - (1 - \mathbf{f}.RRS_x)(1 - \mathbf{f}.RRS_y)(1 - \mathbf{f}.RRS_z)$$

$$\mathbf{f} = .03$$

$$RRS_{x(para)} = \mathbf{a}.RRS_x = 0.16.RRS_x$$

Scheme 5. Additivity of the Substituent Effects on Radical Stabilisation for the Radical $^{\bullet}$CXYZ.

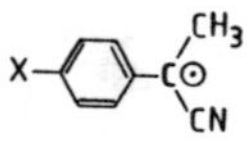

Bz

$RRS_{Ph} = 11.2$
$RRS_{Me} = 2.3$
$RRS_{CN} = 8.6$

$\mathbf{f} = .03$ $\mathbf{a} = .16$

Calculation of RRS_{Bz} for X = H:

- Arithmetic : $11.2 + 2.3 + 8.6 = 22.1$
- Spin dens. like : $\dfrac{1-(1-.336)(1-.069)(1-.258)}{.03} = 18.0$

Calculation or RRS_{Bz} for X = CN:

- Spin dens. like : $\dfrac{1-(1-.03x18.0)(1-.03x0.16x8.6)}{.03} = 18.63$
- % stabil. $(CN_{para}) = \dfrac{18.63 - 18.0}{18.0} \times 100 = 3.5$ %

Scheme 6. Method of estimation of the additivity of substituent effects with RRS_x values. The example of the isomerisation of benzylic radical (Bz) dimers (12b, see p. 304).

X	RRS_x	% stabil. (calc.)	ΔG≠ isomerisation of dimer (exp.) (160°C)	% stabil. (exp.)
H	0	0	33.3	0
Cl	2.4	1.0	32.9	1.2
CN	8.6	3.5	32.8	1.5 } antagonism
COOMe	7.9	3.2	32.8	1.5 }
OMe	4.5	1.8	32.3	3.0
SMe	10.7	4.4	31.8	4.5 } synergism
NH_2	8.4	3.4	30.7	7.8 }

Table XI. Comparison of estimated and experimental stabilisation of p-substituted benzylic radicals (Bz). (Cfr. Scheme 6 for the example X = CN : % stabil. means : stabilisation introduced by p-substitution.

RRS-values are to be multiplied for these estimations by a factor which should be around .03. For substituents directly bound to the radical center the factor **f** = .03 gives approximately the actual spin delocalisation. The RRS-value of phenyl 11.2, for example, gives .03 x 11.3 = 0 .34 corresponding to the fraction of delocalisation in the benzylic radical when calculated by the Fischer relation. A substituent in para position on a benzylic radical requires a supplementary attenuation coefficient **a** : when calculated from σ_α values (29), one obtains **a** = 0.16 (scheme 5).

Table XI illustrates such calculations. As shown before (p. 304), we studied the meso to dl isomerisation of the dimers of the parasubstituted benzylic radicals (Bz), shown in Table II. When X is equal to H one can calculate the RRS value by arithmetic additivity or by Spin density like additivity. For the RRS values of phenyl, methyl and cyano substituents, the first gives 22.1 and the second is 18.0 when using a factor = .03

For the supplementary stabilisation resulting from para-substituents on this benzyl radical the calculated values given in percent are listed in the table. One can compare these calculated data to the lowering effect of the p-substitution on ΔG≠, also given in percent as experimental result. The outcome, as expected, shows synergism for the phenylogous captodative substitution and antagonism for the dicapto substitution.

References

1. (a) H.G. Viehe, R. Merényi, L. Stella, Z. Janousek, Angew. Chem., **91**, 982 (1979), Internat. Ed. Engl., **18**, 917 (1979); (b) H.G. Viehe, Z. Janousek, R. Merényi, L. Stella, Acc. Chem. Res., **18**, 148 (1985).
2. (a) M.J.S. Dewar, J. Am. Chem. Soc., **74**, 3353 (1952); (b) A.T. Balaban, Rev. Roum. Chim., **16**, 725 (1971); (c) R.W.

Baldock, P. Hudson, A.R. Katritzky, F. Soti, J. Chem. Soc., Perkin Trans 1, **1974**, 1422
3. (a) See (1b) and references quoted there; (b) T.H. Koch, J.A. Olesen, J. De Niro, J. Am. Chem. Soc., **97**, 7285 (1975); (c) J.B. Olson, T.H. Koch, J. Am. Chem. Soc., **108**, 756 (1986); (d) W.J. Leigh, D.R. Arnold, R.W.R. Humphreys, P.C. Wong, Can. J. Chem., **58**, 2537 (1980).
This book : see in the index.
4. (a) H. Naarman, M. Beaujean, R. Merényi, H.G. Viehe, Polym. Bull., **2**, 363, 417 (1980); (b) M. Beaujean, S. Mignani, R. Merényi, Z. Janousek, H.G. Viehe, M. Kirch, J.M. Lehn, Tetrahedron, **40**, 4395 (1984); (c) L. Vertommen M. Beaujean, R. Merényi, Z. Janousek, H.G. Viehe, This book, p. 329.
5. (a) S. Mignani, M. Beaujean, Z. Janousek, R. Merényi, H.G. Viehe, Tetrahedron, **37**, (suppl. 1), 111 (1981; (b) S. Mignani, R. Merényi, Z. Janousek, H.G. Viehe, Bull. Soc. Chim. Belg., **93**, 991 (1984); (c) S. Mignani, Z. Janousek, R. Merényi, H.G. Viehe, Bull. Soc. Chim. France, **1985**, 1267; (d) S. Mignani, R. Merényi, Z. Janousek, H.G. Viehe, Tetrahedron, **41**, 769 (1985).
6. (a) C. De Cock, S. Piettre, F. Lahousse, Z. Janousek, R. Merényi, H.G. Viehe, Tetrahedron, **41**, 4183 (1985); (b) K.D. Gundermann, E. Rohrl, Liebigs Ann. Chem., **1974**, 1661; R.M. Moriarty, C.R. Romain, I.L. Kalre, J. Karle, J. Am. Chem. Soc., **87**, 3521 (1965); (d) G. Coppe-Motte et al., This book p. 371; (e) H.K. Hall Jr., Angew. Chem., **95**, 448 (1983), Int. Ed. Engl. **22**, 490 (1983).
7. (a) F. Lahousse, R. Merényi, J.R. Desmurs, H. Allaime, A. Borghese, H.G. Viehe, Tetrahedron Lett., **25**, 3823 (1984); (b) S.-U. Park, S.-K. Chung, M. Newcomb, J. Am. Chem. Soc., **108**, 240 (1986).
8. (a) A. De Mesmaeker, L. Vertommen, R. Merényi, H.G. Viehe, Tetrahedron Lett., **23**, 69 (1982); (b) R. Merényi, A. De Mesmaeker, H.G. Viehe, Ibid., **24**, 2765 (1983); (c) R. Merényi, A. De Mesmaeker, W. Masamba, H.G. Viehe, in preparation.
9. B.S. Rabinovitch, E.W. Schlag, K.B. Wiberg, J. Phys. Chem., **28**, 504 (1958).
10. (a) J.A. Berson, J.M. Balquist, J. Am. Chem. Soc., **90**, 7343 (1968); (b) L.B. Rodewald, C.H. De Puy, Tetrahedron Lett., **1964**, 2951.
11. For an analogous study on p-substituted tetraphenyl cyclopropanes, see : D.R. Arnold, M. Yoshida, J. Chem. Soc. Chem. Commun., **1981**, 1203; Can. J. Chem., **60**, 2313 (1982).
12. (a) R. Merényi, V. Daffe, J. Klein, W. Masamba, H.G. Viehe, Bull. Soc. Chim. Belg., **91**, 456 (1982); (b) J. Klein, R. Merényi, H.G. Viehe, in preparation.
13. M. Van Hoecke, A. Borghese, J. Penelle, R. Merényi, H.G. Viehe, Tetrahedron Lett., submitted.
14. This sequence of relative radical stabilization by substituent couple is the following one : $NR_2/CN > SR/CN > OR/CN > SR/COOR > OR/COOR > CN/COOR > COOR/COOR$. It also corresponds to the results of several unpublished studies.

15. H. Regenstein, W. Ahrens, A. Berndt, Tetrahedron, **31**, 2837 (1975) and references quoted.
16. J. Hine, N.W. Flachskam, J. Am. Chem. Soc., **95**, 1179 (1973). (b) G. Leroy, C. Wilante, D. Peeters, M.M. Uyewa, J. Mol. Struct. (Theochem), **124**, 107 (1985).
17. (a) L. Sylvander, L. Stella, H.G. Korth, R. Sustmann, Tetrahedron Lett., **26**, 749 (1985); (b) see also : this book : R. Sustmann, p. 143, H.G. Korth, P. Lommes, R. Sustmann, L. Sylvander, L. Stella, p. 167.
18. F. Lahousse, R. Merényi, H.G. Viehe, unpublished.
19. G. Coppe-Motte, Ph. D. Thesis, Louvain-la-Neuve.
20. (a) D.C. Nonhebel, J.C. Walton, 'Free-radical Chemistry', Cambridge University Press, 1974); (b) T. Koenig in 'Free Radicals', J.K. Kochi (Ed.), Wiley Interscience, N.Y., Vol. 1, p. 113; (c) F.S. Hyuser, R.M. Van Scoy, J. Org. Chem., **33**, 3524 (1968); (d) C. Walling, D. Bristol, J. Org. Chem., **36**, 733 (1971).
21. I.V. Khudyakov, V.A. Kuzmin, A.I. Yasmenko, W. Smit, J. Salve, C.R.H.I. De Jonge, Int. J. Chem. Kinet., **16**, 1481 (1984).
22. (a) M.G. Alder, J.E. Leffler, J. Am. Chem. Soc., **76**, 1425 (1954); (b) W.G. Bentrude, A.K. Mc Knight, Tetrahedron Lett., **1966**, 3147; (c) E.M.Y. Quinga, G.D. Mendenhall, J. Org. Chem., **50**, 2836 (1985).
23. O. Exner, in 'Progress in Physical Organic Chemistry', A.S. Streitwieser Jr., R.W. Taft Eds; Wiley-Interscience, N.Y. 1973; Vol. 10, p. 510 and references therein.
24. Several reactants leading to "Umpolung" of the reactivity are formed via radical intermediates. See : (a) D. Seebach, Angew. Chem., **91**, 259 (1979); Int. Ed. Engl., **18**, 239 (1979); (b) B. Giese, H. Horler, Tetrahedron, **41**, 4025 (1985); (c) B. Giese, V. Erfort, Angew. Chem., **94**, 133 (1982); Int. Ed. Engl., **21**, 130 (1982).
25. M. Szwarc, J. Chem. Phys., **16**, 128 (1948).
26. (a) S.W. Benson, J. Chem. Educ., **42**, 510 (1965); (b) A.S. Rodgers, M.C.R. Wu, L. Kuita, J. Phys. Chem., **76**, 918 (1972); (c) R.T. Sanderson, J. Org. Chem., **47**, 3835 (1982).
27. (a) G. Leroy, in P.O. Lowdin (Ed.) 'Adv. Quant. Chem.', (Academic Press N.Y.), **17**, 1-95 (1985); (b) G. Leroy,D. Peeters, C. Wilante, J. Mole. Struct. (Theochem), **120**, 91 (1985); (c) G. Leroy, D. Peeters, M. Sana, C. Wilante, This book, p. 1.
28. S. Dinçturk, R.A. Jackson, J. Chem. Soc. Perkin II, **1981**, 1127 and references quoted there.
29. (a) J.M. Dust, D.R. Arnold, J. Am. Chem. Soc., **105**, 1221 (1983); (b) D.R. Arnold, This book, p. 171.
30. (a) X. Creary, J. Org. Chem., **45**, 280 (1980); (b) X. Creary, This book p. 245.
31. J. Timberlake, This book, p. 271, and references quoted there.
32. (a) A. De Mesmaeker, Ph. D. Thesis, Louvain-la-Neuve, 1983; (b) see ref. 9, 10. b; (c) M.C. Flowers, H.M. Frey; Proc. Ray. Soc. (A), **257**, 122 (1960); (d) M. Arai, R.J. Crawford, J. Can. Chem., **50**, 2158 (1972); (e) 1,2-Dimethoxy-3-methylcyclopropane :

W. Kirmse, M. Zeppenfeld, J. Chem. Soc. Chem. Commun., **1977**, 124.

33. (a) C. Rüchardt, H.D. Beckhaus, Topics in Curr. Chem., **130**, 1 (1986); (b) H. Birkhofer, H.D. Beckhaus, C. Rüchardt, This book, p. 199; (c) Data from 33a,b have not been taken in account for the phenyl/vinyl mean value; to have the most reliable average for OMe and CN, we fitted Rüchardt data on Me and Ph to the mean RRS values calculated from other sources.
34. D.C. Nonhebel, J.C. Walton, J. Chem. Soc. Chem. Commun., **1984**, 731
35. (a) D.F. Mc Millen, D.M. Golden, Ann. Rev. Phys. Chem., **33**, 493 (1982); (b) J.A. Hawari, J.M. Kanabus-Kaminska, D.D.M. Wayner, D. Griller, This book, p. 91.
36. See ref. 27(c), p. 14.
37. H. Fischer, Z. Naturforsch. A, **19**, 866 (1964).

$\sigma^{\cdot}$ REVISITED

Richard A. Jackson
School of Chemistry and Molecular Sciences,
University of Sussex, Brighton BN1 9QJ, U.K.

Numerous attempts have been made to quantify the effects of substituents on radical reactions: most are based on extended Hammett equations. The realization that both electron-withdrawing and electron-releasing substituents conjugated with a radical centre can stabilize the radical suggests that polar substituent constants (σ, σ^+, σ^-, etc.) are inappropriate measures of radical stabilization has led to a number of suggestions for $\sigma^{\cdot}$ or equivalent scales. Polar as well as radical stabilizing effects are important in many reactions: many schemes therefore contain both radical and polar components.

Some of these scales are shown in table I. A number of them are flawed for various reasons such as likely side reactions or the inability of substituents to conjugate fully; others would be suitable in principle but have not been tested on a wide enough range of substituents. The last five entries in the table all have significant advantages: the remainder of the discussion is focussed on these scales.

The $\sigma^{\cdot}$ scale we suggested in 1979 on the basis of the thermal decomposition of dibenzylmercurials (designated $\sigma^{\cdot}_J$ here) has the advantage of a transition state which should approximate very closely to a fully formed benzyl radical and thus $\rho^{\cdot}$ values for other reactions based on these $\sigma^{\cdot}$ values should give a quantitative indication of the build-up or loss of radical character in the transition state.

Creary's rearrangement of aryldimethylmethylenecyclopropanes is postulated to go via an intermediate with approximately full benzylic stabilization. Synthetic difficulties are probably of a similar magnitude to those involved in making dibenzylmercurials, and the extent of electron decoupling in the transition state is not certain.

We include data on the reaction of poly(methylmethacrylate) radicals with substituted styrenes as an example of a reaction where the radical character build-up in the transition state is not expected to be complete: this system is rendered more suitable than a number of alternative radical addition or transfer reactions by its very small polar effect, as judged by *meta* substituents.

An attractive scheme is based on the reduction of the α C-H coupling constant by stabilizing substituents in benzyl

H. G. Viehe et al. (eds.), Substituent Effects in Radical Chemistry, 325–328.
© 1986 by D. Reidel Publishing Co.

radicals. The advantages include the relative ease of making the radicals (though nitrobenzyl radicals have proved elusive) and the absence of problems associated with side reactions. Possible disadvantages include the uncertainty of whether stabilization by different types of substituents is linearly related to a_H.

The final scheme considered $|\sigma^{\pm}-\sigma|/n$ is derived simply from Hammett polar substituent constants. The stabilizing effect of a *para* substituent, presumed to be predominantly a resonance effect, is taken as $|\sigma^{+}-\sigma|$ or $|\sigma^{-}-\sigma|$ as appropriate, divided by 2 if the electron in the radical would have to go into an antibonding orbital, and by 1 if the electron is in a non-bonding orbital. The principal advantage of this scale is its ease of construction: values are available for virtually all likely substituents and thus thorough comparisons can be made with all other schemes.

Although Arnold's $\sigma_{\alpha}^{\cdot}$ scheme and the $\sigma_{C}^{\cdot}$ scale derived from Creary's results do not invoke separation into a polar and a radical contribution, we felt that such a separation should be tried. Accordingly, each set of results was re-analysed in terms of a polar contribution defined by *meta* substituents, with the remainder ascribed to a radical stabilization contribution. These new scales (denoted A and C in table II) were correlated against the other proposed measures of radical stabilization. Better correlations were obtained with $\sigma_{\alpha}^{\cdot}$ rather than A, and with $\sigma_{C}^{\cdot}$ rather than C. A and C are therefore excluded from the final analysis, and table III shows the correlations between the remaining five schemes, based on *para* substituents only.

The overall agreement is good, favouring the concept of a radical stabilization parameter which is transferable from one reaction to another. The good agreement of the $\sigma_{\alpha}^{\cdot}$ scale with reaction based scales suggests empirically that lowering of α C-H coupling constants in benzyl radicals is approximately linearly related to stabilization energies. The importance of the $\sigma_{J}^{\cdot}$ and $\sigma_{C}^{\cdot}$ scales is that these should reflect the effects of substituents on an almost completely formed benzylic radical: on this basis, if the transition state for dibenzylmercury decomposition approximates to the complete breakage of the benzyl-Hg bond, then in the aryldimethylmethylenecyclopropane rearrangement, radical formation is ca. 70% complete in the transition state and $\rho^{\cdot}$ values for other reactions give an indication of the amount of build-up of radical character in the transition state.

Finally we note that the $|\sigma^{\pm}-\sigma|/n$ scale gives a slightly better overall correlation with other scales than does any of the other scales considered: since values are available for virtually all likely substituents, this may be the scale of choice for prediction of the effects of substituents for which no experimental data are available.

TABLE I

Designation	Reaction or System	Comments	Ref.
Q/e	Copolymerization data	First scheme. Recalculation needed when new data added.	1
E_R	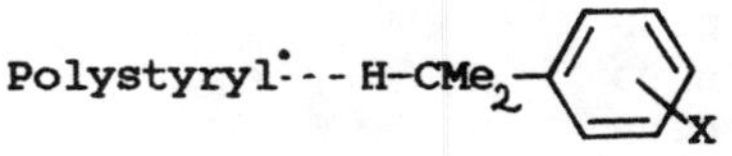	Wide range of substituents. Primary data not reported.	2
E_D	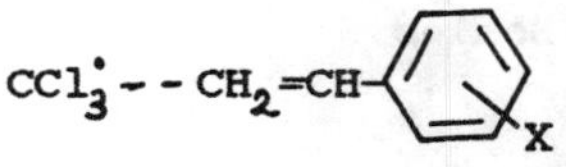	Only five *para* substituents.	3
τ_P	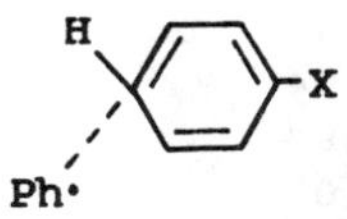	Only five *para* substituents. Likelihood of side reactions.	4
$\sigma^{\bullet}_F$	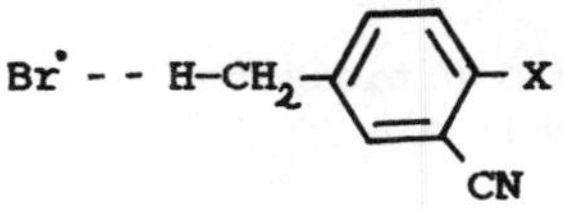	CN reduces polar effect but will interfere with resonance of *para* group.	5
W	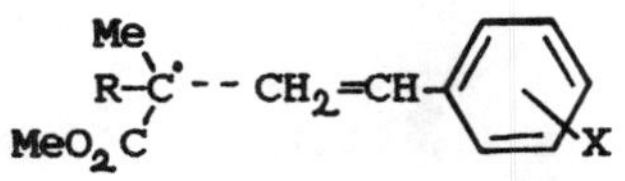	Very small polar effect.	6
$\sigma^{\bullet}_J$	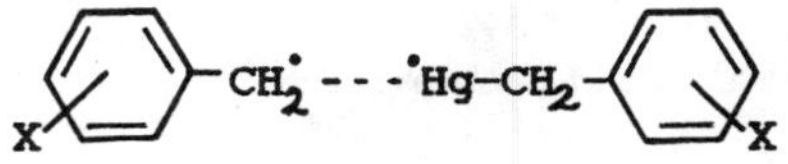	Full benzylic character in TS. Some compounds difficult to make.	7
$\sigma^{\bullet}_C$	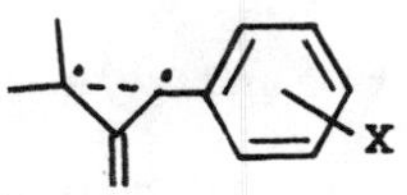	Very small polar effect. Syntheses not trivial.	8
$\sigma^{\bullet}_{\alpha}$	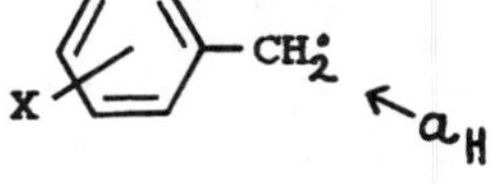	Very wide range of substituents. Stab. α to a_H?	9
$\lvert\sigma^{\pm}-\sigma\rvert/n$	Hammett polar constants	Values available for all likely substituents.	10

TABLE II Correlations for Alternative Scales

	A	$\sigma_{\alpha}^{\cdot}$	C	$\sigma_{C}^{\cdot}$
$\sigma_{J}^{\cdot}$	0.67	0.80	0.71	0.82
C	0.97	0.84	–	–
$\sigma_{C}^{\cdot}$	0.93	0.93	–	–
W	0.72	0.89	0.83	0.93
A	–	–	0.97	0.93
$\sigma_{\alpha}^{\cdot}$	–	–	0.84	0.93
$\lvert\sigma^{\pm}-\sigma\rvert/n$	0.77	0.84	0.77	0.87

TABLE III Correlations Between σ· Type Scales

	$\sigma_{J}^{\cdot}$	$\sigma_{C}^{\cdot}$	W	$\sigma_{\alpha}^{\cdot}$	$\lvert\sigma^{\pm}-\sigma\rvert/n$
$\sigma_{J}^{\cdot}$	1.00	0.82	0.84	0.80	0.98
$\sigma_{C}^{\cdot}$	(8)	1.00	0.93	0.93	0.87
W	(6)	(6)	1.00	0.89	0.94
$\sigma_{\alpha}^{\cdot}$	(7)	(11)	(5)	1.00	0.84
$\lvert\sigma^{\pm}-\sigma\rvert/n$	(10)	(12)	(6)	(11)	1.00

Figures in top right triangle are correlation coefficients between scales. Figures in bottom left triangle are the number of data points in common.

References

1. T. Alfrey, jun. and C. C. Price, *J. Polym. Sci.*, 1947, **2**, 101.
2. T. Yamamoto and T. Otsu, *Chem. Ind.*, 1967, 787.
3. H. Sakurai, S. Hayashi, and A. Hosomi, *Bull. Chem. Soc. Jpn*, 1971, **44**, 1945.
4. R. Itô, T. Migita, N. Morkawa, and O. Simamura, *Tetrahedron*, 1965, **21**, 955.
5. T. H. Fisher and A. W. Meierhoefer, *J. Org. Chem.*, 1978, **43**, 224.
6. C. Walling, E. R. Briggs, K. B. Wolfstirn, and F. R. Mayo, *J. Am. Chem. Soc.*, 1948, **70**, 1537.
7. S. Dinçtürk, R. A. Jackson, and M. Townson, *J. Chem. Soc. Chem. Comm.*, 1979, 172; H. Ağirbaş and R. A. Jackson, *J. Chem. Soc. Perkin Trans. 2*, 1983, 739.
8. X. Creary, *J. Org. Chem.*, 1980, **45**, 280.
9. J. M. Dust and D. R. Arnold, *J. Am. Chem. Soc.*, 1983, **105**, 1221; D. D. M. Wayner and D. R. Arnold, *Can. J. Chem.*, 1985, **63**, 2378.
10. S. Dinçtürk and R. A. Jackson, *J. Chem. Soc. Perkin Trans. 2*, 1981, 1127.

CAPTODATIVE EFFECT ON THE RELATIVE RATES OF H-ABSTRACTION BY TERT-BUTOXY RADICALS

L. Vertommen*, M. Beaujean, R. Merényi, Z. Janousek, H.G. Viehe
Université de Louvain, Laboratoire de Chimie Organique,
Pl. L. Pasteur,1 B-1348 Louvain-la-Neuve, Belgium.
* Actual address : AKZO, Research Laboratories, Velperweg, 76
NL-6824 BM Arnhem, The Netherlands

INTRODUCTION : The synergism in radical stabilisation by captodative (**cd**) substitution is essentially a thermodynamic resonance effect (1). We have been interested whether in different radical reactions a kinetic **cd**-effect would be observable. A rate enhancement by **cd** substitution would agree with synthetic results leading to selectivity in radical addition and H-abstraction reactions (2).

In radical addition reactions such rate acceleration have been shown as a weak effect superimposed to polar interactions when isobutyronitril radical was added to p,p'-substituted 1,1-diphenyl ethylenes (3), and also in radical cyclisation reactions (4).

We report here on results obtained by studying the rate of H-abstraction from several mono- or disubstituted methane derivatives using ditert-butylperoxyde as oxidizing agent.

RESULTS : The method using the thermal degradation of ditert-butylperoxide as internal clock is well known. The generated tert-butoxy radicals can either abstract $H^{\cdot}$ from an adjacent molecule (RH) or they decompose into acetone and methyl radicals.

$$\text{t.But-O-O-t.But.} \xrightarrow{120\text{-}150°\ C} \text{t.But-O}^{\cdot} \begin{cases} \xrightarrow[k_a]{RH} \text{t.But-OH} + R^{\cdot} \\ \xrightarrow{k_d} \;>\!\!=O + H_3C^{\cdot} \end{cases}$$

The [tert-butanol][acetone] ratio, determined after the reaction is a measure of the rate k_a. For substrates RH and R'H relative H-abstraction rates can be calculated when the initial concentrations of [RH] and [R'H] are known.

$$\frac{k_a}{k_d} = \frac{[+OH]}{[\,>\!\!=O]} \cdot \frac{1}{[RH]} \qquad \frac{k_a}{k_{a'}} = \frac{([+OH]/[\,>\!\!=O])_{RH}}{([+OH]/[\,>\!\!=O])_{R'H}} \cdot \frac{[R'H]_i}{[RH]_i}$$

$XYCH_2$-Substrate and ditert-butylperoxyde (molar ratio 10:1) in neat phase was heated at 150° during 10 hrs in a sealed NMR tube. Afterwards tert-butanol/acetone ratios were determined by integration of the 60 MHz NMR spectrum.

H. G. Viehe et al. (eds.), Substituent Effects in Radical Chemistry, 329–330.
© 1986 by D. Reidel Publishing Co.

Compounds $XYCH_2$ are listed in Fig. 1 with the values of k_a relative to $k_{a'}$ ($R'H$ = cyclohexane), and logarithms of $k_a/k_{a'}$ are plotted against the sum of the Hammett σ_p-values of the substituents of the CH_2 groups.

Tert-butoxy radicals are electrophilic. One would therefore expect a strong influence of polar substituents. This is indeed the case for most didonor(**dd**) or dicaptor(**cc**) substituted methylenes. Amines and CH_2 groups with **cd** substitution , however, show enhanced reactivity. The results obtained for amines can be interpreted in terms of electron transfer to the tert-butoxy radical, followed by proton elimination from the aminyl radical-cation.

For **cd** compounds, we believe that enhanced stabilisation of the radical explains the higher $H^{\cdot}$ abstraction rates.

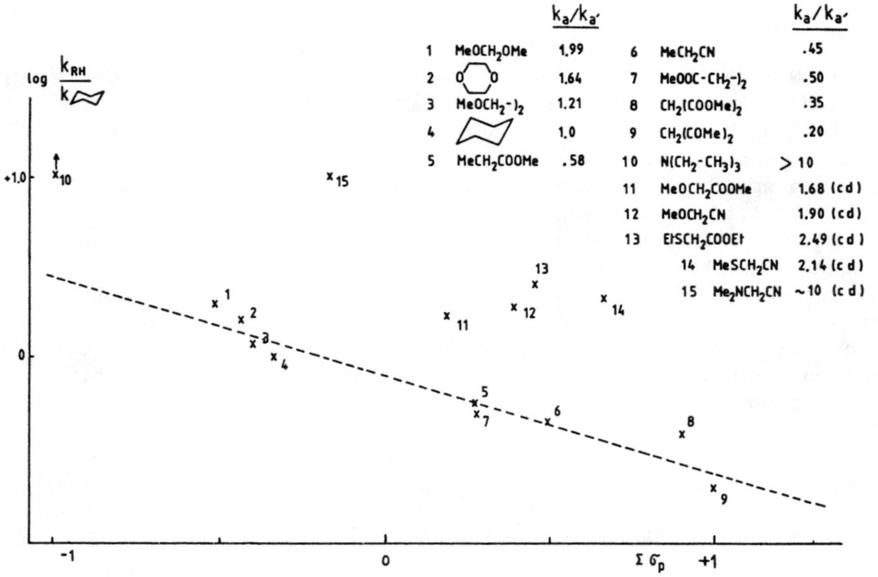

Figure 1. Correlation of relative H-abstraction rates with Hammett constants.

REFERENCES

(1) a) H.G. Viehe, R. Merényi, L. Stella, Z. Janousek, Angew. Chem., **91**, 982 (1979), Internat. Ed. Engl., **18**, 917 (1979); b) H.G. Viehe, Z. Janousek, R. Merényi, L. Stella, Acc. Chem. Res., **18**, 148 (1985).

(2) This book R. Merényi, Z. Janousek, H.G. Viehe, p.301 and references there.

(3) F. Lahousse, R. Merényi, J.R. Desmurs, H. Allaime, A. Borghese, H.G. Viehe, Tetrahedron Lett., **25**, 3823 (1984).

(4) S.U.Park, S.-K. Chung, M. Newcomb, J. Am. Chem. Soc., **108**, 240 (1986).

SUBSTITUENT EFFECTS IN GAS-PHASE THERMAL HYDROXYLATION AND HYDROGENOLYSIS

R. Louw, J.A. Manion and P. Mulder
Center for Chemistry and the Environment
Gorlaeus Laboratories, The State University of Leiden,
P.O. Box 9502
2300 RA LEIDEN, The Netherlands

ABSTRACT

The effect of substituents, especially Cl, on four basic free-radical gas-phase reactions is outlined, viz. H-abstraction from (chloro)benzenes by $^{\bullet}OH$ (1), <u>ipso</u> substitution $ArCl + OH^{\bullet} \longrightarrow ArOH$ (2), hydro-dechlorination $ArCl + H^{\bullet} \longrightarrow ArH + Cl^{\bullet}$ (3) and H-abstraction from vinylchloride $Cl^{\bullet} + CH_2{=}CHCl \longrightarrow CH_2{=}C^{\bullet}Cl + {}^{\bullet}CH{=}CHCl$ (4a,b).

INTRODUCTION

In our studies on slow combustion and on reductive thermolysis of organic chlorine compounds important attacking species are $^{\bullet}OH$ and $H^{\bullet} + Cl^{\bullet}$, respectively. Depending on substrates and conditions, said radicals may add, or abstract hydrogen. Four examples of substituent effects are discussed in this paper.

I. HYDROXYLATION OF CHLOROBENZENE VIA H-ABSTRACTION

Addition and subsequent conversion of intemediate hydroxycyclohexadienyl type radicals is the usual reaction of $^{\bullet}OH$ with benzene and derivatives at ambient temperature, both in solution[1] and in the atmosphere[2]. At elevated temperature this reaction is reversible, which allows H-abstraction to become a major first step in product formation[3,4]. The different site selectivities for both types of reaction (1a), (1b) are exemplified for chlorobenzene:

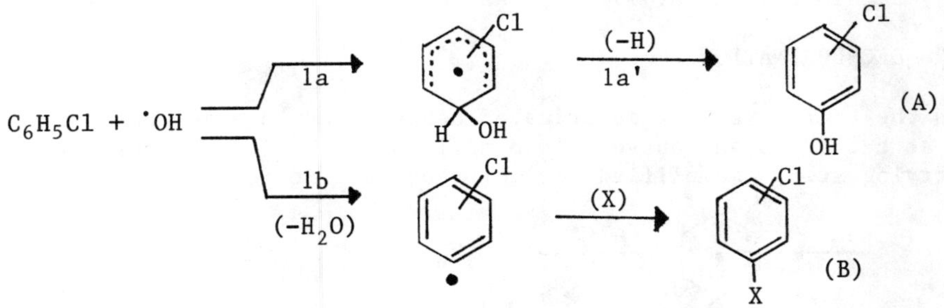

(1)

H. G. Viehe et al. (eds.), Substituent Effects in Radical Chemistry, 331–334.
© 1986 by D. Reidel Publishing Co.

Reaction	Conditions	Isomer Pattern	o/m/p (%)	Ref.
1a	H_2O, 25°C, Ar	A	44.6/23.1/32.3	1b
1a	H_2O, 25°C, O_2	A	37.1/21.8/41.1	1b
1b	G, 295°C, N_2+I_2	B(X=I);	24/52/24	5
1b	G, 276°C,air+tBuOOH	B(X=OH)≡A;	22/56/22	5
	G, 314°C,air+tBuOOH	B(X=OH)≡A;	20/58/22	5

Despite the difficulties encountered in adequate conversion by step 1a'[1], phenolic product A shows the preference for o- + p- attack usual for homolytic substitution. In H-abstraction (1b), m-H seems to be slightly preferred over p-H. This is analogous to the behaviour of Cl$^{\bullet}$, the radical determining the pattern for gas-phase chlorination $C_6H_5Cl \rightarrow C_6H_4Cl_2$. (o/m/p ≅ 13/62/25 at 300°C[6]), although the transition states must involve quite different degrees of C-H bond breaking.

II. HYDROXY-DECHLORINATION

(Reaction 2) is straight-forward and governed by the (relative) rate of *ipso* addition of $^{\bullet}$OH (step 2a)[5].

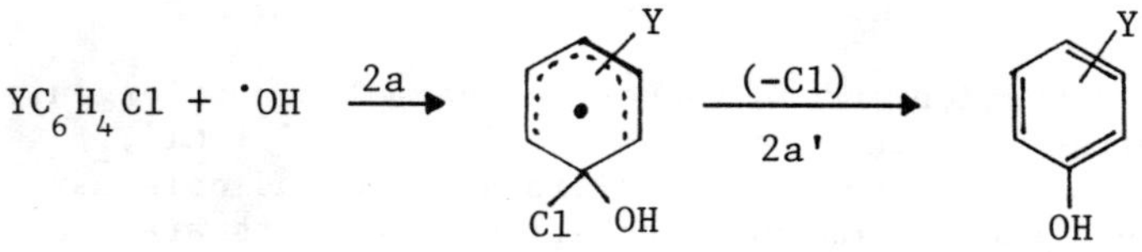

(2)

With C_6H_5Cl at 290°C, in the gas-phase, reaction 2 is 0.3 times as fast as reaction 1b, and on a per-site basis, the rate of displacement of a chlorine in m-$C_6H_4Cl_2$ is as fast as that for conversion $C_6H_5Cl \rightarrow C_6H_5OH$. From competitive runs with o/m/p - $C_6H_4Cl_2$ the relative rates would correspond to an "isomer distribution" at 290°C of 55/25/20; when assuming that deviation from the statistical distribution is due to enthalpy differences only, this pattern extrapolates to 68/15/17 at 25°C. Such ratios tally with those usually observed for free radical additions to chlorobenzene; there may be some steric acceleration in the *ortho* isomer, however. Such an effect is quite prominent in o-$ClC_6H_4CF_3$, which is hydroxy-dechlorinated ca. 19 times as fast as the p-isomer.

III. HYDRO-DECHLORINATION

occurs in the thermolysis of chlorinated benzenes in an atmosphere of hydrogen at ca. 600°C and above[7]. *Ipso* addition of H$^{\bullet}$ is an important chain-carrying step, exemplified for chlorobenzene in eq. 3:

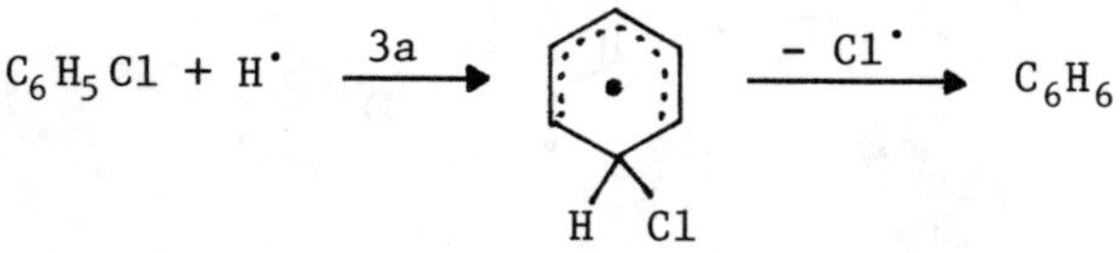

(3)

On the basis of competitive runs with C_6H_5Cl and o/m/p-$C_6H_4Cl_2$, and partial conversions of 1,2,4-$C_6H_3Cl_3$ the following partial rate factors f have been determined (720°C):

o-Cl (1.01); m-Cl (0.74); p-Cl (0.88)

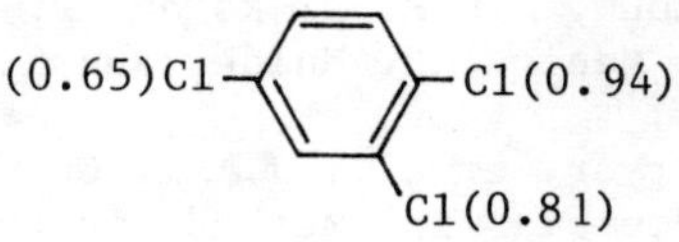

Step (3a) is irreversible and involves a rather early TS; f-values are consonant with the slightly electrophilic character of H$^{\cdot}$ [8] and show some steric acceleration by neighbouring chlorines.

IV. THERMOLYSIS OF CHLOROETHYLENE

Vinylchloride thermolysis, in a 10-15 fold excess of nitrogen, 600-640°C, τ ca. 3s, shows conversions up to 10%. Major products are 2-chloro-1,3-butadiene (C) and acetylene. The latter product is formed, to some extent, via direct HCl-elimination (4c), but prominent steps are (4a) and (4b):

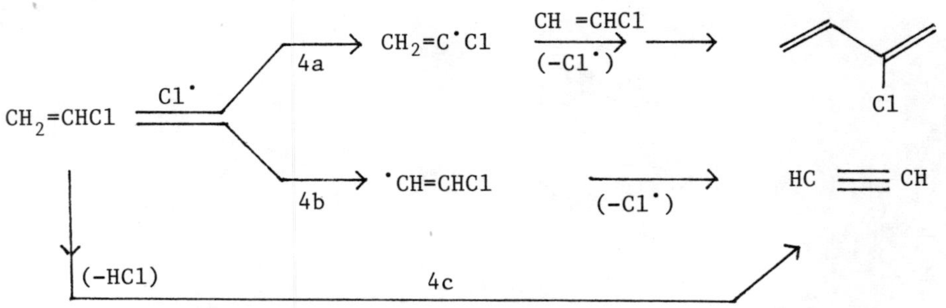

(4)

Our product data lead to $k_{4a}/k_{4b} \cong 5$. This order seems to be at variance with structural effects on H-abstraction from (chloro)alkanes[9]. Whereas in the latter cases usually early transition states are involved, steps (4a, 4b) must have fairly late transition states. This will make the difference in C-H bond strengths in vinylchloride an important factor. The retarding polar effect of Cl will also play a role; hence, compared with abstraction of H from ethylene reaction (4b) is slower, but the retarding effect at C-1 is more than compensated by the stabilization exerted by chlorine in the 1-chlorovinyl radical.

REFERENCES

1. a M.K. Eberhardt et al., J. Am. Chem. Soc., 104, 7069 (1982);
   b M.K. Eberhardt et al., J. Phys. Chem., 77, 589 (1973).
2. J.A. Leone and J.H. Seinfeld, Int. J. Chem. Kinet., 16, 159 (1964).

3. P. Mulder and R. Louw, Recl. Trav. Chim. Pays-Bas, 103, 282 (1984).
4. F.R. Tully, et al. J. Phys. Chem., 85, 2262 (1981).
5. P. Mulder and R. Louw, Recl. Trav. Chim. Pays-Bas, in press.
6. R. Louw and W. Dorrepaal, Int. J. Chem. Kinet. 10, 249 (1978).
7. a R. Louw, J.H.M. Dijks and P. Mulder, Chem. and Ind., 759 (1983).
b J.A. Manion, P. Mulder and R. Louw, Environm. Sc. Techn., 19, 280 (1985).
8. W.A. Pryor, et al., J. Am. Chem. Soc., 95, 6993 (1973).
9. M.L. Poutsma, in 'Methods in Free-Radical Chemistry', Vol 1, p. 138, E.S. Huyser, Editor: Marcel Dekker, New York (1969).

HETEROCYCLIC SUBSTITUENTS IN $S_{RN}1$ REACTIONS

M.P. CROZET and P. VANELLE
Aix-Marseille University ; CNRS UA 109
P.O. Box 80
13397 Marseille cedex 13
France

ABSTRACT - Electron-tranfer substitution reaction (SRN1) is generalized to the heterocyclic electrophile, 1-methyl-2-chloromethyl- 5-nitro-imidazole, and to the heterocyclic nucleophiles, 3-ethyl-5-nitro-tetrahydro-1,3-oxazine salt and 2,2-dimethyl-5-nitro-1,3-dioxane salt. New 1-methyl-5-nitroimidazoles bearing a trisubstituted ethylenic double bond in the 2 position are obtained in high yields. Heterocyclic substituents change very little the substitution reaction but increase dramatically the elimination of nitrous acid.

INTRODUCTION

An important current goal of mechanisms chemistry, particularly of the subfield of radical chemistry, is the establishment of generalizations which govern the structure and properties of paramagnetic species under the influence of experimental variables such as substituents, solvents, etc... Sucessful generalizations of this kind are fundamentally important and practically useful.

Substitution reactions which proceed via radical anion intermediates have been of great interest in recent years (1,2) from both a synthetic and mechanistic point of view. It appears that with few change in the nature of the substituents of both electrophile (acceptor) and nucleophile (donor), the reaction may proceed via a SRN1 pathway involving paramagnetic species rather than via a SN2 pathway. By way of contrast, mechanistic and synthetic investigations involving heterocyclic reactants have received little attention (3, 4). It is of obvious interest to discover whether appropriately substituted heterocycles would undergo analogous substitution reactions owing to the fact that most of the important classes of chemotherapeutic agents and pesticides are heterocycles.

Since the introduction of Metronidazole **1** into the therapy, nitroimidazoles as a class of compounds have attracted much attention (5). As a part of a program directed toward the synthesis of analogues of **1** modified in the position 2 and in order to test whether nitronate anions were able to react by a SRN1 mechanism in this heterocyclic

H. G. Viehe et al. (eds.), Substituent Effects in Radical Chemistry, 335–338.
© 1986 by D. Reidel Publishing Co.

system, we choose 1-methyl-2-chloromethyl-5-nitroimidazole **2** because of its known reactivity with different nucleophiles (3).

We present in this work the results of our investigations on the reaction of **2** with heterocyclic nitronate anions **3** (6) and **4** (7).

RESULTS AND DISCUSSION

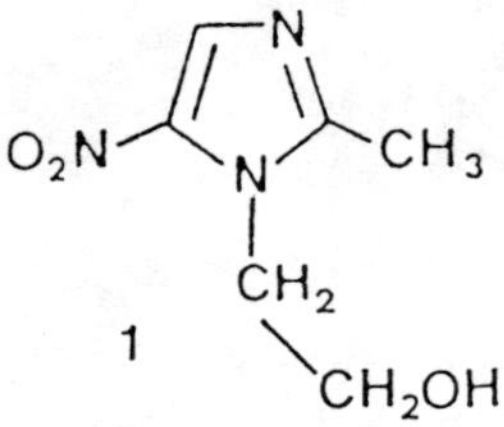

1

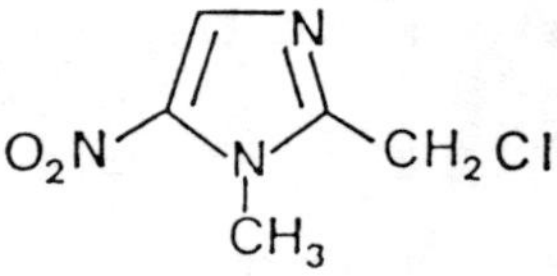

2

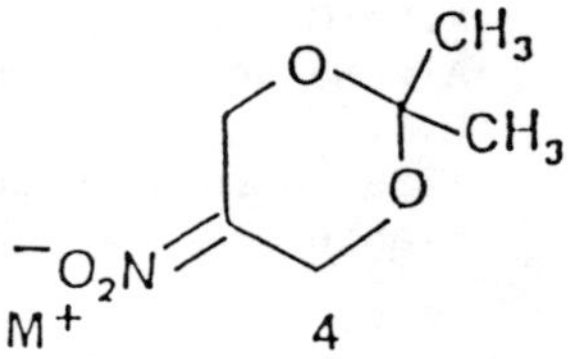

4

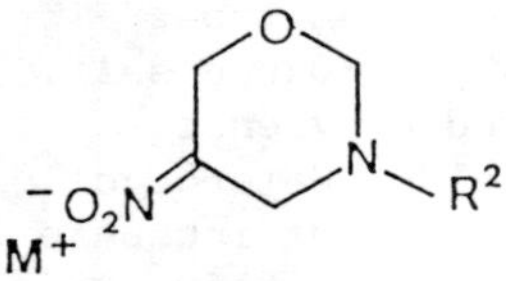

R^2 = Et M = Na

3

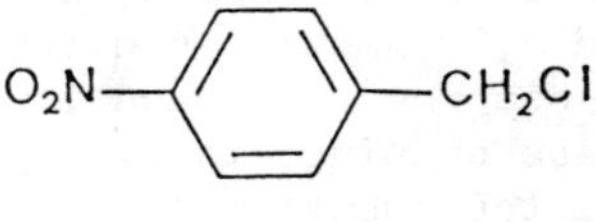

5

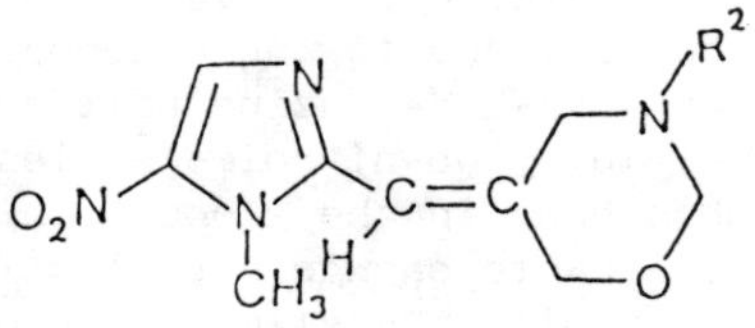

6

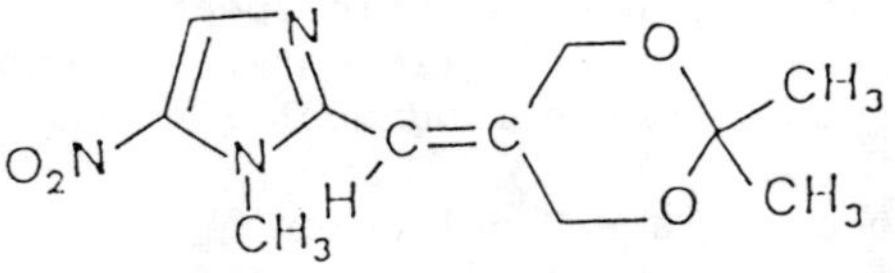

8

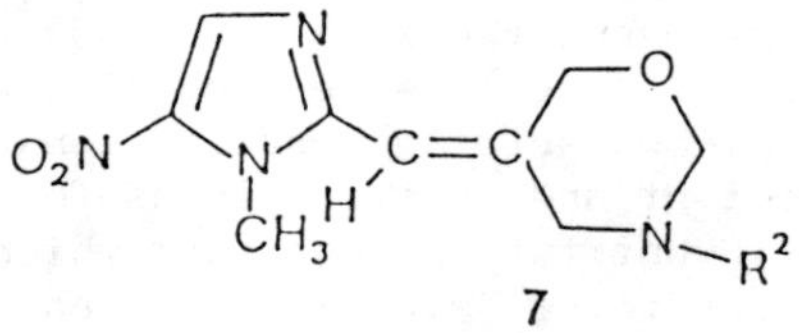

7

TABLE 1 - Reactions[a] between **2** and **5** with **3** and **4**

Anion[b]	Molar equiv. of anion	RX	Yield % : Subst. Prod.	Elim. Prod.
3	1	**5**	63	6[c]
3	2	**5**	15	69[c]
3	3	**5**	5	78[c]
3	2	**2**	-	75[d]
4	2	**5**	85	5
4	2	**2**	-	89

(a) The reaction was allowed to proceed for 24h at room temperature in MeOH under argon (b) Sodium salt. (c) Isomers ratios estimated from NMR of isolated mixtures are 1/1. (d) Isolated isomers : ratio is 1/1.

The present paper is concerned with two questions : (1) the usefulness in synthesis of the reaction of **2** with the salts of heterocyclic nitro compounds and (2) how these results compare with those obtained when the electron-transfer substitution is applied to the model p.nitrobenzyl chloride.

The results are easily summarized : although some by-products and tears require purification of the elimination products formed in reactions of **2** with **3** or **4** by flash chromatography on silica gel, the yields are good when two molar equivalents of anions are used.

We have shown previously (3) that the best yield (88%) is obtained in this reaction with simple nitronate anions in DMF with exposure to the illumination of two ordinary 60-W fluorescent lamps with two molar equivalents of lithium salt after rigorous degassing called freeze-pump-thaw procedure (8).

The mechanism by which the carbon alkylate is obtained appears to be :

$$NI\text{-}CH_2\text{-}Cl + {}^{-}CR^1R^2NO_2 \longrightarrow NI\text{-}CH_2\text{-}Cl^{\overline{\cdot}} + CR^1R^2NO_2$$

$$NI\text{-}CH_2\text{-}Cl^{\overline{\cdot}} \longrightarrow NI\text{-}CH_2^{\cdot} + Cl^{-}$$

$$NI\text{-}CH_2^{\cdot} + {}^{-}CR^1R^2NO_2 \longrightarrow NI\text{-}CH_2\text{-}CR^1R^2NO_2^{\overline{\cdot}}$$

$$NI\text{-}CH_2\text{-}CR^1R^2NO_2^{\overline{\cdot}} + NI\text{-}CH_2\text{-}Cl \longrightarrow NI\text{-}CH_2\text{-}CR^1R^2NO_2 + NI\text{-}CH_2\text{-}Cl^{\overline{\cdot}}$$

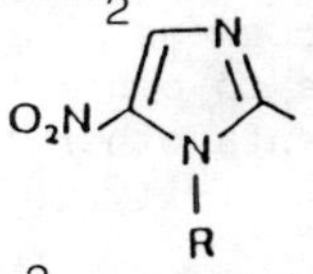
= NI

This mechanism is fully consistent with what is known about electron-transfer substitution reactions of benzylic nitro compounds (1) and a number of observations support the proposed radical anion-free radical chain mechanism.

The nitro group facilites the C-alkylation of nitronate anions ; in no instance did 1-methyl-2-chloromethylimidazole react with nitronate anions to give C-alkylation derivatives under these conditions. The C-alkylation product is the major product and except in intramolecular substitution reaction, there is no example of C-alkylation of nitronate anion which is not an electron-transfer substitution reaction (9). When the reaction is repeated with one equivalent of p-dinitrobenzene which acts as an inhibitor the yield of C-alkylation decreases dramatically.

The second equivalent of anion is involved in the elimiation process

$$NI\text{-}CH_2\text{-}CR^1R^2NO_2 + {}^{-}CR^1R^2NO_2 \longrightarrow NI\text{-}CH{=}CR^1R^2 + HCR^1R^2NO_2$$

When one equivalent of anion is used the yield of olefin is only 43% indicating that the base-promoted nitrous elimination is not much slower than that of the SRN1.

Products resulting of nitrous acid elimination from the C-alkylation products formed from **5** and simple nitronate anions even in excess are generally not observed except if the tertiary nitro group is on a cyclic carbon atom (6). On the other hand, if an electron-attracting group is in β of the nitro group, base-promoted elimination of nitrous acid becomes easy. In our examples these two factors operate for HNO2 elimination and these reactions provide **6, 7** and **8** in high yields.

All the 1-methyl-5-nitroimidazoles with a trisubstituted double bond in the position 2 prepared by electron-transfer substitution reaction followed by nitrous acid elimination are very active against amoebic, trichomonal, giardial and anaerobic infections. An increase of the delocalization of the unsaturated system change little the activity of these compounds in comparative trials with metronidazole but the nature of the substituents of the double bond can dramatically affect this activity (10).

The results of this investigation lead to the following conclusions : electron-transfer substitution reaction (SRN1) can be generalized to nitroheterocyclic electrophiles and to nitro heterocyclic salts. The fact that the methylene group in the position 2 is activated by the nitro heterocyclic system and that the tertiary nitro group is on a cyclic carbon provide an easy elimination of nitrous from the C-alkylation product in sharp contrast with the stability of C-alkylation products in p-nitrobenzylic series. Studies are in progress to generalize these electron-transfer substitution reactions to other heterocyclic systems with electron-withdrawing substituents other than nitro group and with nitro heterocyclic salts whose reactivity in SRN1 reactions is not yet known.

ACKNOWLEDGMENTS

We thank the Centre National de la Recherche Scientifique for support of this research (PIRMED AIP/06931/1173 and PVD AI/06943/8025). We are grateful to RHONE-POULENC SANTE for generous support.

REFERENCES

(1) Kornblum, N. The Chemistry of Functional Groups, Supplement F : 'The Chemistry of Amino, Nitroso and Nitro Compounds and their Derivatives' ; Pataï, S., Ed. ; Wiley : New York, 1982 ; pp 361-384.

(2) Julliard, M. ; Chanon, M. Chem. Rev. 1983, **83**, 425.

(3) Crozet, M.P. ; Surzur, J-M. ; Vanelle, P. ; Ghiglione, C. ; Maldonado, J. Tetrahedron Lett. 1985, **26**, 1023 and references cited therein.

(4) Beadle, C.D. ; Russell Bowman, W. J. Chem. Research (S) 1985, 150.

(5) Nair, M.D. ; Nagarajan, K. 'Nitroimidazoles as Chemotherapeutic agents' in Progress in Drug Research, Jucker, E. ; Ed. ; Birkhauser Verlag : Basel, 1983 ; Vol. **27**, pp. 163-252.

(6) Crozet, M.P. ; Vanelle, P. Tetrahedron Lett. 1985, **26**, 323 and references cited therein.

(7) Crozet, M.P. ; Archaimbault, G. ; Vanelle, P. ; Nouguier, R. Tetrahedron Lett. 1985, **26**, 5133 and references cited therein.

(8) Kornblum, N. ; Carlson, S.C. ; Widmer, J. ; Fifolt, M.J. ; Newton, B.N. ; Smith, R.G. J. Org Chem. 1978, **43**, 1394.

(9) Seebach, D. ; Colvin, E.W. ; Lehr, F. ; Weller, T. Chimia, 1979, **33**, 1.

(10) Crozet, M.P. ; Vanelle, P. ; Rhône-Poulenc Santé to be published in J. Med. Chem.

SUBSTITUENT EFFECTS ON THE OXIDATION OF TOLUENES BY MOLECULAR OXYGEN : THE SURPRISING REACTIVITY OF p.METHOXYTOLUENE

M. COSTANTINI and L. KRUMENACKER
RHONE POULENC
Centre de Recherches de Saint-Fons
69192 - SAINT FONS Cedex
France

ABSTRACT. Some data on the catalysis by Co(II) or Co(II)-Br^- of the oxidation of substituted toluenes to the corresponding aldehydes by dioxygen are described. Most generally, the selectivities into aldehydes decrease when the conversions of the toluenes increase ; however, with some toluenes as p.methoxytoluene or 2-methylnaphtalene, it is possible to obtain high selectivities and conversions. To try to understand this difference of reactivity, a study of the mechanisms, involving the ionisation potentials of toluenes and the oxidation rates of corresponding aldehydes, is presented.

INTRODUCTION. Benzaldehyde and substituted analogues (p.methoxybenzaldehyde, p.t.butylbenzaldehyde, m.phenoxybenzaldehyde,...) are very important products for industrial chemistry. They can be used either on their own or as intermediates.

They are generally obtained by oxidation of the corresponding toluene. Several oxidation processes can be envisaged. Economically, those using dioxygen in the presence of catalysts are the most efficient. However, in most cases, selectivities into aldehydes decrease when toluene conversions increase : this is an inconvenient for industrial processes where high conversions are most interesting.

The results always depend strongly on the substituents of the aromatic ring. In this brief study, we would like to report on some examples of this behaviour.

RESULTS. Reactions were conducted in acetic acid using either cobalt(II) and bromide or cobalt(II) as catalysts (eq.1).

$$XArCH_3 + O2 \xrightarrow[CH_3COOH]{Co^{II}(+\ Br^-)} XArCHO \dashrightarrow XArCOOH \qquad \text{(eq. 1)}$$

With the exception of a few rare cases - which are described here - high aldehyde selectivity is obtained only with partial conversion of toluenes (to avoid formation of the corresponding acids) ; for example, with X = p.chloro, 3-4 dichloro, 2-4 dichloro, p.t.butyl, m.phenoxy, etc..., the highest aldehyde selectivity is about 75 % for a conversion rate of 20-25 %.

H. G. Viehe et al. (eds.), Substituent Effects in Radical Chemistry, 339–342.
© 1986 by D. Reidel Publishing Co.

We found the astonishing behaviour of p.methoxytoluene most interesting as the corresponding aldehyde selectivity is almost independant of the conversion rate of this hydrocarbon, making it exceptional among all the other toluenes of the same type (this is true either with Co(II)-Br^- catalyst or with Co(II) catalyst). In Table 1, we compare the results obtained using p.methoxytoluene and p.t.butyltoluene (the latter chosen as reference).

Hydrocarbons $[XArCH_3]_o = 3,3$ M/1	Catalysts	θ°C	P_{O_2}/b	conversions (%)	"AL" (%) *	ACID (%) *
$p.CH_3OArCH_3$	Co^{II}-Br^-	75	1	25	75	25
$p.CH_3OArCH_3$	Co^{II}-Br^-	75	1	50	73	27
$p.t.BuArCH_3$	Co^{II}-Br^-	75	1	25	75	25
$p.t.BuArCH_3$	Co^{II}-Br^-	75	1	50	50	50
$p.CH_3OArCH_3$	Co^{II}	75	1	70	71	29
$p.t.BuArCH_3$	Co^{II}	115	1	25	5	95
$p.CH_3OArCH_3$	Co^{II}	115	20	79	74	26
$p.t.BuArCH_3$	Co^{II}	125	20	83	0	90
2-CH_3 naphtalene **	Co^{II}	90	1	70	70	30
3,4,5$(CH_3O)_3ArCH_3$	Co^{II}-Br^-	95	1	0	-	-

* Selectivities
** Using traces of Br^- to initiate oxidation

We can notice that 2-methylnaphtalene has the same behaviour as p.methoxytoluene. Furthermore, catalysis using only Co(II) is more selective than that using Co(II)-Br^- in the case of p.methoxytoluene, whereas for p.t.butyltoluene (and the other toluenes studied), the opposite is true.

DISCUSSION. The difference between p.methoxytoluene and other toluenes can only be explained by inversion of the reactivity between the hydrocarbon and the corresponding aldehyde. This inversion is clearly demonstrated by comparing the kinetic studies of stoichiometric oxidations by $Co(OAc)_3$ of the couples $p.CH_3OArCH_3/p.CH_3OCHO$ and $p.ClArCH_3/p.ClArCHO$ successively (Fig. 1).

It is shown that the oxidation kinetics of XArCHO by Co(III) do not depend on the nature of X : the slow step has no electrophilic character but is linked to the direct abstraction of aldehydic hydrogen whose rate is independant of the nature of substituents on the nucleus (eq.2).

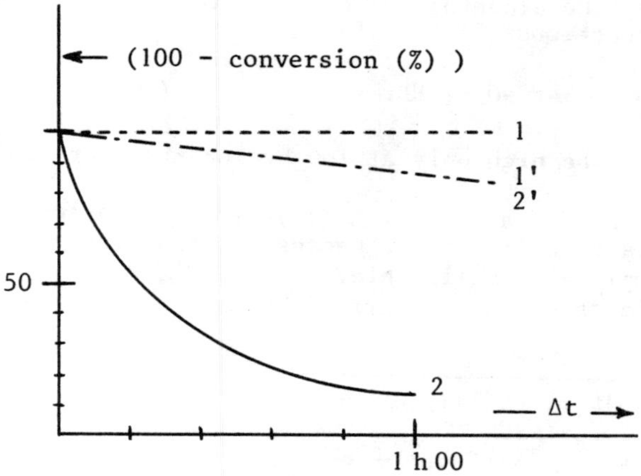

Fig. 1 : Stoichiometric oxidations by $Co(OAc)_3$ of the couples p.$ClArCH_3$(1)/p.ClArCHO(1') and p.CH_3OArCH_3(2)/p.$CH_3OArCHO$(2')

$$XArCHO + Co(III) \xrightarrow{k} XAr\dot{C}{=}O + H^+ + Co(II) \qquad \text{(eq. 2)}$$

On the contrary, the oxidation kinetics of $XArCH_3$ by Co(III) are strongly influenced by the nature of X, the slow step beeing an electrophilic attack (eq.3).

$$XArCH_3 + Co(III) \underset{}{\overset{k'}{\rightleftharpoons}} [XArCH_3]^{+\cdot} + Co(II) \qquad \text{(eq. 3)}$$

Thus the ratio between the reactivity of $XArCH_3$ and XArCHO can be inverted when changing from one substituent to another : in the case of p.methoxytoluene, the analysis of the kinetic data of Fig.1 gives $k'/k \simeq 14$. Because of the electrophilic character of the reaction (3), this reactivity is well correlated to the ionisation potentials (2-CH_3 naphtalene = 7,56 eV, p.CH_3OArCH_3 = 7,9 eV, p.t.Bu$ArCH_3$=8,44 eV, p.$ClArCH_3$ = 8,69 eV, p.$CH_3OArCHO$ = 8,60 eV, p.ClArCHO = 9,6 eV). The great differences between the ionisation potentials of the toluenes and the corresponding aldehydes justify reaction (2).

When using Co(II)-Br^- as catalyst, if the hydrocarbon has a low ionisation potential (p.methoxytoluene, methylnaphtalenes), the oxidation of the Br^- anion by Co(III) is slower than the toluene oxidation and reaction (3) is the main way (selectivities and conversions are increased) . If, on the other hand, the hydrocarbon has a high ionisation potential, the opposite is true and reactions (4) and (5) are the main pathway :

$$Co(III) + Br^- \longrightarrow Co(II) + Br^\cdot \qquad \text{(eq.4)}$$

$$XArCH_3 + Br^\cdot \longrightarrow XArCH_2^\cdot + HBr \qquad \text{(eq.5)}$$

In this latter case, the aldehyde selectivity thus depends on the competition between reactions (5) and (6).

$$XArCHO + Br^{\bullet} \longrightarrow XAr\dot{C}{=}O + HBr \qquad \text{(eq.6)}$$

This selectivity will be high only at low hydrocarbon conversion.

Using p.methoxytoluene, catalysis by Co(II)-Br^- gives a lower selectivity than catalysis by Co(II) alone because of reactions (4) and (6) which are not completely negligeable.

All the preceding facts can be summarised by the two following schemes :

CATALYSIS BY Co(III) ALONE

$XArCH_3 + Co(III) \rightleftharpoons [XArCH_3]^{\ddagger} \longrightarrow XArCH_2^{\bullet}$
$XArCH_2^{\bullet} + O_2 \longrightarrow XArCH_2OO^{\bullet}$
$XArCH_2OO^{\bullet} + Co(II) + H^+ \longrightarrow XArCHO + Co(III) + H_2O$
$XArCHO + Co(III) \longrightarrow XAr\dot{C}{=}O + Co(II) + H^+$
$XAr\dot{C}{=}O \dashrightarrow XArCOOH$

CATALYSIS BY Co(II)-Br^-

$XArCH_3 + Br^{\bullet} \longrightarrow XArCH_2^{\bullet} + HBr$
$XArCH_2^{\bullet} + O_2 \longrightarrow XArCH_2OO^{\bullet}$
$XArCH_2OO^{\bullet} + Co(II) + H^+ \longrightarrow ArCHO + Co(III) + H_2O$
$Co(III) + HBr \longrightarrow Co(II) + H^+ + Br^{\bullet}$
$XArCHO + Br^{\bullet} \longrightarrow XAr\dot{C}{=}O + HBr$
$XAr\dot{C}{=}O \dashrightarrow XArCOOH$

CONCLUSION. In this brief study, we have shown that, in the case of oxidation of toluenes by molecular oxygen in the presence of catalysts, the aldehyde selectivities are closely dependent on the ionisation potentials of toluenes. However, it is important to note that all toluenes having low ionisation potentials do not give high selectivities and conversions (for example, 3,4,5-trimethoxytoluene, which is more oxydisable by Co(III) than p.methoxytoluene, is not oxydised : cf. Table 1) because of the absence of initiation reactions or the occurence of fast reactions between the intermediary radical-cations and the very nucleophilic toluenes (eq. 7) :

$$[XArCH_3]^{\ddagger} + XArCH_3 \dashrightarrow \text{Products} \qquad \text{(eq. 7)}.$$

REFERENCES.

1. R.A. SHELDON and J.K. KOCHI, "Metal-Catalysed Oxidations of Organic compounds", Academic Press, 1981, 122-130, 318-327.
2. A. CLERICI and O. PORTA, Can. J. Chem. 1980, 2177.

FREE RADICAL ADDITIONS OF DICHLOROMETHANE TO δ-ALKENYL t-BUTYL PEROXIDES.

B. Maillard[(a)], M. Agorrody[(b)], E. Montaudon[(b)]
(a) Laboratoire de Chimie Organique et Organométallique, associé au CNRS, UA 35.
(b) Laboratoire de Chimie Appliquée.
Université de Bordeaux 1, 351, cours de la Libération, F 33405, Talence-Cedex, France.

ABSTRACT. The free radical additions of dichloromethane to δ-alkenyl t-butyl peroxides were performed in order to define the limits of the synthetic potential of their induced decomposition as a route to substituted tetrahydrofurans. When the double bond was terminal, five membered cyclic ethers were isolated with yields close to 40 %, meanwhile in the case of t-butyl 5-methyl hex-4-enyl peroxide, a tetrahydropyran was obtained with a similar yield.

In a recent series of articles, we have described the synthetic potential of the induced homolytic decomposition of unsaturated peroxycompounds ; this reaction was found to be efficient for 5-pentan-4-olidation [1] and 2,3-epoxypropanation [2] of hydrogen donor solvents ZH (fig. 1).

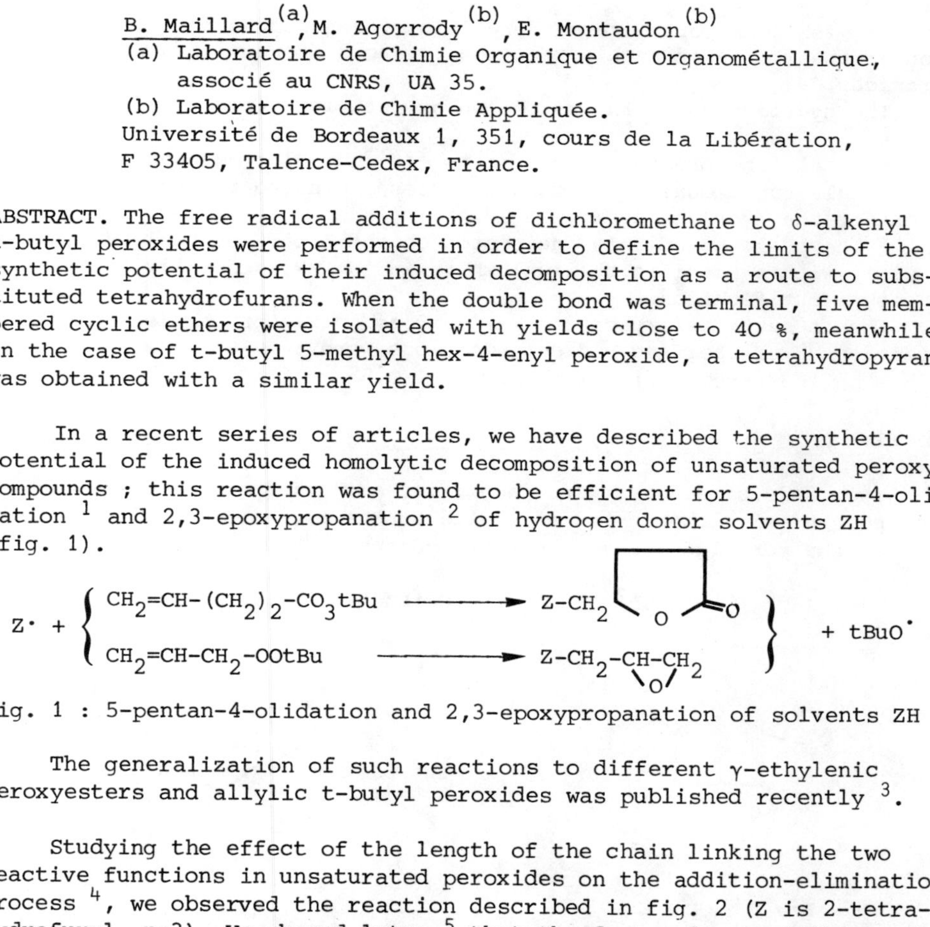

Fig. 1 : 5-pentan-4-olidation and 2,3-epoxypropanation of solvents ZH

The generalization of such reactions to different γ-ethylenic peroxyesters and allylic t-butyl peroxides was published recently [3].

Studying the effect of the length of the chain linking the two reactive functions in unsaturated peroxides on the addition-elimination process [4], we observed the reaction described in fig. 2 (Z is 2-tetrahydrofuryl, n=3). We showed later [5] that the free radical addition of ZH to t-butyl pent-4-enyl peroxide is an efficient route to 2-substituted tetrahydrofurans.

H. G. Viehe et al. (eds.), Substituent Effects in Radical Chemistry, 343–346.
© 1986 by D. Reidel Publishing Co.

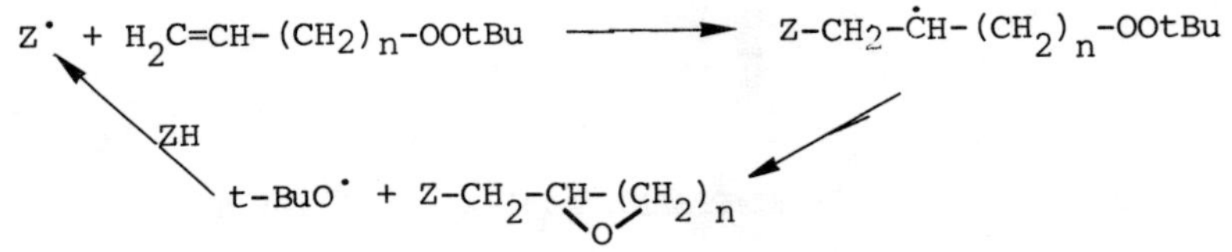

Fig. 2 : Mechanism of the induced decomposition of t-butyl ethylenic peroxides ($1 \leqslant n \leqslant 4$).

We wish to report here our preliminary results on the determination of the limits of such a reaction, using different unsaturated peroxides.

The hydrogen donor ZH was chosen as methylene chloride for the following reasons :

- In all the previous experiments [1,3,5] dichloromethane led only to a single functionalized compound corresponding to the substitution of one of the hydrogen atoms.
- The presence of only one proton on the Z part of the adduct ($CHCl_2$), with a specific chemical shift in ^1H NMR, generally allowed an easy identification.

When t-butyl peroxyacetate was used as initiator, the standard conditions [5] of the free radical additions were :

ZH/peroxide/initiator : 100/1/0,5   110°C ; 10 h .

RESULTS

Two kinds of adducts were identified according to the peroxide studied. Indeed although 1,2 and 3 led to the tetrahydrofurans 1a, 2a and 3a, the peroxide 4 gave the tetrahydropyran 4a.

$R^1_2C{=}CR^2{-}(CH_2)_2{-}CR^3_2{-}OOtBu$

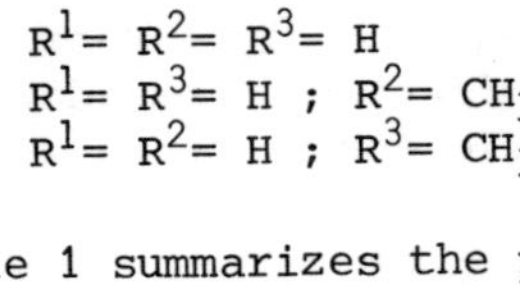

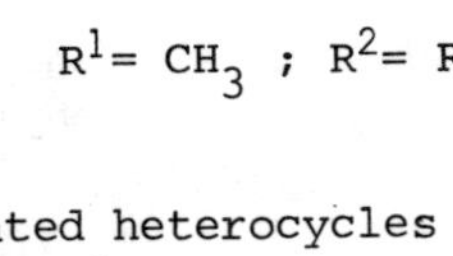

1, 1a $R^1= R^2= R^3=$ H
2, 2a $R^1= R^3=$ H ; $R^2= CH_3$
3, 3a $R^1= R^2=$ H ; $R^3= CH_3$

4, 4a $R^1= CH_3$; $R^2= R^3=$ H

Table 1 summarizes the yields of the isolated heterocycles 1a-4a relative to the peroxide 1-4 involved in the reaction.

Peroxide	1	2	3	4
Adduct	1a	2a	3a	4a
Yield %	42	47	41	39

Table 1 : Free radical additions of dichloromethane to the peroxides 1-4.

DISCUSSION

First of all, let us keep in mind the mechanism of the induced decomposition of unsaturated peroxides that has been determined previously [4]. Indeed, whatever is the value of n ($1 \leq n \leq 4$), the propagation phase of the chain reaction procedes in three steps (fig. 2).

The analysis of the results led us to three observations concerning the isolated cyclic ethers, their yields, and the effect of the reactive chain length.

The size of the cyclic ether seems to be related to the nature of substitution at the point of unsaturation. According to the above mechanism, the orientation of the addition is in good agreement with the previous studies on the free radical additions [6].

The well known inefficiency of free radical additions to trisubstituted double bond contrasts seriously with the yield relative to the formation of 4a. The reversibility of the addition could be invoked to explain them (fig. 3). The reaction of evolution of the radical adduct could be then the determining step ; the faster homolytic displacement ($k_{S_{H_i}} > k_H$ [ZH]) could compete efficiently with the retroaddition (k_{-a}), although it is not possible for the hydrogen transfer. This hypothesis is not contradicted by the apparent non-addition of dichloromethane to the t-butyl 3-methyl but-2-enyl peroxide [3], as we will see later. However, to confirm this hypothesis, the free radical addition of methylene chloride to a trisubstituted alkene (e.g. 2-methyl hex-2-ene) should be studied, but this is not the purpose of this work.

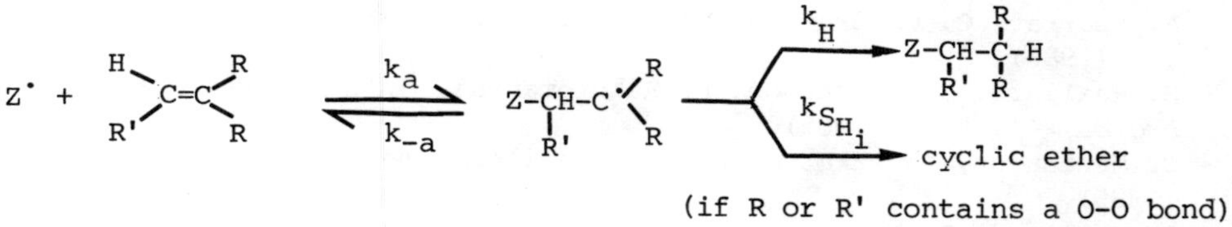

Fig. 3 : Hypothetic mechanism of the free radical additions to trisubstituted alkenes

The comparison of the yields of the different reactions seems to indicate that the substitution of an hydrogen by a methyl do not really affect the efficiency of the addition-elimination reaction to unsaturated peroxides. This differs seriously from the analysis that was effected previously for other peroxydic compounds [3]. To eliminate the ambiguity of the use of a single ZH, we started the generalization of free radical additions of other ZH solvents to 1, 2, 3 and 4 . The initial results with methyl propanoate and acetonitrile confirm this observation.

It should be underlined that, in the case of 1, the chain reaction is very short ; a ratio 1/initiator equal to two was necessary to consume all the peroxide[5]. To confirm the apparent relation between the different results and the peroxide/initiator ratios in the free radical additions of methylene chloride to 1 and t-butyl 3-methyl but-2-enyl peroxide 5, we have performed the last one with a half equimolar amount of t-butyl peroxyacetate relatively to 5 and isolated the oxetane 5a with a yield of 45 %.

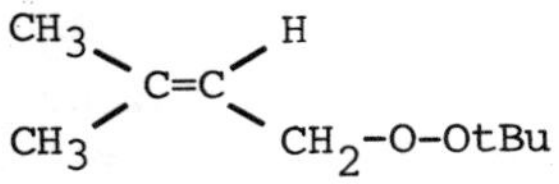

5

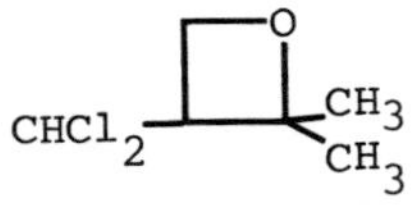

5a

Taking into account the influence of the peroxide/initiator ratio on the yield of adducts for the peroxide 5, the free radical additions of ZH to β-alkenyl t-butyl peroxides under different initiation conditions (peroxide/initiator = 2) should be done. Indeed, beyond the mechanistic point of view, there is also a proper synthetic interest.

In conclusion, the main effect of substituents upon the course of the induced decomposition of a δ-alcenyl t-butyl peroxide appears to occur in the first step. Indeed, depending on the substitution of the double bond, the addition takes place on one or the other carbon leading finally to a tetrahydrofuran or a tetrahydropyran.

LITTERATURE

1a - B. Maillard, A. Kharrat, C. Gardrat, *Tetrahedron,* 40, 3534 (1984);
b - A. Kharrat, C. Gardrat, B. Maillard, *Synth. Comm.,* 13, 1209 (1983);
c - A. Kharrat, C. Gardrat, B. Maillard, *Can. J. Chem.,* 62,2385 (1984);
d - A. Kharrat, C. Gardrat, B. Maillard, *Can. J. Chem.,* 63,2522 (1985);
e - A. Kharrat, C. Gardrat, B. Maillard, *Bull. Chem. Soc. Belg.,* 94, 29, (1985).
2a - B. Maillard, E. Montaudon, F. Rakotomanana, M.J. Bourgeois, *Tetrahedron,* 41, 5039 (1985).
b - E. Montaudon, F. Rakotomanana, B. Maillard, *Tetrahedron,* 41, 2727, (1985).
3 - B. Maillard, A. Kharrat, F. Rakotomanana, E. Montaudon, C. Gardrat, *Tetrahedron,* 41, 4047 (1985).
4 - M.J. Bourgeois, E. Montaudon, B. Maillard, *to be published.*
5 - M. Agorrody, E. Montaudon, B. Maillard, *to be published.*
6a - J.M. Tedder, *Angew. Chem. Int. Ed. Engl.,* 21, 401, (1982).
b - B. Giese, *Angew. Chem. Int. Ed. Engl.,* 22, 783, (1983).

THE NON-CHAIN RADICALOID MECHANISM OF SOME NUCLEOPHILIC SUBSTITUTION REACTIONS

Alan R. Katritzky
Department of Chemistry
University of Florida
Gainesville, FL 32611
USA

ABSTRACT. Evidence is presented for a non-chain radical mechanism for the reactions of N-alkylpyridinium cations with nitronate anions. The first step is considered to be charge-transfer. The extension of this type of reaction to other C-nucleophiles and to N-arylpyridinium cations is discussed.

Nucleophilic substitutions at saturated carbon atoms are one of the most important synthetic reactions of organic chemistry and the study of their mechanism has been of great interest over the past 50 years. A recent objective summary of this work is provided by March.[1] Highlights of the past decades have been Ingold's recognition on the fundamental difference between S_N1 and S_N2 reactions, which was followed by Winstein's realization that the S_N1 mechanism could involve either ion pairs or free carbocations. Later Sneen put forward his much criticised hypothesis that S_N2 reactions occur on intimate ion-pairs formed reversibly in a preequilibrium step. Subsequently, the concept of variable nucleophilic assistance was emphasized. Most recently, it has been amply demonstrated that electron transfer can cause nucleophilic substitution reactions to take place by radical pathways (see ref 1 for leading references).

Most of this classical work on nucleophilic substitution was carried out using negatively charged leaving groups departing from neutral substrates. This means that in the S_N1 type reactions, charge is created in the transition state and such reactions therefore require a highly polar medium in which it is often difficult to distinguish between the role of the solvent as solvent and as nucleophile. Our group has made extensive investigations of nucleophilic substitutions in which substituted pyridines have been utilized as neutral leaving groups.[2] In such reactions, no charge is created on bond heterolysis and the rates of S_N1 reactions are consequently much less affected by the polarity of the solvent. We believe that in this way we have

H. G. Viehe et al. (eds.), Substituent Effects in Radical Chemistry, 347–360.
© 1986 by D. Reidel Publishing Co.

demonstrated conclusively that S_N1 and S_N2 reactions can take place simultaneously. Moreover, we have demonstrated that in some instances the S_N1 reactions involve carbonium ions, and in others reactions on ion pairs.[3] In addition, we have provided evidence that S_N2 reactions can take place either by the classical Ingold mechanism or by a Sneen-type mechanism.[4] This work has been comprehensively reviewed.[2]

The purpose of the present paper is to consider nucleophilic substitutions reactions with neutral leaving groups which take place by initial electron transfer and thus by a radical type mechanism.

1. Preparative Work with Nitronate Anions

Much literature work has shown that the reaction of nitronate anions with alkyl halides or tosylates results in exclusive O-alkylation, although Michael reactions of nitronate anions with alpha,beta-unsaturated nitriles, ketones, <u>etc</u>, are reported to give C-alkylated products (Scheme 1).[5] The only other exceptions to the exclusive O-alkylation of nitronate anions is in reactions that have been demonstrated to proceed through radical chains. The first examples of these were discovered by Kornblum who showed that a very limited number of benzyl substrates (<u>p</u>-nitrobenzyl halides for the most part) react with alpha-nitronate anions to give C-alkylation. This work has been summarized.[6] More recently, Russell showed that alkylmercury halides could C-alkylate nitronate anions.[7]

Scheme 1: NITRONATE ANIONS AS AMBIDENT NUCLEOPHILES

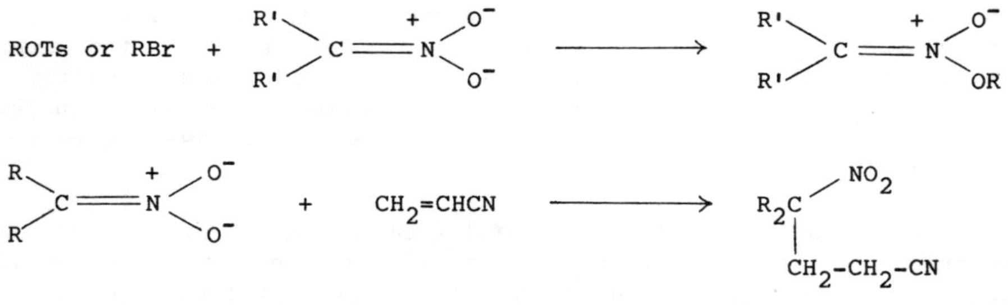

In our investigation of the pyridine mediated conversions of primary amino groups in aliphatic amines to other functionality,[8] we showed that such compounds react with nitronate ions to afford high yields of C-alkylated derivatives.[9] A variety of nitronate anions could be used and C-alkylation was achieved with both primary and secondary alkyl groups (Scheme 2). This was the first indication that radical type mechanisms were involved.

2. Kinetic Evidence for the Occurance of Second-Order Reactions

Numerous experiments have shown that the reaction between a variety of nitronate ions and N-alkylpyridinium alkylating agents is an overall second order reaction, being first order in each constituent.[10]

Scheme 2: C-ALKYLATION OF NITRO-ALKANE ANIONS

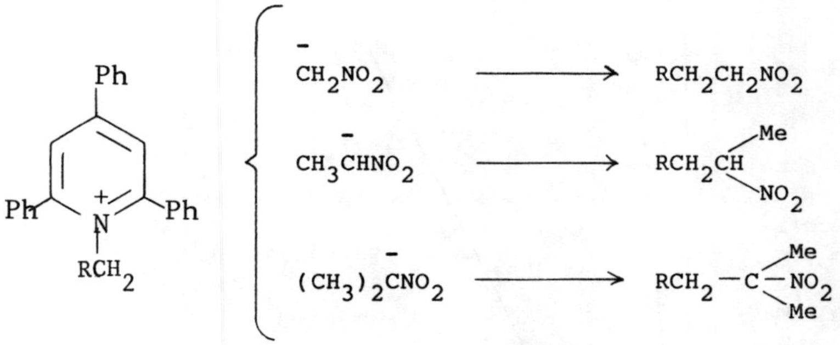

Reaction for R = aryl in EtOH at 80 °C: yields 60-70%
for R = alkyl in AmOH at 130 °C: yields 40-50%

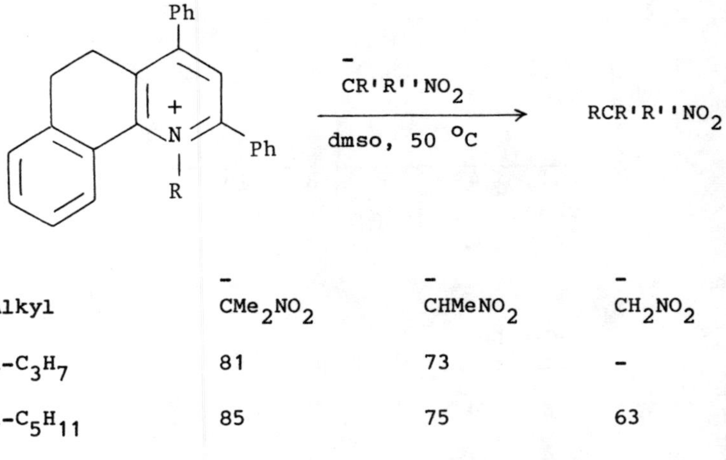

Alkyl	$\bar{C}Me_2NO_2$	$\bar{C}HMeNO_2$	$\bar{C}H_2NO_2$	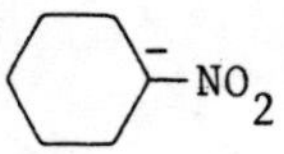
n-C_3H_7	81	73	-	91
n-C_5H_{11}	85	75	63	-
n-C_6H_{13}	85	78	64	-
n-C_7H_{15}	86	74	63	-

Such second order reactivity persists in other solvents (Fig 1). It applies for several different leaving groups (Fig. 2) and for different N-substituents (Fig. 3) as well as for different nitronate anions (Fig. 4).[10]

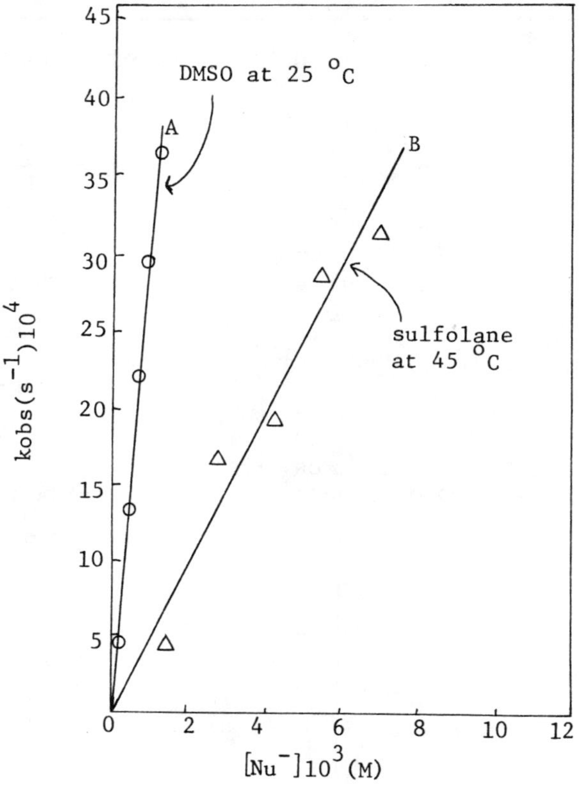

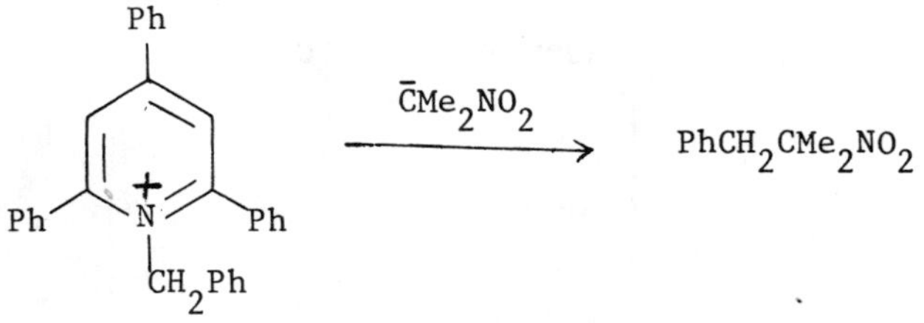

Figure 1: Dependence of k_{obs} on nucleophile concentration for the reaction of 1-benzyl-2,4,6-triphenylpyridinium tetrafluoroborate with sodium 2-nitropropanide in (A) dimethyl sulfoxide at 25 °C, (B) sulfolane at 45 °C.

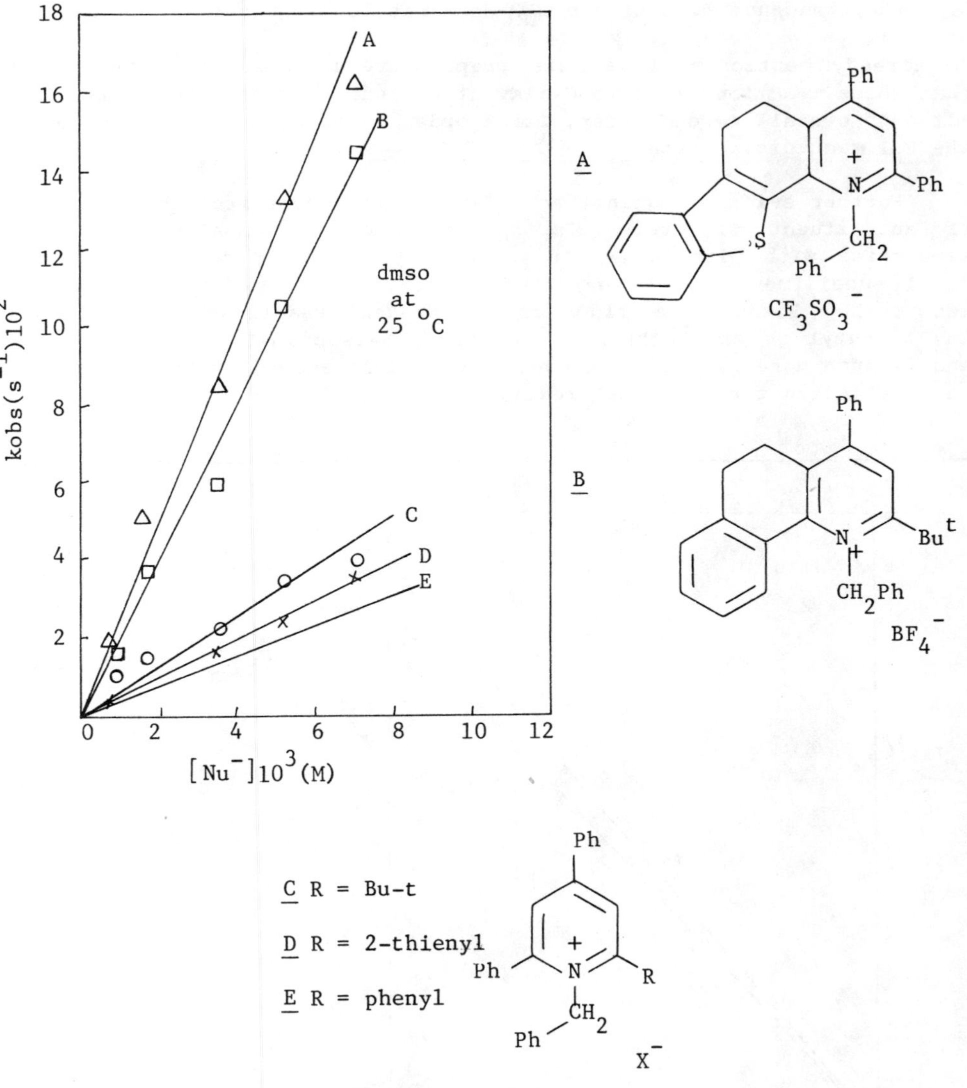

Figure 2: Dependence of k_{obs} on nucleophile concentration for reactions with sodium 2-nitropropanide in DMSO at 25 °C of: (A) 1-benzyl-5,6-dihydro-2,4-diphenyl-1-azonia-1-thiabenzo[a]fluorene, (B) 1-benzyl-2-t-butyl-5,6-dihydro-4-phenylbenzo[h]quinolinium, (C) 1-benzyl-2-(2-thienyl)-4,6-diphenylpyridinium, (D) 1-benzyl-2-t-butyl-4,6-diphenylpyridinium tetrafluoroborates.

3. Other Evidence Against the S_N2 Mechanistic Pathway

As already mentioned above, the preparative results which demonstrate that these reactions lead to C-alkylation indicate that these reactions, although overall second order, first order in each component, are not of the S_N2 mechanistic type.

Further evidence against an S_N2 mechanism comes from considerations of substituent effects. In S_N2 reactions of N-alkyl pyridinium substrates with piperidine, the reactivity sequence depends on the N-alkyl substituent in the expected fashion: N-benzyl > N-methyl > N-isopropyl. However, we find for the present reactions that, although the N-benzyl is still the most reactive, N-isopropyl is not far behind and is much more reactive than N-methyl. This accords with the expected rate variation for a radical reaction.

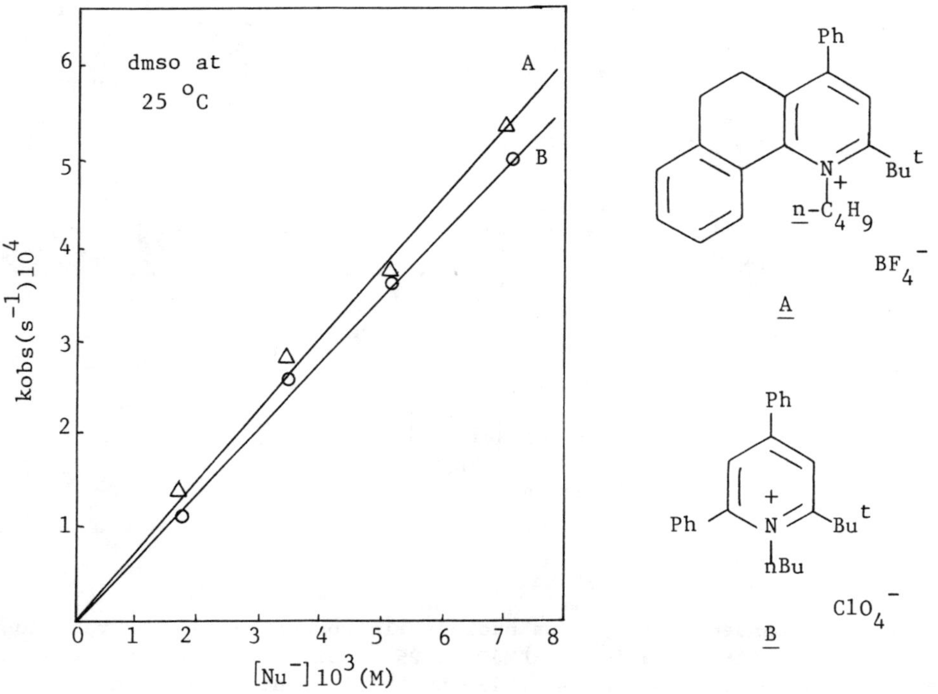

Figure 3: Dependence of k_{obs} on nucleophile concentration for the reactions with sodium 2-nitropropanide at 25 °C in DMSO of (A) 1-*n*-butyl-2-*t*-butyl-5,6-dihydro-4-phenyl-benzo[*h*]quinolinium, (B) 1-*n*-butyl-2-*t*-butyl-4,6-diphenylpyridinium.

Again, we can compare a series of N-(p-substituted benzyl)-2,4,6-triphenylpyridinium cations. Rates for the S_N2 piperidine substitution show the expected rate-enhancement of donor substituents, and are depressed for acceptor substituents. However, the rates for the reactions with nitronate anions are increased by both types of substituents, and especially by the radical-stablizing nitro substituent (see Table 1).

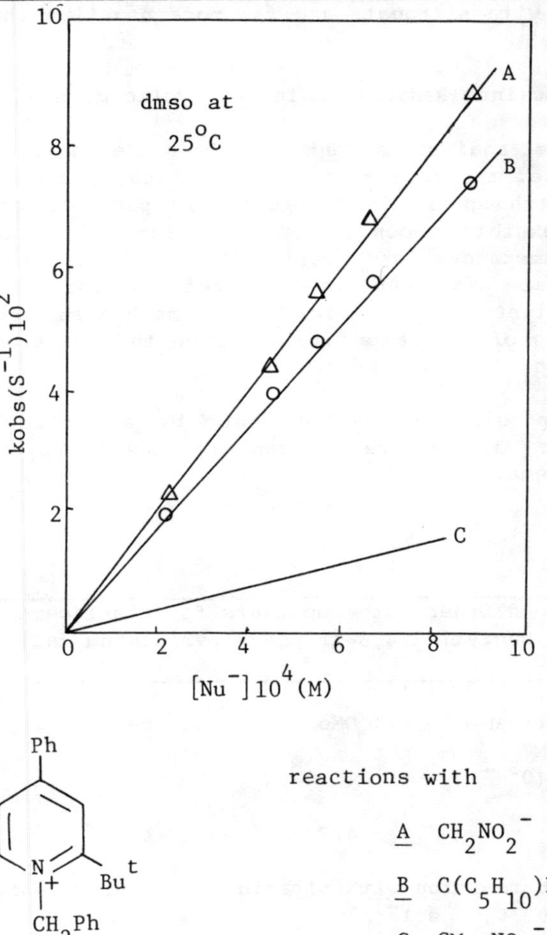

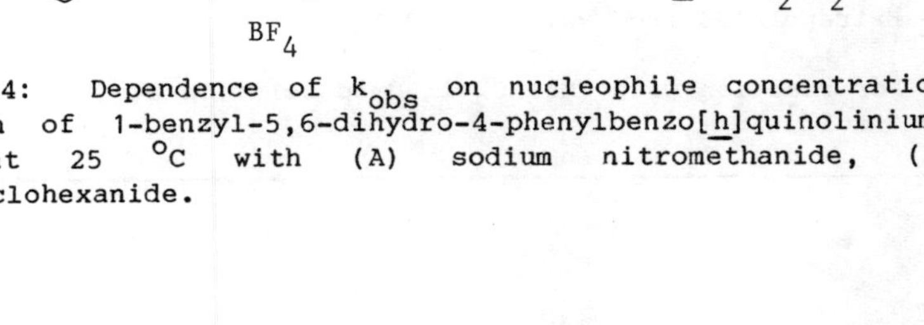

Figure 4: Dependence of k_{obs} on nucleophile concentration for the reaction of 1-benzyl-5,6-dihydro-4-phenylbenzo[h]quinolinium (9a) in DMSO at 25 °C with (A) sodium nitromethanide, (B) sodium nitrocyclohexanide.

Moreover, absolute rate comparisons[11] demonstrate that there is a mechanistic changeover. Rates extrapolated to the same temperature (25 °C) and in the same solvent (dmso) indicate that, given identical mechanisms, the nitronate anion is behaving as a nucleophile 10^4 times more powerful than piperidine for the reaction with 1-benzyl-2,4,6-triphenylpyridinium cation. This is a most unlikely situation.

Furthermore, activation entropy values do not fit in with those expected for a classical S_N2 reaction type: activation entropies for the reactions with nitronate are far more negative than those with, eg, piperidine.[10,11]

4. Evidence Against Radical Chain and Cyclic Mechanisms

Major evidence against a radical chain mechanism has already been discussed: i.e. the fact that kinetics clearly show clean second-order reactions. Although it is theoretically possible for a radical chain reaction to exhibit second order kinetics, this holds only in very special circumstances and certainly not over the wide range of conditions that has been investigated in this instance. Further evidence against the chain type mechanism derives from the ineffectiveness of initiators to speed up the reaction or of inhibitors to slow it down.[11]

Another possible mechanism would be a cyclic transformation as shown in Scheme 3. However, cogent reasons to reject this are listed within the scheme.

Table 1. Second Order Rate Constants for Displacement Reactions of 1-(p-substituted benzyl)-2,4,6-triphenylpyridinium cations

Substituent	OMe	Me	H	NO_2
$k^2 \times 10^3$ [a]	18.0[b]	8.52	4.95	2.96
k^2 [c]	4.2	4.1	3.30	12.4

[a] For reaction with piperidine in chlorobenzene solution at 100 °C, $1\ mol^{-1}s^{-1}$.

[b] Extrapolated from value at 70 °C.

[c] For reaction with CMe_2NO_2 in dmso solution at 25 °C, $1\ mol^{-1}s^{-1}$

Scheme 3: EVIDENCE AGAINST A CYCLIC MECHANISM

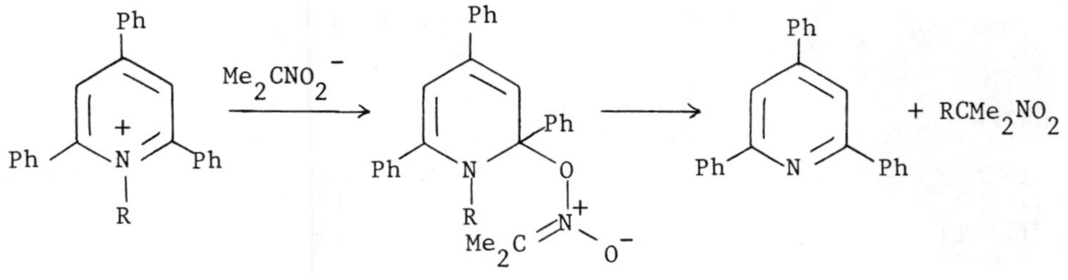

(a) Pyridinium cations normally add nitronate at the 4-position to form a C-C bond.

(b) Unfavorable on orbital grounds

(c) Complexes formed do not show expected nmr.

(d) Bulky groups at alpha-position do not inhibit reaction (cf. the effect of a t-butyl).

---

5. Complex Formation Between Charged Nucleophiles and Pyridinium Derivatives

It is well known that N-substituted pyridinium cations form sigma complexes with nucleophiles by addition at either the 2- or the 4-position. Such additions are facilitated if the ring position to which addition is taking place is initially unsubstituted.[12]

We have investigated complex formation of a variety of anionic nucleophiles with both the 1,2,4-triphenylpyridinium cation and the 1,2,6-triphenylpyridinium cation.[13] We showed that, as expected, complex formation took place in each case by the formation of a sigma-bond with the unsubstituted carbon at either the 6- or the 4-position, respectively (Scheme 4).

With the n.m.r. and electronic spectral characterization of the above mentioned series of sigma-complexes to use as models, we studied the interaction of the 1,2,4,6-tetraphenylpyridinium cation with the same series of nucleophiles (Scheme 4). We thus demonstrated that

Scheme 4: REACTIONS OF N-PHENYLPYRIDINIUM WITH NUCLEOPHILES

(i) With a free alpha or gamma position

(ii) No free alpha or gamma position

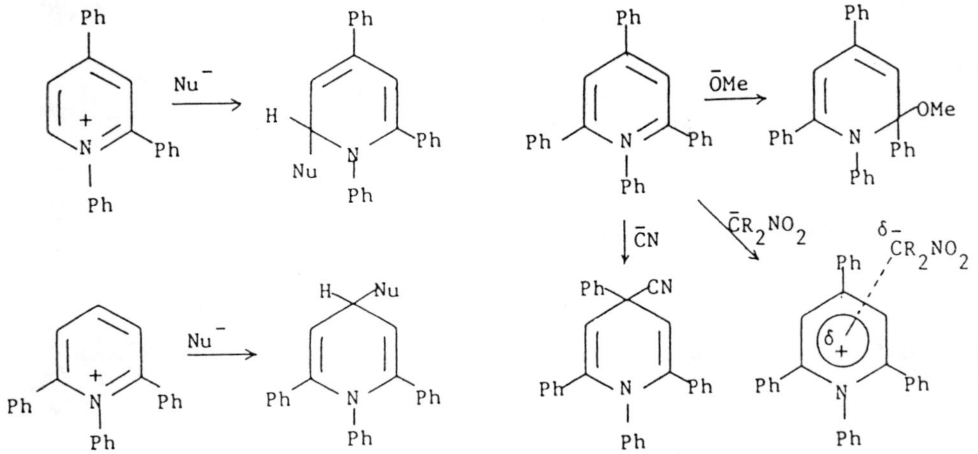

Nu = MeO, PhS, CN, CH_2NO_2, CMe_2NO_2, cyclo-$C_6H_{10}NO_2$

---

methoxide formed a sigma-complex at the 2-position and the cyanide anion, a sigma-complex at the 4-position (Scheme 5). This change in orientation is in accordance with the hard nature of the methoxide and the much softer character of the cyanide anion. However, of much greater interest was the situation with the nitronate anions derived from nitromethane, 2-nitropropane, and nitrocyclohexane. The spectra of the products formed in these cases (Scheme 5) were not in agreement with the formation of sigma-complexes but instead indicated association by charge-transfer interaction. The UV spectral maxima of the complex formed with sodium 2-nitropropanide in different solvents, when plotted against the E_T value gives a linear relationship (Figure 5) similar to those that have been found for such well recognized charge transfer complexes as 1-ethyl-4-cyanopyridinium iodide.

We believe that such charge-transfer complexes are intermediates in the reactions we are now discussing. Indeed, in the case of the N-alkylpyridiniums (as distinct from the N-benzylpyridiniums) we have been able to get evidence for their formation. We believe that the mechanism of the nucleophilic substitution reactions between N-substituted pyridinium cations and nitronate anions that we have been describing are as described in Scheme 6.

Scheme 5: STRUCTURE OF METHOXIDE AND CYANIDE ADDUCTS

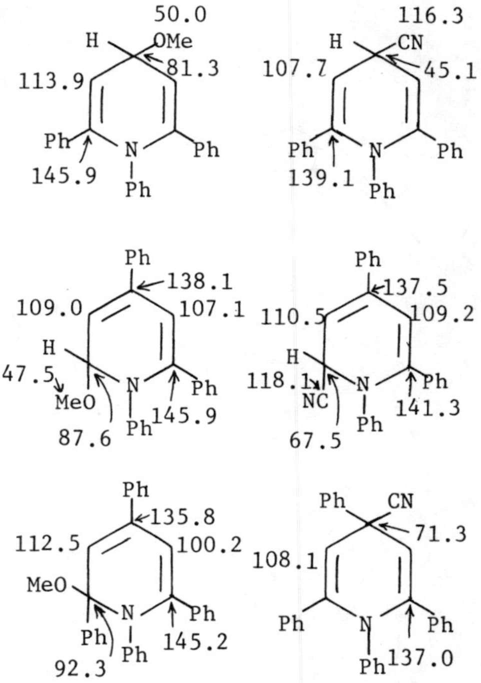

CHARGE TRANSFER NATURE OF NITRONATE ANION-PYRIDINIUM CATION INTERACTION

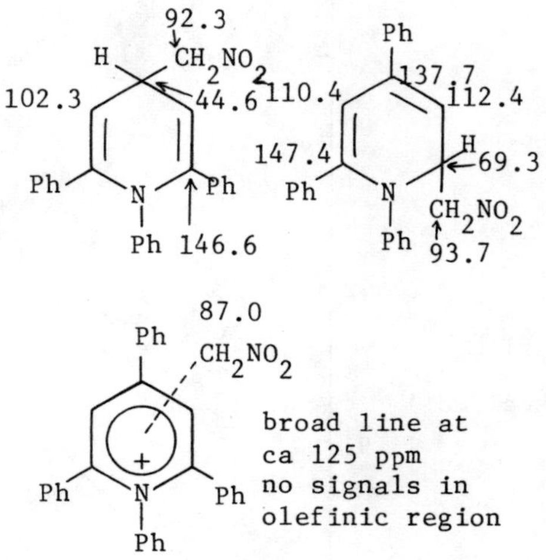

broad line at
ca 125 ppm
no signals in
olefinic region

Scheme 6: PROPOSED RADICALOID MECHANISM

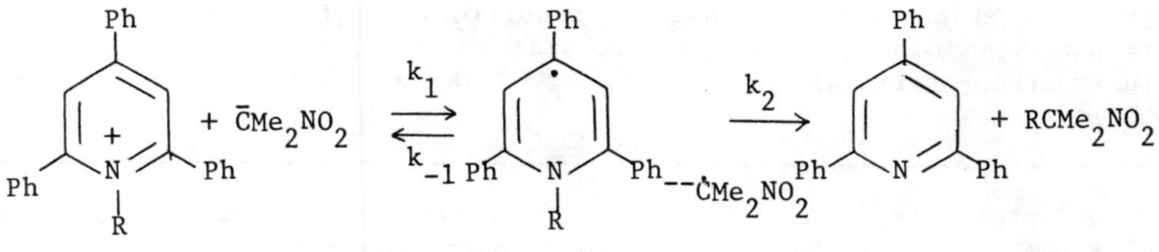

Radical pair or
CT complex

For R = CH_2Ph, k_2 is fast and k_1 is rate determining; For R = n-Bu, $k_{-1} >> k_2$

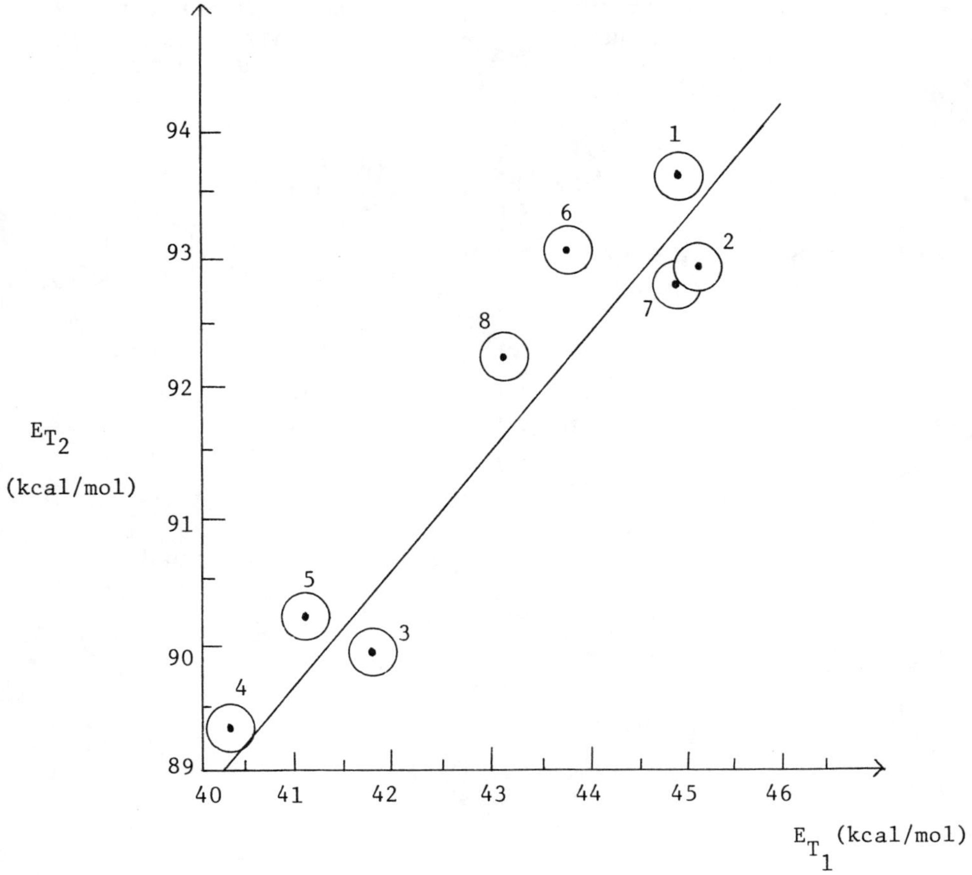

Figure 5: Plot of Charge-Transfer Transition Energies (E_T) of 1,2,4,6-Tetraphenylpyridinium Tetrafluoroborate (E_{T_2}) vs. 1-Ethyl-4-methoxycarbonylpyridinium Iodide (E_{T_2}) as a function of solvent polarity.

Unfortunately, extensive ESR work has given only broad unresolved single peak signals. This is applied to both the complexes derived from the N-phenyl compounds and those from the N-alkyl compounds. Moreover, extensive attempts to obtain CNDIP signals from the nitro compounds produced have failed. However, reaction rates do correlate with radical decay half lives (Scheme 7).[14]

Scheme 7: CORRELATION OF RATES WITH RADICAL DECAY HALF-LIVES

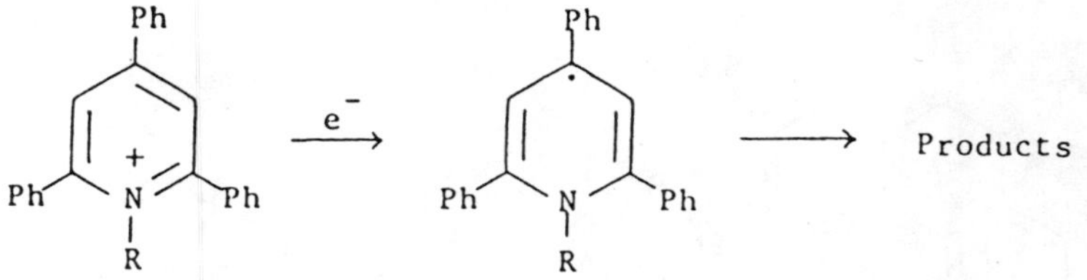

Cyclic voltammetry shows that the half lives of the radical intermediates 14 at 25 ^{o} depends on the N-substituent as follows:

R	CH_2Ph	nBu	iPr	Ph
$t(\frac{1}{2})$ (sec)	<0.01	30	5	>1000
Rate fast	slow	fast	---	

6. Other Examples of Radical Nonchain Mechanisms

Similar experiments with N-aryl substituted pyridinium cations have been carried out and it has been shown that reactions of this type do indeed lead to the transfer of the N-aryl group to the carbon atom of the nitronate anion (Scheme 8).[15] However, the yields of isolated products are at best 30% and usually much lower. Thus the reaction is not useful preparatively, but the results are significant from an unmechanistic point of view and give further evidence for the radical mechanism suggested.

Other C-nucleophiles probably react by a similar mechanism. We have carried out preparative work with highly hindered malonate anions and have shown that these can be C-alkylated under surprisingly mild conditions (Scheme 9).[16] This points to an electron transfer mechanism: further studies are proceeding in this area.

Scheme 8: PREPARATIVE TRANSFER OF N-PHENYL GROUPS TO NITRONATE ANIONS

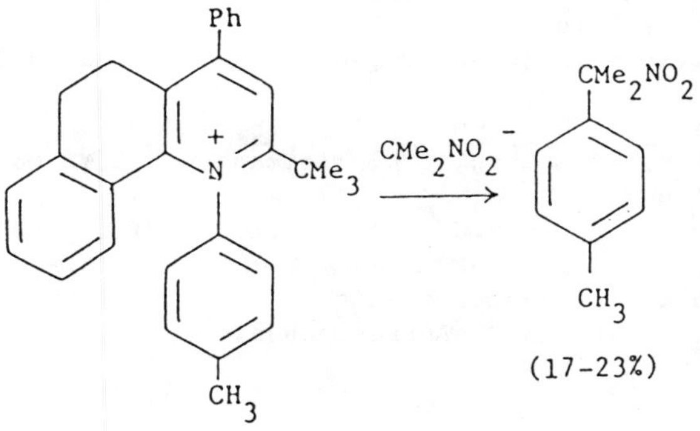

Scheme 9: REACTIONS OF N-ALKYLPYRIDINIUM CATIONS WITH MALONATE ANIONS

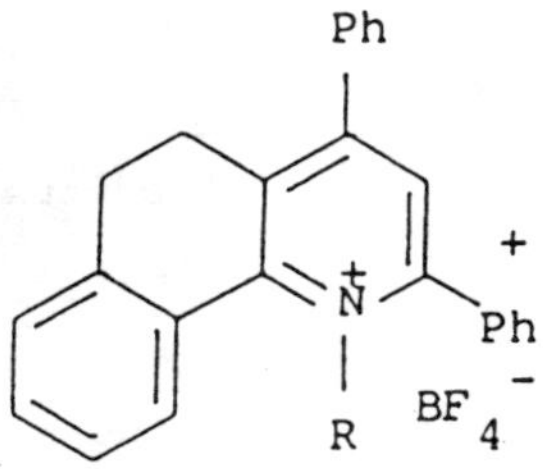

$+ R'\bar{C}(CO_2Et)_2 \xrightarrow{25\,^{o}C} RR'C(CO_2Et)_2$

	R	R'	Yield %
a	$i\text{-}C_3H_7$	Et	44
b	$i\text{-}C_3H_7$	$s\text{-}C_4H_9$	37
c	$2\text{-}C_5H_{11}$	Et	37

Acknowledgements: I wish to thank all my coworkers, whose names are mentioned in the references, for their dedicated and careful work.

References

1. J. March "Advanced Organic Chemistry", J. Wiley, New York, 3rd Edition 1985, p255.
2. Reviews: (a) A.R. Katritzky and G. Musumarra, Chem. Soc. Rev. 1984, 13, 47. (b) A.R. Katritzky, G. Musumarra and K. Sakizadeh, Heterocycles 1985, 23, 1765.
3. A.R. Katritzky, Z. Dega-Szafran, M.L. Lopez-Rodriquez, and R.W. King, J. Am. Chem. Soc. 1984, 106, 557.
4. A.R. Katritzky, K. Sakizadeh, B. Gabrielsen, and W.J. le Noble, J. Am. Chem. Soc. 1984, 106, 1879.
5. D. Seebach, F.W. Colvin, F. Lehr and T. Wellr, Chimia 1979, 33, 1.
6. N. Kornblum, Angew Chem. Internat. Ed. 1975, 14, 734.
7. G.A. Russell, J. Herschberger, and K. Owens J. Am. Chem. Soc. 1979, 101, 1312.
8. Review: A.R. Katritzky and C.M. Marson, Angew. Chem. Internat. Ed., 1984, 23, 420.
9. (a) A.R. Katritzky, G.Z. de Ville and R.C. Patel, Tetrahedron 1981, 37, Supplement 1, 25. (b) A.R. Katritzky, M.A. Kashmiri and D.K. Wittman, Tetrahedron 1984, 40, 1501.
10. A.R. Katritzky, J-L Chen, C.M. Marson, A. Maia and M.A. Kashmiri, Tetrahedron 1985 in press.
11. A.R. Katritzky, M.A. Kashmiri, G.Z. de Ville and R.C. Patel, J. Am. Chem. Soc. 1983, 105, 90.
12. J.W. Bunting Adv. Het. Chem. 1979, 25, 1.
13. A.R. Katritzky, J-L Chen, D.K. Wittman, and C.M. Marson, manuscript in preparation.
14. Unpublished work with R.C. Patel, G.Z. de Ville, and E. Lunt.
15. A.R. Katritzky, D.K. Wittmann, J-L Chen, C.M. Marson, and A. Ossana, manuscript in preparation.
16. Unpublished work with J. Aurrecoechea.

CAPTODATIVE SUBSTITUENT EFFECTS IN CYCLOADDITION REACTIONS

Lucien Stella
Laboratoire de Chimie Organique B
Université d'Aix-Marseille III
F - 13397 Marseille cedex 13
France

ABSTRACT. The captodative olefins, well known for their radicophilic properties, are also good partners for cycloaddition reactions. The focus of that report is to illustrate the remarkable dienophilic reactivity and the synthetic potential of this class of olefins in Diels-Alder reactions.

The captodative effect, the combined action of an electron-withdrawing *(captor)* and an electron-releasing *(donor)* substituent on a radical center, leads to an enhanced stabilisation (1, 2). This effect has two main mechanistic consequences : by weakening of either σ or π bonds it provides molecules activated respectively toward either homolysis *(proradical* property) or radical additions *(radicophilic* property). The thermodynamic stabilization of the captodative radical 2 obtained by homolytic addition to the radicophilic olefins 1 (figure 1) allows selective coupling reactions leading to the formation of double-adducts

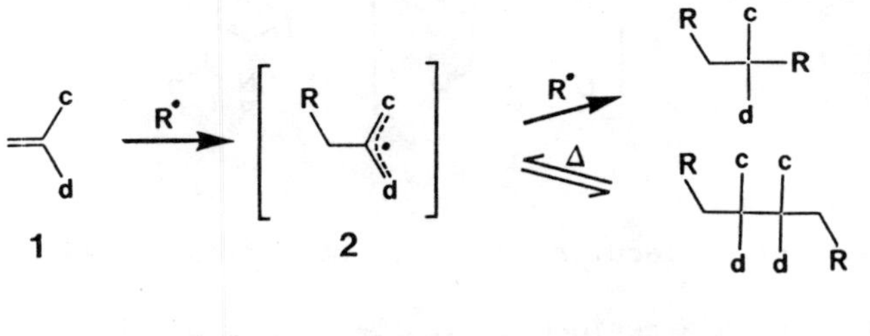

1 2 3 4

Figure 1. Radical addition reactions to captodative olefins.

3 and/or adduct-dimers 4 (3). Furthermore, captodative olefins show in addition reactions higher reactivity than expected from polar factors (4). The easy and efficient twofold CC bond formation process, thermodynamically and even kinetically promoted by the captodative effect, would also be expected in cycloaddition reactions which develop

H. G. Viehe et al. (eds.), Substituent Effects in Radical Chemistry, 361–370.
© *1986 by D. Reidel Publishing Co.*

a biradical intermediate or a weakly perturbed biradical transition state (5). In agreement with this, some of the captodative olefins (see figure 2 for an example) dimerise already at room temperature (6) to head-to-head cyclobutane derivatives (such as 7) via 1,4 diradical type tetramethylene intermediate (such as 6). The comparison with the very

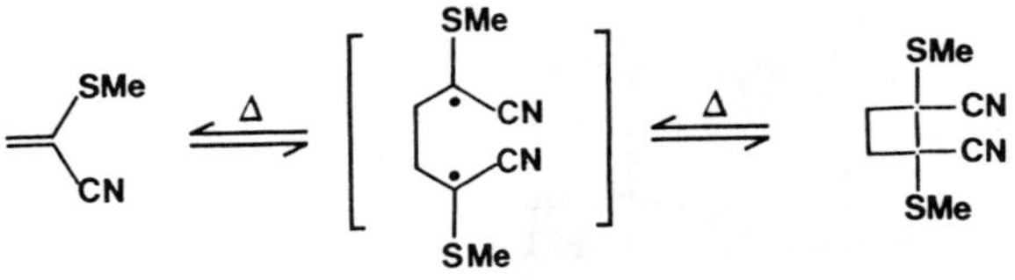

Figure 2. Diradical (2+2) dimerisation reaction of a captodative olefin

sluggish cyclodimerisation of acrylonitrile (7) leads to assume that the captodative stabilization of the 1,4-diradical 6 must be the driving force of the easy two-step process.
Furthermore, captodative olefins are excellent partners in (2+2) thermal cycloadditions to ethylene derivatives having pronounced tendency for this reaction : allenes, methylene cyclopropanes and gem-difluoroethylene derivatives (8). Competition experiments reveal that 2- *tert* -butylthio-acrylonitrile 16 reacts faster than acrylonitrile with chlorotrifluoroethylene and that the cycloaddition rates of the captodative olefins increase with the radical stabilizing power of the couple of substituents, in the following order :
CN/OR < COOR/SeR << COOR/SR < CN/SR ~ CN/SeR ~ CN/NRR'.

The thermal intramolecular (2+2) cycloaddition of the captodative olefin 8 has been reported (9) to occur with exceptional ease (Figure 3).

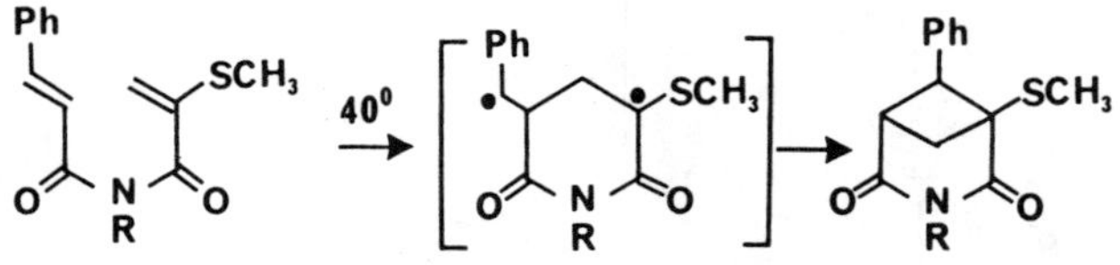

Figure 3. An easy intramolecular (2+2) cycloaddition reaction.

Aldonitrones 11 undergo a regiospecific (3+2) cycloaddition to the 2-morpholino-acrylonitrile 12 (10) whereby the stabilized biradical 13 may be the intermediate (figure 4). Similarly, the 2- *tert* -butylthio-acrylonitrile 16 undergoes a smooth 1,3-dipolar cycloaddition to the C,N-diphenyl-nitrone 17 with formation of 5,5-disubstituted isoxazolidines as the major products (11) (figure 4). On the other hand, the good reactivity of electron-poor *(electrophilic)* olefins in Diels-Alder reaction is a familiar feature. Since it is well established that the two major limitations imposed to the generality of the Diels-Alder reaction are the inherent inefficiency of electron-rich *(nucleophilic)* olefins and its pronounced sensitivity to steric effects (12), than the question about the dienophilicity of the captodative *(radicophilic)* olefins is of great significance : if both

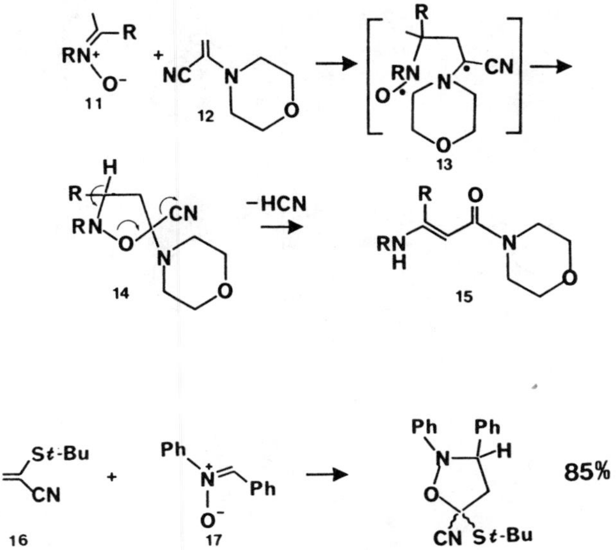

Figure 4. (3+2) cycloaddition reactions of nitrones to captodative olefins.

electronic and steric effects of the donor group are dominant in the Diels-Alder reactions involving the captodative olefins, one might then reasonably expect that the formation of cycloadducts would be disfavored, whereas, if the captodative effect come into play in stabilizing a transition state close to a biradical (5), the dienophilicity of the captodative olefins would resemble more closely that of electrophilic olefins than that of nucleophilic ones.
The Diels-Alder reactivity of 2-morpholino-acrylonitrile 12 has been evaluated toward four conjugated dienes (figure 5). In every cases, isomeric (4+2) cycloadducts are obtained in good yields. The adducts are easily transformed into the corresponding ketones by a mild hydrolysis of the α-amino-nitrile group. It is thus seen that the 1-cyano-enamines can be considered as good ketene equivalents useful for the synthesis of unsaturated cyclohexanones (13).
The dienophilic reactivity of a series of 2-alkylthio-acrylonitriles have been also investigated. With cyclopentadiene the spontaneous cycloadditions occur at room temperature affording adducts in excellent yields (table I). The costumary *endo* -cyano addition rule is found to be unimportant in determining the *endo/exo* ratio of the adducts but increase of size of the thio-alkyl group reduces the *endo* selectivity of that group. This result, related to the variation of the donor

β-steric effect is just the opposite of what is observed when the α-steric effect or the captor β-steric effect are varied (14) (tableI). With slightly activated or really nucleophilic dienes, thermal cycloadditions occur very efficiently (figure 6). These reactions

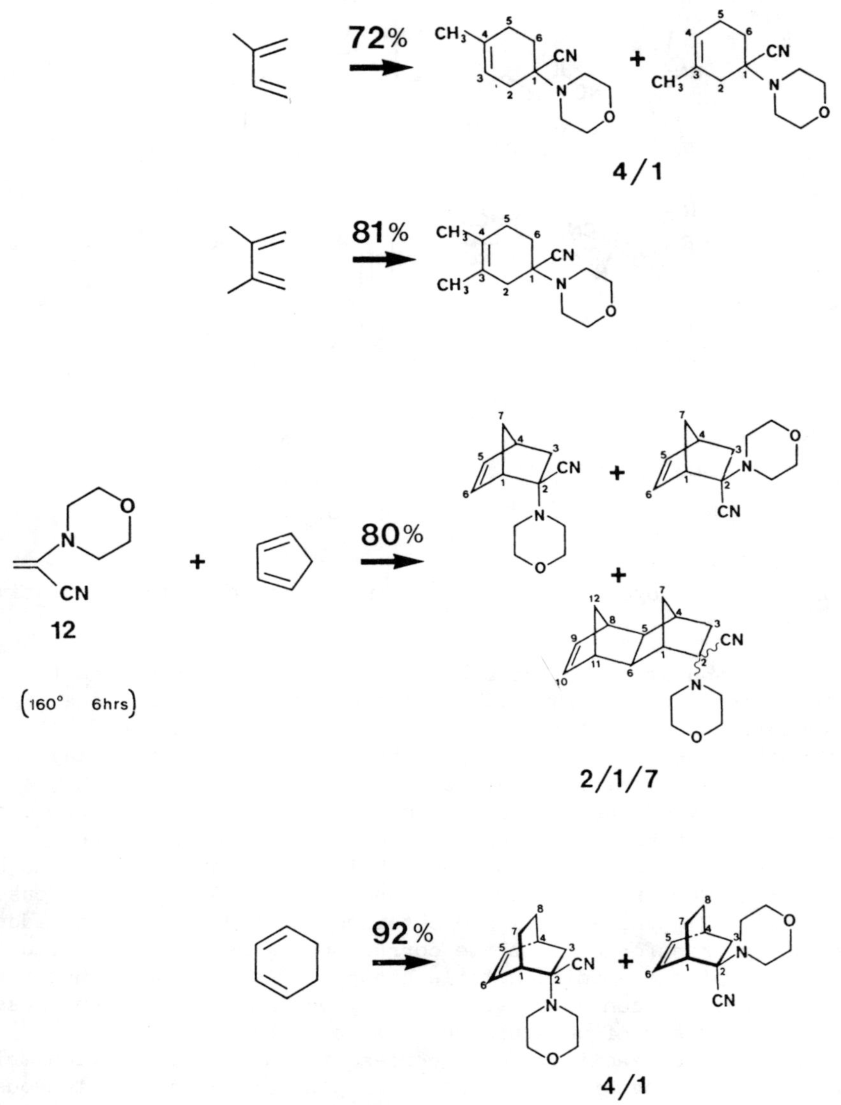

Figure 5. (4+2) cycloaddition reactions to 2-morpholino-acrylonitrile.

TABLE I. α and β steric effects for Diels-Alder reactions of cyclopentadiene with olefins $H_2C{=}CRR'$

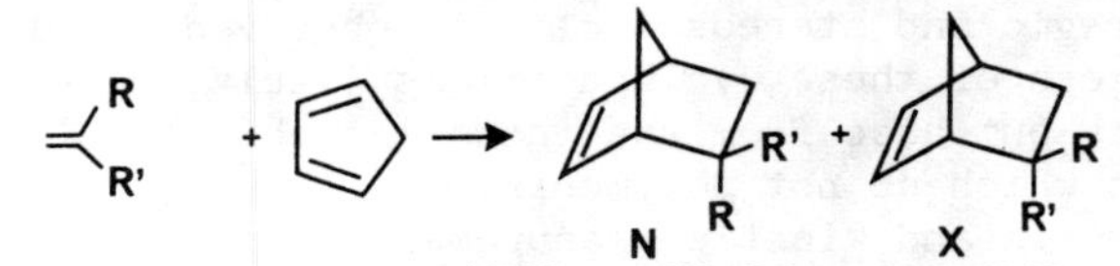

R	R'	t(°C)	Yield	N/X	ref.
H	CN	138		0.8	22
Me	CN	138	86 %	4.9	"
Et	CN	138	52 %	6.1	"
iPr	CN	138	30 %	10.1	"
tBu	CN	138	5 %	49	"
MeS	CN	20	85 %	19	14
$PhCH_2S$	CN	20	96 %	3.5	"
iPrS	CN	20	85 %	3	"
tBuS	CN	20	95 %	1.5	"
Me	CHO	80	90%	3.9	22
Me	COMe	80	73 %	1.4	"
Me	COEt	80	62 %	1.3	"
Me	COiPr	80	50 %	1.1	"

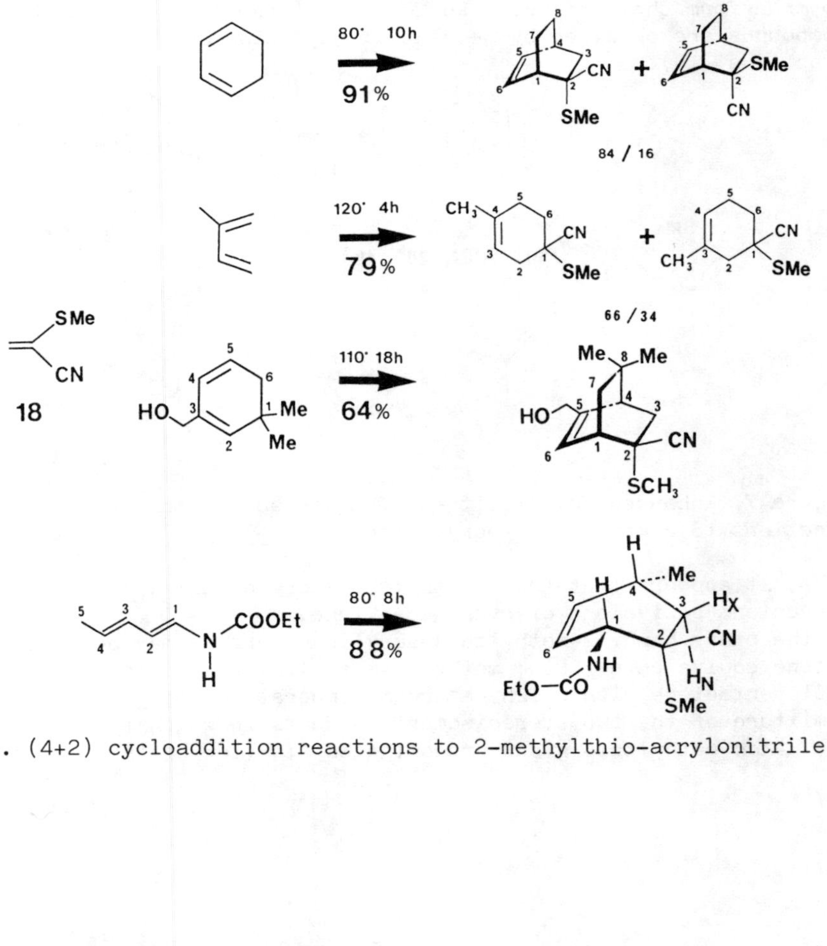

Figure 6. (4+2) cycloaddition reactions to 2-methylthio-acrylonitrile.

provide convenient access to diversely substituted, thioalkyl- and cyano-functionalized cyclo- and bicyclo-hexenes (15). Of particular note are the high regio and stereoselectivity observed in all these reactions. The success of these cycloadditions likely derives from their dienophilicity but also from the thermal stability of the captodative olefins which do not polymerise.

Competition experiments and kinetic measurements show that cycloaddition of 2-methylthio-acrylonitrile 18 to cyclohexadiene is seven times faster than that of acrylonitrile (16). This indicated that the captodative substitution exerts an accelerating effect on the Diels-Alder reaction despite the larger steric bulk of the methylthio group compared to hydrogen. As for the methylthio-ethylene, both the σ and π electron transfer occur in the same direction (17), *i.e.* , from the heteroatom to the adjacent ethylene fragment, it appears that an interpretation of this result based upon the simple additivity of classical polar effects does not hold. The enhanced reactivity of 2-methylthio-acrylonitrile 18 compared with acrylonitrile may be well interpreted by employing a model of a concerted reaction with an unsymetrical transition state having biradical character (5).

Both the regio and stereoselectivities of the cycloaddition to 2-methylthio-acrylonitrile 18 can be enhanced by Lewis acid catalysis as illustrated in figure 7. A further attractive feature of 2-alkylthio-acrylonitriles is the variety of methods which can be used to transform their adducts. For example (figure 8), cyclic dienic compounds are obtained by sequential oxydation and elimination reactions (15).

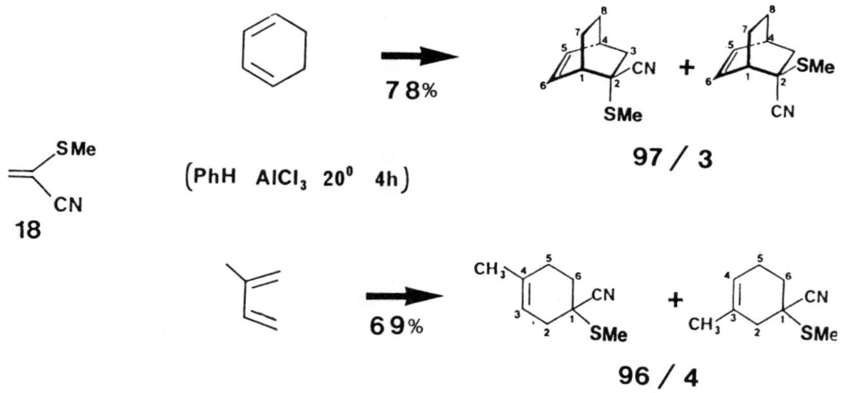

Figure 7. Enhanced selectivities of Lewis acid catalyzed cycloaddition reactions to 2-methylthio-acrylonitrile.

Thus, it appears that the 2-methylthio-acrylonitrile 18 can serve as a convenient equivalent of cyano-acetylene.

On the other hand, 2-methylthio-acrylic esters 19 may be considered as ketene equivalents (18) : methyl 2-methylthio-acrylate 20 reacts with cyclopentadiene within 12h, at room temperature, to give in 94% yield, a mixture of the two stereoisomeric adducts in which the isomer with the *endo* methylthio group predominates. In view of this surprising stereochemistry, the reaction of other alkyl 2-alkylthio-acrylates was examinated (table II). The steric effect of the ester function's alkyl

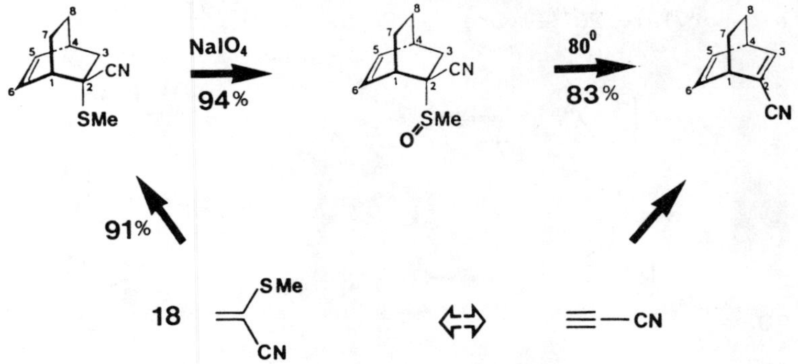

Figure 8. The 2-methylthio-acrylonitrile is an efficient cyano-acetylene equivalent for the Diels-Alder reaction with 1,3-cyclohexadiene.

group is found to be of minor importance in determining the *endo/exo* ratio of the adducts, but the increase of the size of the alkylthio group reduces the *endo* selectivity of that group, much more than for the 2-alkylthio-acrylonitriles (see table I). Such a violation of the Alder rule is not unprecedented for 2-substituted acrylic dienophiles (19) and has been interpreted in terms of either stabilizing (20) or destabilizing (21) Van der Waals interaction.

TABLE II. Steric effects of alkylthio and ester functions on the stereoselectivity of the Diels-Alder reaction.

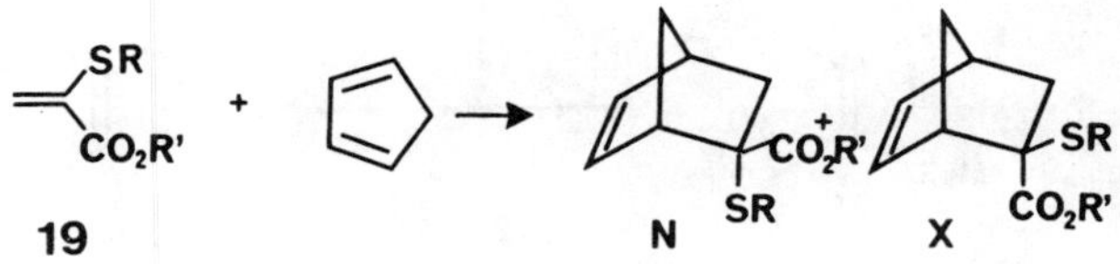

R	R'	Conditions	Yield	N/X Ratio
Me	Me	20°, 12h	94 %	69/31
tBu	Me	" "	90 %	22/78
Me	tBu	" 48h	82 %	70/30
tBu	tBu	" "	88 %	25/75
Me	Me	$AlCl_3$, 20°, 3h	68 %	3/97
tBu	Me	"	58 %	3/97
Me	tBu	"	54 %	55/45
tBu	tBu	"	38 %	20/80

Particularly noteworthy is the increase of the methylester *endo* orientation observed in the Lewis acid catalyzed Diels-Alder reactions of the alkyl 2-alkylthio-acrylates series (table II). Thus, the use of aluminium trichloride catalysis leads to very regio- and stereoselective cycloaddition reactions of the methyl 2-methylthio-acrylate 20 (figure 9).

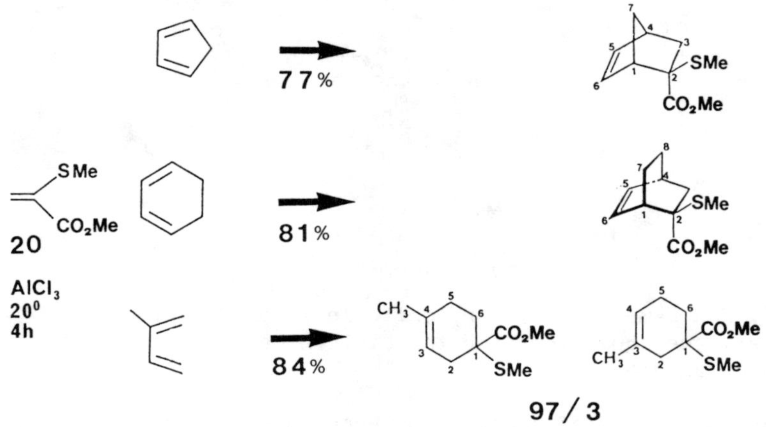

Figure 9. Aluminium trichloride catalyzed cycloadditions of methyl 2-thiomethyl-acrylate.

A significant feature contributing to the synthetic potential of the captodative olefins is the good reactivity of trisubstituted dienophiles such as the methyl 2-methylthio-crotonate 21 : compared to methyl tiglate 22, this olefin, which profits by an increazed functionality, still retains a remarkable dienophilicity (figure 10).

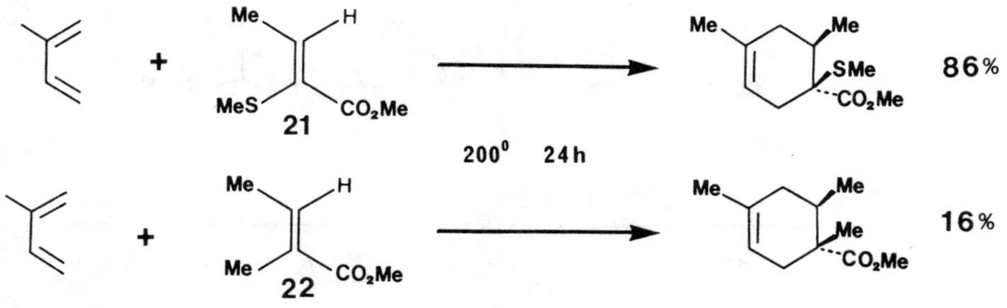

Figure 10. Compared thermal Diels-Alder reactivities of methyl 2-methylthio-crotonate and methyl tiglate.

Replacing the ester function of 21 by the less hindered and the better electron-withdrawing carboxaldehyde group, leads to 2-methylthio-crotonaldehyde 23 which undergoes a very smooth stereoselective cycloaddition reaction to cyclopentadiene (figure 11).

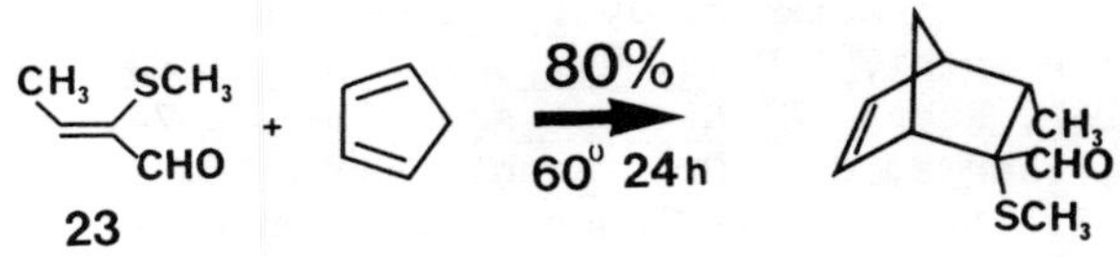

Figure 11. Thermal Diels-Alder reaction of 2-methylthio-crotonaldehyde

Finally, under analogous conditions, the captodative diene, methyl 2-methylthio 2,4-pentadienoate <u>24</u>, reacts as a dienophile, at the less substituted double bond, in the highly site selective cycloaddition with cyclopentadiene (figure 12).

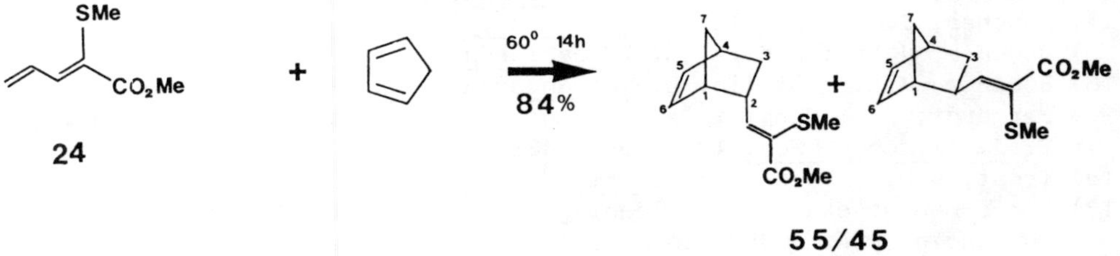

Figure 12. Thermal Diels-Alder reaction of methyl 2-methylthio 2,4-pentadienoate.

This reaction reveals new possibilities for organic synthesis and we anticipate that the future will see additional examples of the synthetic utility of these versatile olefins and dienes classes. Further development of the intramolecular Diels-Alder cycloadditions involving captodative olefins is now in progress in our laboratory.

Acknowledgements. The author wishes to thank Dr. J.L. Boucher for his fundamental contribution and the CNRS for financial support.

REFERENCES

(1) Reviews : (a) Viehe, H.G. ; Merényi, R. ; Stella, L. ; Janousek, Z. Angew. Chem. Int. Ed. Engl. 1979, **18**, 917 ; (b) Viehe, H.G. ; Janousek, Z. ; Merényi, R. ; Stella, L. Acc. Chem. Res. 1985, **18**, 148.

(2) A special effect on polysubstituted radical stabilization has been first suggested by theoretical argument by (a) Dewar, M.J.S. J. Am. Chem. Soc. 1952, **74**, 3353 and then observed experimentally and termed "push-pull stabilization" by (b) Balaban, A.T. Rev. Roum. Chim. 1971, **16**, 725 and "merostabilization" by (c) Baldock, R.W. ; Hudson, P. ; Katritzky, A.R. ; Soti, F. J. Chem. Soc., Perkin Trans.1 1974, 1422.

(3) (a) Stella, L. ; Janousek, Z. ; Merényi, R. ; Viehe, H.G. Angew. Chem. Int. Ed. Engl. 1978, **17**, 691 ; (b) Mignani, S. ; Merényi, R. ; Janousek, Z. ; Viehe, H.G. Tetrahedron, 1985, **41**, 769.

(4) (a) Lahousse, F. ; Merényi, R. ; Desmurs, J.R. ; Allaime, H. ; Borghese, A. ; Viehe, H.G. Tetrahedron Lett. 1984, **25**, 3823 ; (b)

see also : Giese, B. ; Meixner, J. Chem. Ber. 1981, **114**, 2138.
(5) Dewar, M.J.S. ; Pierini, A.B. J. Am. Chem. Soc. 1984, **106**, 203.
(6) Gundermann, K.D. ; Röhrl, E. Liebigs Ann. Chem. 1974, 1661.
(7) Roberts, J.D. ; Sharts, C.M. Org. React. 1962, **12**, 1.
(8) De Cock, Ch. ; Piettre, S. ; Lahousse, F. ; Janousek, Z. ; Merényi, R. ; Viehe, H.G. Tetrahedron, 1985, **41**, 4183.
(9) Alder, A. ; Belluš, D. J. Am. Chem. Soc. 1983, **105**, 6712.
(10) Döpp, D. ; Walter, J. Heterocycles, 1983, **20**, 1055.
(11) Döpp, D. ; Henseleit, M. Chem. Ber. 1982, **115**, 798.
(12) (a) Wollweber, H. Diels-Alder Reactions ; Georg Thieme Verlag : Stuttgart, 1972 ; (b) Sauer, J. Angew. Chem., Int. Ed. Engl. 1966, **5**, 211 ; 1967, **6**, 16.
(13) Boucher, J.L. ; Stella, L. Tetrahedron, 1985, **41**, 875.
(14) Boucher, J.L. ; Stella, L. Bull. Soc. Chim. Fr. in press.
(15) Boucher, J.L. ; Stella, L. Tetrahedron, in press.
(16) Boucher, J.L. ; Stella, L. Tetrahedron Lett. 1985, **26**, 5041.
(17) Bernardi, F. ; Mangini, A. ; Epiotis, N.D. ; Larson, J.R. ; Shaik, S. J. Am. Chem. Soc. 1977, **99**, 7465.
(18) Trost, B.M. ; Tamaru, Y. J. Am. Chem. Soc. 1977, **99**, 3101.
(19) For a recent example and leading references, see : Creary, X. ; Inocencio, P.A. ; Underiner, T.L. ; Kostromin, R. J. Org. Chem., 1985, **50**, 1932.
(20) (a) Kokube, Y. ; Fueno, T. ; Furukawa, J. J. Am. Chem. Soc. 1970, **92**, 6548 ; (b) Kokube, Y. ; Sugimoto, T. ; Furukawa, J. ; Fueno, T. Ibid, 1972, **94**, 3633.
(21) Houk, K.N. ; Luskus, L.J. J. Am. Chem. Soc. 1971, **93**, 4606.
(22) Mellor, J.M. ; Webb, C.F. J. Chem. Soc., Perkin II, 1974, 17.

CAPTODATIVE SUBSTITUTION EFFECT IN [2+2] CYCLOADDITIONS BETWEEN ALLENES AND OLEFINS

G. Coppe-Motte, A. Borghese, Z. Janousek, R. Merényi, H.G. Viehe.
Université de Louvain, Laboratoire de Chimie Organique,
Place L. Pasteur, 1, B 1348 Louvain-la-Neuve, Belgium.

Olefins with captodative substitution are valuable precursors of variously substituted methylenecyclobutanes via [2+2] cycloaddition reactions with different allenes (1). These reactions are regioselective and lead only to 3-methylenecyclobutanes. When unsymmetrically substituted allenes are used, each double bond can participate in cycloaddition. The arising mixture of isomers A and B can be generally separated by column chromatography. It was established that the isomer B can be converted into A upon heating.

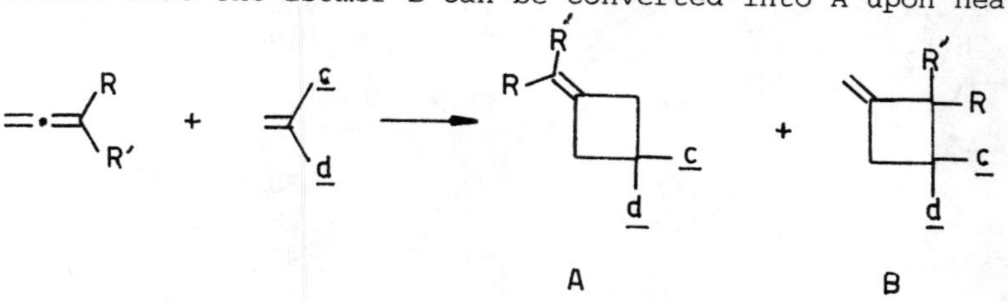

These results are summarized in the following tables.

Table I. Cycloadditions between **cd** olefins and allene or dimethylallene (2,3)

R,R' = -H			R,R' = $-CH_3$				
c	d	Yield %	c	d	Global Yield %	A %	B %
-CN	$-S-t-C_4H_9$	86	-CN	$-S-t-C_4H_9$	76	56	44
-CN	-Cl	66	-CN	$-OCH_3$	42	64	36
$-CO_2CH_3$	$-S-t-C_4H_9$	42	-CN	-Cl	25	25	75

H. G. Viehe et al. (eds.), Substituent Effects in Radical Chemistry, 371–374.
© 1986 by D. Reidel Publishing Co.

Table II. Cyanoallene and methoxyallene cycloadditions(4).

R,R' = -H, -CN			R,R' = -H, $-OCH_3$				
c	d	Yield % A	c	d	Global Yield %	A %	B%
-CN	$-S-t-C_4H_9$	78	-CN	$-S-t-C_4H_9$	67	36	64
-CN	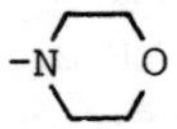	79	-CN	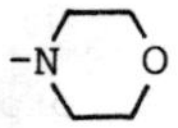	73	-	-
-CN	-Cl	34	-CN	-Cl	95	35	65

Table III. Tert.-butylthioallene cycloadditions

R,R' = -H, -S-t-butyl

c	d	Global Yield %	A %	B %
-CN	$-S-t-C_4H_9$	95	70	30
-CN	$-N(CH_3)_2$	49	82	18
-CN	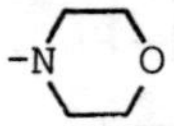	75	88	12
-CN	$-OCH_3$	73	70	30
-CN	-Cl	48	72	28
$-CO_2CH_3$	$-S-t-C_4H_9$	60	57	43
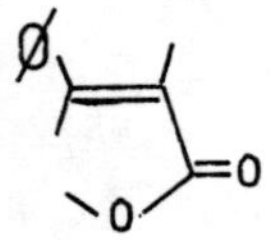		47	88	12
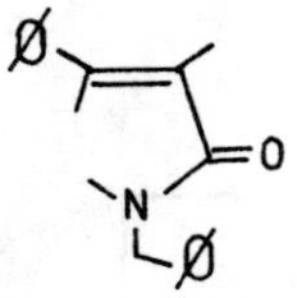		100	100	0

The exclusive formation of 3-methylenecyclobutanes in these [2+2] cycloaddition reactions can be rationalized in terms of a two-step mechanism. **cd**-Substitution apparently favours diradical character of the intermediate even though polar contributions cannot be excluded.

Scheme I. Two-step cycloaddition between allenes and **cd**-olefins.

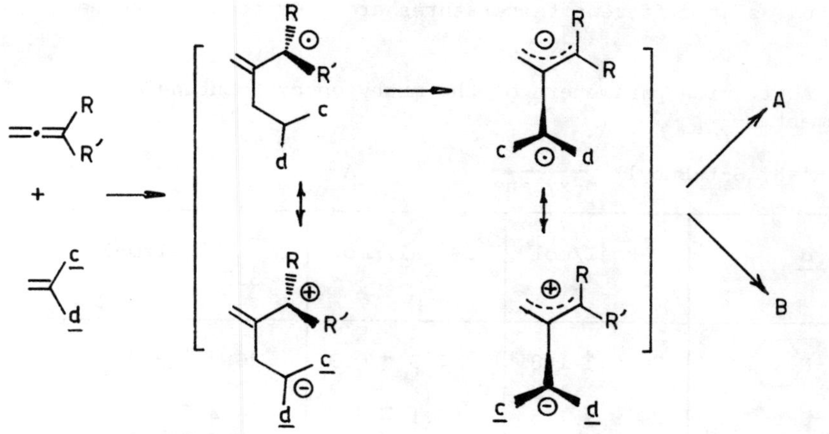

On the other hand, cycloaddition reactions of **cd** allenes with **cd** olefins occur with particular ease at room temperature.

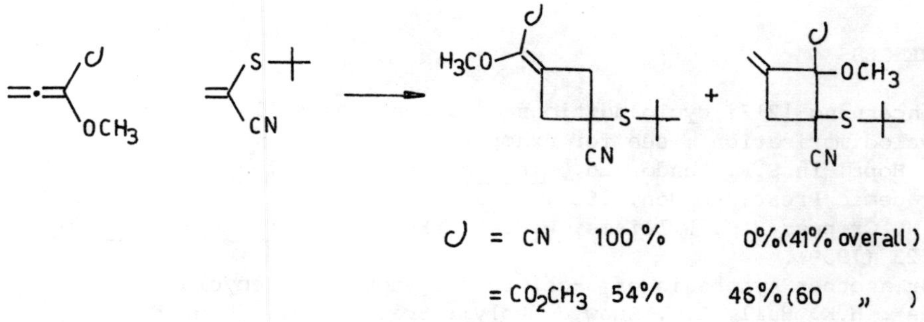

The synthesis of new **cd** allenes is based on the reaction of the metalated donor substituted allenes with different electrophiles according to the following scheme :

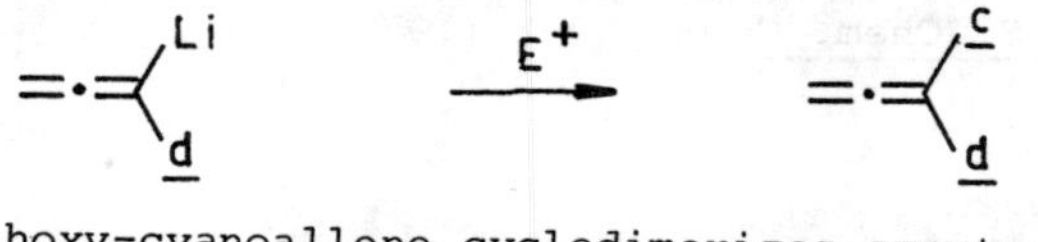

Methoxy-cyanoallene cyclodimerizes spontaneously in solution and leads to a mixture where the 1,2-dimethylenecyclobutane with all the substituents on the cyclobutane ring predominates (1,5).
In the case of tert.-butylthio allene cycloadducts, the methylene-

cyclobutane rearrangement is complete upon equilibration in o-xylene (6). The Arrhenius parameters for this isomerization estimated from kinetic studies at different temperatures are presented in the next table.

Table IV. Activation parameters of the methylenecyclobutane rearrangement.

$$\text{B(R,R'=H, S-t-Butyl)} \xrightarrow[\text{o-xylene}]{\Delta} \text{A}$$

c	d	$\Delta H^{\ddagger}$ kcal/mol	$\Delta S^{\ddagger}$ cal/Kmol	$\Delta G^{\ddagger}_{300}$ kcal/mol
-CN	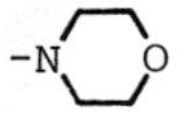	28.6 ± 1.8	7.9 ± 6.9	26.2 ± 3.8
-CN	$-S-t-C_4H_9$	24.9 ± 1.5	-5.1 ± 5.8	26.4 ± 3.2
-CN	$-OCH_3$	44.2 ± 0.6	36.1 ± 2.2	33.4 ± 2.8

These results are in agreement with the proposed biradical mechanism because the relative rates ($\Delta G^{\ddagger}$) follow the order of relative stabilization effect of the **cd** substituent couples involved in each case.

REFERENCES

1. Concerning [2+2] cycloadditions of allenes as well as their cyclodimerisation , see for example : H. Hopf, in S.R. Landor Ed., The Chemistry of Allenes, Vol.2, Academic Press, London, 1982, p. 525.
2. H.N. Cripps, J.K. Williams, W.H. Sharkey, J. Amer. Chem. Soc. **81**, 2723 (1959).
3. For another synthesis of 3-chloro-3-cyanomethylenecyclobutane, see : H.K. Hall, L.G. Snow, J. Polym. Sci., Polym. Chem. Ed., **21**, 1885 (1983).
4. [2+2] Cycloadditions between cyanoallenes and enamines : J.E. Baldwin, R.H. Fleming, D.M. Simmons, J. Org. Chem., **37**, 3963 (1972).
5. Cyclodimerisation of ethyl 2-acetoxybuta-2,3-dienoate M. Verny, R. Vessière, Bull. Soc. Chim. Fr., 1729 (1969).
6. Rearrangements of methylenecyclobutanes, see for example : W. Kirmse, H.R. Murawski, J.C.S. Chem. Comm., 392 (1978).

CAPTODATIVE ALKENES AS DIPOLAROPHILES - KINETIC STUDIES[1]

Dietrich Döpp* and Jutta Walter
Fachgebiet Organische Chemie, FB 6, Universität Duisburg,
Postfach 101629,
D-4100 Duisburg 1
Federal Republic of Germany

ABSTRACT. 1,3-Dipolar cycloadditions of α -(tert-butylthio)acrylonitrile and related captodative alkenes to various dipolesare investigated. Competition constants for selected pairs of dipolarohiles competing for diphenyl diazomethane, benzonitrile oxide and benzylidene methylamine-N-oxide as well as rate constants, enthalpies and entropies of activation for addition of two captodative alkenes to diphenyl diazomethane are quite similar in magnitude to those of other well known 1,3-dipolar cycloadditions.

1. INTRODUCTION

The question whether captodative[2] (c,d-)substitution in olefinic dienophiles or dipolarophiles enhances the rates of Diels-Alder-additions[3] or 1,3-dipolar cycloadditions[4] by stabilizing one terminus of a conceivable diradical like transition state or even intermediate (see scheme below), has been treated recently[3,4].

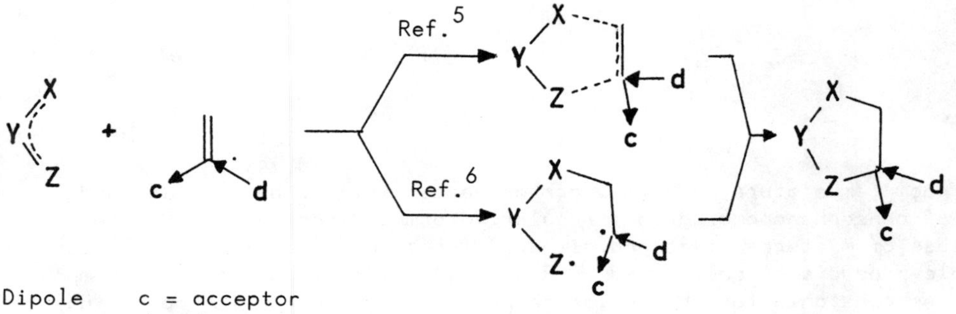

Dipole c = acceptor
d = donor
as in alkenes **1a–c**

H. G. Viehe et al. (eds.), Substituent Effects in Radical Chemistry, 375–378.
© 1986 by D. Reidel Publishing Co.

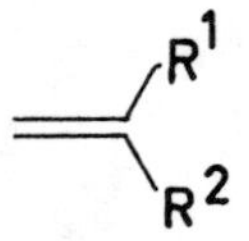

1a–g

	R^1	R^2
a	CN	morpholino
b	$COOCH_3$	$-S-(t-C_4H_9)$
c	CN	$-S-(t-C_4H_9)$
d	CN	CH_3
e	$COOCH_3$	CH_3
f	H	$-S-(t-C_4H_9)$
g	H	$-O-C_4H_9$

Up to now, only few quantitative data have become available[3c,4b], since studies aimed at these may be hampered by either the instability of the adducts[4c] or lack of reactivity of suitable c,d-olefins to certain dipoles or dienes. In this paper, we report on competition experiments using alkenes **1b–g** and various 1,3-dipoles, as well as on rate constants and activation parameters in selected cases. All experimental details will be published elsewhere.

2. COMPETITION STUDIES[7]

2.1 Diphenyl diazomethane as 1,3-dipole

Diphenyl diazomethane (**2**) instantaneously and quantitatively reacts with neat α-(tert-butyl-thio)acrylonitrile (**1c**) at room temperature with evolution of nitrogen. Likewise, α-morpholino-acrylonitrile (**1a**) can be added to **2** in refluxing benzene solution within 12 hours, albeit only in 23% yield.

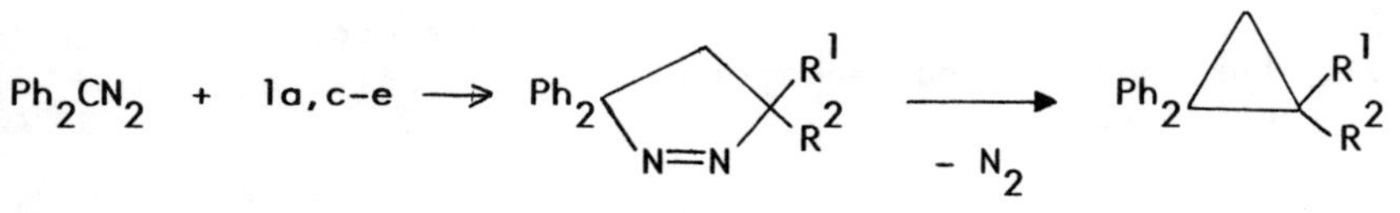

2

3a,c–e

(R^1, R^2 as in **1a,c–e**)

At room temperature, 92% of **2** are converted into **3c** by neat **1c**, and 8% of benzophenone azine are regularly formed. Treatment of **2** with an excess of mixtures of **1c** and either **1d** or **1e** and assaying for all possible products by calibrated HPLC analysis allowed to determine competition constants for the following pairs of dipolarophiles competing for **2**: **1c/1d**, $\varkappa$ = 115, **1c/1e**, $\varkappa$ = 16. Thus, under these conditions, **1c** is added to **2** significantly (but not outstandingly) faster than the sterically slightly less encumbered alkenes **1d,e**.

2.2 Other dipoles

Benzylidene methylamine-N-oxide (**4**) adds methyl methacrylate (**1e**) unidirectionally[8] to form the 5-disubstituted isoxazolidine **5e**, and the same behaviour is observed for alkenes **1c,d**.

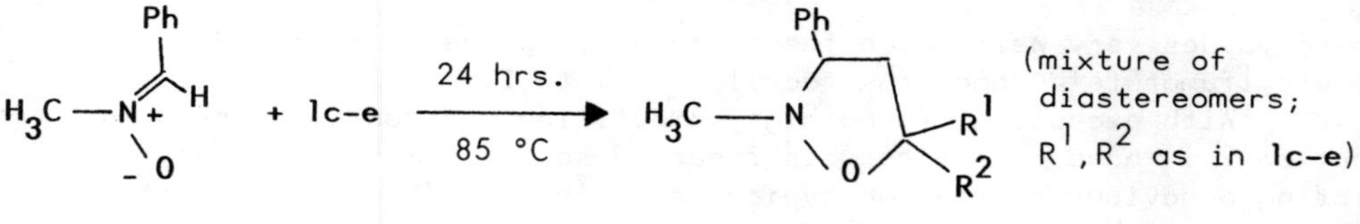

4 **5c-e**

Competition constants, however, were low for the same pairs of dipolarophiles as used before competing for **4**: With **1c/1d**: $\varkappa$ = 2.1 ($\pm$ 0.5), with **1c/1e**: $\varkappa$ = 1.2 ($\pm$ 0.2).

The corresponding 5-substituted 4,5-dihydroisoxazoles are formed from in situ generated benzonitrile oxide[7] and **1e**[9a] or butyl vinyl ether (**1g**)[9b] as well as methyl-(α-tert-butylthio)acrylate (**1b**), **1c**, methacrylonitrile (**1d**), tert-butylvinylsulfide (**1f**) and butylvinylether[9a] (**1g**). Following the procedure of Huisgen[7], competition constants $\varkappa$ for the pairs of dipolarophiles competing for benzonitrile oxide in ether solution at 0 °C could be determined: **1c/1b** 0.13; **1c/1d** 2.1; **1d/1e** 0.25; **1c/1f** 1.17; **1c/1g** 0.44.

3. KINETIC STUDIES FOR ADDITION OF 1b,c TO DIPHENYLDIAZOMETHANE (2).

Rate constants for the cycloaddition of **1b,c** to **2** in acetonitrile solution have been determined photometrically at 520 nm at six different temperatures (see table 1).

TABLE I: Rate constants for addition of **1b,c** to diphenyl diazomethane (**2**) in acetonitrile

Temp.	$10^3 k_2$ [$L mol^{-1} s^{-1}$]	
[K]	**1b**	**1c**
273	2.36 $\pm$ 0.09	1.91 $\pm$ 0.07
283	4.12 $\pm$ 0.20	3.40 $\pm$ 0.09
293	6.10 $\pm$ 0.10	5.96 $\pm$ 0.06
303	10.78 $\pm$ 0.10	10.01 $\pm$ 0.11
313	17.89 $\pm$ 0.30	16.04 $\pm$ 0.39
323	32.6 $\pm$ 0.14	23.24 $\pm$ 0.48

From these constants, the following activation parameters have been derived:

1b + **2** → **3b**: $\Delta H^{\ddagger} = 35.5\ kJmol^{-1}$, $\Delta S^{\ddagger} = -165.5\ JK^{-1}mol^{-1}$

1c + **2** → **3c**: $\Delta H^{\ddagger} = 34.7\ kJmol^{-1}$, $\Delta S^{\ddagger} = -169,3\ JK^{-1}mol^{-1}$

These values very well match the corresponding ones for the addition of dimethylfumarate[10a] and ethyl acrylate [10b] to **2**.

With exception of the high competition constants reported in section 2.1, none of the results reported so far demonstrates any deviating behaviour of the two typical c,d-alkenes (**1b,c**) investigated. They can well be accomodated within the band-width of rate constants and activation parameters known[7,11] for cycloadditions of the dipoles investigated here.

Acknowledgements: This work has been supported by Ministerium für Wissenschaft und Forschung des Landes Nordrhein-Westfalen and Fonds der Chemischen Industrie.

---

1 Taken in part from the Doctoral Thesis of Jutta Walter, University of Duisburg 1985.

2 H. G. Viehe, Z. Janousek, R. Mérenyi, and L. Stella, Acc. Chem. Res. **18**, 148(1985).

3 a) L. Stella and J. L. Boucher, Tetrahedron Lett. **23**, 953 (1982). b) J. L. Boucher and L. Stella, Tetrahedron **41**, 875 (1985). c) ibidem, Tetrahedron Lett. **26**, 5041 (1985).

4 a) D. Döpp and M. Henseleit, Chem. Ber. **115**,798(1982). b) F. Texier, A. Derdour, H. Benhaoua, T. Benabdallah and O. Yebdri, Tetrahedron Lett. **23**, 1893 (1982). - c) D. Döpp and J. Walter, Heterocycles **20**, 1055(1983).- d) See also: H. Horikawa, T. Nishitani, T. Iwasaki and I. Inoue, Tetrahedron Lett. **21**, 2193 (1983).

5 R. Huisgen, J. Org. Chem. **41**, 403 (1976) and earlier papers.

6 R. A. Firestone, Tetrahedron **33**, 2249 (1977) and earlier papers.

7 In analogy to:K. Bast, M. Christl, R. Huisgen and W. Mack, Chem.Ber. **106**, 3312 (1973).

8 R. Huisgen, H. Hauck, R. Grashey and H. Seidl, Chem. Ber. **101**, 2568 (1968).

9 a) M. Christl, R. Huisgen and R. Sustmann, Chem. Ber. **106**, 3275(1973). b) K. Bast, M. Christl, R. Huisgen, W. Mack and R. Sustmann, Chem. Ber. **106**, 3258 (1973).

10 a) R. Huisgen, J. Org. Chem. **33**, 2291 (1968).- b) R. Huisgen, H. Stangl, H. J. Sturm and H. Wagerhofer, Angew. Chem. **73**, 170 (1963).

11 L. Fisera, J. Geittner, R. Huisgen and H.-U. Reissig, Heterocycles **10**, 153 (1978).

CAPTODATIVE ALKENES IN CONCERTED 1,3-DIPOLAR CYCLOADDITIONS[1]

Dietrich Döpp*, Jutta Walter and Sabine Holz
Fachgebiet Organische Chemie, FB 6, Universität Duisburg,
Postfach 101629,
D-4100 Duisburg 1
Federal Republic of Germany

ABSTRACT. The 1,3-dipolar cycloaddition of an E-/Z-pair of captodative alkenes to a cyclic aldonitrone is demonstrated to occur stereospecifically. Additions of α-(tert-butylthio)acrylonitrile to a series of C,N-diphenylnitrones show a change in regioselectivity upon donor substitution in the C-phenyl-ring. Both findings are in accord with well accepted principles of frontier molecular orbital control and a concerted mechanism.

1. INTRODUCTION

Demonstrations of stereospecifical cycloadditions of captodative[2] alkenes to suitably reactive dipoles, while sometimes being hampered by the instability of the primary adducts, are highly desirable to prove the validity of either the concerted or the non-concerted[4] mechanism of 1,3-dipolar cycloadditions also for such alkenes, which carry a radical stabilising array of substituents at one terminus of the double bond. Nitrones as dipoles seem to be especially suitable for this purpose, because a conceivable 1,5-diradical intermediate would show captodative substitution on one terminus and be a (per se) stable aminyl oxide at the other:

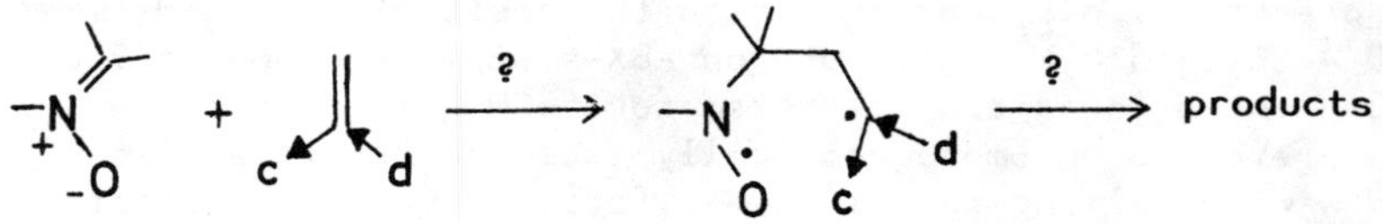

(**c** = acceptor, **d** = donor)

2. RESULTS

The reactive dipole 6-tert-butyl-3,3-dimethyl-3H-indole-1-oxide (**1**)[5] smoothly adds both of the easily accessible[6] dipolarophiles dimethyl 2-(phenylthio)maleate (**2**) and -fumarate (**3**) at room temperature. Compound **1** was added to an excess of the neat esters, and the reaction

H. G. Viehe et al. (eds.), Substituent Effects in Radical Chemistry, 379–381.
© 1986 by D. Reidel Publishing Co.

times (42 hrs. for **2** and 20 hrs. for **3** at room temperature) were chosen to leave 20% of **1** unchanged. The sole products were **4** and **5**, yields are base on converted **1** and refer to pure recrystallized material. Absence of any other isomers including the C3a-epimers in the mother liquors was substantiated[7].

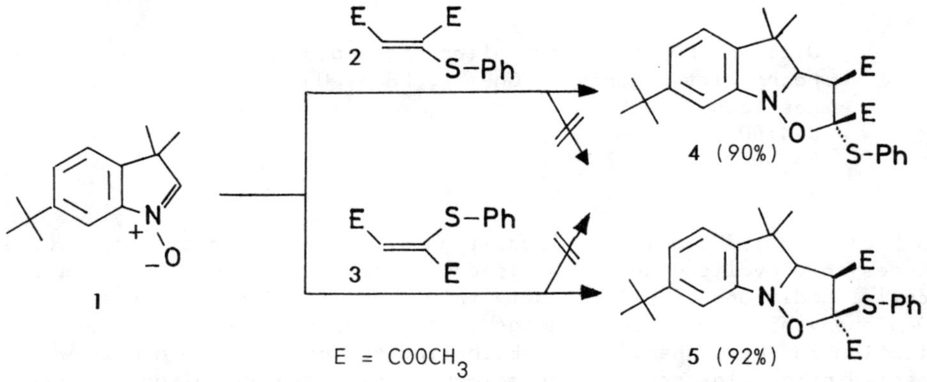

Thus, the additions of **2** and **3** to **1** were both stereospecific and highly diastereo- and regioselective. A diradical intermediate of sufficient lifetime to allow for non-stereospecific addition can be ruled out.

When **7** (usually applied in 50% excess) was added to N-phenyl-C-arylnitrones **6** in benzene solution, a complete reversal of the sense of addition was observed when the more electron poor nitrone **6a** was replaced by the more electron rich **6e**. In principle, both the 5- and the 4-disubstituted isoxazolidines (**8, 9**) may arise as mixture of diastereomers, which in some cases may be partially separated. Regioisomers **8** and **9** are easily distinguishable on the basis of their ^{1}H-NMR-spectra: In **8** 4-CH_2 and 3-H constitute an ABX-system, whereas in **9** a singlet for 3-H and a low-field AB-system for 5-CH_2 are always observed. Yields refer to chromatographically isolated products. Isomers **9** may show wide melting ranges due to facile reversal to starting materials and/or formation of the more stable isomers **8**. We currently investigate the thermolability of **9b-e**, to which the captodative substitution at C-4 certainly contributes.

The change in sense of addition may be interpreted[8] in terms of a change in type of FMO-control: With **6a** (X=CN), LUMO(dipole)-HOMO(dipolarophile) control (Sustmann's Type III[9]) prevails, whereas with **6e** (X=NMe_2), the HOMO(dipole)-LUMO(dipolarophile) control (Sustmann's Type I[9]) becomes dominant[10]. The LUMO of **7** is predicted[11] to have a substantially larger coefficient on β-C than on α-C, which is favourable for interaction with the lobe at oxygen in the nitrone-HOMO.

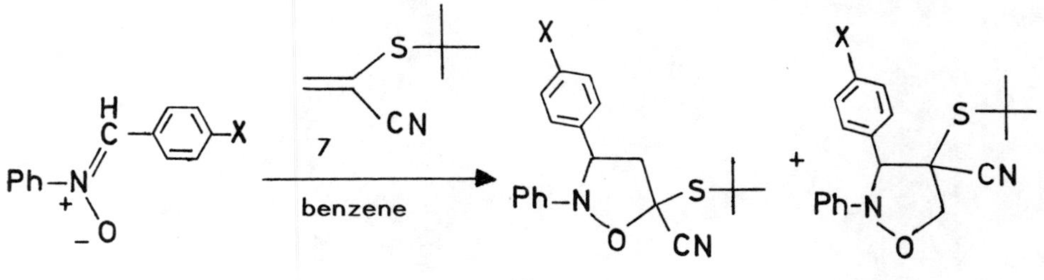

a: X = CN	60 °C, 5 hrs.	**a**: 85%	**a**: -
b: X = Cl	60 °C, 4 hrs.	**b**: 67%	**b**: 6%
c: X = H	80 °C, 1 hr.	**c**: 85%	**c**: 9% (Ref.[12])
d: X = OCH_3	60 °C, 5 hrs.	**d**: 31%	**d**: 10%
e: X = $N(CH_3)_2$	80 °C, 1 hr.	**e**: -	**e**: 80%

Again, all findings are in accord with Huisgen's[3] concerted mechanism for the 1,3-dipolar cycloaddition and do not support Firestone's[4] view of a spin-paired diradical intermediate.

Acknowledgements: This work has been supported by Ministerium für Wissenschaft und Forschung des Landes Nordrhein-Westfalen and Fonds der Chemischen Industrie. We thank Miss P. Blitgen, Mr. M. Pies and Mr. K. Samouris for preliminary experiments.

[1] Taken in part from the Doctoral Thesis of Jutta Walter, University of Duisburg 1985.

[2] H. G. Viehe, Z. Janousek, R. Mérenyi and L. Stella, Acc.Chem. Res. **18**, 148(1985).

[3] R. Huisgen, J. Org. Chem. **41**, 403 (1976) and earlier papers.

[4] R. A. Firestone, Tetrahedron **33**, 2249 (1977) and earlier papers.

[5] D. Döpp, Chem. Ber. **109**, 3849 (1976).

[6] J. A. Kaydos and D. L. Smith, J. Org. Chem. **48**, 1096 (1983).

[7] All experimental details fully support the conclusions drawn and will be published separately.

[8] I. Fleming: 'Frontier Orbitals and Organic Chemical Reactions', J. Wiley & Sons, London 1976.

[9] R. Sustmann, Tetrahedron Lett. **1971**, 2717.

[10] The same type of control is favoured by lowering the FMO's in the dipolarophile: J. Sims and K. N. Houk, J.Am.Chem.Soc. **95**, 5798(1973).

[11] R. Sustmann, private communication.

[12] D. Döpp and M. Henseleit, Chem. Ber. **115**, 798 (1982).

PHOTO-DIELS-ALDER ADDITION OF TWO CAPTODATIVE ALKENES TO 1-ACETONAPHTHONE[1]

Dietrich Döpp and Hamid Reza Memarian
Fachgebiet Organische Chemie, FB 6
Universität Duisburg
Postfach 101629
D-4100 Duisburg 1
Federal Republic of Germany

ABSTRACT. Triplet excited 1-acetonaphthone readily reacts with α-morpholinoacrylonitrile and α-tert-butylthioacrylonitrile under formation of 1,4-ethanonaphthalenes with remarkably high regio- and stereoselectivity, whereas the parent acrylonitrile fails to react at all and tert-butyl vinylsulfide is added both regio- and stereounselectively.

1. INTRODUCTION

A facile 1,4-photocycloaddition of α-morpholinoacrylonitrile (**2a**) to 1-acetonaphthone (**1**) has been reported recently[2]. The main product is a single diastereomer (**3aA**) of the 9-cyano-9-morpholino substituted 1,4-ethanonaphthalene **3a**, the structure of which has been unambiguously established by X-ray crystallography[2]. A bimolecular reaction of triplet excited **1** and **2a** has been established, and the degree of conversion of starting material **1** within fixed time-intervals is independent of the nature of solvent used for photolysis[2].

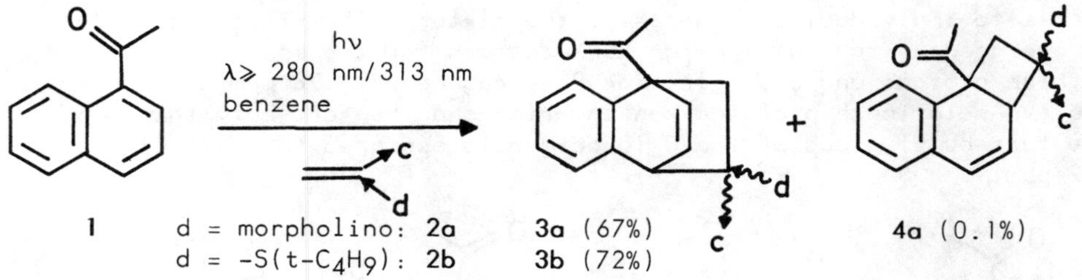

1 d = morpholino: **2a** **3a** (67%) **4a** (0.1%)
d = -S(t-C_4H_9): **2b** **3b** (72%)
c = CN

2. RESULTS

Only recently, a cyclobutane type by-product **4a** was detected and isolated in very low yield by HPLC[3]. While **3a** exhibits UV absorption typical for a phenyl ring and isolated unsaturation[2], **4a** clearly shows conjugation[λ in nm (ε), in chloroform solution: 320(1200), 284(3500),

H. G. Viehe et al. (eds.), Substituent Effects in Radical Chemistry, 383–385.
© 1986 by D. Reidel Publishing Co.

268(8000), 261(7100), 254(5300)] as reported[4] for other (2+2)-adducts of naphthalene[3]. Upon irradiation at 313 nm in solution it reverts to starting materials **1** and **2a**, it may reach a small stationary concentration as long as it is screened by starting material.

As **2a**, α-(tert-butylthio)acrylonitrile (**2b**) quenches[5] the lowest excited triplet state of **1** at a rate which is well below diffusion controlled,

$$k_q = 2.5 \times 10^7 M^{-1} s^{-1}.$$

On a preparative scale, **2b** is added to **1** in the same fashion as **2a**. After 96 hrs. of irradiation with a 150 W high pressure burner through a water cooled Duran-immersion well (λ ≥ 280 nm) under an atmosphere of pure nitrogen in 0.1 M solution in benzene containing an equimolar amount of **2b**, 60% of **1** are converted. After chromatographic workup, a crude yield of 84% of **3b** was obtained, crystallization from hexane gave a 67% yield of pure material, m. p. 106-108 °C[3]. It should be noted that all yields refer to that fraction of starting material **1**, which could not be recovered and tend to be lowered by the sensitivity of **3b** towards silica gel chromatography and heat (the half life of **3b** is 2 days at 50 °C in chloroform solution).

Chromatography and ^{1}H-NMR-spectroscopy of the mother liquors would have led to detection of any other products present. Thus again, only one dominant diastereomer of the only formed regioisomer (**3b**) is obtained, although its relative configuration is still unknown.

It is interesting to compare the behaviour of both acrylonitrile and tert-butyl vinylsulfide (**5**) towards photoexcited **1**. While acrylonitrile was found to be unreactive, **5** could be added to photoexcited **1**. After 50% conversion of **1** was reached, besides small amounts of a dimer of **5** (as suggested by the mass spectrum of that compound), a mixture (**6, 7**) of diastereomeric pairs of the two possible regioisomeric 1,4-ethanonaphthalenes resulted, from which so far only one component (assigned as **6A**) could be partly separated off. Careful analysis of the 300-MHz-^{1}H-NMR spectra, however, allows to assign individual signals to individual components of the mixture. The structure of **6A** was assigned on the basis of an ABPX-pattern for the sp^3-methylene and -methine protons and a singlet for 9 hydrogens at 1.28 ppm attributed to a syn-relationship of the benzene ring and the tert-butylthio group (the tert-butyl group of the C-10 epimer is assigned to a signal at 1.32 ppm)[3].

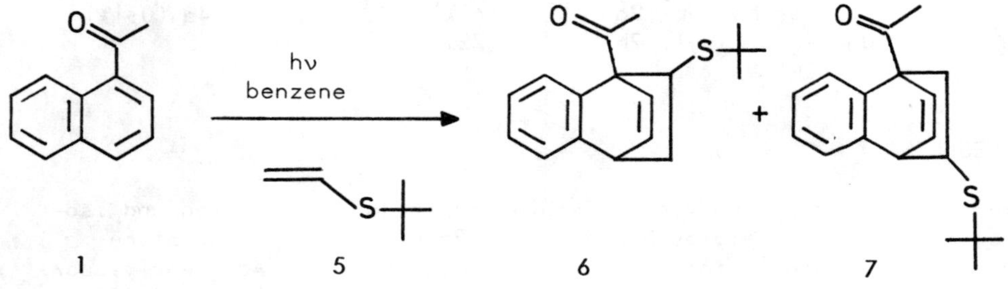

Thus, the addition of **5** to excited **1** lacks both regio- and stereoselectivity in sharp contrast to the photoaddition of **2a,b**. Also, c,d-substitution of the olefin added enhances the overall efficiency. The results are in agreement with either a rapidly collapsing complex anticipating the product geometry (as in **3aA**) or - since an excited triplet state is likely to be involved[2] - a polarized diradical like **8** undergoing rapid and selective bond formation towards a 1,4-ethanonaphthalene after spin pairing. This diradical species may well profit from c,d-stabilization at one end and extensive delocalization at the other. Extensive charge transfer interaction of **2a,b** (donor) with excited **1** (acceptor) can be ruled out due to lack of influence of solvent polarity in the photoaddition of **2a**[2].

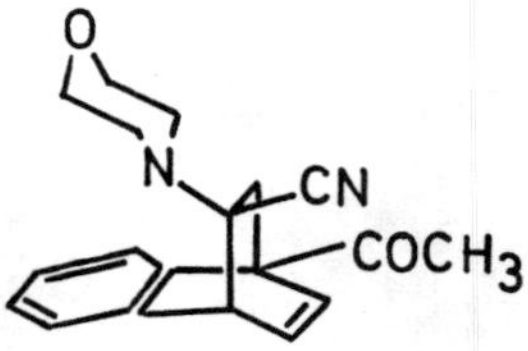

3aA

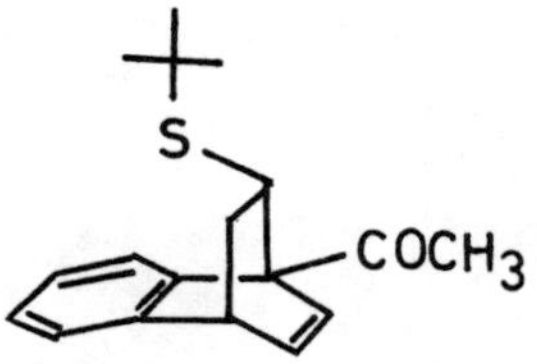

6A

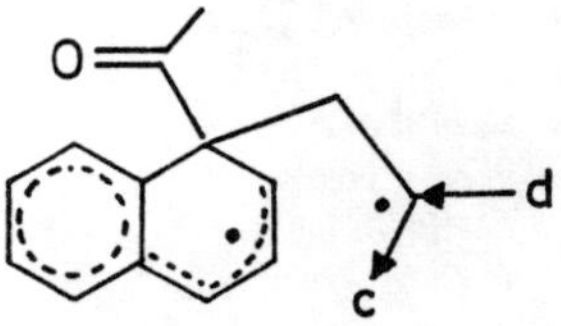

8

Acknowledgements. Support of this research from Ministerium für Wissenschaft und Forschung des Landes Nordrhein-Westfalen, Fonds der Chemischen Industrie and Deutscher Akademischer Austauschdienst (by a stipend to H. R. Memarian) is gratefully acknowledged.

[1] Taken from the planned doctoral thesis of H. R. Memarian, University of Duisburg.

[2] D. Döpp, C. Krüger, H. R. Memarian and Y.-H. Tsay, Angew. Chem. **97**, 1059 (1985); Angew. Chem. Int. Ed. Engl. **24**, 1048 (1985).

[3] All details will be published in a full paper elsewhere.

[4] N. C. Yang, J. Libman and M. F. Savitzky, J. Am. Chem. Soc. **94**, 9226 (1972); N. C. Yang and J. F. Libman, J. Am. Chem. Soc. **94**, 9228 (1972); C. Pac, t. Sugioka, K. Mizuno and H. Sakurai, Bull. Chem. Soc. Jpn. **46**, 238 (1973).

[5] We thank Dr. Helmut Görner, Max-Planck-Institut für Strahlenchemie, D-4330 Mülheim/Ruhr, for these determinations.

FREE RADICAL CYCLIZATION OF UNSATURATED CARBONIC ACID DERIVATIVES

M.D. Bachi* and E. Bosch
Department of Organic Chemistry
The Weizmann Institute of Science
Rehovot 76100
Israel.

Abstract. Phenylselenocarbonate derivatives of homopropargylic alcohols react with tri-n-butylstannane to give α-alkylidene-γ-lactones. Product analysis of the reaction between an unsaturated dithiocarbonate and tri-n-butylstannane indicates that the stannyl radicals attack the thiono-, rather than the thiolo-sulphur atom.

Alcohols can be readily transformed into their carbonic acid derivatives (1). The free-radical reductive degradation of these compounds by trialkylstannanes (Scheme I) has been extensively used for the deoxygenation of alcohols[1,2]. The degradation of selenocarbonates (1a) to the corresponding desoxy compounds (4) probably involves the intermediacy of alkoxycarbonyl radicals (2)[3], while the degradation of thionocarbonates (1b)[4,5] and of thiocarbonylimidazolides (1c)[2] proceeds via triheterosubstituted free-radicals of type (3). For the degradation of dithiocarbonates (1d), Barton[2,6]

Scheme I

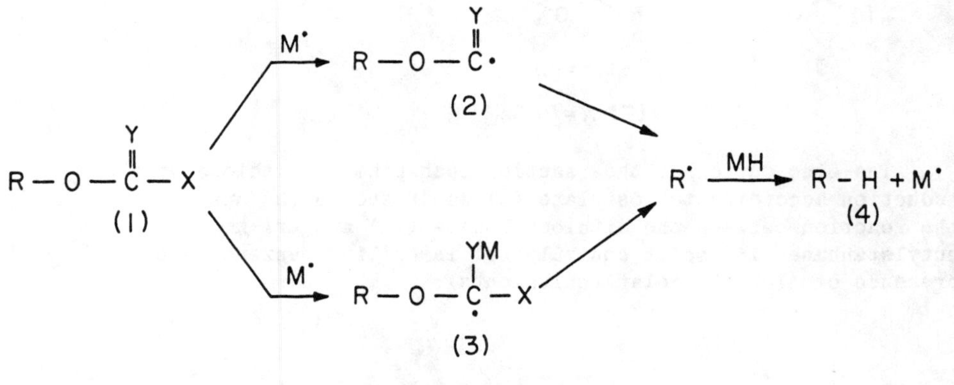

(a) Y = O, X = SePh; (b) Y = S, X = OR′; (c) Y = S, X = Imidazolide; (d) Y = S, X = SR′. M = R''_3 Sn

H. G. Viehe et al. (eds.), Substituent Effects in Radical Chemistry, 387–390.
© 1986 by D. Reidel Publishing Co.

proposed a mechanism involving initial attack of trialkyltin radicals on the sulphur atom of the thiocarbonyl to give an adduct radical (3) which then undergoes β-fission, while Beckwith[7] postulated a homolytic substitution on the sulphide sulphur atom to give an alkylthiocarbonyl radical (2) (Y=S), which undergoes β-fission.

Analysis of possible transformations of transient free-radicals of types (2) and (3) led us to the following two postulates:
(a) A carbonic acid derivative (1) which has a properly postioned multiple bond on the residue R should be subject to cyclization when converted into free-radicals of either types (2) or (3); such a transformation may be useful for the synthesis of lactones.
(b) A multiple bond suitably situated on the residue R′ of dithiocarbonates (1d) may serve as an internal trap to radicals of type (3) but not to radicals of type (2); consequently product analysis of such a transformation may elucidate the reaction mechanism of the reduction of dithiocarbonates with tributylstannane.

Investigation of postulate (a) has provided a new method for the synthesis of α-alkylidene-γ-lactones from homopropargylic alcohols (Scheme II)[8]. Alcohols (5) were converted (phosgene) into the chloroformates (6) and subsequently (PhSeH, pyridine) into the corresponding phenylselenocarbonates (7)[3]. When treated with tri-n-butylstannane and AIBN (azaisobutyronitrile) in boiling benzene, the selenocarbonates (7) were converted, in high yields, into the corresponding α-alkylidene-γ-lactones (9). Some examples are displayed in the Table.

Scheme II

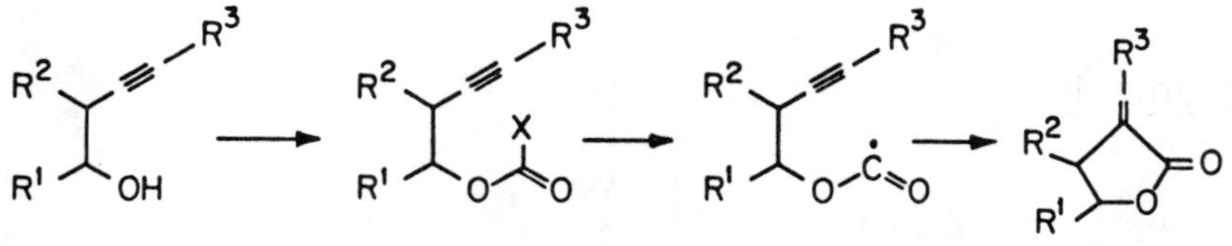

(5) (6) X = Cl (7) X = PhSe (8) (9)

The elucidation of the reaction mechanism of dithiocarbonate reduction according to postulate (b) was based on the analysis of the reaction between the dithiocarbonate (10) and tri-n-butylstannane (25% molar equivalents) in boiling benzene in the presence of AIBN (2% molar equivalents).

The major product isolated from this reaction was the thiolactone (14) (71%, based on Bu_3SnH) and the minor product consisted of the dithiolactone (13) (Scheme III). This result supports the reaction mechanism proposed by Barton[6] which requires, in the first stage, formation of the transient free-radical (11). In contrast to the usual pattern which requires a subsequent β-fission, and in line with our expectations, radical (11) underwent cyclization followed by hydrogen abstraction to give the intermediate ortho compound (12). Hydrolysis ofthe crude product on silica gel afforded the lactones (13) and (14). The alternative mechanism, which would lead to compound (15), COS, and propane is thus not involved in this reaction.

Table: The Reaction of Homopropargylic Selenocarbonates (7) with n-Bu_3SnH and AIBN in Boiling Benzene

	Compounds type (7)	Products (9)	Yield %	E/Z
a	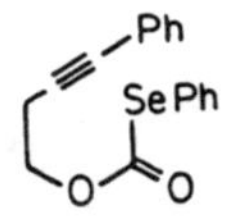	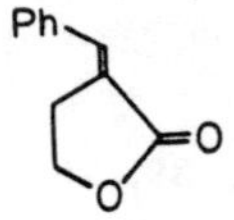	87	E
b	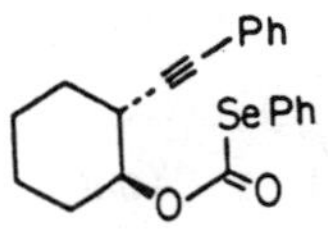	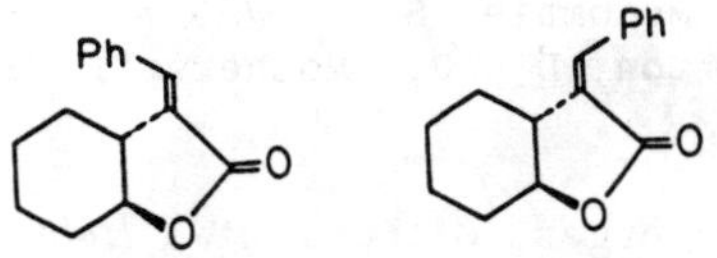	90	5/1
c	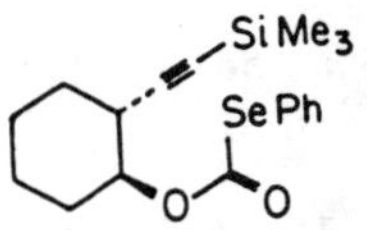	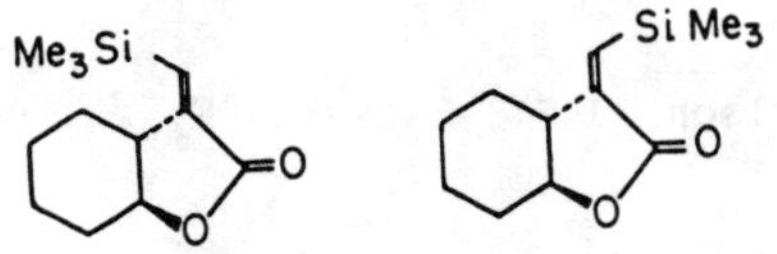	80	1/12
d	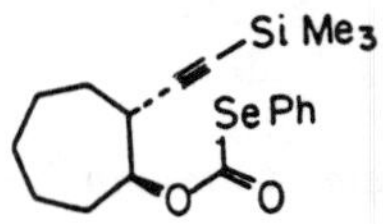	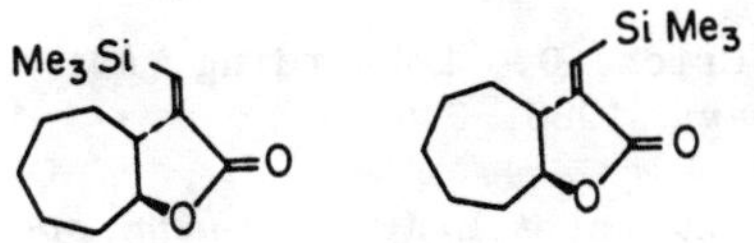	92	1/10

Scheme III

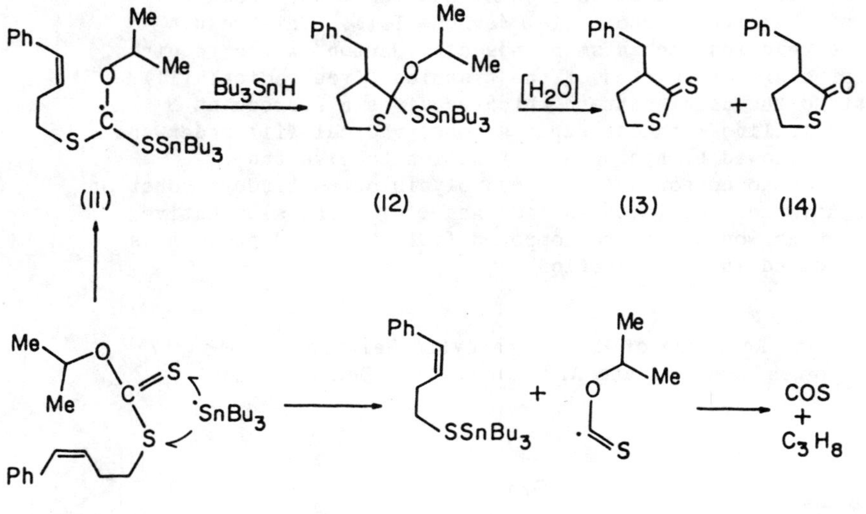

References

1. Review: Hartwig, W. *Tetrahedron* 1983, **39**, 2609.

2. a) Barton, D.H.R.; McCombie, S.W. *J. Chem. Soc., Perkin Trans 1,* 1975, 1574. b) Barton, D.H.R.; Motherwell, W.B.; Stange, A. *Synthesis* 1981, 743.

3. Pfenninger, J.; Heuberger, C.; Graf, W. *Helv. Chim. Acta.* 1980, **63**, 2328.

4. Robins, M.J.; Wilson, J.S.; Hansske, F. *J. Am. Chem. Soc.* 1983, **105**, 4059.

5. Prisbe, E.J.; Martin, J.C. *Synthetic Commun.* 1985, **15**, 401.

6. Barton, D.H.R.; Crich, D.; Löbberding, A.; Zard, S.Z. *J. Chem. Soc., Chem. Commun.* 1985, 646.

7. Barker, P.J.; Beckwith, A.L.J., *J. Chem. Soc., Chem. Commun.* 1984, 683.

8. Preliminary report: Bachi, M.D.; Bosch, E. *Tetrahedron Lett.* 1986 **27**, 641.

SYNTHETIC APPLICATIONS OF THE POLAR EFFECTS OF THE SUBSTITUENTS IN FREE-RADICAL REACTIONS

Francesco Minisci
Dipartimento di Chimica, Politecnico di Milano
Piazza Leonardo da Vinci, 32
20133 Milano
Italy

ABSTRACT - The polar effects can play a dominant role in free-radical reactions, when charged species are involved. A charged radical is as much important as a charged substrate in determining the overall sensitivity to the polar effect. Substituents on the radicals affect their polarity and selectivity and substituents on the substrate influence the polarity and the selectivity also of the intermediate radicals. Reactions of general character and high synthetic value have been developed by taking advantage of the polar effect of the protonated amino radicals with alkanes, alkenes and aromatics or nucleophilic carbon-centered radicals with positively charged unsaturated compounds. Solvent and captodative effects are discussed in relation to the polar effects.

The awareness that polar effects can play a role in free-radical reactions is old, but for long time, with the exception of copolymerization, they were considered a minor aspect of the free-radical reactivity without significant involvements in organic synthesis.

Synthetic organic chemists, with few exceptions, have always considered with suspicion free-radical reactions owing to the low selectivity. Now polar effect, selectivity and synthetic interest are, obviously, strictly connected in ionic reactions. Thus the low incidence of the polar effect in free-radical reactions was considered one of the reasons of the low selectivity. This bad synthetic reputation is still largely diffused.

H. G. Viehe et al. (eds.), Substituent Effects in Radical Chemistry, 391–433.
© 1986 by D. Reidel Publishing Co.

The problem of the substituent effect in free radical reactions can be considered as influence of the substituent on the radical and on the substrate; these two aspects are strictly connected because generally the reaction of a radical with a substrate leads to an intermediate radical arising from the substrate, whose structure affects both the primary interaction and the further evolution of the intermediate radical (eq 1)

$$R\cdot + S \longrightarrow S\cdot \longrightarrow \text{Products} \qquad (1)$$

The enthalpic factor has been for long time considered the major responsible of the reactivity and selectivity of the free-radical reactions. That is actually often the case. However for several free-radical reactions, the reaction enthalpy is a minor factor in determining reactivity and selectivity, which is mainly determined by the polar effects.

Thus alkyl radicals add faster to acrylonitrile (eq 2) than to butadiene (eq 3)[1], even if the allyl radical is more stable than the α-cyanoalkyl radical and that has been utilized for synthetic purposes[1]

$$R\text{-}CH_2\cdot + CH_2 = CH\text{-}CN \longrightarrow R\text{-}CH_2\text{-}CH_2\text{-}\dot{C}H\text{-}CN \qquad k=9.2\times10^5 M^{-1}s^{-1} \qquad (2)$$

$$R\text{-}CH_2\cdot + CH_2=CH\text{-}CH=CH_2 \longrightarrow R\text{-}CH_2\text{-}CH_2\text{-}\dot{C}H\text{-}CH=CH_2 \qquad k=5.3\times10^5 M^{-1}s^{-1} \qquad (3)$$

On the contrary protonated amino radicals add very fast to butadiene (eq 4), a reaction practically controlled by diffusion, and also to unconjugated α-olefins[2] (eq 5), but no reaction occurs with acrylonitrile[2] (eq 6)

$$R_2\dot{\overset{+}{N}}H + CH_2=CH\text{-}CH=CH_2 \longrightarrow R_2\overset{+}{N}H\text{-}CH_2\text{-}\dot{C}H\text{-}CH=CH_2 \qquad k>10^8 M^{-1}s^{-1} \qquad (4)$$

$$R_2\dot{\overset{+}{N}}H + CH_2=CH\text{-}C_4H_9 \longrightarrow R_2\overset{+}{N}H\text{-}CH_2\text{-}\dot{C}H\text{-}C_4H_9 \qquad k>10^6 M^{-1}s^{-1} \qquad (5)$$

$$R_2\dot{\overset{+}{N}}H + CH_2=CH\text{-}CN \not\longrightarrow R_2\overset{+}{N}H\text{-}CH_2\text{-}\dot{C}H\text{-}CN \qquad (6)$$

Similarly t-alkyl[3], acyl[4] and hydroxymethyl[5] radicals react very fast with protonated 4-cyanopyridine (eqs 7-9), but no reaction occurs with benzene[6] (eq 10)

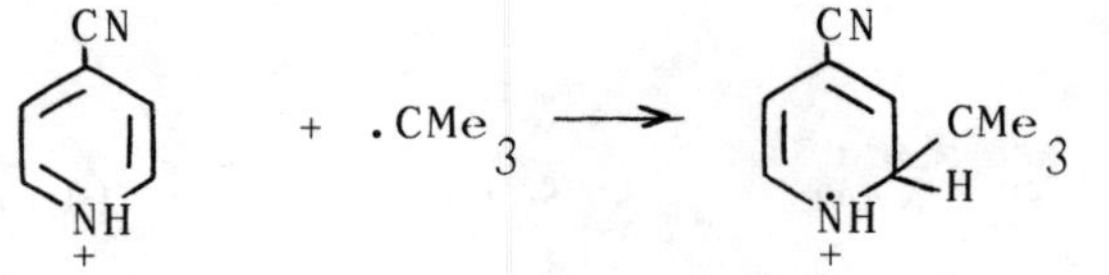

$k = 6.3 \times 10^{7} M^{-1} s^{-1}$ (7)

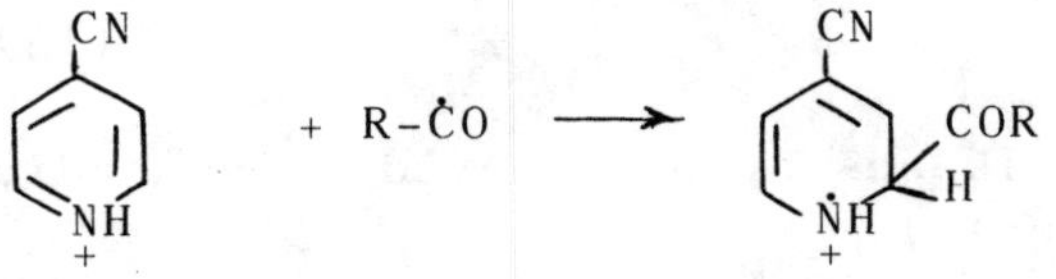

$k > 10^{6} M^{-1} s^{-1}$ (8)

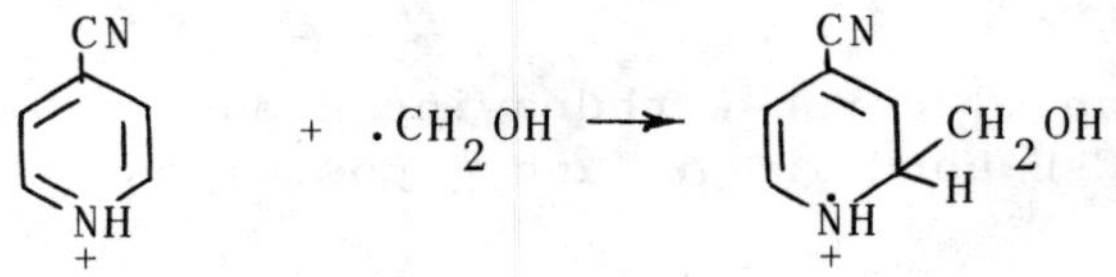

$k > 10^{7} M^{-1} s^{-1}$ (9)

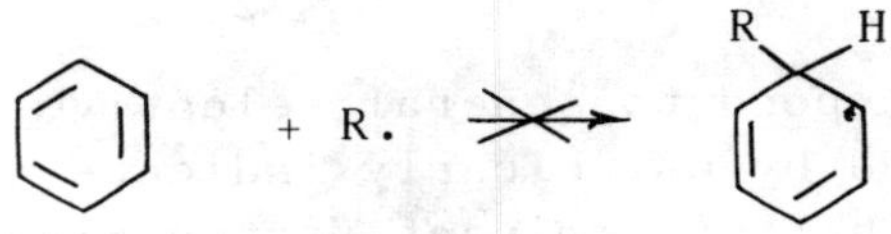

R.=Me_3C., RĊO, .CH_2OH (10)

Protonated amino radicals and carbon-centered radicals, .CH_2-X (X = electronwithdrawing group, COR, CN, NO_2, SO_2R, etc.), react fast with benzene (eqs 11, 12)[7], but no reaction occurs with protonated 4-cyanopyridine (eq 13)

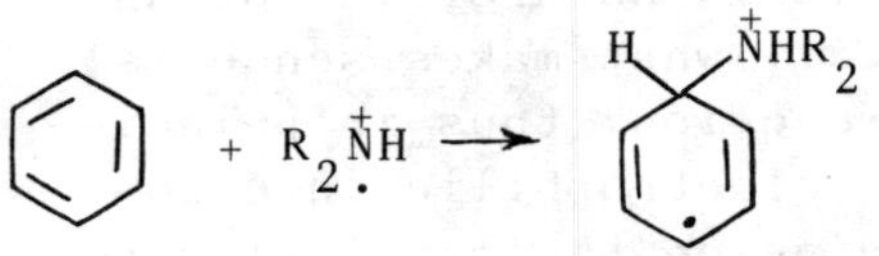

$k > 10^{6} M^{-1} s^{-1}$ (11)

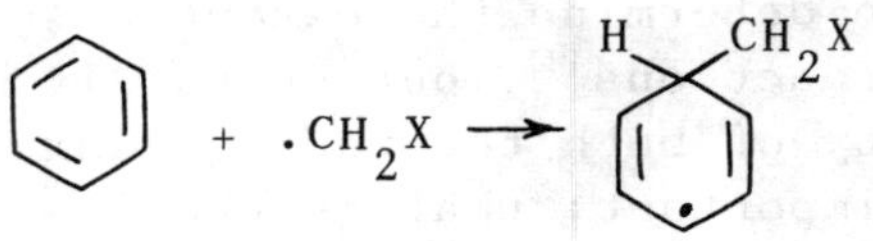

fast (12)

R. = $R_2\dot{N}^{+}H$, $\dot{C}H_2X$ (13)

Alkyl radicals add fast to the diazonium salts[8] (eq 14), but the presence of an electron-withdrawing group directly bound to the radical center completely inhibits the reaction (eq 15)

$$Me_3C\cdot + N{\equiv}\overset{+}{N}{-}A_2 \longrightarrow Me_3C{-}N = \overset{+}{\underset{\cdot}{N}}{-}Ar \qquad k \simeq 10^8 M^{-1}s^{-1} \qquad (14)$$

$$X{-}CH_2\cdot + N{\equiv}\overset{+}{N}{-}Ar \not\longrightarrow X{-}CH_2{-}N = \overset{+}{\underset{\cdot}{N}}{-}Ar \qquad X = COR,\ NO_2,\ CN\ etc. \qquad (15)$$

Hydrogen abstraction from alkanes by protonated amino radicals is a fast reaction[9] (eq 16)

$$R{-}CH_2{-}CH_3 + Me_2\overset{+}{\underset{\cdot}{N}}H \longrightarrow R{-}\underset{\cdot}{C}H{-}CH_3 + Me\overset{+}{N}H_2 \qquad k > 10^6 M^{-1}s^{-1} \qquad (16)$$

however the presence of an electron-withdrawing group completely deactivates the C-H bond in α- and β-positions (eq 17)

$$C_2H_5{-}X + Me_2\overset{+}{\underset{\cdot}{N}}H \not\longrightarrow \cdot C_2H_4{-}X + Me_2\overset{+}{N}H_2 \qquad X{=}COR,\ CN,\ NO_2\ etc. \qquad (17)$$

All these examples, which correspond to general selective free-radical reactions developed by us, clearly indicate not only that the polar effect can be the dominant factor also in the free-radical reactions in determining the reactivity and the consequent possibility of selective syntheses, but also that the extent of the polar effect can be in the most favourable cases similar to that of highly selective ionic reactions.

The polar effect, being kinetic in nature, is not an intrinsic property of a free radical; what makes sense is the polar effect in a particular reaction: thus in principle the same radical can behave as electrophilic, nucleophilic or non-polar species depending on the particular reaction involved[10].

Several factors contribute to determine the extent of the polar effect in free-radical reactions[10], but certainly the polarity and the polarizability of both the radical and the substrate are among the most important; that is the polarity of the substrate is not less important than the polarity of the radical in determining the sensitivity to the polar effect.

Thus polar effects are particularly marked when charged species are involved. The effect is maximum when both the radical and the substrates have opposite charge. However in these cases the reaction generally turns into an electron-transfer process (eqs 18, 19) or into an ionic reaction

$$R^{\dot{+}} + S^{-} \longrightarrow R + S. \qquad (18)$$

$$R^{\dot{-}} + S^{+} \longrightarrow R + S. \qquad (19)$$

From synthetic point of view the reactions, in which only one species, the radical or the substrate, is charged, are much more interesting; also in these cases an electron-transfer is possible, but often the thermodynamic conditions are not suitable for an electron-transfer and the polar effect is still large enough to determine a high selectivity. Reactions, which fall within this picture, will be discussed in this paper.

The polar effect certainly affects the activation energy of the free-radical reactions (see, for instance, the results of Table I). However, recently Ingold and coworkers[11] have reported evidence that the influence can be even larger in the preexponential factor A for some free-radical reactions.

TABLE I - Arrhenius parameters for the addition of primary alkyl radicals to protonated heteroaromatic bases[3a]

Heteroaromatic base	E_a (kcal/mol)	log A
4-Methylpyridine	6.9	9.2
4-Methylquinoline	5.9	9.2
4-Cyanopyridine	4.9	9.3
Quinoxaline	2.8	9.3

Reactions of protonated amino radicals

The protonation strongly increases the polar character of the amino radicals, whose reactions with alkanes, alkenes, alkynes and aromatics are among the most selective free-radical reactions. Often the extent of the polar effect of the reactions of the amino radical cations is comparable to that of the most selective electrophilic reactions.

Alkane halogenation by N-haloamines in acidic medium

In 1966-1968 a highly selective method of halogenation of alkanes by N-haloamines has been developed by us[9b,12-19,22]. An exceptional selectivity in ω-1 positions in 1-substituted n-alkanes has been reported (Table II). A chain mechanism, in which the alkyl amino radical cation is the chain-carrying species, has been suggested (Scheme I).

TABLE II - Isomer distribution of the homolytic chlorination and bromination of substituted alkanes

Radical source	Substituted alkane							Ref.
Me_2NCl	Cl	CH_2	CH_2	CH_2	CH_2	CH_3		12
			2.5	16.7	77.2	3.9		
Me_2NCl	Cl	CH_2	CH_2	CH_2	CH_2	CH_2	CH_3	13
		–	–	6.8	19.4	71.6	2.2	
Cl_2		1.8	9.4	21.3	23.9	26.5	17.2	
	MeOCO	CH_2	CH_2	CH_3				12,14
Me_2NCl		–	86.4	13.6				
	MeOCO	CH_2	CH_2	CH_2	CH_3			12,14
Me_2NCl		–	13.3	79.9	6.7			
Me_2NBr		0.47	6.0	81.3	12.2			15
Cl_2	1.3	5.4	30.1	44.5	18.7			16
Br_2		35.0	19.6	45.4	–			16
	MeOCO	CH_2	CH_2	CH_2	CH_2	CH_3		
Me_2NCl		–	0.7	6.3	87.3	6.7		12,14
Me_2NBr		–	0.41	5.9	87.0	6.8		15
$i\text{-}Bu_2NCl$		–	–	3.0	89.9	7.1		17

	MeOCO	CH_2	CH_2	CH_2	CH_2	CH_2	CH_3	
MeNHCl		-	-	2.2	19.5	71.6	6.3	9b
Me_2NCl		-	-	1.2	14.7	80.3	7.2	9b
$n\text{-}Bu_2NCl$		-	-	1.2	11.8	81.2	5.8	9b
$i\text{-}Bu_2NCl$		-	-	0.8	10.2	82.8	6.6	9b

	MeOCO	CH_2	CH_2	CH_2	CH_2	CH_2	CH_2	CH_2	CH_2	CH_3	
$i\text{-}Bu_2NCl$		-	-	-	1.7	6.7	13.4	19.1	57.7	1.4	18

	MeCOO	CH_2	CH_2	CH_2	CH_2	CH_2	CH_3	
Me_2NCl		-	-	1.2	14.7	80.4	3.7	17
Me_2NBr		-	-	1.7	14.6	80.8	2.9	19
$i\text{-}Bu_2NCl$		-	-	0.8	10.2	84.6	4.4	19

	CH_3	CH_2	CH_2	CH_2	CH_2	CH_2	CH_3	
Me_2NCl	1.1	55.6	29.0	14.3				18
	1	75.8	39.5	39 (per hydrogen atom)				
$i\text{-}Bu_2NCl$	1.3	64.4	22.9	11.3				18
	1	74.3	26.4	26.1 (per hydrogen atom)				

Initiation $$R_2\overset{+}{N}HCl + M^+ \longrightarrow R_2\overset{+}{\dot{N}}H + MCl^+ \quad (M^+ = Fe^{++}, Cu^+)$$

Propagation $$R_2\overset{+}{\dot{N}}H + H\text{-}R' \longrightarrow R_2\overset{+}{N}H_2 + R'^{\bullet}$$

$$R'^{\bullet} + R_2\overset{+}{N}HCl \longrightarrow R' - Cl + R_2\overset{+}{\dot{N}}H$$

Termination $$R_2\overset{+}{\dot{N}}H + M^+ + H^+ \longrightarrow R_2\overset{+}{N}H_2 + M^{++}$$

Scheme I

Steric, enthalpic and polar effects were suggested[12-19] to contribute to the exceptional selectivity. Particularly the polar effect had a magnitude never observed before in hydrogen abstractions. Many evidences, partly summarized in

Table II, supported the mechanism of Scheme I and excluded that a halogen atom could be the chain-carrying species of the halogenation:

(i) The positional selectivity with substituted alkanes were quite different from those observed with Cl. and Br.; the halogenation by N-haloamines was much more sensitive to the polar substituents (Table II).

(ii) The reactivity and selectivity of chlorination and bromination by N-chloro and N-bromoamines were identical, strongly suggesting that the abstracting species is the same in the two cases.

(iii) The substrate selectivity was also very high. Thus high conversions and high yields of monosubstitution were obtained in contrast with the conventional halogenation by Cl_2, where moderate yields of monosubstitution are obtained only at low conversions. The polar effect is the main responsible for this behaviour because the introduction of a halogen atom strongly deactivates the substrate towards further hydrogen abstraction.

(iv) The selectivity, identical for chlorination and bromination, between CH_2 and CH_3 in unsubstituted alkanes[18] is much higher than for the hydrogen abstraction by chlorine atom and close to that of bromine. This fact is very significant because the energy of the bond $Me_2\overset{+}{N}H$-H has been evaluated[13] in the range of 94-101 kcal/mole and therefore the hydrogen abstraction is either thermoneutral or exothermic; that is confirmed by the fact that the reaction is fast at room temperature also with N-bromoamines (it is completed in few minutes when an effective initiator is utilized). The hydrogen abstraction by bromine is an endothermic process and it is much slower at room temperature. Thus, in our opinion[21], not only the enthalpic, but also the polar factor contributes to determine the CH_2/CH_3 selectivity of hydrogen abstraction by amino radical cations.

(v) The CH_2 group in position 2 of n-heptane is more reactive than the CH_2 groups in the positions 3 and 4. This selectivity was ascribed to a high sensitivity to steric effects and it increases with the bulk of the alkyl groups in the dialkylamino radical.

(vi) The intramolecular chlorination of N-chloro-n-hexylamine in acidic medium mainly leads to pyrrolidines, whereas the intermolecular chlorination of n-hexylammonium salt by $Me_2\overset{+}{N}HCl$ mainly gives piperidines[21], thus clearly showing the difference of the two processes.

(vii) The chlorination of methyl 2-hexenoate[14] by $Me_2\overset{+}{N}HCl$ gives the attack of the position 5 even higher than that of the position 4, although this last position is allylic; that is the polar effect is more important than the enthalpic effect.

In spite of this great deal of evidences Tanner[23] in 1969 questioned the mechanism (Scheme I) suggesting that the chlorination did not proceed via an amino radical cation, but rather via a chlorine atom chain (Scheme II), likewise the Goldfinger mechanism of bromination by N-bromosuccinimide.

$$R\cdot + Cl_2 \longrightarrow R\text{-}Cl + Cl\cdot$$

$$Cl\cdot + R\text{-}H \longrightarrow R^{\cdot} + HCl$$

$$HCl + R'_2\overset{+}{N}HCl \longrightarrow R'_2\overset{+}{N}H_2 + Cl_2$$

Scheme II

Tanner stated: "It was immediately apparent that an inconsistency in the proposed mechanism (Scheme I) existed: the variability of product distribution with the initiator used (Fe^{2+}, Cu^{+})".

Actually in concentrated sulphuric acid the reported variability was very small as the results of Table III indicate. It was really surprising that such a small variability, almost within the limits of the experimental errors, could cancel the great deal of evidences provided by the points (i)-(vii).

To explain our selectivity Tanner stated, without any experimental evidence, that our results were not genuine, but due to the selective solvolysis of the haloderivatives by sulphuric acid. Such a statement was still more surprising because it was very easy to verify by treating the mixture of chloroderivatives under the reaction conditions and observing the change of isomer distribution.

TABLE III - Isomer distribution for the chlorination of methyl hexanoate by Me_2NCl in 96% H_2SO_4 (initiator effect)[14]

Initiator	MeOCO	$-CH_2$	$- CH_2$	$- CH_2$	$- CH_2$	$- CH_3$
Fe^{++}		-	0.7	6.3	87.3	5.7
Cu^{+}		-	0.3	4.7	89.4	5.6

In all cases investigated we did not observe significant changes of the isomer distributions, due to the reaction medium, even because the reactions are completed in few minutes at room temperature by effective metal salt initiators (Fe^{2+}, Cu^{+}), which are preferable to the photochemical initiation. In any case an incidental solvolysis should decrease and not increase the selectivity as suggested by Tanner: the solvolysis of a -CHCl- group is faster than that of a $-CH_2Cl$ group and that should be reflected in a lower CH_2/CH_3 selectivity; moreover, as Deno[24] has rightly evidenced, the solvolysis rate of the -CHCl- groups increases with the distance from the electron-withdrawing substituents and that should be reflected in a lower ω-1 selectivity.

Paradoxically, it was just the strong polar effect, contested by Tanner, which misled this author because he investigated a substrate, 1-chloropropane, too deactivated for polar reasons towards the attack of the amino radical cations, whereas we had already reported[12] that the less deactivated 1-chlorobutane, 1-chloropentane and 1-chlorohexane were halogenated with a high selectivity, absolutely inconsistent with a halogen atom chain.

Two research groups, Ingold[25] and Deno[28], subsequently confirmed our mechanism reporting similar results when the same substrates were investigated (Table IV).

In less acidic medium the reaction is more complex and the low selectivity reported by us and Deno[29] at low acidity has been ascribed by us[30] to a superimposition of a chlorine carrying chain in these conditions, without, however, identifying the mechanism of the influence of the acidity on the Cl_2 formation. More recently we[31] have recognized a new chain process of the protonated N-chloroamines (Scheme III) which well explains the acidity effect and the formation of Cl_2 .

TABLE IV - Isomer distribution for the chlorination and bromination of 1-chlorobutane and methylhexylether

Radical source							Ref.
	Cl - CH_2	- CH_2	- CH_2	- CH_3			
Me_2NCl	-	6.0	88.5	5.5			Minisci[12] (1967)
Me_2NCl	traces	6.6	87.8	5.4			Minisci[13] (1970)
Me_2NCl	1.2	6.1	87.3	5.4			Ingold[25] (1970)
Cl_2	7.8	23.5	49.2	19.5			Walling[26] (1959)
Br_2	23	21.7	55.3	traces			Tedder[27] (1960)
	MeO-CH_2-	CH_2-	CH_2-	CH_2-	CH_2-	CH_3	
Me_2NCl	-	-	0.9	11.3	82.5	4	Minisci[17] (1968)
Me_2NBr	-	-	0.9	11.7	83.0	4	Minisci[19] (1969)
i-PrNCl	-	-	-	-	92	8	Deno[28] (1971)

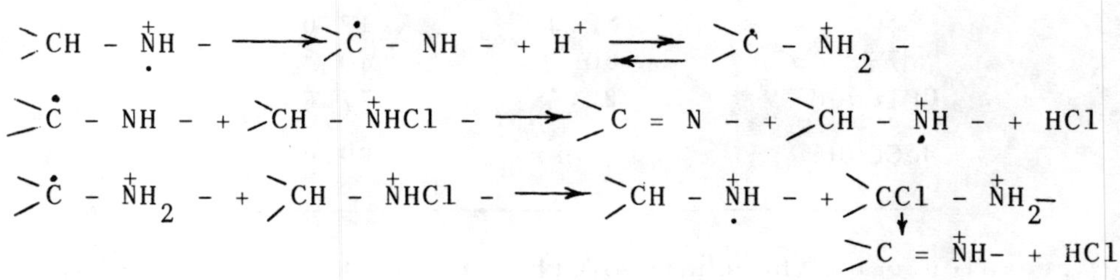

Scheme III

Thus with dicyclohexyl-N-chloroamine the intramolecular hydrogen abstraction, which is slower compared with the acyclic alkylamine because the cyclohexyl group must be forced to the boat conformation, competes with the irreversible loss of the α-proton (Scheme IV) and the ratio of the reaction products 1 and 2 is strongly affected by the acidity (Table V).

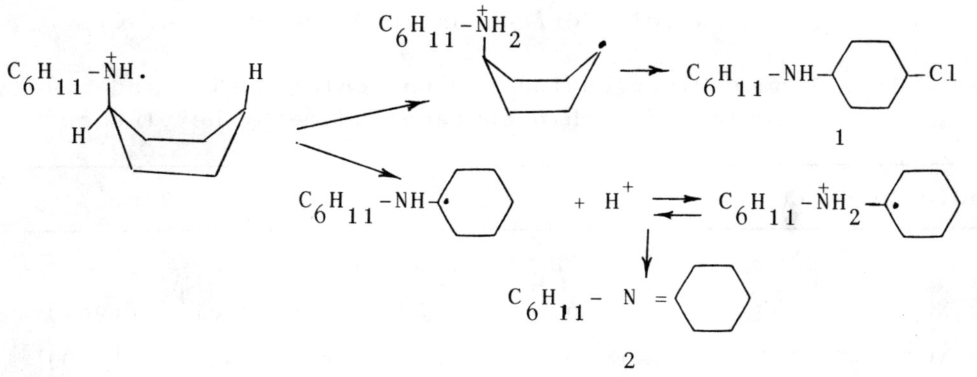

Scheme IV

TABLE V - Reaction products of the free-radical decomposition of dicyclohexyl-N-chloroamine[31]

H_2SO_4 %	1 %	2 %
96	97	3
92.5	93.4	6.6
90	91.9	8.1
80	82.4	17.6
75	60	40
CF_3COOH	22.5	77.5
MeCOOH	traces	99.9

The prevalence of the chain of the Scheme III at low acidity leads to HCl, responsible for the Cl_2 formulation (eq 20)

$$R_2\overset{+}{N}HCl + HCl \longrightarrow R_2\overset{+}{N}H_2 + Cl_2 \qquad (20)$$

Actually, the reaction (20) does not appear to be fast compared with the free-radical chain, but it becomes important with substrates deactivated by polar groups, as 1-chloropropane. Thus the incursion of chlorine atom chains in the chlorination by N-chloroamines must be mainly ascribed to three factors: (a) A low acidity favours the chain of Scheme III and the possibility to form Cl_2. (b) A low reactivity of the substrate, often associated with a low solubility in the reaction medium, also favours the chain of Scheme III. (c) A less effective initiator can also contribute to the formation of Cl_2 according to eq 20 because it increases the reaction times, thus explaining the small differences observed in our results[14] (Fe(II) and Cu(I) salts are both effective initiators) and the larger differences subsequently observed with less effective metal salt initiators at lower acidity by Ingold[25].

The intramolecular chlorination (Hoffman-Löefler reaction) is less sensitive to the incursion of chlorine atom chain compared with the intermolecular process for problems of solubility, which is low with substrates such as haloalkanes in the reaction medium, and reactivity. Thus di-n-butyl-N-chloroamine gives only the intramolecular process in the presence of 1-chlorobutane[25]; however in the presence of methyl heptanoate the intermolecular mode competes with the intramolecular process[9b] (Table II). In this last case the higher reactivity expected for an intramolecular process is counterbalanced by the lower polar deactivation of the position 6 of the heptanoate compared with the position 4 of the protonated butylamino radical and by the CH_2/CH_3 selectivity. Since the intramolecular hydrogen abstraction by the amino radical cations has been evaluated[9a] $>10^6 s^{-1}$, the intermolecular hydrogen abstraction from non-deactivated CH_2 group must be very fast.

A very recent paper of Tanner[32] does not consider any more the hydrogen abstraction by amino radical cations a phantom process, but he claims that for the first time his results show how the intra- and intermolecular chlorinations of alkylamonium derivatives are different, completely ignoring that almost 20 years ago we[22] have shown that the intramolecular chlorination of N-chlorohexylamines mainly gives pyrrolidines, whereas the intermolecular chlorination of hexylammonium salts by $Me_2\overset{+}{N}HCl$ mainly gives piperidines and that these results have been reported in reviews of large diffusion[33]. Moreover Tanner in this paper has not yet realized that his previous results were due to the deactivated substrate investigated, 1-chloropropane, never utilized by us, but he ascribes his different results to the different procedure of synthesis of the N-chloroamine, which actually is irrelevant. What has relevance in determining the incursions of chlorine chains are the factors (a)-(c) above discussed, completely ignored by Tanner, whereas the presence of inhibitor impurities does not substantially influence the selectivity when an effective initiator is utilized in concentrated H_2SO_4.

Amination of aromatics by amino radical cations

Amino radical cations add very fast to non-deactivated aromatic rings giving highly selective aromatic substitutions. Simple amino radical, $\dot{N}H_3^{+}$, mono- and dialkylaminoradicals, arising from N-haloamines, hydroxylamines and hydroxylamine-O-sulphonic acids, have been utilized.[34]

With activated aromatic substrates, such as phenol, anisole, resorcinol dimethyl ether a strongly acidic medium is not required because the addition of the protonated alkylamino radical to the aromatic ring is in any case faster than the loss of the α-proton according to the Scheme III. A strongly acidic medium is not even necessary with the simple amino radical, $\dot{N}H_3^{+}$, because there is no possible competition of loss of an α-proton. With benzene or slightly activated (i.e. alkyl-benzene) or slightly deactivated aromatics (i.e. halobenzene derivatives) a strongly acidic medium is necessary in order to overcome the competitive chain of Scheme III. Thus the aromatic amination of benzene by methyl-benzyl-N-chloroamine gives the substitution product (3) and the corresponding imines (4 and 5) in ratios strongly affected by the acidity[31] (Table VI) (Scheme V)

TABLE VI - Reaction products of the free-radical decomposition of methyl-benzyl-N-chloroamine in benzene[31]

H_2SO_4 %	3 %	4 %
96	98.4	1.6
90	98.2	1.8
85	95.9	4.1
80	84.9	15.1
70	traces	99.9

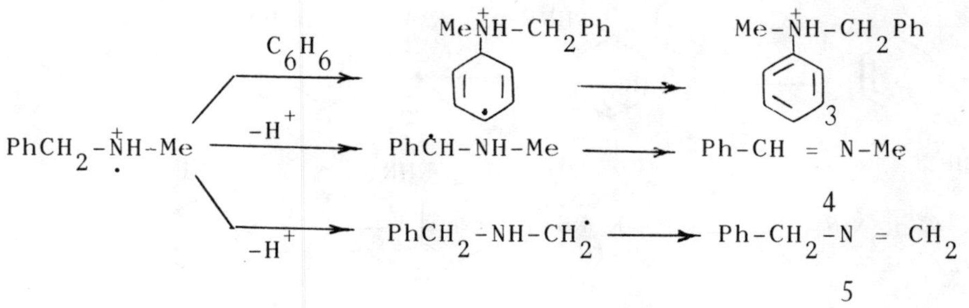

Scheme V

The results of Tables V and VI explain another obscure aspect of the chemistry of the amino radical cations: the fact that when these radicals were generated by other radical sources (chemical or electrochemical oxidation of amines, photolysis of N-nitroso amines, redox decomposition of alkylhydroxylamines, alkylhydroxylamine-O-sulphonic acids and N-oxides of t-alkylamines) the hydrogen abstraction from C-H bonds and the addition to benzene did not occur, whereas the addition to non-deactivated olefins and to activated aromatics took place in all cases. Compared with N-haloamines the main difference was constituted by the acidity of the used medium. Now the overall behaviour is clear: the addition of the alkylamino radical cations to olefins or activated aromatics is in all cases faster or at least competitive, than the loss of the α-proton, whereas the hydrogen abstractions or the additions to non-activated aromatics are slower than the loss of the α-proton in non strongly acidic medium.

The stoichiometry of the amination by N-chloroamines is shown by eq 21

$$Ar\text{-}H + R_2\overset{+}{N}HCl \longrightarrow Ar\text{-}\overset{+}{N}HR_2 + HCl \qquad (21)$$

Since the yields of the aromatic amination are high, the reaction of HCl with $R_2\overset{+}{N}HCl$ leading to Cl_2 (eq 20) appears to be much slower than the free-radical chain leading to the aromatic substitution (Scheme VI)

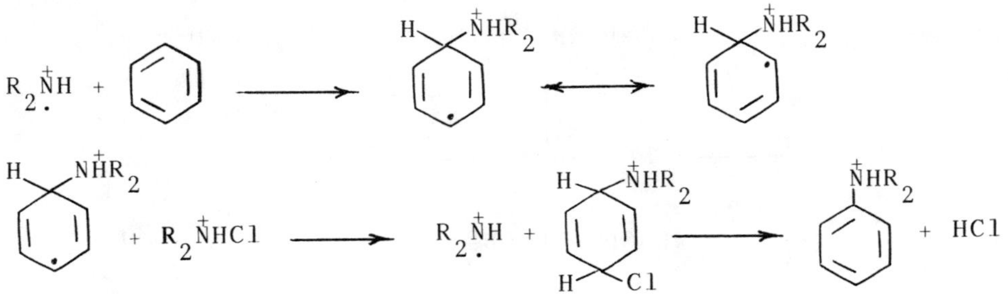

Scheme VI

Qualitative thermodynamic considerations led us to exclude in strongly acid medium an electron-transfer process (eq 22) in the rearomatization step because the protonation considerably increases the redox potential of the aminocyclohexadienyl radical

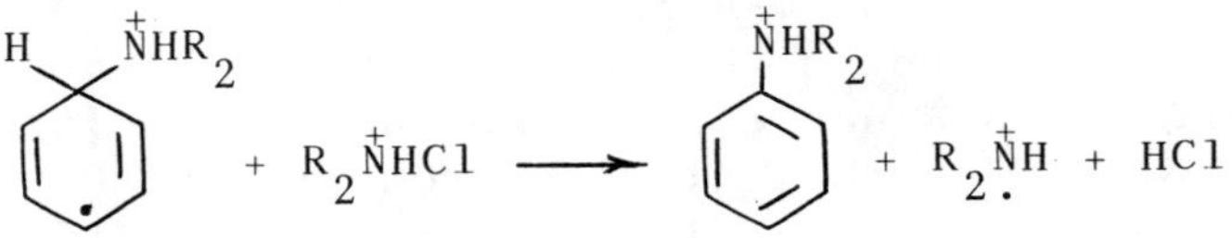

(22)

Such a mechanism, however, cannot be excluded when electron-releasing groups (OH, OR, NHCOR) in the aromatic ring decrease the redox potential of the radical adduct and allow to work at low acidity. The exceptional polar effect is shown by the regioselectivity with activated aromatic derivatives and polycyclic derivatives (Tables VII and Table VIII) and by the fact that electron-withdrawing substituents (NO_2, CN, COR, $R_3\overset{+}{N}$ etc.) completele inhibit the reaction. That is a limitation, but it is also useful from synthetic point of view because one of the main complications of the homolytic aromatic substitutions is the low substrate selectivity, which often leads to a complex mixture of polysubstituted products as the conversions increase. In the homolytic amination the introduction of a protonated amino group completely deactivates the aromatic ring towards further attack so that complete conversions of the aromatic ring occurs without significant formation of disubstituted products. In some cases, as with naphtalene, the polar effect of the protonated amino group deactivates both the aromatic rings and only traces of disubstituted derivatives are formed at complete conversion of naphtalene.

TABLE VII - Homolytic amination of activated aromatic derivatives

Aromatic substrate	N-chloroamine	Orientation (%)	Yield (%)	Ref.
Acetanilide	Me_2NH	p (100)	93	35
"	Piperidine	p (100)	98	35
"	Methylbenzyl amine	p (100)	88	35
Phenol	Piperidine	o (9) p (91)	97	36
Anisole	"	o (8.9) p (91.1)	85	37

TABLE VIII - Homolytic amination of polycyclic aromatic compounds

Aromatic substrate	N-chloroamine	Orientation (%)	Yield (%)	Ref.
4-Nitrobiphenyl	Me_2NH	4' (100)	86	38
"	Piperidine	4' (100)	85	38
4-Biphenylsulphonic acid	Me_2NH	4' (100)	86	38
4-Chlorobiphenyl	"	4' (100)	84	38
4-Aminobiphenyl	"	4' (100)	96	38
Fluorenone	"	2 (100)	98	38
Oxindole	"	5 (100)	86	37
8-Methoxyquinoline	"	5 (100)	89	39

The very high sensitivity to the inductive polar effect is shown by the partial rate factor of the meta position in the amination of toluene, which is one of the highest values ever observed even among the electrophilic aromatic substitutions (Table IX). The reaction is also very sensitive to conjugative polar effects, as shown by the partial rate factors of naphtalene (Table X) comparable with the most selective electrophilic species, and anisole (Table XI). Thus a large change separation must be working in the transition state (eq 23),

similar to a change-transfer complex

(23)

TABLE IX - Partial rate factors for the meta position of toluene in the homolytic amination and electrophilic substitutions

Reaction	Partial rate factor	Ref.
Amination by $Me_2\overset{+}{N}H\cdot$	8	40
Nitration	1.2-3	41
Sulphonation	5.7	"
Chlorination	4-9.1	"
Bromination	2.5-10	"
Ethylation	1.6	"
Acetylation	4.8	"

TABLE X - Partial rate factors for the homolytic amination and electrophilic substitutions of naphtalene

Reaction	Partial rate factor		Ref.
	α	β	
Amination by $Me_2\overset{+}{N}H\cdot$	9.6×10^4	2.9×10^3	42
Nitration	470	50	41
Chlorination	9.1×10^4	-	"
Bromination	2.1×10^5	2.1×10^3	"

TABLE XI - Partial rate factors for the para position in the amination of x-substituted benzene derivatives[43]

X	$\cdot\overset{+}{N}H_3$	$Me_2\overset{+}{N}H\cdot$
OMe	20.1	10^3
Me	4.2	51
H	1	1
CN	$2.5x10^{-2}$	No reaction

The polarity of the protonated radical certainly is a fundamental factor of the sensitivity to the polar effect, but also the reaction enthalpy is a basic factor in influencing the polar selectivity. That is illustrated by the partial rate factors in the amination by $\overset{+}{\underset{\cdot}{N}}H_3$ and $Me_2\overset{+}{\underset{\cdot}{N}}H$ radicals (Table XI). The radical $\overset{+}{\underset{\cdot}{N}}H_3$ has a somewhat higher electron affinity that the radical $Me_2\overset{+}{\underset{\cdot}{N}}H$, however it is a much less electrophilic species for enthalpic reasons. The bond energies[20] for $\overset{+}{N}H_3$-H and $Me_2\overset{+}{N}H$-H have been evaluated respectively in the ranges of 123-128 and 94-101 kcal/mol and therefore a more exothermic process occurs for the aromatic addition of $\overset{+}{\underset{\cdot}{N}}H_3$ compared with $Me_2\overset{+}{\underset{\cdot}{N}}H$. This higher exothermicity is reflected in an earlier transition state with less charge-transfer character. The high rate of addition of protonated amino radicals to benzene is emphasized by the fact that with N-chloro-n-butylamine and N-chloro-n-butylmethylamine the addition to benzene is competitive with the intramolecular hydrogen abstraction (Scheme VII), in spite of the low solubility of benzene in the reaction medium[7a]

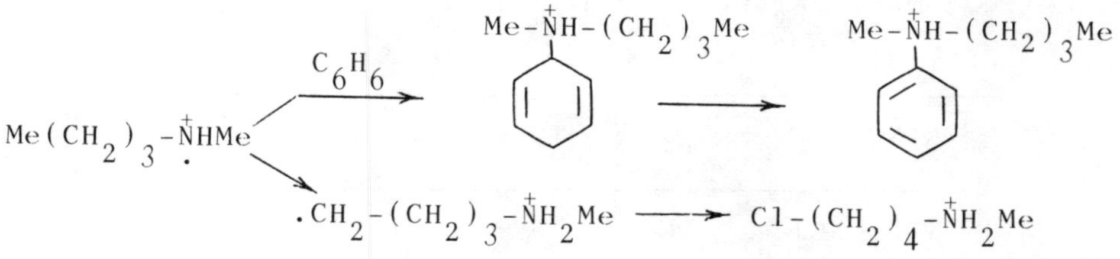

Scheme VII

Benzylic halogenation by N-haloamines

Combined steric and polar effects have allowed to develop the most selective method of benzylic chlorination. Alkylbenzenes can give with protonated amino radicals addition to the aromatic ring and benzylic hydrogen abstraction. Both reactions are strongly influenced by polar effects, but also by steric effects. These last are, however, much more marked in the addition to the aromatic ring than in hydrogen abstraction because in the first case the nitrogen-centered radical must approach the nuclear carbon atom, in the other case it must approach a hydrogen atom in the transition states. Thus bulky dialkylamino radicals (diisopropyl, diisobutyl, dicyclohexyl) only give hydrogen abstraction, whereas less hindered amino radicals ($\dot{N}^{+}H_3$, $R-\dot{N}^{+}H_2$, $Me_2\dot{N}^{+}H$, pyrrolidine, piperidine, morpholine) only give nuclear addition; examples of intermediate steric effects, in which both the reactions simultaneously occur, are obviously possible. The results of benzylic chlorination of m-xylene to m-dichloromethyl-benzene, particularly investigated for practical purposes[44], illustrates the method (Table XII). With the utilized N-chloroamines no trace of nuclear addition was observed for steric reasons and the polar effect plays a dominant role in determining the high selectivity because the introduction of a chlorine atom completely deactivates the chloromethyl group towards further chlorination. No other known method of halogenation provides a selectivity comparable with these values. The utilized amines are very cheap, commercial products, they however, can be easily recovered and recycled, making the overall process particularly unexpensive.

TABLE XII - Selectivity of the reaction[44]

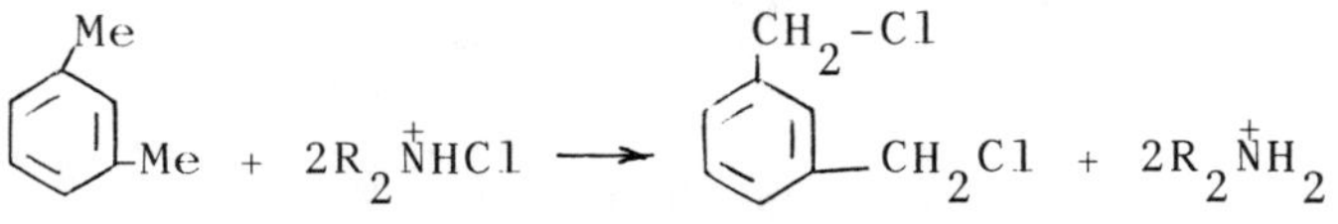

R_2NCl	Selectivity %
$i\text{-}Pr_2NCl$	91.9
$i\text{-}Bu_2NCl$	92.2
$(c\text{-}C_6H_{11})_2NCl$	96.1

Reactions of carbon-centered radicals with positively charged unsaturated systems

The addition of nucleophilic carbon-centered radicals to unsaturated systems bearing a positive charge is characterized by large polar effects, which have allowed to develop new reactions of high synthetic value. Protonated heteroaromatic bases and diazonium salts have been extensively investigated among the unsaturated systems, but recently pyrilium and iminium salts have shown quite a similar behaviour, suggesting a large contribution of charge separation to the transition states of all these systems (eq 24)

$$R\cdot \quad X = Y^{+} \longleftrightarrow R^{+} \quad X = \dot{Y} \qquad (24)$$

Thus we have previously strongly increased the polar character of the amino radicals by protonation, in these cases the polarity of the radicals is generally low, but the polarity of the substrates is strongly increased very often by protonation. However, the overall polar effect is quite similar in the two cases.

Substitutions of heteroaromatic bases by nucleophilic radicals

The substitution of protonated heteroaromatic bases by nucleophilic carbon-centered radicals is a general reaction[45] of[46] high synthetic interest in that it reproduces most of the numerous aspects of the Friedel-Crafts aromatic substitution, but with opposite reactivity and selectivity. As the Friedel-Crafts substitutions, also these homolytic reactions can be divided in generalized sense into alkylation and acylation, which include reactions of considerable diversity.

The great synthetic interest of the heteroaromatic substitution results from several factors:

(i) A large variety of radical sources allows to generate nucleophilic radicals corresponding to most of the electrophilic species utilized in the Friedel-Crafts substitutions. Several of the most important classes of organic compounds (alkanes, alkenes, alkylbenzenes, alkyl halides, alcohols, ethers, aldehydes, ketones, carboxylic acids, esters, amides, amines, peroxides, N-chloroamines, oxaziridines, etc.) have been successfully utilized.

(ii) A very large variety of carbon-centered radicals proved to be suitable for the selective heteroaromatic substitutions. The substituents on the radical center play a fundamental role in determining the polar character of the radical and the consequent reactivity and selectivity. The main synthetic limitation occurs when an electron-withdrawing group is directly bound or conjugated to the radical center because the decreased nucleophilic character completely inhibits the reaction. Thus, for example, the radicals $.CH_2COOEt$ generated in several solvents from ethyl iodoacetate and benzoyl peroxide do not attack the protonated bases, but in benzene solution add to the solvent[7b] (eq 25)

$$C_6H_6 + I\text{-}CH_2COOEt + (PhCOO)_2 \longrightarrow Ph\text{-}CH_2COOEt + Ph\text{-}I + PhCOOH + CO_2 \qquad (25)$$

Benzoyl peroxide is a very selective source of alkyl radicals from alkyl iodides because the rates of iodine abstraction by the phenyl radical are close to the diffusion controlled limit[47] (eq 26)

$$Ph. + I\text{-}CH_2COOEt \xrightarrow{k} Ph\text{-}I + .CH_2COOEt \quad k \simeq 10^9 M^{-1} s^{-1} \qquad (26)$$

However, it is sufficient that the electronwithdrawing group is in the β position and the alkyl radical has enough nucleophilic character for the selective alkylation of the heteroaromatic bases. Thus the same reaction with ethyl β-iodopropionate in benzene solution (benzene : lepidine 50:1) leads only to the selective attack of the heteroaromatic base (eq 27) without significant addition to the solvent[7b]

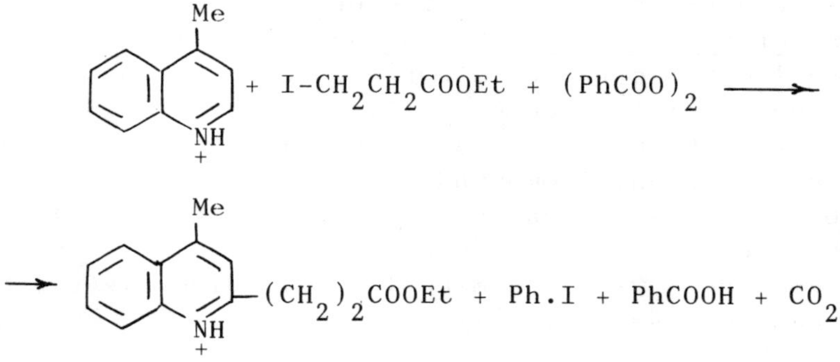

(27)

(iii) The regio- and chemoselectivity are exceptionally high. Carbon-centered radicals are reactive species; they undergo a variety of reactions: hydrogen and halogen abstraction, additions to unsaturated systems, oxidation, reduction, isomerization, dimerization, disproportionation, etc. Thus a high substrate selectivity is the primary condition for the synthetic success and for the utilization of a large variety of radical sources.

The results of Table XII indicate that the rates of addition of alkyl and acyl radicals to protonated heteroaromatic bases can be much higher than those of most of the possible competitive reactions of the same radicals. Moreover, these results clearly emphasize two fundamental aspects of these reactions: (a) t-Alkyl, hydroxymethyl and acyl radicals can be even more reactive than primary alkyl radicals, in spite of the unfavourable steric and enthalpic factors, thus showing the dominant role of the polar effect, which is also emphasized by the high regioselectivity. Actually with the most nucleophilic radicals (t-alkyl, acyl, hydroxymethyl) the rate constants of Table XIII must be considered minima values because they were evaluated by competitive methods, in which the "radical clocks" are irreversible processes, whereas the radical addition to the heterocyclic rings are reversible. (b) Phenyl is a highly reactive, unselective radical, in agreement with the Reactivity-Selectivity Principle; this is a generally valid criterion for free-radical reactions when the reactivity is mainly governed by the reaction enthalpy; however, when polar effects are important the Principle can be reversed[3a,10]. Phenyl radical is actually in several reactions much more reactive and less selective than alkyl and acyl radical; that occurs also in addition to benzene. However, with protonated heteroaromatic bases alkyl and acyl radicals are always much more selective, but they can be even more reactive than the phenyl radical. On the other hand when polar effects are important in free-radical reactions it has less sense to consider the reactivity of the free radicals in general without specifying the particular kind of reaction and the particular selectivity involved.

TABLE XIII - Absolute rate constants ($M^{-1}s^{-1}$) (25-60°) for the addition of carbon-centered radicals to benzene and protonated heteroaromatic bases

Aromatic substrate	Radical				
	n-Bu.[3a]	t-Bu.[3b]	$.CH_2OH$[5]	t-BuĊO[4]	Ph.[7b]
Benzene	3.8×10^2	no react.[6]	no react.	no react.	1.03×10^6
4-Acetyl-pyridine	6.8×10^5	4.6×10^6	-	-	-
4-Cyano-pyridine	8.9×10^5	6.7×10^6	$\sim10^7$	$\sim10^6$	6.10×10^6
Quinoline	8.5×10^5	4.1×10^6	$\sim10^7$	$\sim10^6$	-
Quin-oxaline	2.7×10^7	-	-	3.6×10^6	-
Benzo-thiazole	9.9×10^5	-	-	7.1×10^5	-

(iv) The variety of heteroaromatic bases is very large. All the heteroaromatic bases, with at least one α or γ free position, are in general suitable for these substitutions. Sometimes also the ipso-substitution can occur in α and γ positions.

(v) The rearrangement of the carbon-centered radicals in a minor pathway. Contrary to the carbonium ions involved in Friedel-Crafts reaction the corresponding radicals do not generally undergo similar rearrangements during the substitutions because they have less tendency than the carbonium ions to rearrange and the fast addition to the heterocyclic ring sometimes overcomes the possible rearrangement. This has been observed in the heteroaromatic substitution only in particular cases, but the phenomenon is very limited.

(vi) Simple and mild experimental conditions, unexpensive reagents and catalysts can be often used providing the conditions also for commercial applications.

(vii) Highly unstable radicals can be utilized. Thus t-acyl radicals, t-alkyl-ĊO, whose decarbonylation to t-alkyl radical and CO is a fast process[49] (10^4-$10^5 s^{-1}$), have been successfully utilized for the heteroaromatic substitution[4].

(viii) The rearomatizaion step is easy and selective.

All these aspects have been described in reviews[45], recent articles[46] and patents[48]. In this paper the point (viii) will be discussed in relation to the substituent effect on the radical reactivity, the synthetic involvements, the solvent effect and a possible captodative effect, the study of which is still in progress.

Rearomatization step, substituent effect and synthetic involvements

A factor which affects the synthetic success of the homolytic aromatic substitutions is the rearomatization step of the aromatic radical adduct. The substitutions of protonated aromatic bases with a large variety of radical sources and nucleophilic radicals are very selective. Now the question arises why the rearomatization of the intermediate radical adduct is not selective in benzene series when the same radical sources and identical reaction conditions are employed as with protonated heteroaromatic bases. From the induced decomposition of several radical sources we[50] have suggested that the easy and selective rearomatization is also due to polar effects according to the Scheme VIII

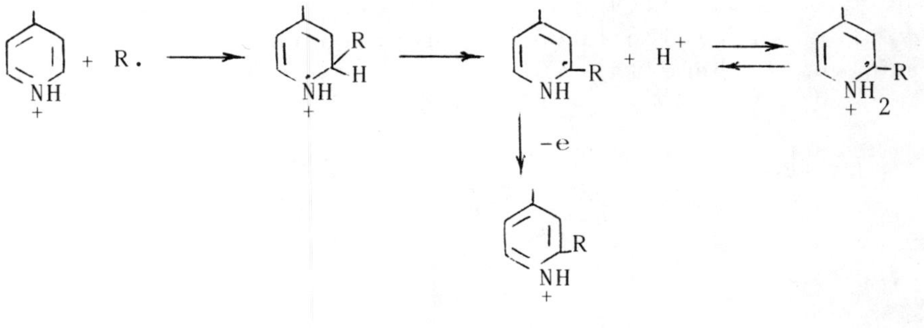

Scheme VIII

The radical adduct can irreversibly lose a proton from the α C-H bond giving a pyridinyl radical (the process is similar to that described by the Scheme III with N-chloroamines). Pyridinyl radicals are very weakly basic (pka 2-3) compared with the corresponding dihydropyridine (pka 7) and the unprotonated radical must be present in significant amount at the equilibrium in non strongly acidic medium. Moreover, the ionization potential of the α-aminoalkyl radicals (5.4 eV) are the lowest observed for any organic and organometallic species[51], close to that of Li and Na. It is therefore not surprising that very mild oxidants can fast and selectively oxidize the pyridinyl radicals.

This interpretation has allowed to develop new synthetic methods and different selectivities of the heteroaromatic substitution[46]. All the other conditions being equal, the reactivity and the selectivity of the reaction increase by increasing the nucleophilic character of the radical. Actually, a too marked nucleophilic character can entail synthetic limitations due to two factors:

(a) The increased nucleophilic character is generally associated with an increased stability and oxidability of the radical: in particular oxygen and nitrogen atoms in α position (i.e. radicals from alcohols, ethers, amines, N-alkylamides) increase the nucleophilicity, the stability and the oxidability of the alkyl radicals (eq 28, 29)

$$>\dot{C}-O- \longleftrightarrow >\bar{C}-\dot{O}^{+}- \qquad (28)$$

$$>\dot{C}-N< \longleftrightarrow >\bar{C}-\dot{N}^{+}< \qquad (29)$$

The increased stability can be reflected in the reversibility of the addition of the radical to the heterocyclic nucleus (Scheme IX)

Scheme IX

The oxidations of R. and of the pyridinyl radical are in general irreversible and the equilibrium of the Scheme IX is governed by the relative values of k_1, k_{-1}, k_2 and k_3. On the other hand an oxidizing medium is required for the selective rearomatization of the radical adduct.

(b) A too nucleophilic character of the radical can determine favourable conditions for an electron-transfer process, particularly with bases of high electron affinity, thus preventing reaction from substitution (eq 30)

(30)

Thus, for example, the α-hydroxymethylation of heteroaromatic bases with methanol gives good results with a variety of oxidants (H_2O_2, t-BuOOH, $S_2O_8^{2-}$, perborate, percarbonate, NH_2OSO_3H), whereas with ethanol only traces of α-hydroxyethylation were obtained under the same conditions and no substitution was observed with isopropanol. The difficulties with ethanol have been overcome mainly by influencing the rate constant k_2 of the Scheme IX by the use of a very mild oxidizing radical source, the redox couple $\overset{+}{N}H_3OH/Ti(III)$ (eq 31)

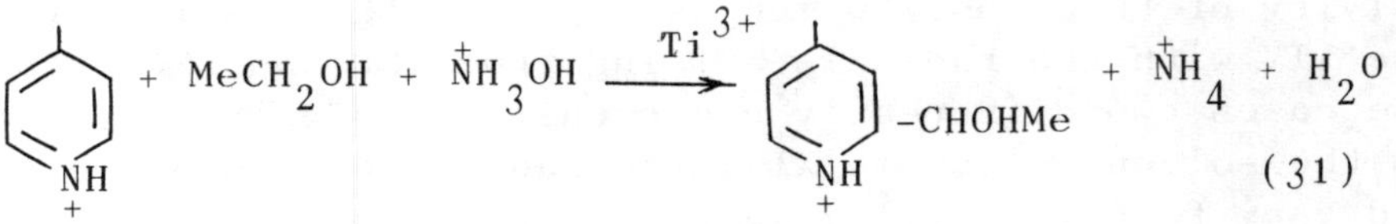

(31)

The mechanism of the reaction is shown by the Scheme X. The oxidation rate of the α-hydroxyethyl radical (k_2) is relatively low by Ti(IV) salt, which has however sufficient oxidizing character for the selective oxidation of the pyridinyl radical (k_4); thus the equilibrium of the Scheme X is shifted on the right ($k_1 > k_2$).

$$\overset{+}{N}H_3OH + Ti^{3+} \longrightarrow .\overset{+}{N}H_3 + Ti^{4+} + OH^-$$

$$.\overset{+}{N}H_3 + MeCH_2OH \longrightarrow \overset{+}{N}H_4 + Me\dot{C}HOH$$

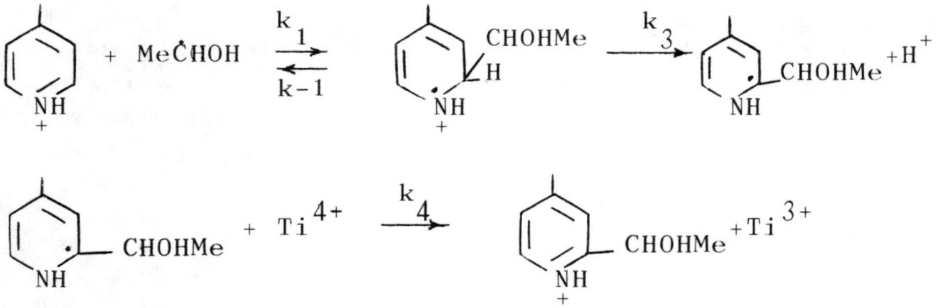

$$Me\dot{C}HOH + Ti^{4+} \xrightarrow{k_2} MeCHO + Ti^{3+} + H^+$$

Scheme X

With the t-α-hydroxylalkyl radical from isopropanol, however, no substitution was observed with all the investigated radical sources. An electron-transfer process according to eq 30 is probably responsible (the redox potentials of $.CH_2OH$, $Me\dot{C}HOH$, $Me_2\dot{C}OH$ have been evaluated respectively -0.98, -1.18 and -1.30 V).

The oxidizing character of the radical source, which affects the value of the rate constant k_2, can also influence the selectivity of the substitution as shown by the results of Table XIV with the radicals arising from THF and DMF. In these cases two differently nucleophilic radicals are formed from the solvents (Scheme XI): the radical 6 and 8 are oxidized very fast by Fe(III) salt, but not by Ti(IV) salt and the overall selectivity of the heteroaromatic substitution is quite different with the two radical sources, $\overset{+}{N}H_3OH/Ti(III)$ and $\overset{+}{N}H_3OSO_3^-/Fe(II)$, although the intermediate abstraction radical is identical.

TABLE XIV - Substitution at the position 2 of lepidine by the radicals 6, 7, 8, 9 obtained by THF, DMF and $.\overset{+}{N}H_3$

Radical source	Metal salt (mol %)	Solvent	Reaction 6 : 7	Products 8 : 9
$\overset{+}{N}H_3OSO_3^-/Fe(II)$	-	THF	18	-
"	Fe(III) (3)	"	12	-
"	Fe(III) (7)	"	6	-
"	Fe(III) (15)	"	2.5	-
$\overset{+}{N}H_3OH/Ti(III)$	-	"	32	-
"	Ti(IV) (30)	"	31	-
$\overset{+}{N}H_3OSO_3^-/Fe(II)$	-	DMF	-	13.3
"	Fe(III) (1)	"	-	5.3
"	Fe(III) (3)	"	-	2.3
"	Fe(III) (5)	"	-	1.8
"	Fe(III) (7)	"	-	1.4
"	Fe(III) (14)	"	-	0.7
$\overset{+}{N}H_3OH/Ti(III)$	-	"	-	60
"	Ti(IV) (30)	"	-	57

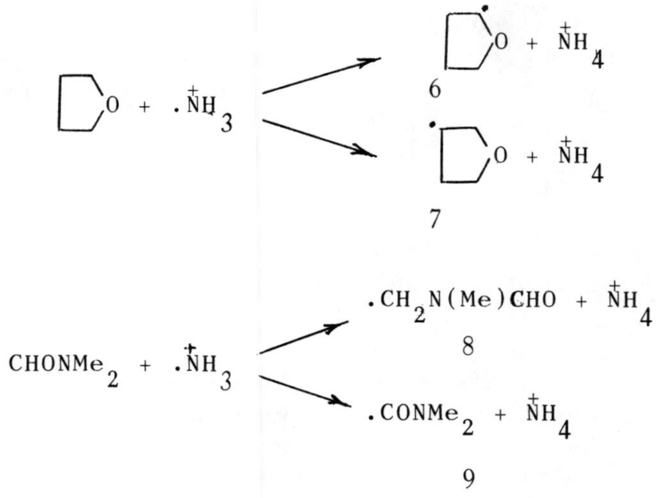

Scheme XI

Substituent and solvent effects

Few cases are known where solvents do have a significant effect on the rates and selectivity of free-radical reactions. The best known and striking example is in chlorine atom reactions, where selectivity is markedly increased by aromatic solvents, CS_2 and SO_2. Perhaps it is not strictly correct to consider these examples true solvent effects, at least in the classical sense given to the solvent effect of ionic reactions: the chlorine atom reacts very fast with benzene ($6x10^9$ $M^{-1}s^{-1}$) to form a well defined new radical species identified as the C_6H_6Cl. π-complex[52], which has rates and selectivity different from those of the "free" chlorine atom. Thus the π-complex must be considered more an intermediate radical than a solvated species.

The results of Table XV show a new exmple[53] of large solvent effect in free-radical reactions, which affects the regioselectivity of the substitution of the α and γ-positions of the protonated pyridine, whereas the β-position is not attacked, with the exception of the phenyl radical.

Two facts are clearly evinced from these results: (i) The regioselectivity is substantially independent of the radical source, but it only depends on the solvent (uniform temperature). The solvent dependence of the regioselectivity is small with the phenyl radical, the less nucleophilic, and it progressively increases as the nucleophilicity increases from methyl to primary, secondary, tertiary and α-tetrahydrofuranyl radicals. The dioxanyl radical is less nucleophilic than the α-tetrahydrofuranyl radical because the electron-releasing effect of the α-oxygen atom is in part counterbalanced by the inductive electron-withdrawing effect of the β-oxygen atom and that is reflected in the solvent effect.

The increased nucleophilic character is, however, associated with all the radicals of Table XV to an increased stability and a consequent increased reversibility of the radical addition. Thus considering the general mechanism of the reaction (Scheme XII) the solvent effect can be working either (i) in the addition of the radical to the heteroaromatic ring as consequence of the positively charged substrate, the nucleophilic character of the radical and of the solvation of the polar transition states (different values of k_1 and k_3 in different solvents) or (ii) to the different stability of the radical adducts 10 and 11, which can be reflected to the different values of k_{-1} and k_{-3} or still (iii) to different rates of the proton loss (k_2 and k_4) in the different solvents.

TABLE XV - Solvent effect on the regioselectivity of the positions α and γ in homolytic substitution of protonated pyridine by nucleophilic radicals

Radical	Radical source	Solvent	α %	γ %
Ph.	PhCOOH	Water	63.8	31.9
"	$(PhCOO)_2$	Benzene	70.2	23.8
Me.	MeCOOH	Water	62.3	37.7
"	t-BuOOH	Benzene	73.2	26.8
n-Pr.	n-PrCOOH	Water	57.6	42.4
n-Bu.	n-BuCOOH	Water	56.3	43.7
"	n-BuI	MeCN	64.5	35.5
"	n-BuI	Benzene	74.6	25.4
i-Pr.	i-Pr-COOH	Water	31.7	68.3
"	i-PrI	DMSO	46.1	53.9
"	i-PrI	MeCN	56.3	43.7
"	i-PrI	Benzene	71.9	28.1
t-Bu.	t-BuCOOH	Water	23.0	77.0
"	"	Benzene	71.4	28.6
-THF.	THF	THF+water	16.7	83.3
"	"	THF	91.3	8.7
Dioxanyl	Dioxane	Dioxane+water	25.5	74.5
"	"	Dioxane	76.3	23.7

The dilemma is therefore: can the solvent effect be mainly related to the nucleophilicity of the radical or to the reversibility of the addition?

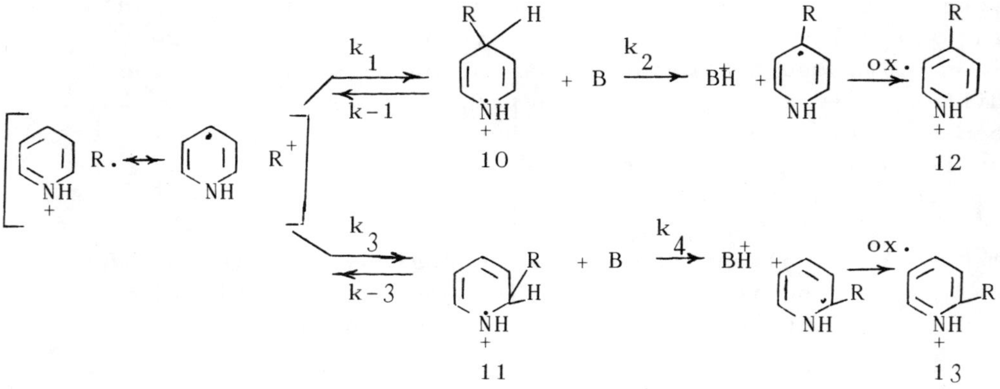

Scheme XII

In an attempt to recognize the step in which the solvent effect is working, the deuterium isotope effect has been determined in water and benzene with the isopropyl, α-tetrahydrofuranyl (α-THF) and phenyl radicals. With the phenyl radical no isotope effect has been observed, supporting the fact that the addition to the pyridine ring is irreversible. With isopropyl and α-THF radicals a significant isotope effect has been observed for the α and γ positions in both solvents (Table XVI) indicating that the C-H bond is broken in a rate determining step. The isotope effect is not much discriminating between the α and γ positions and that agrees with the fact that the isomer distributions are identical from C_6H_5N and C_6D_5N in both the solvents, but it is significantly higher in water than in benzene. That means that the k_{-1}/k_2 $[B^{\cdot}]$ and k_{-3}/k_4 $[B^{\cdot}]$ ratios are higher in water than in benzene. Thus a possible explanation is that the solvent effect influences the reversibility of the radical addition. If we assume that the process is more reversible in water than in benzene, in agreement with the isotope effect, and that the reversibility is higher for 11 than for 10 (even if the isotope effect does not reveal this difference) the increasing amount of 12 with the increasing stability of the nucleophilic radical would be justified in water. A lower reversibility in benzene would minimize this effect. The effect of the temperature actually supports a marked influence of the reversibility on the regioselectivity in water (Table XVII): an increase of the temperature determines an increase of the isomer 12.

TABLE XVI - Deuterium isotope effect in the substitution of protonated pyridine

Radical	Solvent	k_{2H}/k_{2D}	k_{4H}/k_{4D}
Ph.	Water	1.0	1.0
i-Pr.	Water	3.8	4.1
i-Pr.	Benzene	1.7	1.9
α-THF.	Water	5.0	5.1
α-THF.	Benzene	2.5	2.6

TABLE XVII - Effect of the temperature on the regioselectivity for the substitution of protonated pyridine

Radical	Radical source	Temperature (°C)	Solvent	α%	γ%
t-Bu.	t-BuCOOH	20	Water	33.5	66.5
"	"	60	"	28.6	71.4
"	"	100	"	23	77
"	"	60	Benzene	71.4	28.6
Dioxanyl	Dioxane	25	Dioxane + water	46	54
"	"	100	"	36.1	63.9
"	"	100	Dioxane	76.3	23.7

At present we have not sufficient evidence to establish if the influence of the solvent on the reversibility, connected with the rate constants k_2 and k_4, is the only factor or a more complex effect (i.e. solvation of polar transition states of charged species) is involved.

The solvent effect appears to be a general phenomenon in these reactions; it has been qualitatively observed in other substrates with two non-equivalent positions of high nucleophilic reactivity (quinoline, pyrimidine, pyridazine) and also in the relative rates of the competitive reactions between different substrates. From synthetic point of view the results are interesting with the most nucleophilic radicals because it is possible to obtain in large prevalence one isomer of the other depending on the solvent utilized.

Substituent and captodative effects[54]

According to the interpretation of the Schemes IX and XII a captodative effect could be a significant factor in three aspects of the general reaction:

(i) Captodative effect in the rearomatization step. Acyl, RĊO, carbamoyl, R_2NĊO and carboxyl, ROĊO radicals are nucleophilic and add fast to protonated bases giving substitutions of high synthetic value[44,45]. Since the addition exclusively occurs at the α and γ positions the rearomatization step caused by the irreversible loss of the α or γ proton (eq 32) should be particularly favoured by the fact that the allylic captodative radical 14 is formed

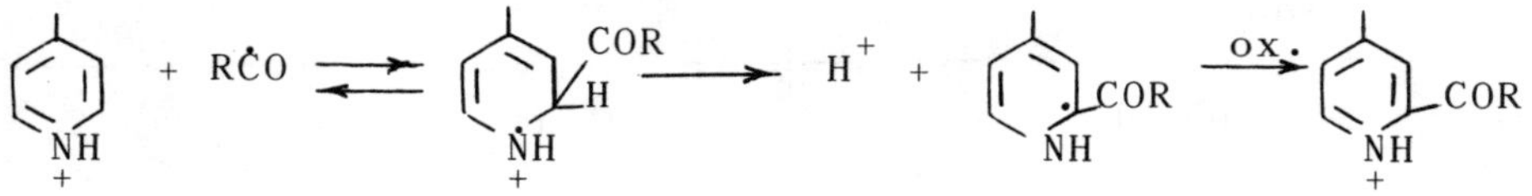

14 (32)

The addition of the acyl radical is a reversible process and the fast irreversible loss of the α proton allows to successfully utilize in the heteroatomatic substitution very unstable acyl radicals, such as pivaloyl radical, which decomposes very fast[49] to t-Bu. radical and CO (10^4-10^5 s^{-1}).

(ii) Captodative effect in the transition state. When the homolytic acylation, carbamoylation or carboxylation takes place in heteroaromatic rings with more α and γ free positions it is difficult to arrest the reaction at monosubstitution even at low conversions. That means that the monosubstituted derivative is much more reactive than the starting base. The polar effect, due to charge separation, could be exalted by the contribution of a captodative effect in the transition state (eq 33)

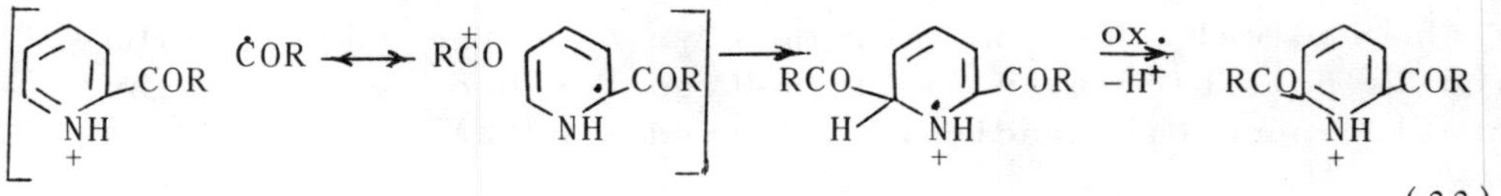

(33)

From synthetic point of view it is, however, possible to obtain monosubstitution in these reactions by taking advantage of the fact that the monosubstituted derivatives are always less basic than the starting base and by adjusting the medium acidity in order that the starting base is, at least in part, protonated and the monosubstituted derivative is unprotonated. The protonated species is more reactive and monosubstitution can be obtained with good selectivity.

The captodative effect should be general in the reactions of every kind of nucleophilic radical with heteroaromatic bases bearing an electron-withdrawing group in α or γ positions. A confirmation of such possibility could arise by comparison of the reactivity of the position 6 in 3- and 4-substituted pyridines (15 and 16)

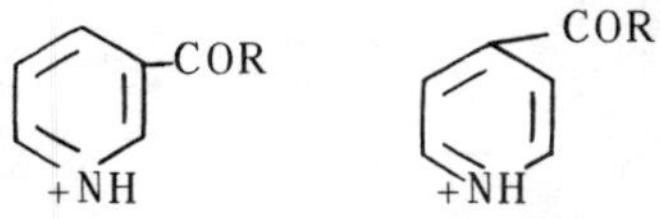

15 16

The activation of the position 6 should be higher in 15 than in 16 (higher polar effect in para than in meta position of the substituent);. an opposite behaviour should be an evidence that a captodative effect, possible in 16 but not in 15, is working in the transition state. Preliminary investigation in the alkylation by t-Bu. radical has given contradictory results because the relative rates determined by the competitive method, are not reliable due to the reversibility of the radical addition and the solvent effect.

(iii) <u>Captodative effect and electron-transfer processes</u>. The complete electron transfer of eq 30 is the limit case of the polar effect determined by charge separation in the transition states of eq 33 and Scheme XII. It occurs with strongly nucleophilic carbon-centered radicals (α-hydroxy-, α-aminoalkyl radicals) and bases of high electron affinity. If a captodative effect is working in the transition states of the radical addition ever more so an effective contribution is possible in an electron-transfer process, where an actual captodative radical is formed (eq 34)

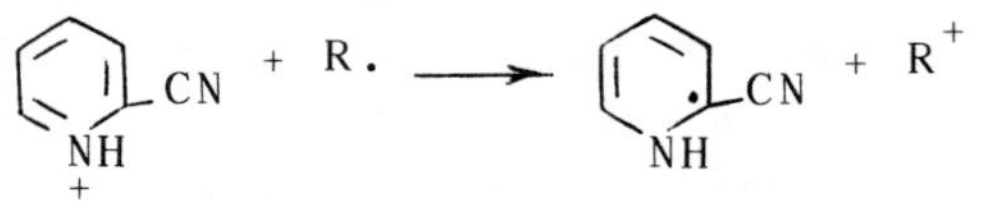

(34)

Reactions of carbon-centered radicals with diazonium salts.

Diazonium salts are very effective traps of nucleophilic alkyl radicals because of large polar effects (eq 35)

$$R\cdot \;\; N\equiv\overset{+}{N}-Ar \longleftrightarrow R^{+} \;\; \dot{N}=N-Ar \longleftrightarrow R-N=\underset{\bullet}{\overset{+}{N}}-Ar \qquad (35)$$

Several radical sources have allowed to develop the "Free-Radical Diazocoupling", a reaction of high synthetic value[8]. An example is shown by eq 36, in which alkyl iodides are used as radical sources.

$$2ArN_2^{+} + RI + 2Ti^{3+} \longrightarrow Ar-N=N-R + ArI + N_2 + 2Ti^{4+} \qquad (36)$$

The mechanism of the reaction is illustrated by the Scheme XIII

$$ArN_2^{+} + Ti^{3+} \longrightarrow Ar\cdot + N_2 + Ti^{4+}$$

$$Ar\cdot + RI \longrightarrow ArI + R\cdot$$

$$R\cdot + N\equiv\overset{+}{N}-Ar \xrightarrow{k} R-N=\underset{\bullet}{\overset{+}{N}}-Ar$$

$$R-N=\underset{\bullet}{\overset{+}{N}}-Ar + Ti^{3+} \longrightarrow R-N=N-Ar + Ti^{4+}$$

Scheme XIII

The higher electron affinity of the diazonium salts compared with protonated heteroaromatic bases makes even more striking the polar effect but at the same time it implies larger synthetic limitations. Thus t-alkyl radicals are considerably more reactive ($k \simeq 10^8\ M^{-1}sec^{-1}$) than primary alkyl radicals ($k \simeq 10^6\ M^{-1}s^{-1}$). However, still more nucleophilic free-radicals, such as $.CH_2OH$, $Me\dot{C}HOH$, $R\dot{C}O$, $.CH_2N(CH_3)CHO$, $.CONMe_2$, α-tetrahydrofuranyl etc., do not add to the diazonium salt because they are fast oxidized by electron-transfer processes (eq 37), whereas the same radicals easily and selectively add to protonated heteroaromatic bases

$$ArN_2^+ + .CH_2OH \longrightarrow Ar. + N_2 + CH_2O + H^+ \qquad (37)$$

The other general limitation is determined by carbon-centered radicals with electron-withdrawing groups directly bound to the radical center. The great importance of the polar effect is emphasized by the example[55] of the Scheme XIV

$$ArN_2^+ + Me_2C = CHCOMe \xrightarrow{Ti^{3+}} \begin{cases} Ar\text{-}N = N\text{-}CMe_2\text{-}CHAr\text{-}COMe \\ Ar\text{-}CMe_2\text{-}CH_2\text{-}COMe \end{cases}$$

Scheme XIV

The phenyl radical adds to the unsaturated ketone with low regioselectivity, giving two carbon-centered radicals with different polar character: the nucleophilic radical gives the diazocoupling reaction, whereas the electrophilic radical is selectively reduced by Ti(III) salt (Scheme XV)

$$ArN_2^+ + Ti^{3+} \longrightarrow Ar\cdot + N_2 + Ti^{4+}$$

$$Ar\cdot + Me_2C{=}CHCOMe \longrightarrow Ar{-}CMe_2{-}\dot{C}HCOMe \quad ; \quad Ar{-}CH(COMe){-}\dot{C}Me_2$$

$$ArCMe_2\dot{C}HCOMe + Ti^{3+} + H^+ \longrightarrow Ar{-}CMe_2{-}CH_2COMe + Ti^{4+}$$

$$Ar{-}CH(COMe)\dot{C}Me_2 + N{\equiv}\overset{+}{N}{-}Ar \longrightarrow ArCH(COMe){-}Me_2C{-}N{=}\dot{\overset{+}{N}}{-}Ar$$

$$ArCH(COMe){-}Me_2C{-}N{=}\dot{\overset{+}{N}}{-}Ar + Ti^{3+} \longrightarrow ArCH(COMe){-}Me_2C{-}N{=}N{-}Ar + Ti^{4+}$$

Scheme XV

<u>Reactions of carbon-centered radicals with pyrilium salts</u>

The pyrilium salts are activated towards nucleophilic species even more than the protonated bases. They react very fast with alkyl radicals giving selective substitution products[56] (Scheme XVI)

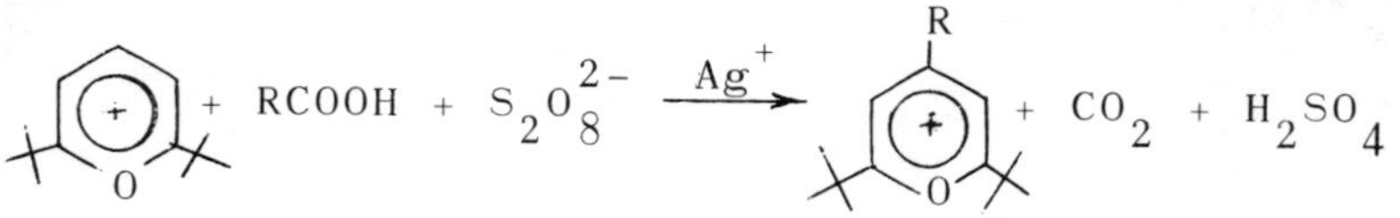

70-90% R = Me, t-Bu., Et_3C.

Scheme XVI

Also in this case a transition state similar to a charge-transfer complex (eq 38) very likely accounts for the selective substitution

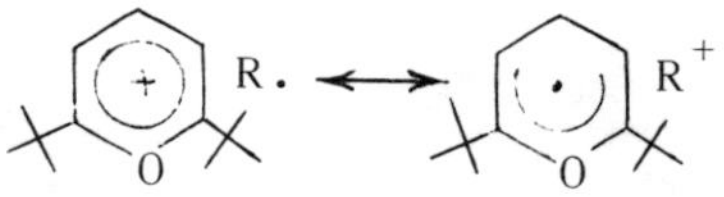

(38)

The mechanism, very probably reproduces the same steps of the redox chain suggested for the substitution of the heteroaromatic bases by the same radical source (Scheme XVIII)

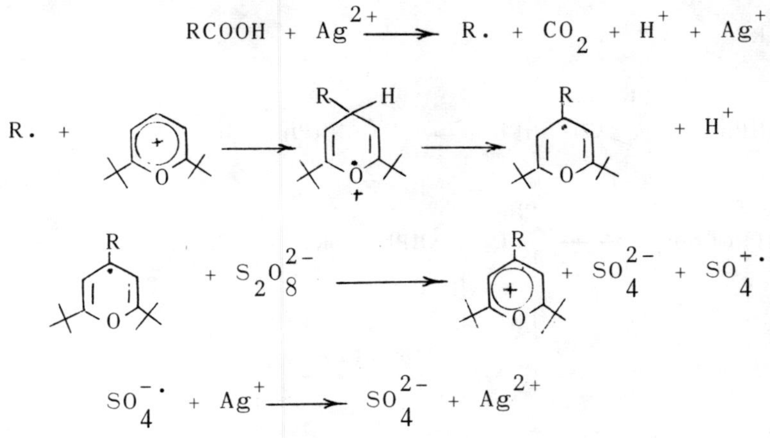

Scheme XVII

The range of application has not yet been investigated. The higher electron affinity of the pyrilium salts compared with protonated heteroaromatic bases suggest that electrophilic carbon-centered radicals will not react and that with the most nucleophilic radicals the electron-transfer process can more easily prevail over the nuclear addition.

<u>Reactions of carbon-centered radicals with iminium salts.</u>

The iminium salts also seem to be considerably activated towards nucleophilic carbon-centered radicals. A free-radical chain similar to those above described takes place, in our opinion, in the alkylation with acyl peroxides[53] (Scheme XVIII)

The synthetic possibilities have not yet been extensively investigated. The low stability of the iminium salts to the hydrolysis limits the radical sources and the suitable experimental conditions.

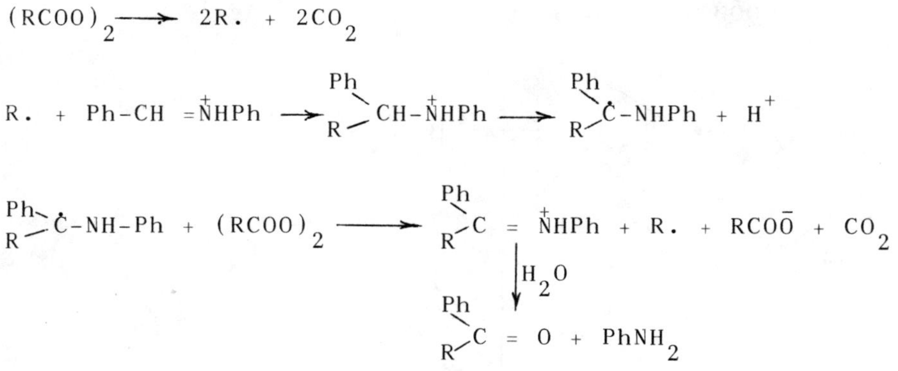

Scheme XVII

Conclusion

Polar effects are particularly important when charged species, the radical or the substrate, are involved in free-radical reactions. The polar effect of the substituents influences the polarity of the free radicals and of the substrates, the selectivity and the synthetic involvements. General synthetic methods of high value have been developed on the ground of the polar effect. Solvent and captodative effects can be important in relation to the nature of the substituents bound to the radical center.

REFERENCES

1. F. Minisci, R. Galli, M. Cecere, V. Malatesta and T. Caronna, Tetrahedron Letters, 5609 (1968); F. Minisci, P. Zammori, R. Bernardi, M. Cecere and R. Galli, Tetrahedron, 26, 4153 (1970); A. Citterio, A. Arnolid and F. Minisci, J.Org.Chem., 44, 2674 (1979).
2. F. Minisci, La Chimica e l'Industria, 49, 705,(1967); F. Minisci and R. Galli, La Chimica e l'Industria, 46, 546 (1964); F. Minisci. R. Galli and M. Cecere, Tetrahedron Letters, 3163 (1966); R.S. Neale, Tetrahedron Letters, 483 (1966).
3. (a) A. Citterio, F. Minisci, O. Porta and G. Sesana, J. Am.Chem.Soc., 99, 7960 (1977); (b) A. Citterio, F. Minisci and F. Franchi, J.Org.Chem., 45, 4752 (1980).
4. M. Bellatti, T. Caronna, A. Citterio, F. Minisci, J.Chem. Soc. Perkin Trans II, 1835 (1976); Gazz.Chim.Ital., 107, 491 (1977).
5. A. Citterio, A. Gentile, F. Minisci, M. Serravalle and S. Ventura, Tetrahedron, 41, 617 (1985).
6. Actually t-Bu. reacts with benzene in particular conditions with a rate constant 10^7 lower than with protonated cyanopyridine (H. Fischer, personal communication).
7. (a) F. Minisci and V. Malatesta, La Chimica e l'Industria, 53, 1154 (1971); (b) F. Minisci, E. Vismara, F. Fontana, G. Morini, M. Serravalle and C. Giordano, submitted to J. Am.Chem.Soc.
8. A. Citterio and F. Minisci, J.Org.Chem., 47, 1759 (1982).
9. (a) J. Spanswick and K.U. Ingold, Can J.Chem., 48, 554 (1970); (b) F. Minisci, R. Galli, M. Perchinunno and R. Bernardi, La Chimica e l'Industria, 50, 453 (1968).
10. F. Minisci and A. Citterio, Adv.Free-Radical Chem., 6, 65 (1980).
11. C. Chatgilialogly, K.U. Ingold, J.C. Scaiano, J.Am.Chem. Soc., 104, 5123 (1982); C. Chatgilialogly, in the press, personal communication.
12. F. Minisci; R. Galli and R. Bernardi, La Chimica e l'Industria, 49, 594 (1967).
13. F. Minisci, G. Gardini and F. Bertini, Can.J.Chem., 48, 544 (1970).
14. F. Minisci, R. Galli, A. Galli and R. Bernardi, Tetrahedron Letters, 2207 (1967).
15. F. Minisci, R. Galli and R. Bernardi, Chem.Comm., 903 (1967).

16. H. Singh and J.M. Tedder, J.Chem.Soc., 4737 (1964).
17. F. Minisci, R. Galli, R. Bernanrdi and M. Perchinunno, La Chimica e l'Industria, 59, 328 (1968).
18. R. Bernardi, R. Galli and F. Minisci, J.Chem.Soc., B 324 (1968).
19. F. Minisci, R. Galli, R. Bernardi and M. Perchinunno, La Chimica e l'Industria, 51, 280 (1969).
20. D.H. Auc, H.M. Webband and M.T. Boxers, J.Am.Chem.Soc., 94, 4726 (1972); R.H. Stanley, M. Taagepera, W.G. Henderson, I. Koppel, J.L. Beauchamp and R.W. Taft, J.Am.Chem. Soc., 99, 326 (1977).
21. Ref.10, page 75
22. F. Minisci, R. Galli and R. Rossetti, La Chimica e l'Industria, 49, 947 (1967).
23. D.D. Tanner and M.W. Mosher, Can.J.Chem., 47, 715 (1969).
24. N.C. Deno and D.G. Pohl, J.Org.Chem., 40, 380 (1975).
25. J. Spanswick and K.U. Ingold, Can.J.Chem., 48, 546 (1970).
26. C. Walling and M.F. Mayahi, J.Am.Chem.Soc., 81, 1485 (1059).
27. P.S. Friedricks and J.M. Tedder, J.Chem.Soc., 144 (1960).
28. N.C. Deno, W.E. Billups, R. Fishbein, C. Pierson, R. Whalen and J.C. Wyckoff, J.Am.Chem.Soc., 93, 438 (1971).
29. N.C. Deno in Methods in Free-Radical Chemistry, 3 (ed. E.S. Huyser), Dekker, New York, 1972, p.147
30. Ref.10, page 82
31. S. Auricchio,M. Bianca, A. Citterio, F. Minisci and S. Ventura, Tetrahedron Letters, 3373 (1984).
32. D.D. Tanner, R. Arhart and C.P. Meintzer, Tetrahedron, 41, 4261 (1985).
33. F. Minisci, Synthesis, 21 (1973); Ref.9, page 82.
34. F. Minisci, Top.Curr.Chem., 62, 1 (1976).
35. F. Minisci, R. Galli and M. Cecere, La Chimica e l'Industria, 48, 1324 (1966).
36. F. Minisci, R. Galli and M. Perchinunno, Org.Prep.Proced., 1, 87 (1969).
37. F. Minisci, R. Galli, M. Cecere and V. Trabucchi, La Chimica e l'Industria, 48, 1147 (1966).
38. F. Minisci and M. Cecere, La Chimica e l'Industria, 49, 1333 (1967).
39. F. Minisci, R. Galli, M. Cecere and V. Trabucchi, La Chimica e l'Industria, 48, 1147 (1966).
40. F. Minisci, M. Perchinunno and O. Porta, J.Chem.Soc.Perkin Trans II, 416 (1974).

41. R.O.C. Norman and R. Taylor 'Electrophilic Substitution in Benzenoid Compounds', London 1965.
42. Ref.34, page 14
43. A. Citterio, A. Gentile, F. Minisci, V. Navarrini, M. Serravalle and S. Ventura, J.Org.Chem., 49, 4479 (1984).
44. F. Minisci and E. Platone, La Chimica e l'Industria, 64, 787 (1982); Ital.Pat. ANIC (15.7.1980).
45. Reviews in the subject: (a) F. Minisci, Refs.10,33,34; (b) E. Vismara, La Chimica e l'Industria, 65, 34 (1983).
46. Most recent results in the field: G. Castoldi, F. Minisci, V. Tortelli and E. Vismara, Tetrahedron Letters, 3897 (1984); A. Citterio, A. Gentile, F. Minisci, M. Serravalle and S. Ventura, J.Org.Chem., 49, 3364 (1984); F. Minisci, A. Citterio, E. Vismara and C. Giordano, Tetrahedron, 41, 4157 (1985); F. Minisci, E. Vismara and U. Romano, Tetrahedron Letters, 4803 (1985); C. Giordano, F. minisci, E. Vismara and S. Levi, J.Org.Chem., in the press; F. Minisci, E. Vismara et al., J.Org.Chem., in the press; Refs.5 and 7b.
47. R.G. Kryger, J.P. Lorand, N.R. Stevens, N.R. Herron, J. Am.Chem.Soc., 99, 7589 (1977).
48. F. Minisci, C. Giordano, E. Vismara, S. Levi and V. Tortelli, Ital.pat. 22700 (1984); 23798 A/84 (1984).
49. H. Schuh, E.J. Hamilton, H. Paul and H. Fischer, Helv. Chim.Acta, 57, 2011 (1974).
50. F. Minisci, C. Giordano, E. Vismara, S. Levi and V. Tortelli, J.Am.Chem.Soc., 106, 7146 (1984).
51. T. Burkey, A. Castellano, D. Griller, F.P. Lossing, J.Am. Chem.Soc., 105, 4701 (1983).
52. N.J. Bunce, K.U. Ingold, J.P. Landers, J. Lusztyk and J. C. Scaiano, J.Am.Chem.Soc., 107, 5464 (1985).
53. F. Minisci, E. Vismara et al., unpublished results.
54. H.G. Viehe, Z. Janousek, R. Merenyi and L. Stella, Acc. Chem.Res., 18,148 (1985).
55. A. Citterio, F. Minisci, E. Vismara, J.Org.Chem., 47, 81 (1982).
56. G. Doddi and G. Ercolani "Symposium of Organic Chemistry", Sirmione (Italy), Abstracts, p.66 (1985).

SELECTIVITY IN HOMOLYTIC AROMATIC SUBSTITUTIONS. ALKYLATIONS OF PYRIDINES.

M. Tiecco
Institute of Organic Chemistry
Faculty of Pharmacy
University of Perugia
Italy

ABSTRACT. Factors controlling positional selectivity in homolytic aromatic substitutions are discussed. Selective reactions leading to different products can be effected using the same substrate and changing the polarity of the attacking radicals as well as using the same radical and changing the polarity of the substrates.

It is well documented that homolytic aromatic substitution reactions occur with the addition-elimination mechanism involving the formation of a σ-complex radical intermediate (path a); in some circumstances a charge transfer between the radical and the substrate preceeds the formation of the σ-complex intermediate (path b):(1-3)

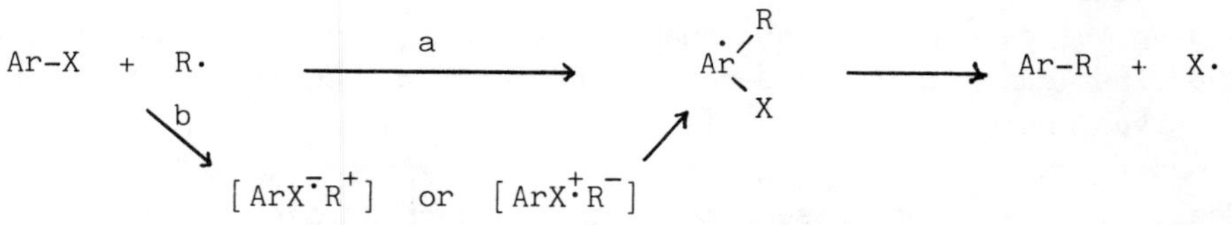

X = H
X = NO_2, COR, SO_2R (ipso substitutions)

Experimental evidences indicate that the course of the reaction is determined by the polarity of both the substrate and the attacking radical. In general one observes that

H. G. Viehe et al. (eds.), Substituent Effects in Radical Chemistry, 435–442.
© 1986 by D. Reidel Publishing Co.

path a i) occurs with scarcely polar radicals or scarcely polar substrates or both
ii) the positional selectivity is governed by the relative stabilities of the σ-complexes intermediates
iii) positional selectivity is usually very poor

path b i) occurs with polar radicals and polar substrates
ii) the positional selectivity is governed by the charge distribution in the substrate
iii) positional selectivity is usually very high.

Thus, when polar effects can operate the homolytic substitution generally leads to the formation of a single isomer and has therefore synthetic importance. When polar effects cannot operate the homolytic substitution leads to mixtures of isomers and its synthetic importance is therefore very limited. In the latter case selective homolytic substitutions can be obtained only with substrates having particular structures so that the σ-complex intermediate deriving from the addition at one nuclear position presents a stability much greater than that of the intermediates deriving from the addition at the other nuclear positions.

It can also be expected that, if conditions are found in which polar effects can operate, substrates possessing these prerequisites should suffer the addition at a different nuclear position to afford a different product or an isomer. In other words one can expect to observe a change in positional selectivity on the same substrate as a consequence of the fact that polar effects can or cannot intervene in the homolytic substitution process.

This problem has been approached in two ways:

i) using the same substrate and changing the polarity of the attacking radicals

ii) using the same radical and changing the polarity of the substrates.

In the first case reactions were carried out with benzene or thiophene derivatives, holding electron-withdrawing substituents, with the polar 1-adamantyl radical (Ad·) and with the scarcely polar methyl or phenyl radicals. In the second case reactions were carried out with the dioxanyl (D·) or the cyclohexyl (C·) radicals using protonated and unprotonated pyridine derivatives as substrates.

The first two examples refer to the reactions of 1,3,5-trinitrobenzene (1) and 1,2,4-trinitrobenzene (4) with adamantyl and phenyl radicals (4). In every case the reactions are very selective. The adamantyl radical adds to the carbon atoms holding the nitro group to afford the products of ipso substitution, (2) and (5), whereas the phenyl radical adds to the nuclear position from which the most stable σ-complexes are formed to give (3) and (6).

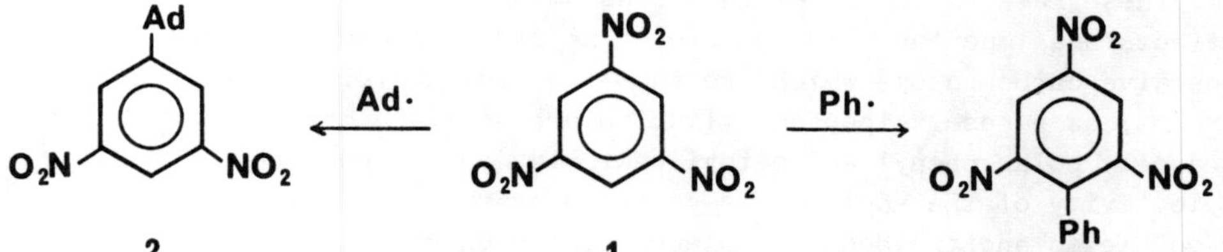

2 1 3

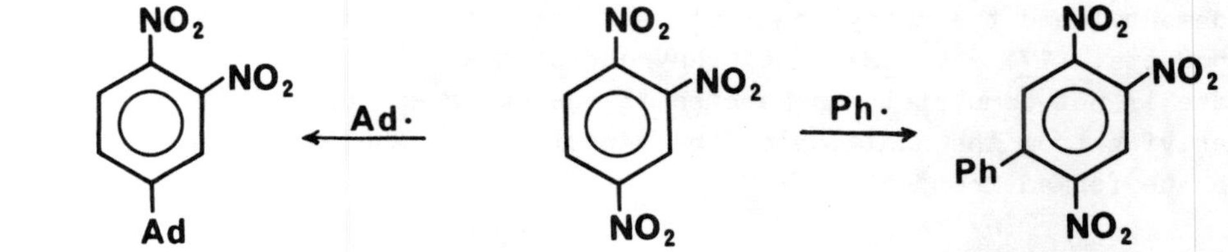

5 4 6

A similar different behaviour was observed (5,6) in the reactions of the 2-carbomethoxy-5-nitrothiophen (7), 2-carbomethoxy-3,5-dinitrothiophen (10), and the 2-ethylsulfonyl-5-acetylthiophen (13); the adamantyl radical gives rise to the ipso substitution products (8), (11) and (14), whereas the methyl radical affords the products deriving by the addition at the unsubstituted nuclear positions (9), (12), and (15):

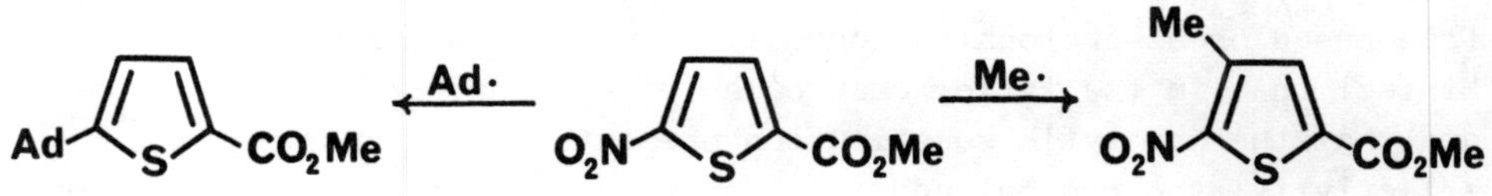

8 7 9

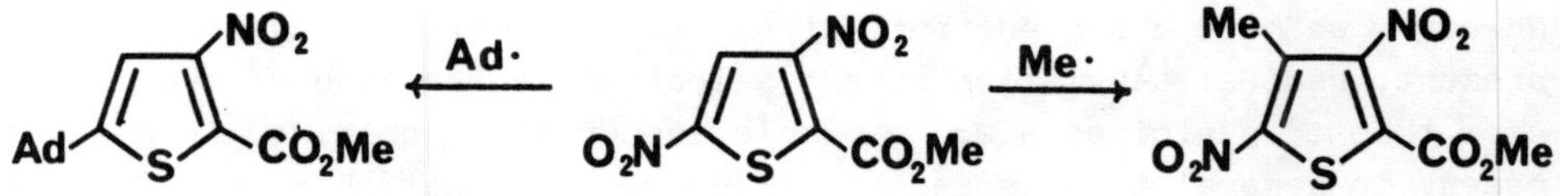

11 10 12

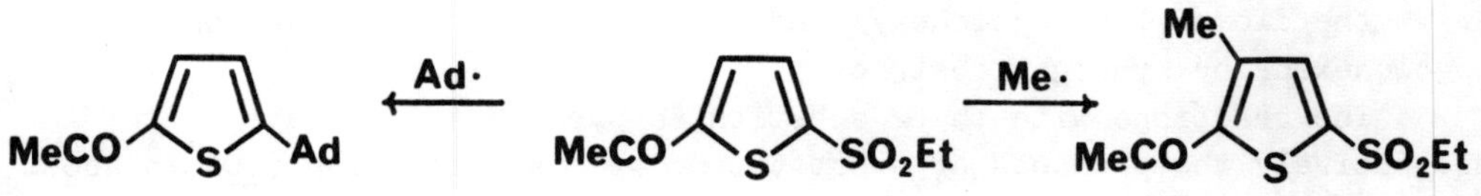

14 13 15

All these results can be explained assuming that with the Ad· polar effects are important and therefore the addition occurs at the most positive carbon atoms which are those holding the electron-withdrawing groups; as a result ipso substitution products are obtained. With the scarcely polar phenyl and methyl radicals, on the contrary, the selectivity of the addition is governed by the relative stability of the σ-complexes and the addition occurs at the unsubstituted nuclear positions.

In the case of the 2-carbomethoxy-4,5-dinitrothiophen (16) both the adamantyl and the methyl radicals give rise to the ipso substitution products, (17) and (18). The behaviour of the methyl radical in this case is not completely unexpected if one observes that the σ-complex deriving from the addition at the 5-position is the most stable which can be formed from (16).

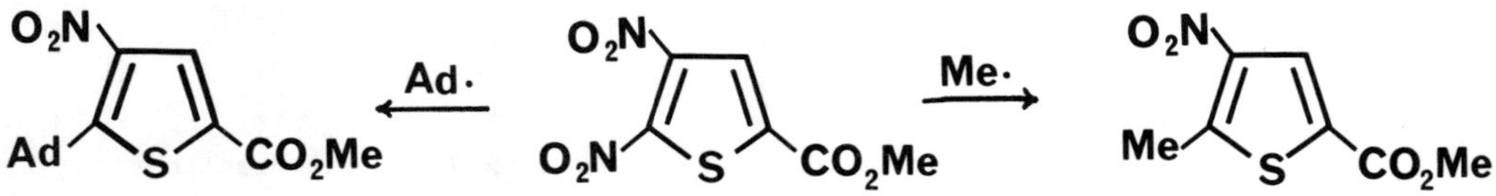

17 16 18

Pyridine, as well as other nitrogen containing heteroaromatic compounds, can be allowed to react with different radicals either as free bases or as protonated ions. It is well known from the works of Minisci that in the latter case polar effects have a relevant importance and that the homolytic substitution presents a great positional selectivity; the radical addition occurs at the nuclear positions which are conjugated with the positive nitrogen atom. Unprotonated pyridine, on the contrary, gives rise to a mixture of the three possible isomers. However, we have observed that, if an electron-withdrawing group is present, radical alkylation became selective in the case of the unprotonated pyridines also, a single isomer being obtained in most cases. In general this isomer is different from that obtained when the same substrates are alkylated in the presence of acids.

This kind of investigations was carried out with pyridines holding an acetyl, a carboethoxy or a cyano group in the 2-, 3- or 4-positions with the dioxanyl or cyclohexyl radical produced from dioxane or cyclohexane by hydrogen abstraction (7).

The reactions with the 2-substituted pyridines (19) gave selectively the products of substitution at the 5-position (20) in about 50% yield. The same reactions carried out in the presence of sulphuric acid afforded instead a mixture of the 6- (21) and 4-substituted compounds (22) (small amounts of the 4,6-disubstituted products (23) were also obtained in some cases).

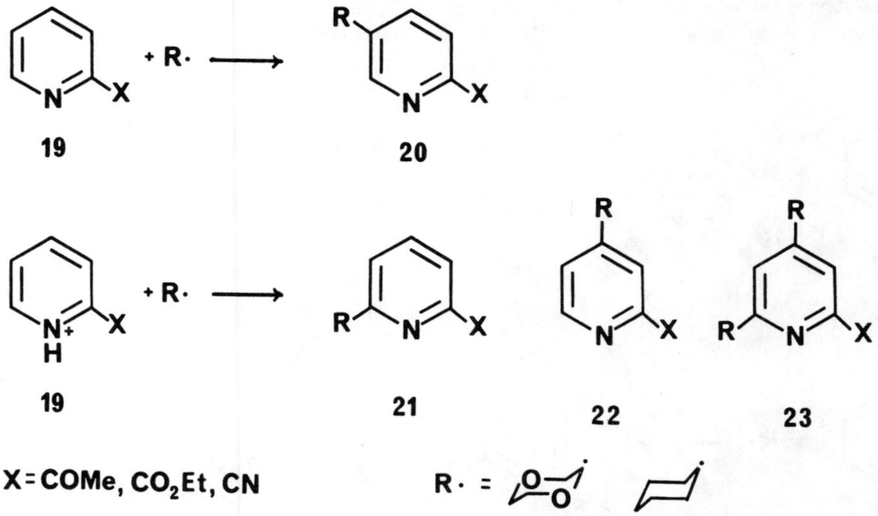

The reaction of the dioxanyl radical with the 3-acetylpyridine (24) gave the 3-acetyl-6-dioxanyl pyridine (25) as the sole reaction product. In this case the same compound is also obtained in the presence of acid; under these conditions however, even at low conversions, this compound was always accompanied by the disubstituted product, 3-acetyl-4,6-bis-dioxanylpyridine (26):

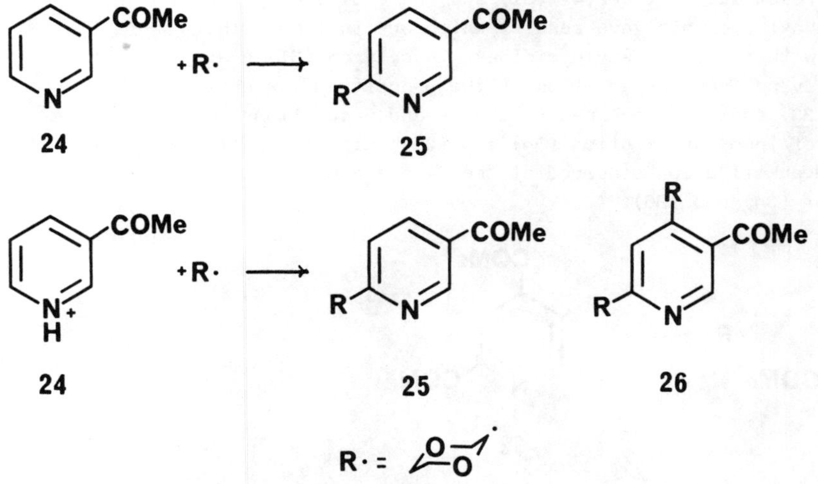

The alkylations of the 4-substituted pyridines (27) were not completely selective. Addition occurred at the 3-position (28) but small amounts of the 2-substituted product (29) were also obtained. The reactions carried out in the presence of acid gave the 2-substituted products (29) contaminated by some 2,6-disubstituted compounds (30):

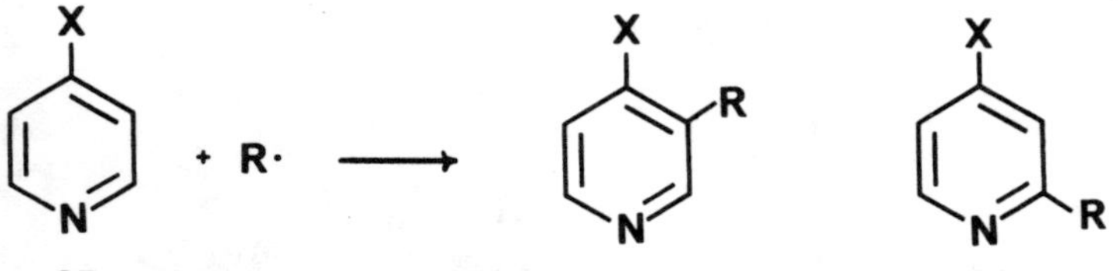

27 28 29

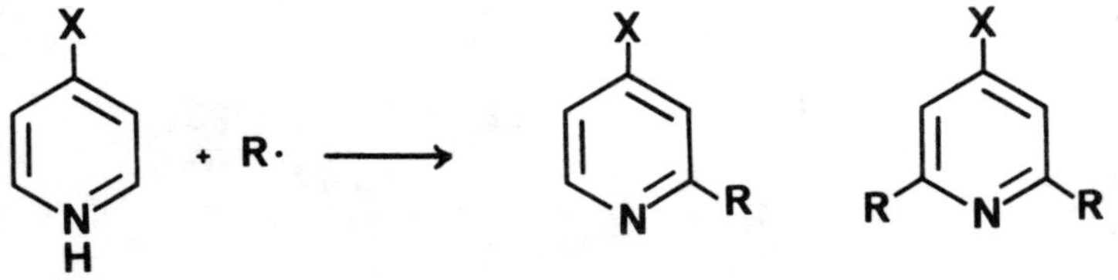

27 29 30

X = COMe, CO_2Et, CN R· = 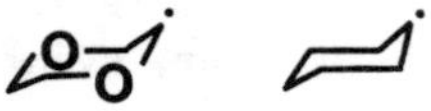

The reactions of the 2,4- (31) and the 2,5-diacetylpyridine (33) with dioxanyl radicals gave results which are in line with those observed with the monoacetylpyridines. Thus, from (31) a single product (32), deriving from the addition at the 5-position was obtained. In the case of (33) addition occurred at the 2- and 5-positions to give the two products of ipso substitution (34) and (35); with the protonated substrate substitution occurred at the 2- and 4-positions to give a mixture of (34) and (36).

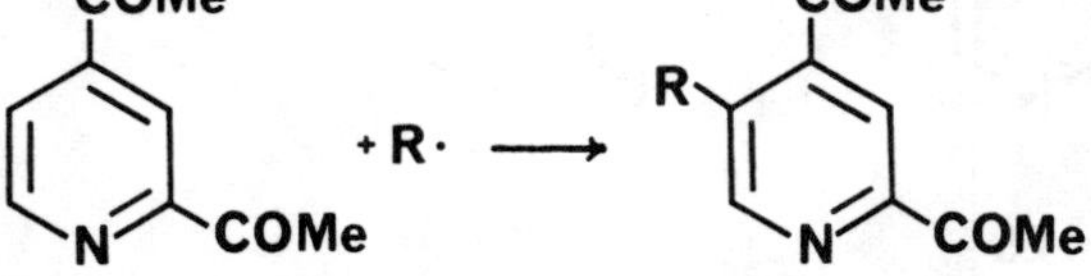

31 32

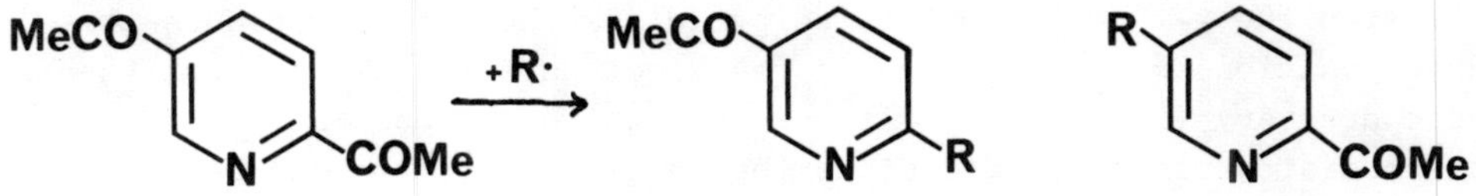

33 34 35

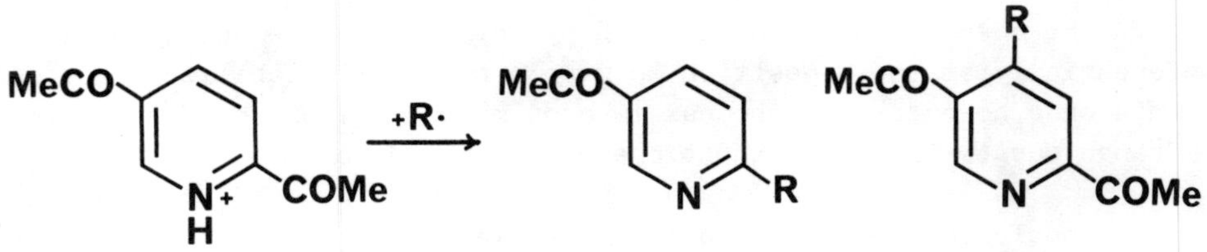

33 34 36

The change in positional selectivity on passing from unprotonated to protonated pyridines can be interpreted on the basis of a change in the nature of the transition state. With protonated pyridines the attack occurs at the most positive sites of the molecule which are the carbon atoms in the α and γ positions. With the unprotonated substrates the positional selectivity is governed by the relative stabilities of the σ-complexes intermediates.

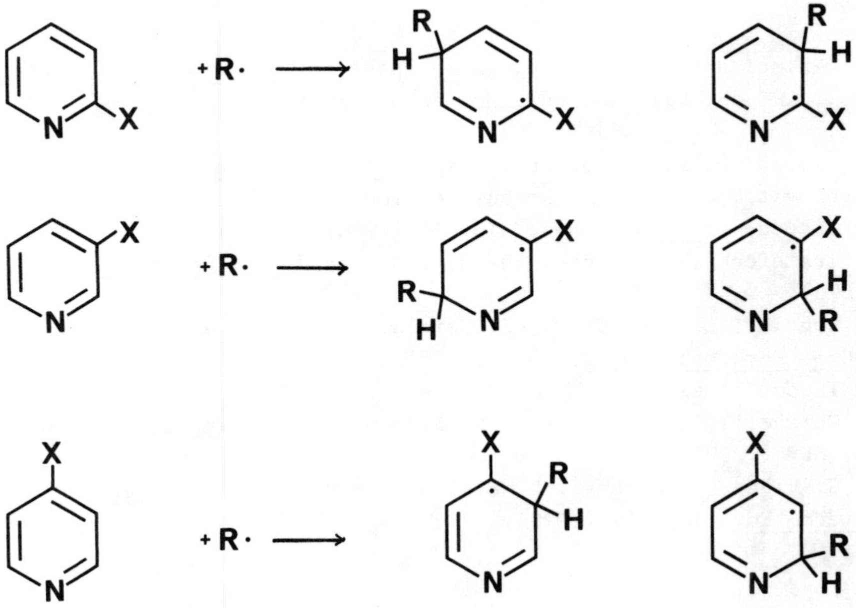

With the 2- and 3-substituted pyridines the addition occurs at those positions which are para to the electron-withdrawing substituents; these are very likely the positions from which the most stable intermediates can be formed since the unpaired electron can be delocalized into the substituents. One can also look at these radicals as capto-dative radicals; as demonstrated by Viehe a particular stabilization is associated with radicals possessing these structures (8). In the case of the 4-substituted pyridines also the addition occurs preferentially at the 3-position to give a resonance-stabilized radical; in this case however this process is made more difficult by the steric hindrance due to the substituent present in the 4-position.

In conclusion selective homolytic aromatic substitutions can be effected also in those cases in which polar effects do not operate, provided appropriate substrates and radicals are selected. The change on positional selectivity, as a function of the change of the mechanism of the addition step, has been demonstrated both in the case in which the same substrate has been allowed to react with radicals possessing different polarities and in the case in which the same radical reacts with substrates with different polarities. This latter case has been studied with pyridine derivatives and has also some synthetic importance since different isomers are obtained when the reactions are carried out on protonated or unprotonated substrates.

REFERENCES

1) F. Minisci and A. Citterio, in Advances in Free Radical Chemistry, Vol. 6, G. H. Williams ed., Heyden and Sons, London (1979), pp 65 - 153.
2) M. Tiecco and L. Testaferri, in Reactive Intermediates, Vol. 3, R. A. Abramovitch ed., Plenum Press, New York (1983), pp 61 - 111.
3) M. Tiecco, Acc. Chem. Res., **13**, 51 (1980).
4) L. Testaferri, M. Tiecco, and M. Tingoli, J.C.S. Perkin Trans. 2, 469 (1979).
5) P. Cogolli, L. Testaferri, F. Maiolo, M. Tiecco, and M. Tingoli, J.C.S. Perkin Trans. 2, 1331 (1980).
6) M. Tiecco et al., unpublished results.
7) D. Chianelli, L. Testaferri, M. Tiecco, and M. Tingoli, Tetrahedron, **38**, 657 (1982).
8) H. G. Viehe, R. Merényi, L. Stella, and Z. Janousek, Angew. Chem. Intern. Ed., **18**, 917 (1979).

THE INVENTION OF HIGH YIELDING RADICAL CHAIN REACTIONS

Derek H.R. Barton and Samir Z. Zard
Institut de Chimie des Substances Naturelles, C.N.R.S.
91190 Gif-sur-Yvette
France

Abstract - The design of new Radical chain reactions useful in the synthesis or modification of Natural Products is discussed. Such reactions can give good yields, and show a selectivity which complements perfectly the ionic reactions in more general use.
A flexible system for the production of many kinds of carbon radicals has been invented. It should be applicable for the generation of other kinds of radicals based on elements other than carbon. In the design of these reactions a vital role is played by a disciplinary group (usually the thiocarbonyl group), an idea which is of general application in the invention of radical chain reactions.

New reactions useful in synthesis are usually discovered by accident. It is, however, possible to invent reactions by conception[1]. Most of this article is devoted to the invention of new radical chain reactions to give good yields of products. This can be done with the aid of the disciplinary group concept (vide infra).

Radical reactions have, of course, an enormous importance in the synthesis of polymers. However, we are concerned here with their use in the chemical synthesis of homogeneous, low molecular weight molecules. In this area of scientific research, radical reactions are not often used, because they are considered to be unselective and to give poor yields of products. It is the purpose of this article to show that well designed radical reactions can give high yields of single products and play an important role in organic synthesis.

Radicals react with various functional groups at very different rates which can vary over many powers of ten. By a judicious choice of reagents, solvents etc... one can devise a system capable of effecting a highly selective transformation. Radical-radical interactions however (coupling, disproportionation etc...) are extremely

H. G. Viehe et al. (eds.), Substituent Effects in Radical Chemistry, 443–461.
© 1986 by D. Reidel Publishing Co.

fast processes and therefore much more difficult to control. If the sequence of radical reactions is conceived so as to constitute a linear chain process where the propagating steps are so fast that the concentration of radical species remains very small, radical-radical interactions can be largely avoided. Under such controlled conditions, clean high yielding reactions become possible. Moreover radical and radical chain reactions have several advantages over conventional ionic processes : neutral conditions; lower steric effects; lower polar effects; lower tendancy to unwanted elimination reactions; tolerance of many functional groups that have to be protected in ionic chemistry. In ionic reactions β-elimination is a serious problem in the carbohydrate and aminoglycoside antibiotic fields. These important compounds are heavily functionalised and selective deoxygenation or deamination presented a challenging problem. We shall first consider the problem of deoxygenation.

Many years ago, Van der Kerk[2] discovered accidentally the facile reduction of alkyl halides by stannanes. The mechanism involved is that of a typical radical chain reaction. The stannyl radicals, generated photochemically, thermally or by a chemical initiator (e.g. azoisobutyronitrile - A.I.B.N.) abstract the halogen to give a carbon radical. This latter abstracts a hydrogen atom from the hydride to give the alkane and a stannyl radical, thus propagating the chain (Scheme 1). This reaction has developped into a useful synthetic tool, not only for reducing halides but also for making carbon-carbon bonds by interception of the intermediate carbon radical.

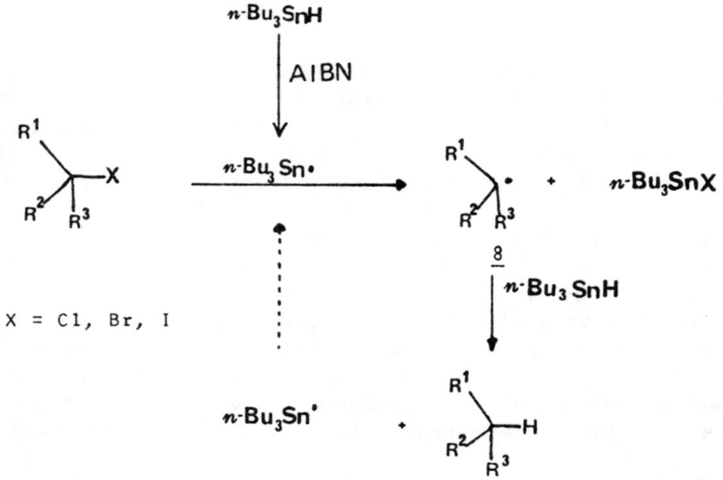

Scheme 1

Our conception for the radical deoxygenation of secondary alcohols is summarised in Scheme 2. In the presence of stannyl radicals, thionoester derivatives of the alcohol such as 1 would react to give a carbon radical 2. This radical can, of course, react with the stannane to give 3. If, however, the temperature is sufficiently high, fragmentation of 2 may take place to form a carbonyl derivative 4 and another carbon radical 5. This latter can now be reduced by the stannane to give the desired alkane 6. The chain is propagated by the stannyl radical also produced. The driving force for the process is the conversion of a thiocarbonyl into a carbonyl and the increase of entropy produced by the fragmentation. In this, and all other reactions where tin hydrides are involved, it is the weakness of the tin-hydrogen bond which produces the disciplinary effect on the radicals and thus prevents random reactivity.

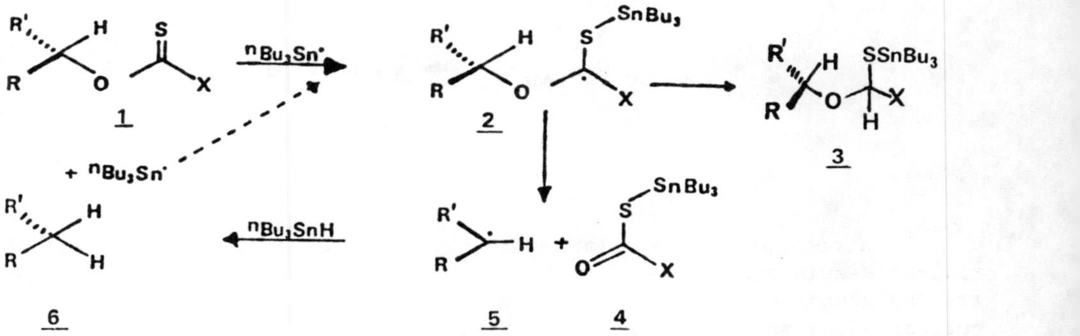

7 X = Ph 8 X = Imidazole 9 X = SMe

Scheme 2

After some experimentation, we found that thionobenzoates 7, thionoimidazolides 8 and especially xanthates 9 were suitable substrates[3]. For secondary alcohols, moderate temperatures (80-110°C) are sufficient to cause the fragmentation of the intermediate radical 2 and yields are generally excellent. In the case of primary alcohols, higher temperatures are required and yields are moderate to good[4]. With tertiary alcohols problems were encountered in the preparation of the various thionoesters. These derivatives underwent Chugaev elimination too readily to survive the reaction conditions. We found, however, that thionoformates were, thermally, stable enough to allow the deoxygenation to take place in good yields, but these derivatives were relatively inaccessible[5].

Despite these various limitations, this deoxygenation reaction has found widespread use especially, as originally intended, in the carbohydrate and aminoglycoside field, where most hydroxy groups belong to the secondary type. The example shown in Scheme 3 illustrates the exceptional tolerance of other functional groups and absence of β-elimination[6].

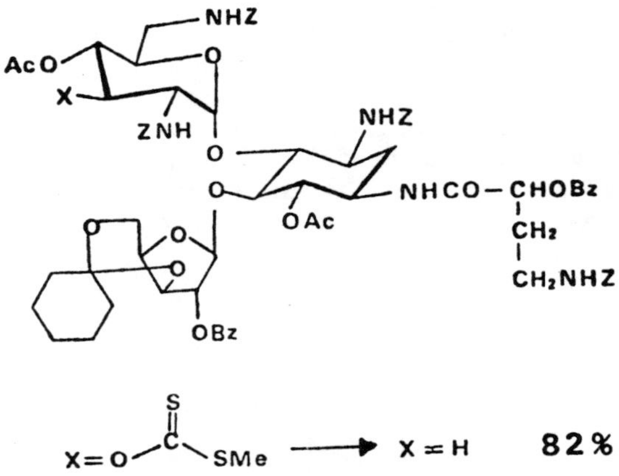

Scheme 3

In a conceptually similar approach, we developped an efficient radical deamination method based on the reaction of isocyanides with tributylstannane. Isocyanides are easily accessible from the corresponding amine and undergo, on treatment with the stannane, a smooth fission of the carbon-nitrogen bond[7] (Scheme 4). As in the case of thionoesters, the carbon radical formed is reduced to the alkane. Although this reaction was independently discovered by Saegusa[9], the yields reported were only moderate and the synthetic potential underestimated. Isothiocyanates and isoselenocyanates also undergo a similar reduction. This deamination reaction was successfully tested on a variety of substrates including a dipeptide pointing to possible applications in the peptide field. Even the 6-amino group in penicillins could be removed without harm to the sensitive β-lactam moiety[9]. As expected, the order of reactivity is tertiary > secondary > primary isocyanide. Consequently, selective deamination is possible by simple adjustment of the reaction temperature. We have thus been able to prepare a number of deaminated neamine derivatives[10].

Another important problem that attracted our attention is the radical decarboxylation of carboxylic acids. Existing methods lack mildness and generality and hence are ill-suited for complex and often fragile natural products. We were aware that a carboxylic radical would not be formed by β-elimination from an alicyclic or aliphatic radical.

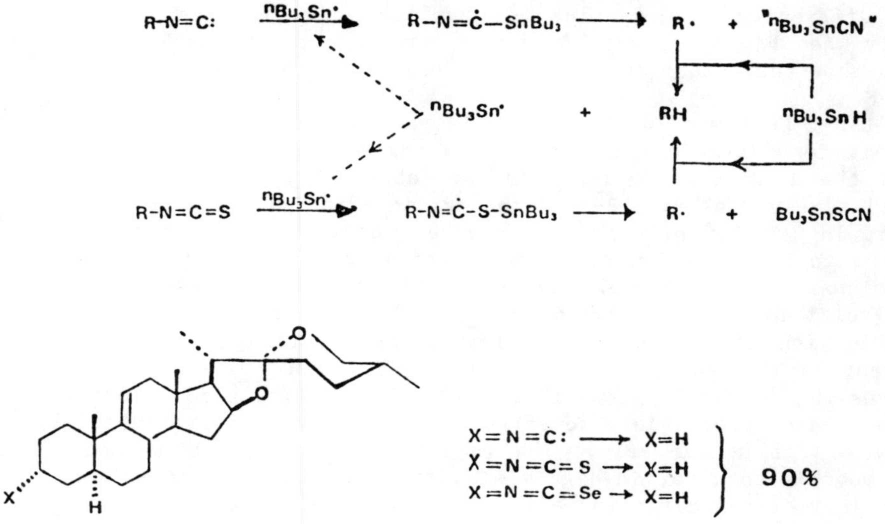

Scheme 4

We, therefore, designed a molecule where the double bond formed is incorporated in an aromatic array, as in the case of dihydrophenanthrene derivatives (Scheme 5). With this aromatisation driving force, good yield of nor-alkanes are obtained by β-elimination[11].

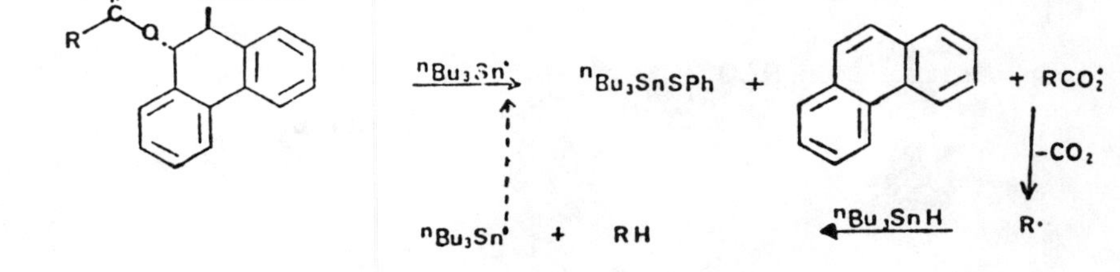

Scheme 5

The difficulty in preparing the various ester intermediates, especially when hindered, was however a drawback to this method. We conceived that the esters (mixed anhydrides) of thiohydroxamic acids of type -C(=S)-N(-)-O-CO-R should show a facile radical fragmentation based on the same considerations that had been taken into account in the design of the deoxygenation reaction (see above). Thus these esters have a relatively weak (N-O) β-bond making, in principle, fragmentation easy. Drawing on our previous experience, we incorporated the C-N double bond to be formed by the fragmentation into an aromatic system for additional driving force. Specifically, we examined esters 10 derived from N-hydroxy thiopyridone 11a. This compound exists in equilibrium with its tautomer, 2-thiopyridine-N-oxide 11b, but the derived esters are entirely in the form shown in 10. These derivatives 10 are easily accessible from acids or from acid chlorides (Scheme 6). A non-negligible advantage is the commercial availability of 11 and of its sodium salt. The reaction with tributylstannane occured smoothly to give, in high yield, the corresponding nor-alkane from a variety of aliphatic and alicyclic acids[12] (Scheme 6).

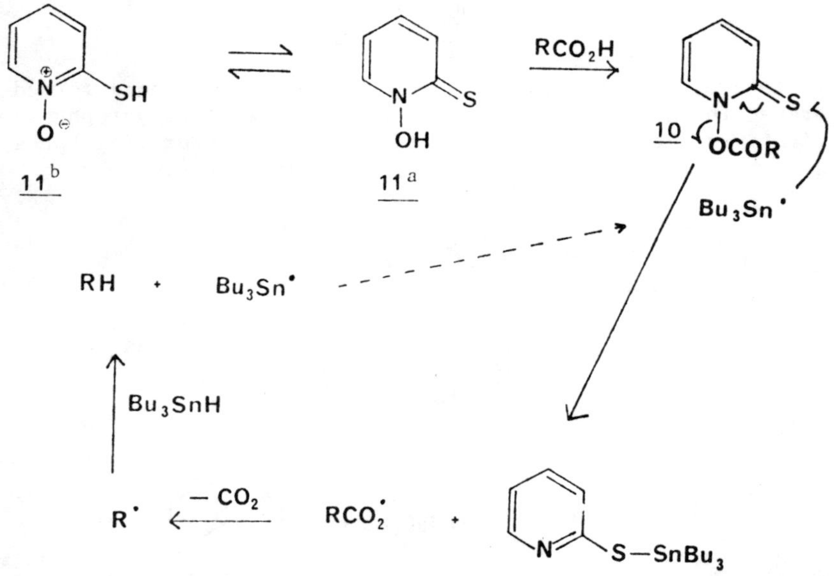

Scheme 6

In the case of primary acids, however, we were surprised to find that reactions were faster and gave a higher yield in benzene at 80°C than in toluene at 110°C. Examination of the reaction mixture in a particular case showed the presence of another compound, identified as the sulphide 12. This is slowly reduced by the stannane to the same nor-alkane (Scheme 7). Apparently, at the higher temperature of refluxing toluene the formation of sulphide can compete with the "normal" reduction. Indeed, in the absence of the reducing agent, the sulphide is the exclusive product. We found this novel decarboxylative rearrangement to be a general high yielding reaction proceeding by the simple radical chain mechanism depicted in Scheme 8. The reaction occurs with other thiohydroxamic esters having an appropriate structure such as 13[13]. Furthermore, the decarboxylation may be promoted thermally or photochemically at room temperature or lower.

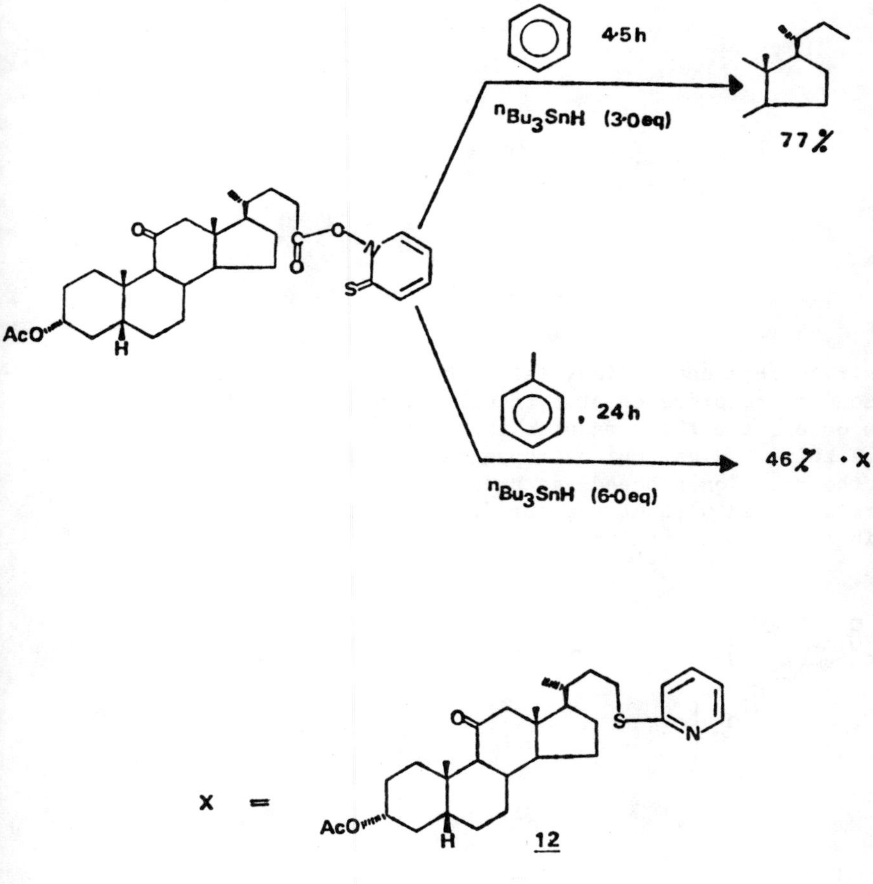

Scheme 7

Having at hand a convenient source of carbon radicals, we considered intercepting them by various reagents and thus diverting the reaction from its normal course of rearrangement. However, in order to retain all the advantages of radical reactions we had to maintain the chain mechanism using an appropriate propagating (chain carrying) radical.

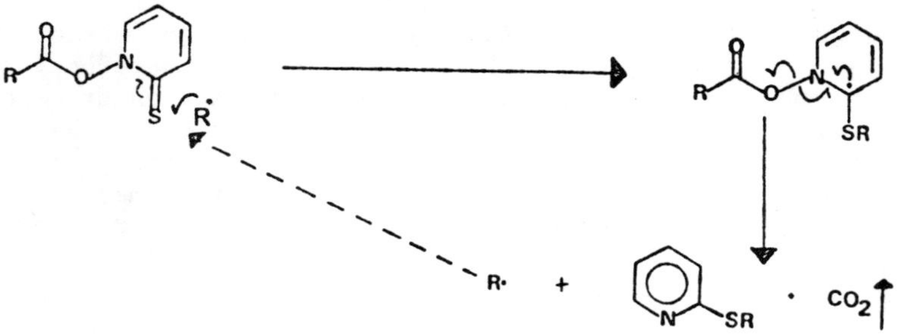

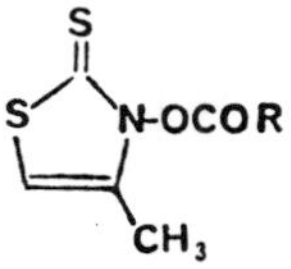

13

Scheme 8

To illustrate this conception, let us consider performing the decarboxylation in the presence of a thiol. Being an excellent hydrogen atom donor, the thiol reduces the radical to the alkane with formation of a thiyl radical which propagates the chain (Scheme 9). In practice, the reaction proceeds as predicted. High yields of nor-alkane are obtained without the purification problems usually encountered in stannane reductions[12].

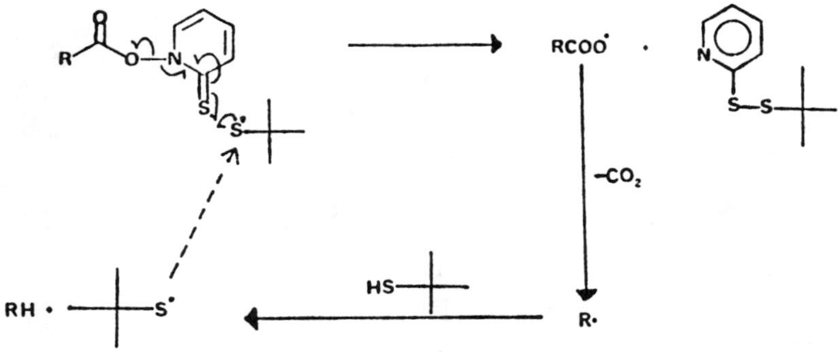

Scheme 9

As a consequene of similar mechanistic considerations, we have succeeded in preparing chlorides, bromides and iodides in excellent yield simply by operating in the presence of carbon tetrachloride, bromotrichloromethane and iodoform respectively[14] (Scheme 10). In terms of mildness of conditions, generality and yields this method is far superior to the classical Hunsdiecker reaction and its variants, practically all based on the use of heavy metal salts. Furthermore, by incorporating an initiator (AIBN) we have been able to extend the reaction to the more difficult case of aromatic and α,β unsaturated acids[15], even those with electron releasing substituents, which usually undergo ring bromination when subjected to Hunsdiecker conditions. As a demonstration of the mildness of our conditions, we can note the successful decarboxylative bromination of the heavily functionalised acid 14 by Prof. Ikegami using our method[16]. All other classical methods failed in this case.

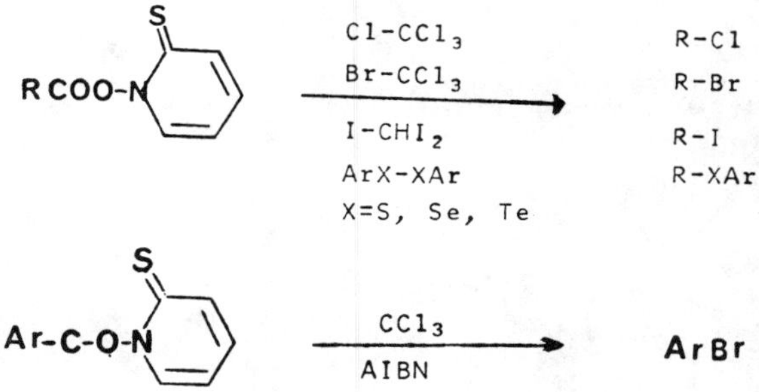

Scheme 10

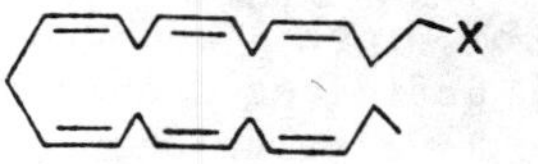

$X=CO_2H \longrightarrow X=Br$ (75%)

14

Reductive chalcogenation can similarly be achieved leading to sulphides, selenides and even tellurides. The carbon radical in this case is trapped with a diaryl disulphide, diselenide or a ditelluride respectively[17] (Scheme 10).

We have also found it possible to intercept the carbon radical with oxygen. The reaction has to be performed in the presence of a thiol to reduce the intermediate hydroperoxy radical to the hydroperoxide and to furnish the thiyl (propagating) radical. The hydroperoxide can be reduced further to the corresponding alcohol by a phosphite or transformed into an aldehyde or a ketone depending on whether the starting acid is primary or secondary[18] (Scheme 11).

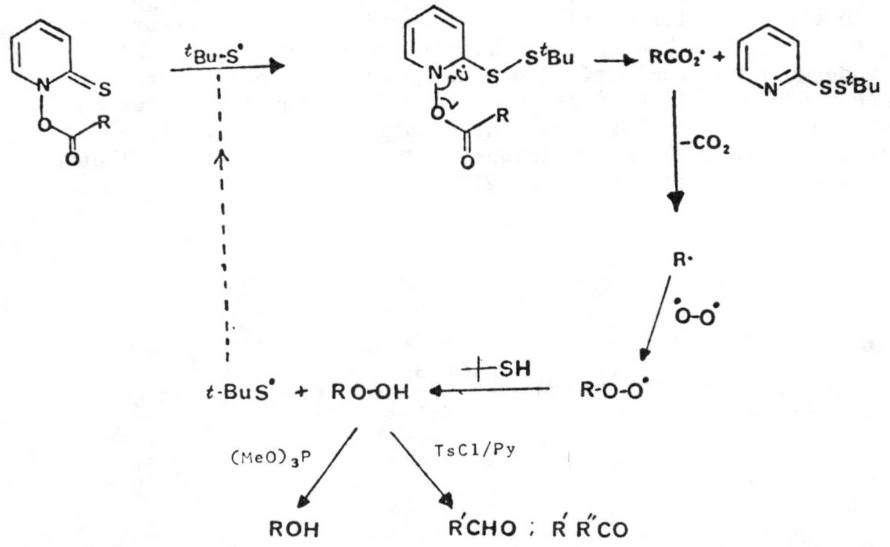

Scheme 11

We next examined a transformation of crucial importance to organic synthesis : the formation of carbon-carbon bonds. We were encouraged in this respect by the work of Giese[19] and others on the addition of carbon radicals to electron deficient double bonds. We found that similar addition could be achieved with our system if the olefin was sufficiently activated for the addition to compete successfully against the background decarboxylative rearrangement[20] (Scheme 12). Yields were variable being highest for reactive, not easily polymerised olefins. An interesting reagent is the acrylic ester derivative 15. The sulphur moiety, by acting as a leaving group, prevents polymerisation and efficiently propagates the chain. The products 16 have furthermore considerable synthetic potential[21].

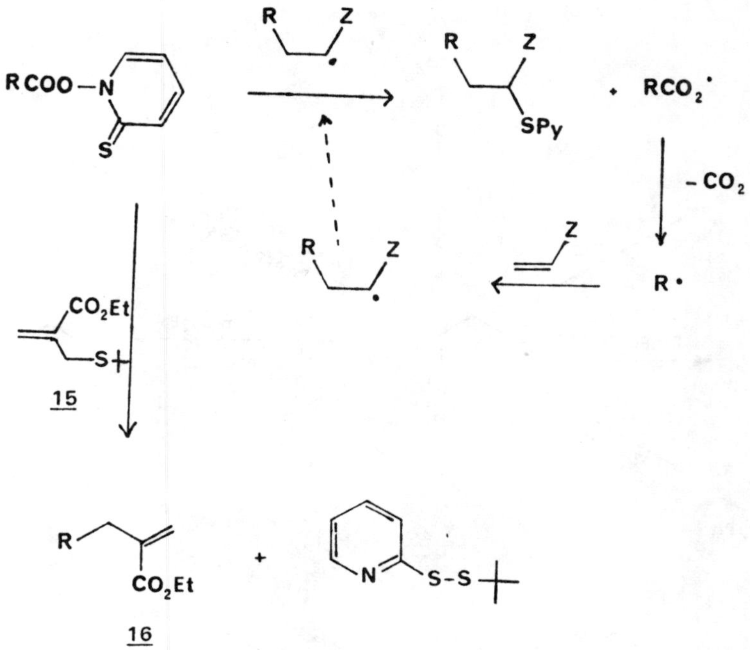

Scheme 12

An additional feature of the decarboxylation is that the normally neutral medium can be made acidic without loss of efficiency. This effectively suppresses complications arising from anionic polymerisations induced by traces of base as is sometimes the case with strongly electrophilic yet non-radical polymerisable olefins such as nitroalkenes, vinylsulphones and vinylphosphonium salts. We have shown that these reactive olefins do give high yields of radical addition and furthermore, that the corresponding adducts are versatile synthetic intermediates easily converted into an array of useful products[22] (Scheme 13). As a further illustration of the synthetic potential of this carbon-carbon bond forming reaction we have succeeded in constructing the important 25-hydroxy- cholesteryl side chain of vitamin D_3 metabolites from a bile acid derivative in essentially two steps[22] (Scheme 13).

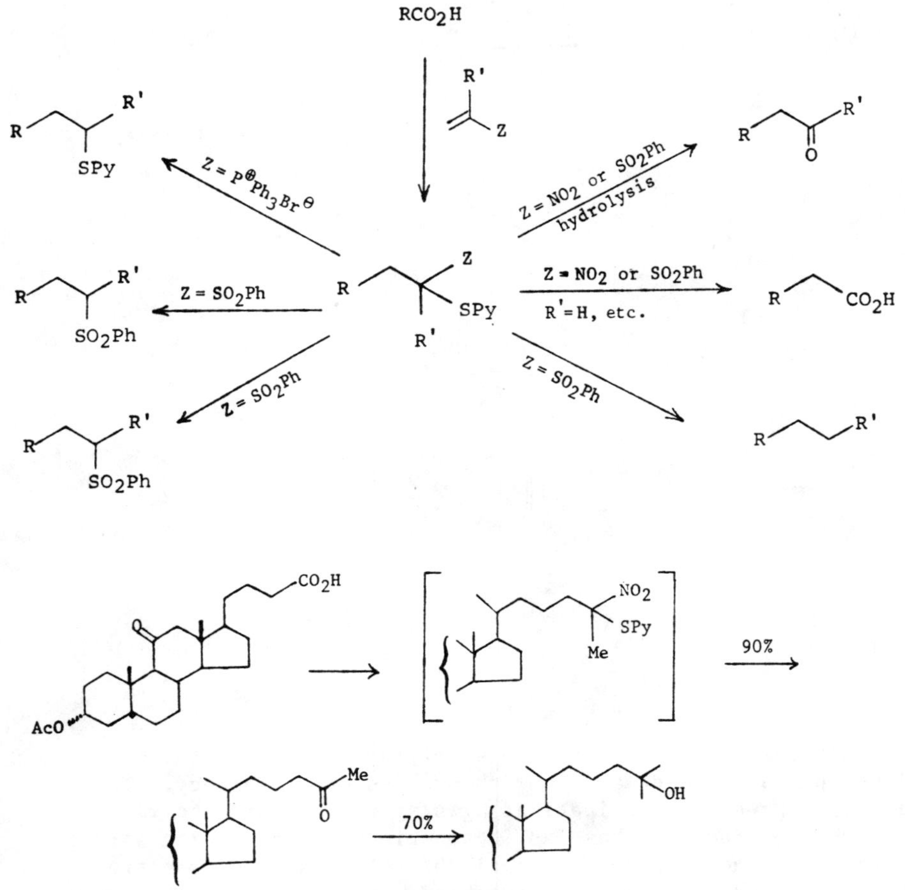

Scheme 13

Somewhat inevitably, our work on the decarboxylation of acids provided us with the key to the so far inadequately solved problem of radical deoxygenation of tertiary alcohols. If a carboxyl radical derived from the hemi-ester of oxalic acid such as 17 can be produced it will, by loss of two molecules of carbon dioxide, give a tertiary carbon radical which can be reduced as before by a thiol (Scheme 14). This conception was readily reduced to practice and a variety of tertiary alcohols were deoxygenated in good yield[23].

Interception of the tertiary radical with the acrylate reagent 23 results in the formation of a quaternary centre directly from an alcohol[24] (Scheme 14). The synthetic potential of this transformation is considerable given the difficulties usually encountered in creating quaternary centres.

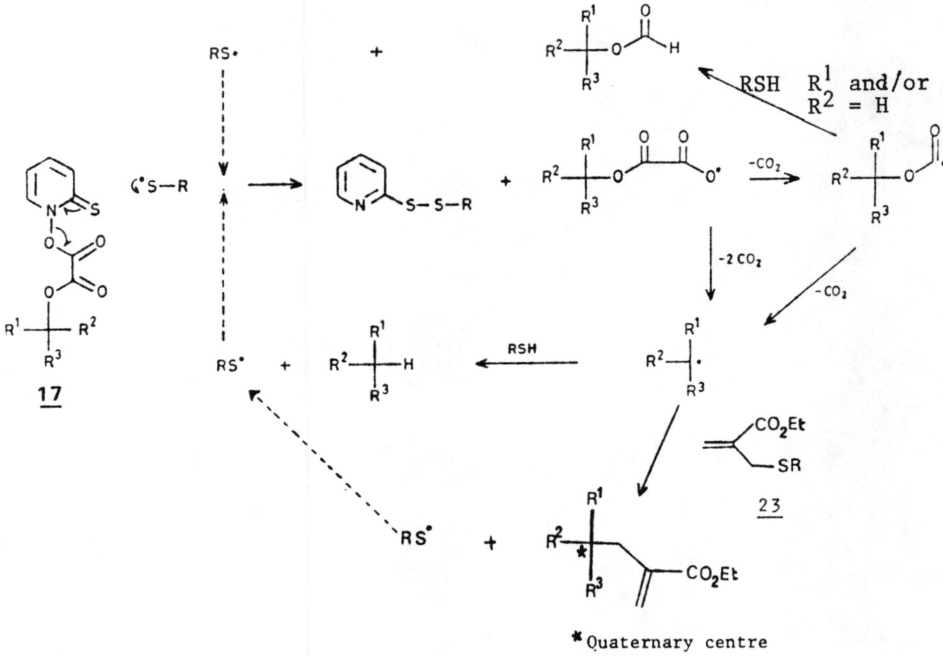

Scheme 14

This type of radical chemistry is well suited for the manipulation of amino- acids and peptides. By using a mixed anhydride coupling procedure N-hydroxy-2-thiopyridone esters of N-protected amino-acids or peptides can be synthesised and decarboxylated in very high yield (Scheme 15) using a thiol as hydrogen atom transfer reagent. Alcoholic, phenolic and even indolic groups do not need protection in this reaction. Equally important is the manipulation of side chain carboxyl groups made possible when the α-carboxyl is appropriately protected[25]. As an example, we carried out the synthesis of optically active vinylglycine 18, an important amino acid, from glutamic acid[26] (Scheme 16).

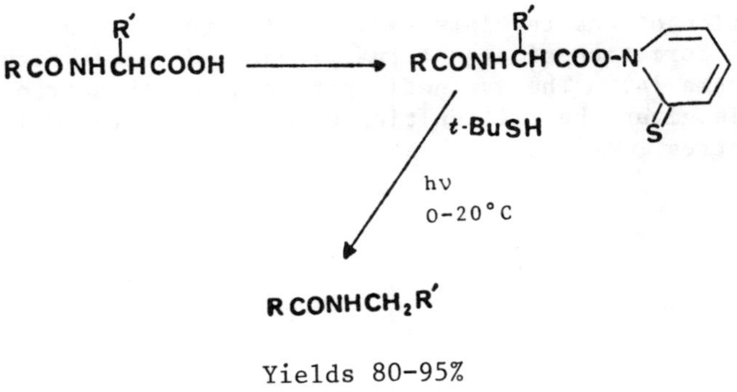

Scheme 15

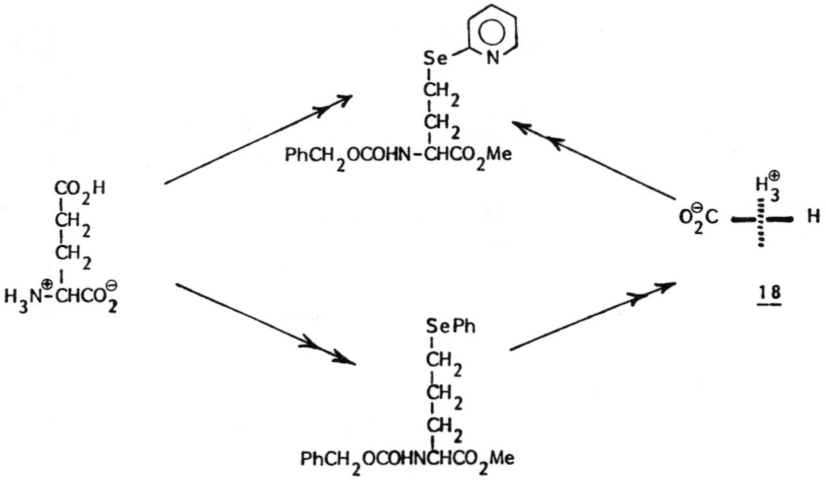

Scheme 16

In the reactions that we have described above, based on the esters of thiohydroxamic esters (mixed anhydrides), the disciplinary group, which prevents random reactions, is the thione function. Another feature is the elevated yields which are seen when a thiyl radical as propagating radical reacts (soft-soft) with the thione function. We conceived that we could exploit this reactivity of thiyl radical to induce carbon radical reactions with other elements such as metalloids and transition metal elements. The following examples illustrate that this idea opens up a new chapter in the preparation of M-R (M = metal or metalloid) bonds. Thus (Scheme 17) one can conceive that a radical generated in the usual way could react with a species $(PhS)_nM$ 19 to give with elimination of a PhS• radical a derivative $R\text{-}M(SPh)_{n-1}$ 20. The chain would then be propagated.

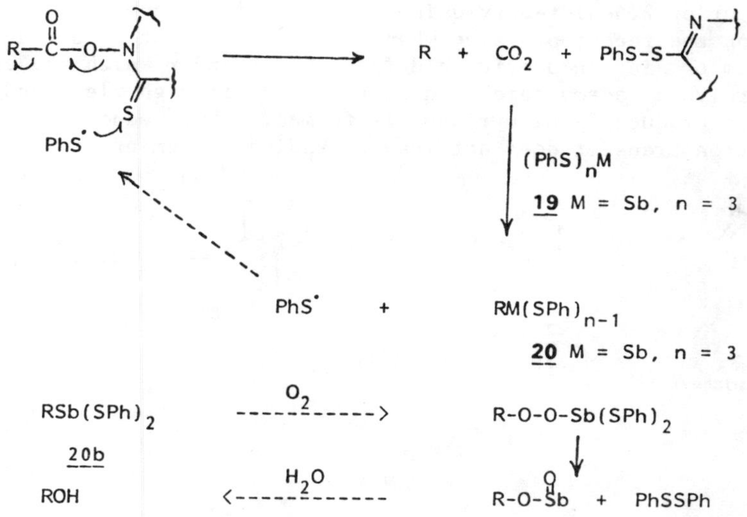

Scheme 17

In fact, in the case where M = antimony and n is 3 we have been able to develop a high yielding reaction which converts an acid into the corresponding nor-alcohol. The reaction works so well because 19 and its congeners are sensitive to oxygen and produce spontaneously thiyl radicals. After hydrolysis excellent yields of nor-alcohol are obtained. This is surely the best method available for the change $R\text{-}CO_2H \rightarrow R\text{-}OH$ under very mild conditions. It works equally well with primary, secondary and tertiary acids in the aliphatic and alicyclic series. Simply stirring a mixture of the ester and $(PhS)_3Sb$ under air at room temperature is sufficient to bring about the desired conversion[27].

The compatability of strong, anhydrous acid with the radical chemistry of thiohydroxamic acid esters (mixed anhydrides) has been referred to above. In basic compounds the addition of such an acid can greatly enhance the radicophilicity of the system. Such rate enhancement for protonated pyridines had already been observed in the masterly work of Minisci and his colleagues.[28] We have recently shown[29] that pyridine and lepidine camphorsulphonate salts in methylene dichloride at room temperature give very high yields on irradiation with a tungsten lamp of the appropriate thiohydroxamic esters. In the absence of the camphorsulphonic acid yields were much lower.

The mechanism of this reaction is probably as indicated in Scheme 18. The first part of this Scheme involves the same radical cation intermediate 21 as in the mechanism already proposed by Minisci[28]. The intermediate 21 is normally stabilised by electron transfer and proton loss. We do not think that this step applies to our system because the addition of the electron acceptors 1,3-dinitrobenzene (1 equiv.) or of 2,6-di-*t*-butylquinone (0.5 equiv.) have very little effect on the reaction. Also when the adamantyl ester is reduced with magnesium or zinc in acetic acid (conditions under which the ester is not hydrolysed) adamantoic acid is recovered in high yield and the reduction product 2-thiopyridone is formed. Thus reduction of esters by electron transfer does not lead to radical fragmentation.

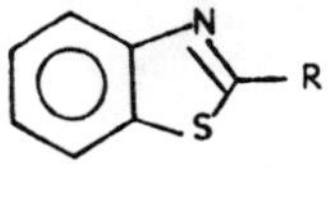

22, R=H
23, R=1-adamantyl

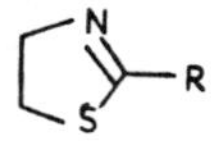

24, R=H
25, R=1-adamantyl

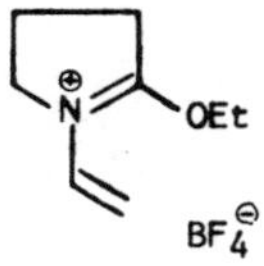

26

Ad-(1)-CH_2CHO

27

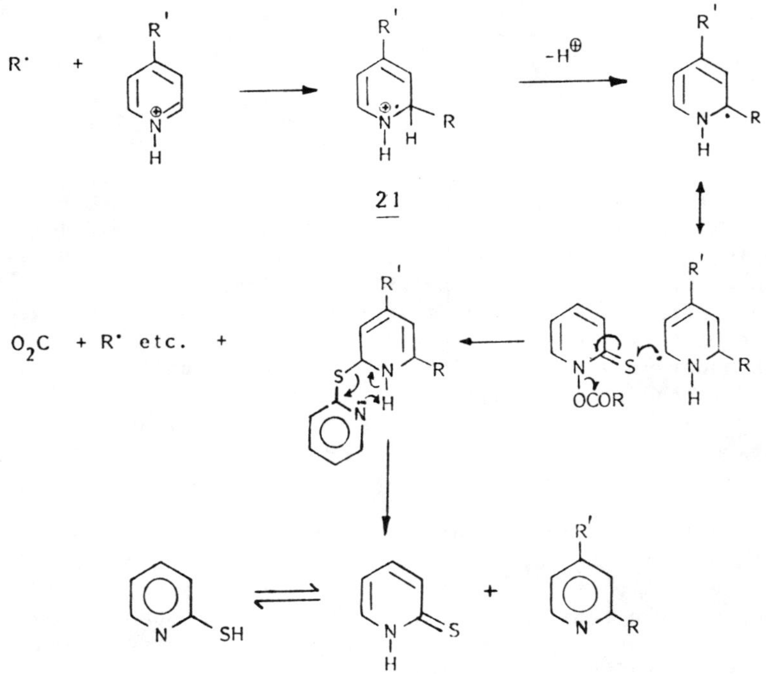

Scheme 18

We have briefly examined radical additions to other heterocyclic systems with encouraging preliminary results. Thus irradiation of the adamantyl ester and the protonated benzothiazole 22 gave the 2-substituted derivative 23 in 52% yield. The simple thiazoline 24 similarly gave the corresponding adduct 25 (37%). This adduct could be converted back to adamantanoic acid in quantitative yield by exposure to sodium nitrite in aqueous acetic acid. If the starting thiazoline 24 was labelled in the 2-position the overall sequence would have led to labelled adamantanoic acid, a transformation which is otherwise difficult to perform under such mild conditions.

Preliminary experiments on the addition of tha adamantyl radical to dimehtylvinylsulphonium tetrafluoroborate and to the N-vinyl-iminium salt[30] 26 gave on work up the corresponding homo-aldehyde 27 in yields which have not yet been optimised. The mechanism of aldehyde formation is presumed to be that shown in Scheme 19.

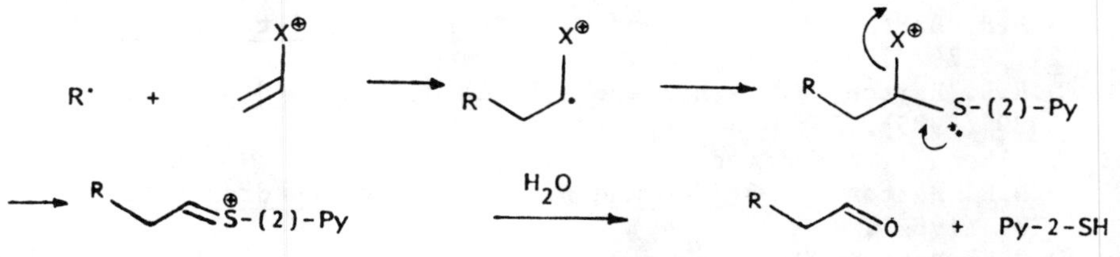

Scheme 19

Acknowledgements. We thank all our colleagues who have participated in this work and whose names are cited in the References. Dr. David Crich played the pioneering role in the early work. We thank also Roussel-Uclaf for their generous support.

REFERENCES

1. D.H.R. Barton and W.B. Motherwell, Pure and Appl. Chem., 53, 15 (1981); idem, in "Organic Synthesis, Today and Tommorrow", p.1, Ed. by B.M. Trost and G.R. Hutchinson, Pergamon Press, 1981.
2. Van der Kerk, G.J.M. Noles and J.G.A. Luitjen, J. Appl. Chem., 7, 356 (1957).
3. D.H.R. Barton and S.W. McCombie, J. Chem. Soc., Perkin Trans 1, 1574 (1975); W. Hartwig, Tetrahedron, 39, 2609 (1983); see also: M.J. Robins, J.S. Wilson and F. Hansske, J. Amer. Chem. Soc., 105, 4059 (1983).
4. D.H.R. Barton, W.B. Motherwell and A. Stange, Synthesis, 743 (1981).

5. D.H.R. Barton, W.B. Motherwell, R.S. Hay-Motherwell and W. Hartwig, Tetrahedron Lett., 23, 2019 (1982).
6. T. Hayashi, T. Iwaoka, N. Takeda and E. Ohki, Chem. Pharm. Bull., 26, 1786 (1978).
7. D.H.R. Barton, G. Bringmann, G. Lamotte, R.S. Hay-Motherwell and W.B. Motherwell, Tetrahedron Lett., 2291 (1979).
8. T. Saegusa, S. Kobayashi, T. Ito and N. Yasuda, J. Am. Chem. Soc., 90, 4182 (1968).
9. D. Ivor John, E.J. Thomas and N.D. Tyrrell, J. Chem. Soc., Chem. Commun., 345 (1979).
10. D.H.R. Barton, G. Bringmann and W.B. Motherwell, J. Chem. Soc., Perkin Trans 1., 2665 (1980).
11. D.H.R. Barton, H. Dowlatshahi, W.B. Motherwell and D. Villemin, J. Chem. Soc., Chem. Commun., 732 (1980).
12. D.H.R. Barton, D. Crich and W.B. Motherwell, J. Chem. Soc., Chem. Commun., 939 (1983).
13. D.H.R. Barton and G. Kretzschmar, Tetrahedron Lett., 24, 5889 (1983).
14. D.H.R. Barton, D. Crich and W.B. Motherwell, Tetrahedron Lett., 24, 4979 (1983).
15. D.H.R. Barton, B. Lacher and S.Z. Zard, Tetrahedron Lett., 26, 5939 (1985).
16. S. Ikegami, private communication.
17. D.H.R. Barton, D. Bridon and S.Z. Zard, Tetrahedron Letts., 25, 5777 (1984).
18. D.H.R. Barton, D. Crich and W.B. Motherwell, J. Chem. Soc., Chem. Commun., 242 (1984).
19. B. Giese, Angew. Chem. Int. Ed., 22, 753 (1983), and references therein.
20. D.H.R. Barton, D. Crich and G. Kretzschmar, Tetrahedron Lett., 25, 1055 (1984).
21. D.H.R. Barton and D. Crich, Tetrahedron Lett., 25, 2787 (1984).
21. D.H.R. Barton and D. Crich, Tetrahedron Lett., 25, 2787 (1984).
22. D.H.R. Barton, H. Togo and S.Z. Zard, Tetrahedron, 41, 5507 (1985); D.H.R. Barton, H. Togo and S.Z. Zard, Tetrahedron Letts., 26, 6349 (1985).
23. D.H.R. Barton and D. Crich, J. Chem. Soc., Chem. Commun., 774 (1984).
24. D.H.R. Barton and D. Crich, Tetrahedron Lett., 757 (1985).
25. D.H.R. Barton, Y. Hervé, P. Potier and J. Thierry, J. Chem. Soc., Chem. Commun., 1298 (1984).
26. D.H.R. Barton, D. Crich, Y. Hervé, P. Potier and J. Thierry, Tetrahedron, 41, 4347 (1985).
27. D.H.R. Barton, D. Bridon and S.Z. Zard, J. Chem. Soc., Chem. Commun., 1066 (1985).
28. a) F. Minisci, A. Citterio and C. Giordano, Acc. Chem. Res., 16, 27 (1983); b) F. Minisci, Top. Curr. Chem., 62, 1 (1976); c) F. Minisci and A. Citterio, "Advances in Free Radical chemistry", Vol. VI, G.H. Williams, Ed., 1980, p. 65; d) F. Minisci, E. Vismara and V. Romano, Tetrahedron Letts., 26, 4803 (1985); e) F. Minisci, A. Citterio, E. Vismara and C. Giordano, Tetrahedron, 41, 4157 (1985) and references there cited; f) J. Seeman, L.E. Clawson and H.V. Secor, Synthesis, 953 (1985).

29. D.H.R. Barton, B. Garcia, H. Togo and S.Z. Zard, Tetrahedron Lett, in press.
30. M.B. Smith and H.N. Shroff, J. Org. Chem., 49, 2900 (1984).

EPILOGUE

Robert Louw*
Center for Chemistry and the Environment,
Gorlaeus Laboratories, The University of Leiden, The Netherlands

Indeed, an absolutely superb meeting, to quote "anchorman" Keith U. Ingold; radical chemistry is alive and well, as he also concluded.

Disciplined radicals - whatever the cause of their constructive behavior - play an increasingly important part in synthesis. This is exemplified by the work of Sir Derek Barton,the first speaker at this conference in Louvain-la-Neuve, where Heinz Günter Viehe and his co-workers organized this very successful NATO Advanced Workshop.

Led by the concept of captodative stabilization - in fact a rediscovery and an actualisation of "merostabilization" or push-pull-interaction (Dewar, Katritzky, Balaban), Viehe et al. have shown how such an inspiration can bring about creative contributions in synthesis through selective free radical addition or abstraction reactions. Moreover, when searching for the "truth behind", they went on performing advanced mechanistic studies by comprehensive measurements on structural effects on rates and equilibria.

The growing interest in the fundamentals of free radical properties and behavior, is reflected by a number of other contributions to this conference, ranging from experimental structural chemistry and dynamics to theoretical approaches.

To illustrate this, consider the papers of D.R. Arnold, K.U. Ingold and R. Sustmann, and observe, inter alia, the continuing importance of ESR in determinations of and measurments on free radicals. Gradually moving from experimentalists to "calculationists" we may meet X. Creary, J.W.Timberlake, Ch. Rüchardt, B. Giese, H.Fischer P. von R. Schleyer and G. Leroy.

Hard workers in the chemical salt mines, struggling with tough matter such as bond strengths or radical stabilization energies, may become jealous when observing Professor Leroy's beautiful, ordered landscape which he was able to lay out using a good computer and some experimental data for picketing. Will experimentalists soon become as outdated as firemen on electric trains today? Indeed, a PC with number-crunching facility costs only a fraction of a modern spectrometer or of proper equipment to perform state-of-the-art kinetic work on rapidly reacting species. When working at structure determination or reaction dynamics, will it remain good practice to say "let us do the

* With thanks to David Griller and Paul von R. Schleyer, who did as I requested and provided letters with their fresh impressions.

H. G. Viehe et al. (eds.), Substituent Effects in Radical Chemistry, 463–466.
© 1986 by D. Reidel Publishing Co.

experiment"? And how about syntheses : will computers design or prescribe the method of choice, to be conducted in a fully automated laboratory? No more "Malaise" reactions (see B. Fraser Reid, Accounts of Chemical Research, Nov. 1985) to fill, or not fill, test tubes, bottles, tankers?

In any case, the overlap area between radical and ionic chemistry, and the interplay between even and odd electron species, for instance via single-electron transfers continues to constitute a challenging territory worth further exploration ; compare, for instance, the contributions by Minisci and Katritzky.

Furthermore, understanding chemical reactivity (a headline of the recent "Pimentel Report") entails, for the time being, more than computing. Structure-reactivity-selectivity correlations, pertaining to free radical reactions, are being further developed. Steric, electronic, and special free radical features play their role in placing facts in a row, as is attempted in Hammett-type or other linear free-energy relations.

Understanding of the factors determining rates and selectivities for radical addition to double bonds, or for (H) abstraction is now reaching a state of maturity.

A major concern, in this respect, is the magnitude of the stabilization (or destabilization) brought about by functional groups,bond at a - say, carbon - centered - radical, at a neighbouring atom or even further away from the unpaired electron site.

At least as interesting is the question : what is the result when two groups are placed on a free radical center? Is their effect additive or not? And how is such an effect defined and determined?

Suppose, for a moment, that we understand, in a thermodynamic sense, the elements affecting rate processes, transition state properties, leading to or involving such functionalized radicals. And let us be so smart not to get involved in solvent effects. As free radicals are generally transient species, direct measurement of thermodynamic properties ($\Delta H_f^{\cdot}$, $S^{\cdot}$) by determination of heats of combustion, for example, is impossible. Relative data can, however, be obtained by equilibrium measurements, as exemplified by Griller who used alkyl iodides ($R_1^{\cdot} + R_2I \rightleftarrows R_1I + R_2^{\cdot}$). Obviously, for adequate interpretation the thermodynamic data on the two iodoalkanes must be known.
Likewise $\Delta H_f^{\cdot}$ and $S^{\cdot}$ of the simplest carbon-centered radical, $^{\cdot}CH_3$, are the "outcome" of measurements on bond dissociation processes such as $CH_3I \rightarrow {}^{\cdot}CH_3 + I^{\cdot}$ or, most basically, $CH_4 \rightarrow {}^{\cdot}CH_3 + {}^{\cdot}H$. Clearly, $\Delta H_f^{\cdot}$-values for radicals are meaningless numbers, unless we compare these data with those of the parent molecules.

Still, common knowledge among common chemists on quantitative (thermodynamic) structural effects in molecules - let alone, radicals - is very limited. They, and, I hope, advanced undergraduate students, will "know" that a branched hydrocarbon, such as isooctane (2,2,3,3-tetramethylbutane, $\Delta H_f^{\circ}(g) = -54$ kcal/mol) is more stable than n-octane, $\Delta H_f^{\circ} = -49.8$ kcal/mol. They may also recall that R-H (and R-Methyl) bond strengths decrease in the order R = primary, secondary, tertiary. To quote today's values : tBu-H = 94, nBu-H = 100 kcal/mol. But how many are aware of the fact - let alone are able to explain - that replacement of one or two hydrogens on a carbon free radical center by methyl results in ΔH_f° changes which are different from those associated with analogous displacements on an sp^3 carbon?

If we look at more interesting cases, where functional groups such as OMe, CN, Cl, COOMe, etc. are involved, it usually turns out that R-Z bond dissociation energies remain nearly constant for R = primary, secondary and tertiary alkyl. Here, increased stabilization, due to interaction between R and Z about compensates for the "stabilization" in $R^{\cdot}$.

Furthermore, structural effects on ΔH_f°, or group increments, are often far from additive. Consider for example the equilibria of the isodesmic "reactions"

$$CH_4 + CH_2(OCH_3)_2 \rightleftarrows 2\ CH_3OH;\ \Delta H^{\circ} = +\ 13 \text{ kcal/mol}$$

$$(CH_3)_2CH_2 + CH_2(CN)_2 \rightleftarrows CH_3CH_2CN;\ \Delta H^{\circ} = -\ 11 \text{ kcal/mol.}$$

Roughly speaking, two OMe groups on one sp^3 carbon are "OK" whereas two cyano groups in that case, are "bad".

Such mutual interactions can be huge in normal molecules, but more significant, they may well differ greatly when the same groups are bound to other centers, including carbon radicals.

Those experimentalists who (still) like to determine thermodynamic properties of functionalized radicals - and, more specifically, or capto-dative substituted radicals - now are fully aware of the associated snags, ambiguities and confidence limits.

In a brainstorming (or, given that we were just within French-speaking Belgian territory, "Impromptu") discussion it was generally agreed that we need more accurate data on heats of formation of properly functionalized molecules which are also suited for measurements on bond dissociation energies, hence providing definitive data on the specified radicals therefrom. In this respect there is a need for new measurements, e.g. on heats of combustion, but a critical review of earlier data is also worthwhile.

However, who are those who are able and willing to do this (and who will pay the bill)? And again : can calculations or (other) estimations based on correlation with some physical property be accepted instead?

Meanwhile, synthesis may continue to take advantage of a rationale, a concept such as captodative interaction. Whatever we really know today about the magnitude of such interactions in a thermodynamic sense, - with or without the involvement of orbital (SOMO) schemes - we can take the intriguing behavior of captodatively substituted radicals for granted; these species, normally, refuse to add to alkenes even if such alkenes are, in turn, **cd**-substituted !

Again, according to all participants, a great conference. Hats off for the organizers, who not only were able to bring together most of the authoritative chemists from NATO and non-NATO countries in a well-suited ambiance, but also succeeded in offering a conference dinner of an unprecedented quality and style : a captodative synergism between French haute-cuisine and Belgian savoir-vivre.

What will be the result, the impact of this conference? A meeting of scientists is quite unlike a congress of a political party. There were no resolutions (except to publish this book as a document !) nor was a planned action program adopted. But most participants will have left satisfied, and inspired to design and develop their experiments or their models, consonant with the code of conduct : do the best you can - eventually, in fair competition, free and radical !

PARTICIPANTS

Agranat Israel, Department of Chemistry, The Hebrew University of Jerusalem, 91904 Jerusalem, Israel

Arnold Donald D., Department of Chemistry, Dalhousie University, Halifax, Nova Scotia, Canada B3H 4J3

Aurich Hans G., Fachbereich Chemie der Philipps-Universität, Hans-Meerwein Strasse, D-3550 Marburg, Germany

Bachi Mario D., The Weizmann Institute of Science, Department of Organic Chemistry, 76100 Rehovot, Israel

Balaban A.T., Str.Poenaru Bordea 6, Apt13, R-70502 Bucharest Roumania.

Barton Sir Derek, Institut de Chimie des Substances Naturelles, CNRS, F-91190 Gif-sur-Yvette, France

Beckhaus Hans-Dieter, Institut für Organische Chemie und Biochemie, Universität Freiburg, Albertstrasse 21, D-7800 Freiburg 1, Germany

Chanon Michel, Université de Droit, d'Economie et des Sciences Naturelles d'Aix-Marseille, Laboratoire de Chimie Inorganique Moleculaire, rue H. Poincaré, F-13397 Marseille Cedex 13, France

Citterio Attilio, Politecnico di Milano, Istituto di Chimica Piazzo L. de Vinci 32 I-20133 Milano, Italy

Clark Timothy, Friedrich-Alexander Universität Erlangen-Nürnberg, Institut für Organ. Chemie, Henkestrasse 42, D-8520 Erlangen, Germany

Costantini M., Rhône-Poulenc Industries, Centre de Recherches de St. Fons, BP62, F-69190 Saint-Fons, France

Creary Xavier, Department of Chemistry, University of Notre Dame, Notre Dame IN 46556 U.S.A.

Crozet Michel P., Université de Droit, d'Economie et des Sciences Naturelles d'Aix-Marseille III. CNRS VA 109, Rue H. Poincaré, F-13397 Marseille Cedex 13, France

Dale Johannes, Department of Chemistry, University of Oslo, P.O.Boks 1033, Blindern, Oslo 3, Norway

Döpp Dietrich, Universität Duisburg, Fachgebiet Organische Chemie Postfach 101629, D-4100 Duisburg 1, Germany

Fischer Hanns, Physikalisch-Chemisches Institut der Universität Zürich, Winterthurerstr.190, 8057 Zürich, Switzerland

Fossey Jacques, L.E.S.C.O.-C.N.R.S. 2 A 8, rue Henri Dunant, BP 28, F-94320 Thiais, France

Giese Bernd, Technische Hochschule Darmstadt, Institut für Organische Chemie, Petersenstrasse 22, D-6100 Darmstadt, Germany

Griller David, National Research Council of Canada, Division of Chemistry, Ottawa, Canada K1A 0R9

Ingold Keith U., National Research Council of Canada, Division of Chemistry, Ottawa, Canada K1A 0R9

Jackson Richard A., School of Chemistry and Molecular Sciences, University of Sussex, Brighton BN1 9QJ, Great Britain

Janousek Zdenek, Université de Louvain, Laboratoire de Chimie Organique, Place Louis Pasteur, 1, B-1348 Louvain-la-Neuve, Belgium

Katritzky Alan R., University of Florida, Department of Chemistry, Gainesville, FL Florida 32611, U.S.A.

Koch Tad H., University of Colorado, Department of Chemistry and Biochemistry, Boulder Co 80309-0215, U.S.A.

Korth H.G., Institut für Organische Chemie, Universität Essen, GHS, D-4300 Essen 1, Germany

Kost Daniel, Ben Gurion University of the Negev, Department of Chemistry, Beer-Sheva 84105, Israel

Leroy George, Université de Louvain, Laboratoire de Chimie Quantique, Place Louis Pasteur, 1, B-1348 Louvain-la-Neuve, Belgium

Louw Robert, Center for Chemistry and the Environment Gorleaus Labs, The State University of Leiden, P.O. Box 9502, NL-2300 RA Leiden, The Netherlands

Lwowski Walter, New Mexico State University, Department of Chemistry, P.O.Box 3C Las Cruces, N.M. 88003 U.S.A.

Maillard Bernard, CNRS VA 35, Laboratoire de Chimie Organique et Organométallique, Université de Bordeaux 1, 351 Cours de la Libération, F-33405 Talence-Cedex, France

Merényi Robert, Université de Louvain, Laboratoire de Chimie Organique, Place Louis Pasteur, 1, B-1348 Louvain-la-Neuve, Belgium

Minisci Francesco, Politechnico di Milano, Dipartimento di Chimica, Piazza L. da Vinci, 32, I-20133 Milano, Italy

Neumann Wilhelm P., Universität Dortmund, Otto-Hahn-Strasse, Postfach 50 05 00, D-4600 Dortmund 50, Germany

Rassat A., Université Scientifique et Médicale de Grenoble, Bâtiment 52, Chimie-Recherche, BP 68, F-38402 Saint-Martin-d'Hères Cédex, France

Roth W., Lehrstuhl für Organische Chemie I, Ruhr-Universität Bochum, Postfach 10 21 48, Universitätsstrasse 150, D-4630 Bochum 1, Germany

Rüchardt Christoph, Institut fur Organische Chemie und Biochemie, Albertstrasse, 21, D-7800 Freiburg 1 , Germany

Stella Lucien, Université de Droit, d'Economie et des Sciences Naturelles d'Aix-Marseille III, Laboratoire de Chimie Organique B, Rue Henri Poincaré, F-13397 Marseille Cedex 13, France

Surzur Jean-Marie, Université de Droit, d'Economie et des Sciences d'Aix-Marseille III, Laboratoire de Chimie Organique B, Rue Henri Poincaré, F-13397 Marseille Cedex 13, France

Sustmann Reiner, Institut für Organische Chemie, Universität-GHS Essen Universitätsstrasse 5, D-4300 Essen 1, Germany

Tedder Lord John M., Department of Chemistry, University of St. Andrews, St. Andrews Fife, KY16 9ST, United Kingdom

Tiecco Marcello, Istituto di Chimica Organica, Facolta di Farmacia, Universita di Perugia , I-06100 Perugia, Italy

Timberlake Jack W., University of New Orleans, Department of Chemistry, Lakefront, New Orleans, LA 70122, U.S.A.

Tordo Paul, Université de Provence - Structure et Réactivité des Espèces Paramagnétiques, C.N.R.S. UA 126, Rue H. Poincaré, F-13397 Marseille Cedex 4, France

Vertommen Luc, AKZO, Research Laboratories, Velperweg, 76, NL-6824 BM Arnhem, The Netherlands

Viehe Heinz Gunter, Laboratoire de Chimie Organique, Université de Louvain, Place Louis Pasteur, 1, B-1348 Louvain-la-Neuve, Belgium

von Ragué Schleyer Paul, Friedrich-Alexander-Universität Erlangen-Nürnberg, Institut für Organische Chemie, Henkestrasse 42, 8520 Erlangen, Germany

Walton John C., Department of Chemistry, University of St. Andrews, St. Andrews Fife, KY16 9ST, United Kingdom

INDEX